T0192820

LONDON MATHEMATICAL SOCIETY LECTURE NOTE SERIES

Managing Editor: Professor M. Reid, Mathematics Institute,
University of Warwick, Coventry CV4 7AL, United Kingdom

The titles below are available from booksellers, or from Cambridge University Press
at www.cambridge.org/mathematics

London Mathematical Society Lecture
Note Series: 392

Surveys in Combinatorics 2011

Edited by

ROBIN CHAPMAN
University of Exeter

CAMBRIDGE
UNIVERSITY PRESS

CAMBRIDGE
UNIVERSITY PRESS

University Printing House, Cambridge CB2 8BS, United Kingdom

One Liberty Plaza, 20th Floor, New York, NY 10006, USA

477 Williamstown Road, Port Melbourne, VIC 3207, Australia

314-321, 3rd Floor, Plot 3, Splendor Forum, Jasola District Centre, New Delhi - 110025, India

103 Penang Road, #05-06/07, Visioncrest Commercial, Singapore 238467

Cambridge University Press is part of the University of Cambridge.

It furthers the University's mission by disseminating knowledge in the pursuit of education, learning and research at the highest international levels of excellence.

www.cambridge.org
Information on this title: www.cambridge.org/9781107601093

© Cambridge University Press 2011

First published 2011

A catalogue record for this publication is available from the British Library

ISBN 978-1-107-60109-3 Paperback

Contents

Preface

The Twenty-Third British Combinatorial Conference was organised by the University of Exeter. It was held in Exeter in July 2011. The British Combinatorial Committee had invited nine distinguished combinatorialists to give survey lectures in areas of their expertise, and this volume contains the survey articles on which these lectures were based.

In compiling this volume I am indebted to the authors for preparing their articles so accurately and professionally, and to the referees for their rapid responses and keen eye for detail. I would also like to thank Roger Astley, Silvia Barbina and Clare Dennison at Cambridge University Press for their advice, assistance and patience. Finally, without the previous efforts of editors of earlier *Surveys* and the guidance of the British Combinatorial Committee, the preparation of this volume would have been quite impossible.

Robin Chapman

University of Exeter

January 2011

Counting planar maps, coloured or uncoloured

Mireille Bousquet-Mélou

Abstract

We present recent results on the enumeration of q-coloured planar maps, where each monochromatic edge carries a weight ν. This is equivalent to weighting each map by its Tutte polynomial, or to solving the q-state Potts model on random planar maps. The associated generating function, obtained by Olivier Bernardi and the author, is differentially algebraic. That is, it satisfies a (nonlinear) differential equation. The starting point of this result is a functional equation written by Tutte in 1971, which translates into enumerative terms a simple recursive description of planar maps. The proof follows and adapts Tutte's solution of *properly* q-coloured triangulations (1973-1984).

We put this work in perspective with the much better understood enumeration of families of *uncoloured* planar maps, for which the recursive approach almost systematically yields algebraic generating functions. In the past 15 years, these algebraicity properties have been explained combinatorially by illuminating bijections between maps and families of plane trees. We survey both approaches, recursive and bijective.

Comparing the coloured and uncoloured results raises the question of designing bijections for coloured maps. No complete bijective solution exists at the moment, but we present bijections for certain specialisations of the general problem. We also show that for these specialisations, Tutte's functional equation is much easier to solve that in the general case.

We conclude with some open questions.

1 Introduction

A planar map is a proper embedding in the sphere of a finite connected graph, defined up to continuous deformation. The enumeration of these objects has been a topic of constant interest for 50 years, starting with a series of papers by Tutte in the early 1960s; these papers were mostly based on recursive descriptions of maps (e.g. [103]). The last 15 years have witnessed a new burst of activity in this field, with the development of rich bijective approaches [98, 39], and their applications to the study of random maps of large size [78, 85]. In such enumerative problems, maps are usually *rooted* by orienting one edge. Figure 1 sets a first exercise in map enumeration.

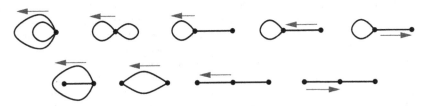

Figure 1: There are 9 rooted planar maps with two edges.

1

Planar maps are not only studied in combinatorics and probability, but also in theoretical physics. In this context, maps are considered as random surfaces, and constitute a model of 2-dimensional *quantum gravity*. For many years, maps were studied independently in combinatorics and in physics, and another approach for counting them, based on the evaluation of certain matrix integrals, was introduced in the 1970s in physics [42, 18], and much developed since then [55, 88]. More recently, a fruitful exchange started between the two communities. Some physicists have become masters in combinatorial methods [35, 37], while the matrix integral approach has been taken over by some probabilists [71].

From the physics point of view, it is natural to equip maps with additional structures, like particles, trees, spins, and more generally classical models of statistical physics. In combinatorics however, a huge majority of papers deal with the enumeration of bare maps. There has been some exceptions to this rule in the past few years, with combinatorial solutions of the Ising and hard-particle models on planar maps [34, 38, 39]. But there is also an earlier, and major, exception to this rule: Tutte's study of properly q-coloured triangulations (Figure 2).

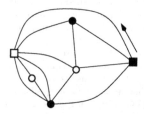

Figure 2: A (rooted) triangulation of the sphere, properly coloured with 4 colours.

This ten years long study (1973-1984) plays a central role in this paper. For a very long time, it remained an isolated *tour de force* with no counterpart for other families of planar maps or for more general colourings, probably because the corresponding series of papers [110, 108, 107, 109, 111, 112, 113, 114, 115, 116] looks quite formidable. Our main point here is to report on recent advances in the enumeration of (non-necessarily properly) q-coloured maps, in the steps of Tutte. In the associated generating function, every monochromatic edge is assigned a weight ν: the case $\nu = 0$ thus captures proper colourings. In physics terms, we are studying the q-state Potts model on planar maps. A third equivalent formulation is that we count planar maps weighted by their Tutte polynomial — a bivariate generalisation of the chromatic polynomial, introduced by Tutte, who called it the dichromatic polynomial. Since the Tutte polynomial has numerous interesting specialisations, giving for example the number of trees, forests, acyclic orientations, proper colourings of course, or the partition function of the Ising model, or the reliability and flow polynomials, we are covering several models at the same time.

We shall put this work in perspective with the (much better understood) enumeration of uncoloured maps, to which we devote Sections 3 and 4. We first present in Section 3 the robust recursive approach found in the early work of Tutte. It applies in a rather uniform way to many families of maps, and yields for their generating functions functional equations that we call *polynomial equations with one catalytic variable.* A typical example is (3.1). It is now understood that the solutions of these equations are always algebraic, that is, satisfy a polynomial equation. For instance, there are $2 \cdot 3^n \binom{2n}{n}/((n+1)(n+2))$ rooted planar maps with n edges, and their generating function, that is, the series

$$M(t) := \sum_{n \geq 0} \frac{2 \cdot 3^n}{(n+1)(n+2)} \binom{2n}{n} t^n,$$

satisfies

$$M(t) = 1 - 16t + 18tM(t) - 27t^2 M(t)^2.$$

Thus algebraicity is intimately connected with (uncoloured) planar maps. In Section 4, we present two more recent bijective approaches that relate maps to plane trees, which are algebraic objects *par excellence*. Not only do these bijections give a better understanding of algebraicity properties, but they also explain why many families of maps are counted by simple formulas.

In Section 5, we discuss the recursive approach for q-coloured maps. The corresponding functional equation (5.3) was written in 1971 by Tutte —who else?—, but was left untouched since then. It involves two "catalytic" variables, and it has been known for a long time that its solution is not algebraic. The key point of this section, due to Olivier Bernardi and the author, is the solution of this equation, in the form of a system of differential equations that defines the generating function of q-coloured maps. This series is thus *differentially algebraic*, like Tutte's solution of properly coloured triangulations. Halfway on the long path that leads to the solution stands an interesting intermediate result: when $q \neq 4$ is of the form $2 + 2\cos(j\pi/m)$, for integers j and m, the generating function of q-coloured planar maps is algebraic. This includes the values $q = 2$ and $q = 3$, for which we give explicit results. We also discuss certain specialisations for which the equation becomes easier to solve, like the enumeration of maps equipped with a bipolar orientation, or with a spanning tree.

Since we are still in the early days of the enumeration of coloured maps, it is not surprising that bijective approaches are at the moment one step behind. Still, a few bijections are available for some of the simpler specialisations mentioned above. They are presented in Section 6. We conclude with open questions, dealing with both uncoloured and coloured enumeration.

This survey is sometimes written in an informal style, especially when we describe bijections. Proofs are only given when they are new, or especially simple and illuminating. The reference list, although long, is certainly not exhaustive. In particular, the papers cited in this introduction are just examples illustrating our topic, and should be considered as pointers to the relevant literature. More references are given further in the paper. Two approaches that have been used to count maps are utterly absent from this paper: methods based on characters of the symmetric group and symmetric functions [68, 69], which do not exactly address the same range of problems, and the matrix integral approach, which is powerful [55], but is not always fully rigorous. The Potts model has been addressed via matrix integrals [51, 56, 123]. We refer to [15] for a description our current understanding of this work.

2 Definitions and notation

2.1 Planar maps

A *planar map* is a proper embedding of a connected planar graph in the oriented sphere, considered up to orientation preserving homeomorphism. Loops and multiple edges are allowed. The *faces* of a map are the connected components of its complement. The numbers of vertices, edges and faces of a planar map M, denoted by $v(M)$, $e(M)$ and $f(M)$, are related by Euler's relation $v(M) + f(M) = e(M) + 2$. The *degree* of a vertex or face is the number of edges incident to it, counted with multiplicity. A map is *m-valent* if all its vertices have degree m. A *corner* is a sector delimited by two consecutive edges around a vertex; hence a vertex or face of degree k defines k corners. The *dual* of a map M, denoted M^*, is the map obtained by placing a vertex of M^* in each face of M and an edge of M^* across each edge of M; see Figure 3.

For counting purposes it is convenient to consider *rooted* maps. A map is rooted by orienting an edge, called the *root-edge*. The origin of this edge is the *root-vertex*. The face that lies to the right of the root-edge is the *root-face*. In figures, we take the root-face as the infinite face (Figure 3). This explains why we often call the root-face the *outer* (or: *infinite*) face, and its degree the *outer degree*. The other faces are said to be *finite*. From now on, every map is *planar* and *rooted*. By convention, we include among rooted planar maps the *atomic* map m_0 having one vertex and no edge. The set of rooted planar maps is denoted $\mathcal{M}$.

A map is *separable* if it is atomic or can be obtained by gluing two non-atomic maps at a vertex. Observe that both maps with one edge are non-separable.

Figure 3: A rooted planar map and its dual (rooted at the dual edge).

2.2 Power series

Let A be a commutative ring and x an indeterminate. We denote by $A[x]$ (resp. $A[[x]]$) the ring of polynomials (resp. formal power series) in x with coefficients in A. If A is a field, then $A(x)$ denotes the field of rational functions in x, and $A((x))$ the field of Laurent series[1] in x. These notations are generalised to polynomials, fractions and series in several indeterminates. We denote by bars the reciprocals of variables: that is, $\bar{x} = 1/x$, so that $A[x, \bar{x}]$ is the ring of Laurent polynomials in x with coefficients in A. The coefficient of x^n in a Laurent series $F(x)$ is denoted

---

[1] A *Laurent series* is a series of the form $\sum_{n \geq n_0} a(n)x^n$, for some $n_0 \in \mathbf{Z}$.

by $[x^n]F(x)$. The *valuation* of a Laurent series $F(x)$ is the smallest d such that x^d occurs in $F(x)$ with a non-zero coefficient. If $F(x) = 0$, then the valuation is $+\infty$. If $F(x;t)$ is a power series in t with coefficients in $A((x))$, that is, a series of the form

$$F(x;t) = \sum_{n \geq 0, i \in \mathbf{Z}} f(i;n)x^i t^n,$$

where for all n, almost all coefficients $f(i;n)$ such that $i < 0$ are zero, then the *positive part* of $F(x;t)$ in x is the following series, which has coefficients in $xA[[x]]$:

$$[x^>]F(x;t) := \sum_{n \geq 0, i > 0} f(i;n)x^i t^n.$$

We define similarly the non-negative part of $F(x;t)$ in x.

A power series $F(x_1, \ldots, x_k) \in \mathbb{K}[[x_1, \ldots, x_k]]$, where $\mathbb{K}$ is a field, is *algebraic* (over $\mathbb{K}(x_1, \ldots, x_k)$) if it satisfies a polynomial equation $P(x_1, \ldots, x_k, F(x_1, \ldots, x_k)) = 0$. The series $F(x_1, \ldots, x_k)$ is *D-finite* if for all $i \leq k$, it satisfies a (non-trivial) linear differential equation in x_i with coefficients in $\mathbb{K}[x_1, \ldots, x_k]$. We refer to [81, 82] for a study of these series. All algebraic series are D-finite. A series $F(x)$ is *differentially algebraic* if it satisfies a (non-necessarily linear) differential equation with coefficients in $\mathbb{K}[x]$.

2.3 The Potts model and the Tutte polynomial

Let G be a graph with vertex set $V(G)$ and edge set $E(G)$. Let ν be an indeterminate, and take $q \in \mathbf{N}$. A *colouring* of the vertices of G in q colours is a map $c : V(G) \to \{1, \ldots, q\}$. An edge of G is *monochromatic* if its endpoints share the same colour. Every loop is thus monochromatic. The number of monochromatic edges is denoted by $m(c)$. The *partition function of the Potts model* on G counts colourings by the number of monochromatic edges:

$$P_G(q, \nu) = \sum_{c:V(G) \to \{1, \ldots, q\}} \nu^{m(c)}.$$

The Potts model is a classical magnetism model in statistical physics, which includes (for $q = 2$) the famous Ising model (with no magnetic field) [120]. Of course, $P_G(q, 0)$ is the chromatic polynomial of G.

If G_1 and G_2 are disjoint graphs and $G = G_1 \cup G_2$, then clearly

$$P_G(q, \nu) = P_{G_1}(q, \nu) \, P_{G_2}(q, \nu). \tag{2.1}$$

If G is obtained by attaching G_1 and G_2 at one vertex, then

$$P_G(q, \nu) = \frac{1}{q} \, P_{G_1}(q, \nu) \, P_{G_2}(q, \nu). \tag{2.2}$$

The Potts partition function can be computed by induction on the number of edges. If G has no edge, then $P_G(q, \nu) = q^{|V(G)|}$. Otherwise, let e be an edge of G. Denote by $G \backslash e$ the graph obtained by deleting e, and by G/e the graph obtained by contracting e (if e is a loop, then it is simply deleted). Then

$$P_G(q, \nu) = P_{G \backslash e}(q, \nu) + (\nu - 1) \, P_{G/e}(q, \nu). \tag{2.3}$$

Indeed, it is not hard to see that $\nu\, P_{G/e}(q,\nu)$ counts colourings for which e is monochromatic, while $P_{G\backslash e}(q,\nu) - P_{G/e}(q,\nu)$ counts those for which e is bichromatic. One important consequence of this induction is that $P_G(q,\nu)$ is always a *polynomial* in q and ν. We call it the *Potts polynomial* of G. Since it is a polynomial, we will no longer consider q as an integer, but as an indeterminate, and sometimes evaluate $P_G(q,\nu)$ at real values q. We also observe that $P_G(q,\nu)$ is a multiple of q: this explains why we will weight maps by $P_G(q,\nu)/q$.

Up to a change of variables, the Potts polynomial is equivalent to another, maybe better known, invariant of graphs, namely the *Tutte polynomial* $T_G(\mu,\nu)$ (see e.g. [19]):

$$T_G(\mu,\nu) := \sum_{S\subseteq E(G)} (\mu-1)^{c(S)-c(G)}(\nu-1)^{e(S)+c(S)-v(G)},$$

where the sum is over all spanning subgraphs of G (equivalently, over all subsets of edges) and v(.), e(.) and c(.) denote respectively the number of vertices, edges and connected components. For instance, the Tutte polynomial of a graph with no edge is 1. The equivalence with the Potts polynomial was established by Fortuin and Kasteleyn [62]:

$$P_G(q,\nu) = \sum_{S\subseteq E(G)} q^{c(S)}(\nu-1)^{e(S)} = (\mu-1)^{c(G)}(\nu-1)^{v(G)}\, T_G(\mu,\nu), \qquad (2.4)$$

for $q = (\mu-1)(\nu-1)$. In this paper, we work with P_G rather than T_G because we wish to assign real values to q (this is more natural than assigning real values to $(\mu-1)(\nu-1)$). However, one property looks more natural in terms of T_G: if G and G^* are dual connected planar graphs (that is, if G and G^* can be embedded as dual planar maps) then

$$T_{G^*}(\mu,\nu) = T_G(\nu,\mu).$$

Translating this identity in terms of Potts polynomials thanks to (2.4) gives:

$$\begin{aligned}
P_{G^*}(q,\nu) &= q(\nu-1)^{v(G^*)-1}\, T_{G^*}(\mu,\nu) \\
&= q(\nu-1)^{v(G^*)-1}\, T_G(\nu,\mu) \\
&= \frac{(\nu-1)^{e(G)}}{q^{v(G)-1}}\, P_G(q,\mu), \qquad (2.5)
\end{aligned}$$

where $\mu = 1 + q/(\nu-1)$ and the last equality uses Euler's relation: $v(G) + v(G^*) - 2 = e(G)$.

3   Uncoloured planar maps: the recursive approach

In this section, we describe the first approach that was used to count maps: the recursive method. It is based on very simple combinatorial operations (like the deletion or contraction of an edge), which translate into non-trivial functional equations defining the generating functions. A recent theorem, generalising the so-called *quadratic method*, states that the solutions of all equations of this type are algebraic. Since the recursive method applies to many families of maps, numerous algebraicity results follow.

3.1 A functional equation for planar maps

Consider a rooted planar map, distinct from the atomic map. Delete the root-edge. If this edge is an isthmus, one obtains two connected components M_1 and M_2, and otherwise a single component M, which we can root in a canonical way (Figure 4). Conversely, starting from an ordered pair (M_1, M_2) of maps, there is a unique way to connect them by a new (root) edge. If one starts instead from a single map M, there are $d+1$ ways to add a root edge, where $d = \mathrm{df}(M)$ is the degree of the root-face of M (Figure 5).

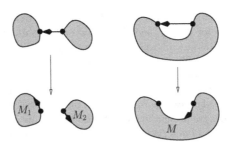

Figure 4: Deletion of the root-edge in a planar map.

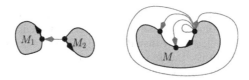

Figure 5: Reconstruction of a planar map.

Hence, to derive from this recursive description of planar maps a functional equation for their generating function, we need to take into account the degree of the root-face, by an additional variable y. Hence, let

$$M(t; y) = \sum_{M \in \mathcal{M}} t^{\mathrm{e}(M)} y^{\mathrm{df}(M)} = \sum_{d \geq 0} y^d M_d(t)$$

be the generating function of planar maps, counted by edges and outer-degree. The series $M_d(t)$ counts by edges maps with outer degree d. The recursive description of maps translates as follows:

$$\begin{aligned}
M(t; y) &= 1 + y^2 t M(t; y)^2 + t \sum_{d \geq 0} M_d(t)(y + y^2 + \cdots + y^{d+1}) \\
&= 1 + y^2 t M(t; y)^2 + t y \frac{y M(t; y) - M(t; 1)}{y - 1}.
\end{aligned} \tag{3.1}$$

Indeed, connecting two maps M_1 and M_2 by an edge produces a map of outer-degree $\mathrm{df}(M_1) + \mathrm{df}(M_2) + 2$, while the $d + 1$ ways to add an edge to a map M such that $\mathrm{df}(M) = d$ produce $d + 1$ maps of respective outer degree $1, 2, \ldots, d + 1$, as can be seen on Figure 5. The term 1 records the atomic map.

The above equation was first written by Tutte in 1968 [105]. It is typical of the type of equation obtained in (recursive) map enumeration. More examples will be given in Section 3.2. One important feature in this equation is the divided difference

$$\frac{yM(t; y) - M(t; 1)}{y - 1},$$

which prevents us from simply setting $y = 1$ to solve for $M(t; 1)$ first, and then for $M(t; y)$. The parameter $\mathrm{df}(M)$, and the corresponding variable y, are said to be *catalytic* for this equation — a terminology borrowed to Zeilberger [122].

Such equations do not only occur in connection with maps: they also arise in the enumeration of polyominoes [24, 59, 101], lattice walks [31, 3, 52, 76, 96], permutations [25, 28, 121]... The solution of these equations has naturally attracted some interest. The "guess and check" approach used in the early 1960s is now replaced by a general method, which we present below in Section 3.3. This method implies in particular that *the solution of any (well-founded) polynomial equation with one catalytic variable is algebraic*. It generalises the *quadratic method* developed by Brown [46] for equations of degree 2 that involve a single additional unknown series (like $M(t; 1)$ in the equation above) and also the *kernel method* that applies to linear equations, and seems to have first appeared in Knuth's *Art of Computer Programming* [76, Section 2.2.1, Ex. 4] (see also [2, 31, 96]).

Contraction vs. deletion. Before we move to more examples, let us make a simple observation. Another natural way to decrease the edge number of a map is to contract the root-edge, rather than delete it (if this edge is a loop, one just erases it). When one tries to use this to count planar maps, one is lead to introduce the degree of the root-vertex as a catalytic parameter, and a corresponding variable x in the generating function. This yields the same equation as above:

$$M(t; x) = 1 + x^2 t M(t; x)^2 + t \sum_{d \geq 0} M_d(t)(x + x^2 + \cdots + x^{d+1}).$$

As illustrated by Figure 6, the term 1 records the atomic map, the second term corresponds to maps in which the root-edge is a loop, and the third term to the remaining cases. In particular, the sum $(x + x^2 + \cdots + x^{d+1})$ now describes how to

Figure 6: Contraction of the root-edge in a planar map.

distribute the adjacent edges when a new edge is inserted. Given that the contraction operation is the dual of the deletion operation, it is perfectly natural to obtain the same equation as before. The reason why we mention this alternative construction is that, when we establish below a functional equation for maps weighted by their Potts (or Tutte) polynomial, we will have to use simultaneously these two operations, as suggested by the recursive description (2.3) of the Potts polynomial. This will naturally result in equations with *two* catalytic variables x and y.

3.2   More functional equations

The recursive method is extremely robust. We illustrate this by a few examples. Two of them — maps with prescribed face degrees, and Eulerian maps with prescribed face degrees — actually cover infinitely many families of maps. Some of these examples also have a colouring flavour.

Maps with prescribed face degrees.   Consider for instance the enumeration of *triangulations*, that is, maps in which all faces have degree 3. The recursive deletion of the root-edge gives maps in which all finite faces have degree 3, but the outer face may have any degree: these maps are called *near-triangulations*. We denote by T the set of near-triangulations. The deletion of the root-edge in a near triangulation gives either two near-triangulations, or a single one, *the outer degree of which is at least two* (Figure 7). In both cases, there is unique way to reconstruct the map we started from. Let $T(t; y) \equiv T(y)$ be the generating function of near-triangulations, counted by edges and by the outer degree:

$$T(t; y) = \sum_{M \in T} t^{e(M)} y^{df(M)} = \sum_{d \geq 0} y^d T_d(t).$$

Figure 7: Deletion of the root-edge in a near-triangulation.

The above recursive description translates into

$$T(y) = 1 + t y^2 T(y)^2 + t \frac{T(y) - T_0 - y T_1}{y}, \tag{3.2}$$

where $T_0 = 1$ counts the atomic map. We have again a divided difference, this time at $y = 0$. Its combinatorial interpretation ("it is forbidden to add an edge to a map of outer degree 0 or 1") differs from the interpretation of the divided difference occurring in (3.1) ("there are multiple ways to add an edge"). Still, both equations are of the same type and will be solved by the same method. Note that we have omitted the variable t in the notation $T(y)$, which we will do quite often in this paper, to avoid heavy notation and enhance the catalytic parameter(s).

Consider now *bipartite* planar maps, that is, maps that admit a proper 2-colouring (and then a unique one, if the root-vertex is coloured white). For planar maps, this is equivalent to saying that all faces have an even degree. Let $B(t; y) = \sum_{d \geq 0} B_d(t) y^d$ be the generating function of bipartite maps, counted by edges (variable t) and by half the outer degree (variable y). Then the deletion of the root-edge translates as follows (Figure 8):

$$
\begin{aligned}
B(y) &= 1 + ty B(y)^2 + t \sum_{d \geq 0} B_d(y + y^2 + \cdots + y^d) \\
&= 1 + ty B(y)^2 + ty \frac{B(y) - B(1)}{y - 1}.
\end{aligned}
\tag{3.3}
$$

This is again a quadratic equation with one catalytic variable, y.

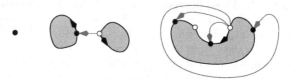

Figure 8: Deletion of the root-edge in a bipartite map.

More generally, it was shown by Bender and Canfield [6] that the recursive approach applies to any family of maps for which the face degrees belong to a given set D, provided D differs from a finite union of arithmetic progressions by a finite set. In all cases, the equation is quadratic, but may involve more than a single additional unknown function. For instance, when counting near-quadrangulations rather than near-triangulations, Eq. (3.2) is replaced by

$$
Q(y) = 1 + ty^2 Q(y)^2 + t \frac{Q(y) - Q_0 - yQ_1 - y^2 Q_2}{y^2},
$$

where Q_i counts near-quadrangulations of outer degree i. Bender and Canfield solved these equations using a theorem of Brown from which the quadratic method is derived, proving in particular that the resulting generating function is always algebraic. Their result only involves the edge number, but, when D is finite, it can be refined by keeping track of the vertex degree distribution [29].

Eulerian maps with prescribed face degrees. A planar map is Eulerian if all vertices have an even degree. Equivalently, its faces admit a proper 2-colouring (and a unique one, if the root-face is coloured white). Of course, Eulerian maps are the duals of bipartite maps, so that their generating function (by edges, and half-degree of the root-vertex) satisfies (3.3). But we wish to impose conditions on the face degrees of Eulerian maps (dually, on the vertex degrees of bipartite maps). This includes as a special case the enumeration of (non-necessarily Eulerian) maps with prescribed face degrees, discussed in the previous paragraph: indeed, if we require that all black faces of an Eulerian map have degree 2, each black face can be contracted into a single edge, leaving a standard map with prescribed (white) face degrees.

Generally speaking, it is difficult to count families of maps with conditions on the vertex degrees *and* on the face degrees (and being Eulerian is a condition on vertex degrees). However, it was shown in [29] that the enumeration of Eulerian maps such that all black faces have degree in $D_\bullet$ and all white faces have degree in $D_\circ$ can be addressed by the recursive method when $D_\bullet$ and $D_\circ$ are finite. This is also true when $D_\bullet = \{m\}$ and $D_\circ = m\mathbf{N}$ (such maps are called *m-constellations*).

Let us take the example of Eulerian near-triangulations. All finite faces have degree 3, while the infinite face, which is white by convention, has degree $3d$ for some $d \in \mathbf{N}$. In order to decompose these maps, we now delete all the edges that bound the black face adjacent to the root-edge (Figure 9). This leaves 1, 2 or 3 connected components, which are themselves Eulerian near-triangulations, and which we root in a canonical way. Let $E(z; y) \equiv E(y) = \sum_{d \geq 0} E_d(z) y^d$ be the generating function of Eulerian near-triangulations, counted by black faces (variable z) and by the outer degree, divided by 3 (variable y). The above decomposition gives:

$$E(y) = 1 + zyE(y)^3 + 2zE(y)(E(y) - E_0) + z(E(y) - E_0) + z\,\frac{E(y) - E_0 - yE_1}{y}.$$

This is a cubic equation with one catalytic variable, which is routinely solved by the method presented below in Section 3.3.

Figure 9: Decomposition of Eulerian near-triangulations.

The enumeration of Eulerian triangulations is often presented as a colouring problem [36, 54], for the following reason: a planar triangulation admits a proper 3-colouring of its vertices if and only if it is (properly) face-bicolourable, that is, Eulerian[2]. Moreover, fixing the colours of the endpoints of the root-edge determines completely the colouring. More generally, let us say that a q-colouring is *cyclic* if around any face, one meets either the colours $1, 2, \ldots q, 1, 2, \ldots, q$, in this order, or $q, q-1, \ldots, 1, q, q-1, \ldots, 1$. Then for $q \geq 3$, a planar map admits a cyclic q-colouring if and only if it is Eulerian and all its face degrees are multiples of q. In this case, it has exactly $2q$ cyclic colourings. The m-constellations defined above are of this type (with $m = q$).

Other families of maps. Beyond the two general enumeration problems we have just discussed, the recursive approach applies to many other families of planar maps:

---

[2]It is easy to see that the condition is necessary: around a face, in clockwise order, one meets either the colours 1, 2, 3 in this order, or 3, 2, 1, and all faces that are adjacent to a 123-face are of the 321-type. The converse is easily seen to hold by induction on the face number, using Figure 9.

loopless maps [8, 119], maps with higher connectivity [43, 47, 67], dissections of a regular polygon [44, 45, 103], triangulations with large vertex degrees [13], maps on surfaces of higher genus [5, 7, 65]... The resulting equations are often fruitfully combined with *composition equations* that relate the generating functions of two families of maps, for instance general planar maps and non-separable planar maps (see. e.g., [104, Eq. (6.3)] or [103, Eq. (2.5)]).

3.3 Equations with one catalytic variable and algebraicity theorems

In this section, we state a general theorem that implies that the solutions of all the functional equations we have written so far are algebraic. We then explain how to solve in practice these equations. The method extends the *quadratic method* that applies to quadratic equations with a unique additional unknown series [68, Section 2.9].

Let $\mathbb{K}$ be a field of characteristic 0, typically $\mathbf{Q}(s_1, \ldots, s_k)$ for some indeterminates $s_1, \ldots, s_k$. Let $F(y) \equiv F(t; y)$ be a power series in $\mathbb{K}(y)[[t]]$, that is, a series in t with rational coefficients in y. Assume that these coefficients have no pole at $y = 0$. The following divided difference (or discrete derivative) is then well-defined:

$$\Delta F(y) = \frac{F(y) - F(0)}{y}.$$

Note that

$$\lim_{y \to 0} \Delta F(y) = F'(0),$$

where the derivative is taken with respect to y. The operator $\Delta^{(i)}$ is obtained by applying i times Δ, so that:

$$\Delta^{(i)} F(y) = \frac{F(y) - F(0) - yF'(0) - \cdots - y^{i-1}/(i-1)! \, F^{(i-1)}(0)}{y^i}.$$

Now

$$\lim_{y \to 0} \Delta^{(i)} F(y) = \frac{F^{(i)}(0)}{i!}.$$

Assume $F(t; y)$ satisfies a functional equation of the form

$$F(y) \equiv F(t; y) = F_0(y) + t \, Q\Big(F(y), \Delta F(y), \Delta^{(2)} F(y), \ldots, \Delta^{(k)} F(y), t; y \Big), \quad (3.4)$$

where $F_0(y) \in \mathbb{K}(y)$ and $Q(y_0, y_1, \ldots, y_k, t; y)$ is a polynomial in the $k + 2$ indeterminates $y_0, y_1, \ldots, y_k, t$, and a rational function in the last indeterminate y, having coefficients in $\mathbb{K}$. This equation thus involves, in addition to $F(y)$ itself, k additional unknown series, namely $F^{(i)}(0)$ for $0 \le i < k$.

Theorem 3.1 ([29, 15]) *Under the above assumptions, the series $F(t; y)$ is algebraic over $\mathbb{K}(t, y)$.*

In practice, one proceeds as follows to obtain an algebraic system of equations defining the k unknown series $F^{(i)}(0)$. An example will be detailed further down. Write (3.4) in the form

$$P(F(y), F(0), \ldots, F^{(k-1)}(0), t; y) = 0, \quad (3.5)$$

for some polynomial $P(y_0, y_1, \ldots, y_k, t; y)$, and consider the following equation in Y:

$$\frac{\partial P}{\partial y_0}(F(Y), F(0), \ldots, F^{(k-1)}(0), t; Y) = 0. \tag{3.6}$$

On explicit examples, it is usually easy to see that this equation admits k solutions $Y_0, \ldots, Y_{k-1}$ in the ring of Puiseux series in t with a non-negative valuation (a Puiseux series is a power series in a fractional power of t, for instance a series in $\sqrt{t}$). By differentiating (3.5) with respect to y, it then follows that

$$\frac{\partial P}{\partial y}(F(Y), F(0), \ldots, F^{(k-1)}(0), t; Y) = 0. \tag{3.7}$$

Hence the following system of $3k$ algebraic equations holds: for $i = 0, \ldots, k-1$,

$$
\begin{aligned}
P(F(Y_i), F(0), \ldots, F^{(k-1)}(0), t; Y_i) &= 0, \\
\frac{\partial P}{\partial y_0}(F(Y_i), F(0), \ldots, F^{(k-1)}(0), t; Y_i) &= 0, \\
\frac{\partial P}{\partial y}(F(Y_i), F(0), \ldots, F^{(k-1)}(0), t; Y_i) &= 0.
\end{aligned}
\tag{3.8}
$$

This system involves $3k$ unknown series, namely Y_i, $F(Y_i)$, and $F^{(i)}(0)$ for $0 \leq i < k$. The fact that the series $F^{(i)}(0)$ are derivatives of F plays no particular role. Observe that the above system consists of k times the *same* triple of equations, so that elimination in this system is not obvious [29] (and will often end up being very heavy). When $k = 1$, however, obtaining a solution takes three lines in Maple. Consider for instance the equation (3.2) we have obtained for near-triangulations. Eq. (3.6) reads in this case

$$Y = t + 2tY^3 T(Y),$$

and it is clear that it has a unique solution, which is a formal power series in t with constant term 0. Indeed, the coefficient of t^n in Y can be determined inductively in terms of the coefficients of T. Then (3.7) reads

$$T(Y) = 1 + 3tY^2 T(Y)^2 - tT_1.$$

These two equations, combined with the original equation (3.2) taken at $y = Y$, form a system of three polynomial equations involving $Y, T(Y)$ and T_1, from which Y and $T(Y)$ are readily eliminated by taking resultants. This leaves a polynomial equation for the unknown series T_1, which counts near-triangulations of outer degree 1:

$$T_1 = t^2 - 27t^5 + 30\,t^3 T_1 + t(1 - 96\,t^3)T_1{}^2 + 64\,t^5 T_1{}^3.$$

One can actually go further and obtain simple expressions for the coefficients of T_1. The above equation admits rational parametrisations, for instance

$$t^3 = X(1 - 2X)(1 - 4X), \quad tT_1 = \frac{X(1 - 6X)}{1 - 4X},$$

and the Lagrange inversion formula yields the number of near-triangulations of outer degree 1 having $3n + 2$ edges (hence $n + 2$ vertices) as

$$\frac{2 \cdot 4^n (3n)!!}{n!!(n+2)!},$$

where $n!! = n(n-2)(n-4)\cdots(n - 2\lfloor \frac{n-1}{2} \rfloor)$. The existence of such simple formulas will be discussed further, in connection with bijective approaches (Section 4).

Algebraicity results. The general algebraicity result for solutions of polynomial equations with one catalytic variable (Theorem 3.1), combined with the wide applicability of the recursive method, implies that many families of planar maps have an algebraic generating function. In the following theorem, the term *generating function* refers to the generating function by vertices, faces and edges (of course, one of these statistics is redundant, by Euler's formula).

Theorem 3.2 ([6, 29]) *For any set* $D \subset \mathbf{N}$ *that differs from a finite union of arithmetic progressions by a finite set, the generating function of maps such that all faces have their degree in D is algebraic. If D is finite, this holds has well for the refined generating function that keeps track of the number of i-valent faces, for all $i \in D$.*

For any finite sets $D_\circ$ and $D_\bullet$ in $\mathbf{N}$, the generating function of face-bicoloured maps such that all white (resp. black) faces have their degree in $D_\circ$ (resp. $D_\bullet$) is algebraic. This holds as well for the generating function that keeps track of the number of i-valent white and black faces, for all $i \in \mathbf{N}$.

Finally, the generating function of face-bicoloured planar maps such that all black faces have degree m, and all white faces have their degree in $m\mathbf{N}$, is algebraic.

Where is the quadratic method? To finish this section, let us briefly sketch why the above procedure for solving equations with one catalytic variable generalises the quadratic method. The first two equations of (3.8) show that $y_0 = F(Y_i)$ is a double root of $P(y_0, F(0), \ldots, F^{k-1}(0), t; Y_i)$. Hence $y = Y_i$ cancels the discriminant of $P(y_0, F(0), \ldots, F^{k-1}(0), t; y)$, taken with respect to y_0. When P has degree 2 in y_0, it is easy to see that the third equation of (3.8) means that Y_i is actually a double root of the discriminant [29, Section 3.2]: this is the heart of the quadratic method, described in [68, Section 2.9]. That each series Y_i is a multiple root of the discriminant actually holds for equations of higher degree, but this is far from obvious [29, Section 6].

4 Uncoloured planar maps: bijections

So far, we have emphasised the fact that many families of planar maps have an algebraic generating function. It turns out that many of them are also counted by remarkably simple numbers, which have a strong flavour of tree enumeration. Both observations raise a natural question: is it possible to explain the algebraicity and/or the numbers more combinatorially, via bijections that would relate maps to trees?

We present in this section two bijections between planar maps and some families of trees that allow one to determine very elegantly the number of planar maps having n edges. The first bijection also explains combinatorially why the associated generating function is algebraic. The second one has other virtues, as it allows to record the distances of vertices to the root-vertex. This property has proved extremely useful is the study of random maps of large size and their scaling limit [49, 78, 80, 86, 85]. Both bijections could probably qualify as *Proofs from The Book* [1]. Both are robust enough to be generalised to many other families of maps, as discussed in Section 4.2.

4.1 Two proofs from The Book?

Both types of bijections involve families of maps with bounded vertex degrees (or, dually, bounded face degrees). So let us first recall why planar maps are equivalent to planar 4-valent maps (or dually, to quadrangulations).

Take a planar map, and create a vertex in the middle of each edge, called *e-vertex* to distinguish it from the vertices of the original map. Then, turning inside each face, join by an edge each pair of consecutive e-vertices (Figure 10). The e-vertices, together with these new edges, form a 4-valent map. Root this map in a canonical way. This construction is a bijection between rooted planar maps with n edges and rooted 4-valent maps with n vertices.

Figure 10: A planar map with n edges and the corresponding 4-valent planar map with n vertices (dashed lines).

Four-valent maps and blossoming trees. The first bijection, due to Schaeffer [98], transforms 4-valent maps into *blossoming trees*. A blossoming tree is a (plane) binary tree, rooted at a leaf, such that every inner node carries, in addition to its two children, a *flower* (Figure 11). There are three possible positions for each flower. If the tree has n inner nodes, it has n flowers and $n + 2$ leaves. Flowers and leaves are called *half-edges*.

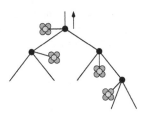

Figure 11: A blossoming tree with $n = 4$ inner nodes.

One obtains a blossoming tree by *opening* certain edges of a 4-valent map (Figure 12). Take a 4-valent map M. First, cut the root-edge into two half-edges, that become leaves: the first of them will be the root of the final tree. Then, start walking around the infinite face in counterclockwise order, beginning with the root edge. Each time a non-separating[3] edge has just been visited, cut it into two half-edges: the first one becomes a flower, and the second one, a leaf. Proceed until all edges

---

[3]An edge is *separating* if its deletion disconnects the map, *non-separating* otherwise; a map is a tree if and only if all its edges are separating.

are separating edges; this may require to turn several times around the map. The final result is a blossoming tree, denoted $\Psi(M)$.

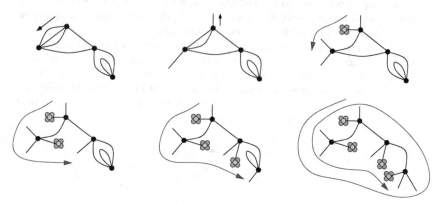

Figure 12: Opening the edges of a 4-valent map gives a blossoming tree.

Conversely, one can construct a map by matching leaves and flowers in a blossoming tree T as follows (Figure 13, left). Starting from the root, walk around the infinite face of T in counterclockwise order. Each time a flower is immediately followed by a leaf in the cyclic sequence of half-edges, merge them into an edge in counterclockwise direction; this creates a new finite face that encloses no unmatched half-edges. Stop when all flowers have been matched. At this point, exactly two leaves remain unmatched (because there are n flowers and $n + 2$ leaves). Observe that the same two leaves remain unmatched if one starts walking around the tree from another position than the root. The tree T is said to be *balanced* if one of the unmatched leaves is the root leaf. In this case, match it to the other unmatched leaf to form the root-edge of the map $\Phi(T)$. Again, the complete procedure may require to turn several times around the tree. We discuss further down what can be done if T is *not* balanced.

Proposition 4.1 ([98]) *The map Ψ is a bijection between 4-valent planar maps with n vertices and balanced blossoming trees with n inner nodes. Its reverse bijection is Φ.*

With this bijection, it is easy to justify combinatorially the algebraicity of the generating function of 4-valent maps.

Corollary 4.2 *The generating function of 4-valent planar maps, counted by vertices, is*

$$M(t) = T(t) - tT(t)^3$$

where $T(t)$, the generating function of blossoming trees (counted by inner nodes), satisfies

$$T(t) = 1 + 3tT(t)^2.$$

Proof By decomposing blossoming trees into two subtrees and a flower, it should be clear that their generating function $T(t)$ satisfies the above equation. Via the bijection Ψ, counting maps boils down to counting balanced blossoming trees. Their generating function is $T(t) - U(t)$, where $U(t)$ counts *unbalanced* blossoming trees. Consider such a tree, and look at the flower that matches the root leaf (Figure 13). This flower is attached to an inner node. Delete this node: this leaves, beyond the flower, a 3-tuple of blossoming trees. This shows that $U(t) = tT(t)^3$.   □

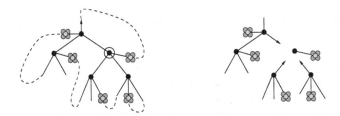

Figure 13: An unbalanced blossoming tree gives rise to three blossoming trees.

This bijection was originally designed [98] to explain combinatorially the simple formulas that occur in the enumeration of maps in which all vertices have an even degree — like 4-valent maps.

Corollary 4.3 *The number of 4-valent rooted planar maps with n vertices is*

$$\frac{2 \cdot 3^n}{(n+1)(n+2)} \binom{2n}{n}.$$

Proof We will prove that the above formula counts balanced blossoming trees of size n. Clearly, the total number of blossoming trees of this size is

$$t_n = \frac{3^n}{n+1} \binom{2n}{n}$$

(because binary trees are counted by the Catalan numbers $\binom{2n}{n}/(n+1)$). Marking a blossoming tree at one of its two unmatched leaves is equivalent, up to a re-rooting of the tree, to marking a balanced blossoming tree at one of its $n+2$ leaves. This shows that $2t_n = (n+2)b_n$, where b_n counts balanced blossoming trees, and the result follows.   □

A more general construction. A variant $\overline{\Phi}$ of the above bijection sends pairs (T, ϵ) formed of a (non-necessarily balanced) blossoming tree T and of a sign $\epsilon \in \{+, -\}$ onto rooted 4-valent maps with a distinguished face. This construction works as follows. In the tree T, one matches flowers and leaves as described above. The two unmatched leaves are then used to form the root edge, the orientation of which is chosen according to the sign ϵ. This gives a 4-valent rooted map. One then marks the face of this map located to the right of the half-edge where T is rooted. For example, the two maps associated with the (unbalanced) tree of Figure 13 are shown in Figure 14.

Figure 14: The two maps associated to the tree of Figure 13 via the map $\overline{\Phi}$.

This construction is bijective. Since a 4-valent map with n vertices has $n+2$ faces, it proves that the number m_n of such maps satisfies $(n+2)m_n = 2 \cdot 3^n \binom{2n}{n}/(n+1)$.

The bijection Φ described earlier can actually be seen as a specialisation of $\overline{\Phi}$: If T is balanced, and one chooses to orient the root edge of the map in such a way it starts with the root half-edge of the tree, the map M one obtains satisfies $\Psi(M) = T$. The distinguished face is in this case the root-face, and is thus canonical.

Quadrangulations and labelled trees. The second bijection starts from the duals of 4-valent maps, that is, from quadrangulations. It transforms them into *well labelled trees*. A *labelled tree* is a rooted plane tree with labelled vertices, such that:
– the labels belong to $\{1, 2, 3, \ldots\}$,
– the smallest label that occurs is 1,
– the labels of two adjacent vertices differ by $0, \pm 1$.
The tree is *well labelled* if, in addition, the root vertex has label 1.

This bijection was first found by Cori & Vauquelin in 1981 [50], but the simple description we give here was only discovered later by Schaeffer [49, 99]. As above, there are two versions of this bijection: the most general one sends rooted quadrangulations with n faces and a distinguished (or: *pointed*) vertex v_0 onto pairs (T, ϵ) formed of a labelled tree with n edges T and of a sign $\epsilon \in \{+, -\}$. Equivalently, it sends rooted quadrangulations with a pointed vertex v_0 *such that the root edge is oriented away from v_0* (in a sense that will be explained below) to labelled trees. The other bijection is a restriction, which sends rooted quadrangulations (pointed canonically at their root-vertex) onto well labelled trees.

So let us describe directly the more general bijection $\overline{\Lambda}$. Take a rooted quadrangulation Q with a pointed vertex v_0, such that the root-edge is oriented away from v_0. By this, we mean that the starting point of the root-edge is closer to v_0 than the endpoint, in terms of the graph distance[4]. Label all vertices by their distance to v_0. The labels of two neighbours differ by ± 1. If the starting point of the root-edge has label ℓ, then the endpoint has label $\ell + 1$. The labelling results in two types of faces: when walking inside a face with the edges on the left, one sees either a cyclic sequence of labels of the form $\ell, \ell + 1, \ell, \ell + 1$, or a sequence of the form $\ell, \ell + 1, \ell + 2, \ell + 1$. In the former case, create an edge in the face joining the two corners labelled $\ell + 1$. In the latter one, create an edge from the "first" corner labelled $\ell + 1$ (in the order described above) to the corner labelled $\ell + 2$. See Figure 15 for an example. The set

---
[4]The fact that Q is a quadrangulation, and hence a bipartite map, prevents two neighbour vertices to be at the same distance from v_0.

of edges created in this way forms a tree, which we root at the edge created in the outer face of Q, oriented away from the endpoint of the root-edge of Q (Figure 16). This tree, $\overline{\Lambda}(Q)$, contains all vertices of Q, except the marked one.

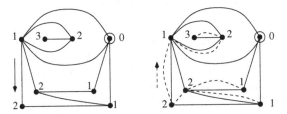

Figure 15: From a rooted quadrangulation with a pointed vertex to a labelled tree (in dashed lines).

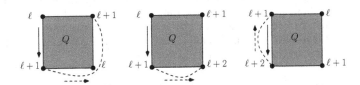

Figure 16: How to root the tree $\overline{\Lambda}(Q)$. The only vertices that are shown are those of the root-face of Q.

The reverse bijection $\overline{V}$ works as follows (see Figure 17 for an example). Start from a labelled tree T. Create a new vertex v_0, away from the tree. Then, visit the corners of the tree in counterclockwise order. From each corner labelled ℓ, send an edge to the next corner labelled $\ell - 1$ (or to v_0 if $\ell = 1$). This set of edges forms a quadrangulation Q. Each face of Q contains an edge of T. Choose the root-edge of Q in the face containing the root-edge of T, according to the rules of Figure 16.

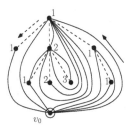

Figure 17: From a labelled tree (dashed lines) to a rooted quadrangulation with a pointed vertex.

Proposition 4.4 *The map $\overline{\Lambda}$ sends bijectively pointed rooted quadrangulations with* *n faces such that the root-edge is oriented away from the pointed vertex to labelled trees with n edges. The reverse bijection is $\overline{V}$.*

When specialised to rooted quadrangulations pointed canonically at their root-vertex, $\overline{\Lambda}$ induces a bijection Λ between rooted quadrangulations and well labelled trees.

Let us now discuss the consequences of this bijection, in terms of algebraicity and in terms of closed formulas. It is possible to use Λ to prove that the generating function of rooted quadrangulations is algebraic [50], and satisfies the system of Corollary 4.2, but this is not as simple as the proof of Corollary 4.2 given above. What *is* simple is to use $\overline{\Lambda}$ to count quadrangulations, and hence recover Corollary 4.3.

This alternative proof works as follows. First, observe that there are $3^n C_n$ labelled trees with n edges, where $C_n = \binom{2n}{n}/(n+1)$ counts rooted plane trees with n edges. Indeed, $3^n C_n$ is clearly the number of trees labelled 0 at the root vertex, such that the labels are in $\mathbf{Z}$ and differ by at most 1 along edges. If ℓ_0 denotes the smallest label of such a tree, adding $1 - \ell_0$ to all labels gives a labelled tree, and this transformation is reversible. Now, the above proposition implies that $3^n C_n = (n+2)q_n/2$, where q_n is the number of quadrangulations with n faces. Indeed, there are $n+2$ ways to point a vertex in such a quadrangulation, and half of these pointings are such that the root-edge is oriented away from the pointed vertex.

4.2   More bijections

Even though it is difficult to invent bijections, the two bijections presented above have now been adapted to many other map families, including the two general families described in Section 3.2: maps with prescribed face degrees (or, dually, prescribed vertex degrees), and Eulerian maps with prescribed face degrees (dually, bipartite maps with prescribed vertex degrees).

On the "blossoming" side, Schaeffer's bijection [98] was originally designed, not only for 4-valent maps, but for maps with prescribed vertex degrees, *provided these degrees are even*. The case of general degrees was solved (bijectively) a few years later by a trio of theoretical physicists, Bouttier, Di Francesco and Guitter [35]. The equations they obtain differ from those obtained by Bender & Canfield via the recursive method [6]. See [29, Section 10] for the correspondence between the two solutions. The case of *bipartite* maps with prescribed vertex degrees was then solved by Schaeffer and the author [34]. The special case of m-constellations was solved earlier in [33].

On the "labelled" side, the extensions of the Cori-Vauquelin-Schaeffer bijection (which applied to quadrangulations) to maps with prescribed face degrees, and to *Eulerian* maps with prescribed face degrees, came in a single paper, again due to Bouttier *et al.* [37].

Other bijections of the blossoming type exist for certain families of maps that are constrained, for instance, by higher connectivity conditions, by forbidding loops, or for dissections of polygons [63, 94, 95]. On the labelled side, there exist bijections for non-separable maps [53, 74], for d-angulations with girth d [17], and for maps of higher genus [48]. But the trees are then replaced by more complicated objects, namely one-face maps of higher genus.

All these bijections shed a much better light on planar maps, by revealing their hidden tree-like structure. As already mentioned, they often preserve important statistics, like distances to the root-vertex. In terms of proving algebraicity results, two restrictions should be mentioned:

- when the degrees are not bounded, it takes a bit of algebra to derive, from the system of equations that describes the structure of trees, polynomial equations satisfied by their generating functions,

- these bijections usually establish the algebraicity of the generating function of maps that are *doubly marked* (like rooted maps with a distinguished vertex, or with a distinguished face). The argument used to prove Corollary 4.2 has in general no simple counterpart.

5 Coloured planar maps: the recursive approach

We have now reviewed two combinatorial approaches (one recursive, one bijective) for the enumeration of families of planar uncoloured maps. We now move to the central topic of this paper, namely the enumeration of coloured planar maps, and compare both types of problems.

A first simple observation is that algebraicity will no longer be the rule. Indeed, it has been known for a long time [91] that the generating function of planar maps, weighted by their number of spanning trees (which is the specialisation $\mu = \nu = 1$ of the Tutte polynomial [102]) is:

$$\sum_{M \in \mathcal{M}} t^{e(M)} \, \mathrm{T}_M(1,1) = \sum_{n \geq 0} \frac{1}{(n+1)(n+2)} \binom{2n}{n} \binom{2n+2}{n+1} t^n.$$

The asymptotic behaviour of the nth coefficient, being $\kappa \, 16^n n^{-3}$, prevents this series from being algebraic [60]. The transcendence of this series implies that it cannot be described by a polynomial equation with one catalytic variable (Theorem 3.1). However, it is not difficult to write an equation with *two* catalytic variables for maps weighted by their Tutte (or Potts) polynomial. This equation is based on the recursive description (2.3). We present this equation in Section 5.1, and another one, for triangulations, in Section 5.2.

The whole point is now to solve equations with two catalytic variables. Much progress has been made in the past few years on the *linear* case. The equations for coloured maps are not linear, but they become linear (or quasi-linear, in a sense that will be explained) for certain special cases, like the enumeration of maps equipped with a spanning tree or a bipolar orientation. Sections 5.3 and 5.4 are devoted to these two simpler problems. They show how the *kernel method*, which was originally designed to solve linear equations with one catalytic variable [2, 31, 96], can be extended to equations with two catalytic variables. Sections 5.3 and 5.4 actually present two variants of this extension.

We then return to the general case. Following the complicated approach used by Tutte to count properly coloured triangulations [116], we obtain two kinds of results:

- when $q \neq 4$ is of the form $2 + 2\cos j\pi/m$, for integers j and m, the generating function of q-coloured maps satisfies also an equation with a *single* catalytic variable, and is thus algebraic. Explicit results are given for $q = 2$ and $q = 3$;

– in general, the generating function of q-coloured maps satisfies a non-linear differential equation.

These results, due to Olivier Bernardi and the author [15, 14], are presented without proof in Sections 5.5 and 5.6. We do not make explicit the differential equation satisfied by the generating function of q-coloured maps, but give an (explicit) *system* of differential equations, which we hope to simplify in a near future.

5.1 A functional equation for coloured planar maps

Let $\mathcal{M}$ be the set of rooted maps. For M in $\mathcal{M}$, recall that $\mathrm{dv}(M)$ and $\mathrm{df}(M)$ denote respectively the degrees of the root-vertex and root-face of M. We define the *Potts generating function* of planar maps by:

$$M(x,y) \equiv M(q,\nu,t,w;x,y) = \frac{1}{q} \sum_{M \in \mathcal{M}} t^{\mathrm{e}(M)} w^{\mathrm{v}(M)-1} x^{\mathrm{dv}(M)} y^{\mathrm{df}(M)} \mathrm{P}_M(q,\nu). \quad (5.1)$$

Since there is a finite number of maps with a given number of edges, and $\mathrm{P}_M(q,\nu)$ is a multiple of q, the generating function $M(x,y)$ is a power series in t with coefficients in $\mathbf{Q}[q,\nu,w,x,y]$. Keeping track of the number of vertices allows us to go back and forth between the Tutte and Potts polynomial, thanks to (2.4).

Proposition 5.1 *The Potts generating function of planar maps satisfies:*

$$\begin{aligned} M(x,y) &= 1 + xywt\left(qy + (\nu-1)(y-1)\right)M(x,y)M(1,y) \\ &\quad + xyt(x\nu-1)M(x,y)M(x,1) \\ &\quad + xywt(\nu-1)\frac{xM(x,y)-M(1,y)}{x-1} + xyt\frac{yM(x,y)-M(x,1)}{y-1}. \end{aligned} \quad (5.2)$$

Observe that (5.2) characterises $M(x,y)$ entirely as a series in $\mathbf{Q}[q,\nu,w,x,y][[t]]$ (think of extracting recursively the coefficient of t^n in this equation). Note also that when $\nu = 1$, then $\mathrm{P}_M(q,\nu) = q^{\mathrm{v}(M)}$, so that we are essentially counting planar maps by edges, vertices, and by the root-degrees dv and df. The variable x is no longer catalytic: it can be set to 1 in the functional equation, which becomes an equation for $M(1,y)$ with a single catalytic variable y.

Proof This equation is not difficult to establish using the recursive definition of the Potts polynomial (2.3) in terms of deletion and contraction of edges. Of course, one chooses to delete or contract the root-edge of the map. Let us sketch the proof to see where each term of the equation comes from. Equation (2.3) gives

$$M(x,y) = 1 + M_{\backslash}(x,y) + (\nu-1)M_{/}(x,y),$$

where the term 1 is the contribution of the atomic map m_0,

$$M_{\backslash}(x,y) = \frac{1}{q} \sum_{M \in \mathcal{M}\backslash\{m_0\}} t^{\mathrm{e}(M)} w^{\mathrm{v}(M)-1} x^{\mathrm{dv}(M)} y^{\mathrm{df}(M)} \mathrm{P}_{M\backslash e}(q,\nu),$$

and

$$M_{/}(x,y) = \frac{1}{q} \sum_{M \in \mathcal{M}\backslash\{m_0\}} t^{\mathrm{e}(M)} w^{\mathrm{v}(M)-1} x^{\mathrm{dv}(M)} y^{\mathrm{df}(M)} \mathrm{P}_{M/e}(q,\nu),$$

where $M \backslash e$ and M/e denote respectively the maps obtained from M by deleting and contracting the root-edge e.

A. The series $M_\backslash$. We consider the partition $\mathcal{M} \setminus \{m_0\} = \mathcal{M}_1 \uplus \mathcal{M}_2 \uplus \mathcal{M}_3$, where $\mathcal{M}_1$ (resp. $\mathcal{M}_2$, $\mathcal{M}_3$) is the subset of maps in $\mathcal{M} \setminus \{m_0\}$ such that the root-edge is an isthmus (resp. a loop, resp. neither an isthmus nor a loop). We denote by $M^{(i)}(x, y)$, for $1 \le i \le 3$, the contribution of $\mathcal{M}_i$ to the generating function $M_\backslash(x, y)$, so that

$$M_\backslash(x, y) = M^{(1)}(x, y) + M^{(2)}(x, y) + M^{(3)}(x, y).$$

• Contribution of $\mathcal{M}_1$. Deleting the root-edge of a map in $\mathcal{M}_1$ leaves two maps M_1 and M_2, as illustrated in Figure 4 (left). The Potts polynomial of this pair can be determined using (2.1). One thus obtains

$$M^{(1)}(x, y) = qxy^2tw\, M(1, y)M(x, y),$$

as the degree of the root-vertex of M_1 does not contribute to the degree of the root-vertex of the final map.

• Contribution of $\mathcal{M}_2$. Deleting the root-edge of a map in $\mathcal{M}_2$ leaves two maps M_1 and M_2 attached by their root vertex, as illustrated in Figure 6. The Potts polynomial of this pair can be determined using (2.2). One thus obtains

$$M^{(2)}(x, y) = x^2yt\, M(x, 1)M(x, y),$$

as the degree of the root-face of M_1 does not contribute to the degree of the root-face of the final map.

• Contribution of $\mathcal{M}_3$. Deleting the root-edge of a map in $\mathcal{M}_3$ leaves a single map M. If the outer degree of M is d, there are $d + 1$ ways to add a new (root-)edge to M, as illustrated in Figure 5 (right). However, a number of these additions create a loop, and thus their generating function must be subtracted. One thus obtains

$$M^{(3)}(x, y) = xt \sum_{d \ge 0} M_d(q, \nu, t, w; x)(y + y^2 + \cdots + y^{d+1}) - xyt\, M(x, 1)M(x, y),$$

where $M_d(q, \nu, w, t; x)$ is the coefficient of y^d in $M(x, y)$. This gives

$$M^{(3)}(x, y) = xyt\frac{yM(x, y) - M(x, 1)}{y - 1} - xyt\, M(x, 1)M(x, y),$$

and finally

$$M_\backslash(x, y) = qxy^2tw\, M(1, y)M(x, y) + x(x - 1)yt\, M(x, 1)M(x, y)$$
$$+ xyt\,\frac{yM(x, y) - M(x, 1)}{y - 1}.$$

B. The series $M_/$. The study of this series is of course very similar to the previous one, by duality. One finds:

$$M_/(x, y) = x^2yt\, M(x, 1)M(x, y) + xy(y - 1)tw\, M(1, y)M(x, y)$$
$$+ xytw\,\frac{xM(x, y) - M(1, y)}{x - 1}.$$

Adding the series 1, $M_\backslash$ and $(\nu - 1)M_/$ gives the functional equation. □

Remark. Equation (5.2) is equivalent to an equation written by Tutte in 1971:

$$\tilde{M}(x,y) = 1 + xyw(y\mu - 1)\tilde{M}(x,y)\tilde{M}(1,y) + xyz(x\nu - 1)\tilde{M}(x,y)\tilde{M}(x,1)$$

$$+ xyw\left(\frac{x\tilde{M}(x,y) - \tilde{M}(1,y)}{x - 1}\right) + xyz\left(\frac{y\tilde{M}(x,y) - \tilde{M}(x,1)}{y - 1}\right), \quad (5.3)$$

where $\tilde{M}(x,y)$ counts maps weighted by their Tutte polynomial [106]:

$$\tilde{M}(x,y) \equiv \tilde{M}(\mu, \nu, w, z; x, y) = \sum_{M \in \mathcal{M}} w^{v(M)-1} z^{f(M)-1} x^{dv(M)} y^{df(M)} T_M(\mu, \nu).$$

We call the above series *the Tutte generating function of planar maps*. The relation (2.4) between the Tutte and Potts polynomials and Euler's relation ($v(M) + f(M) - 2 = e(M)$) give

$$M(q, \nu, t, w; x, y) = \tilde{M}\left(1 + \frac{q}{\nu - 1}, \nu, (\nu - 1)tw, t; x, y\right),$$

from which (5.2) easily follows.

5.2   More functional equations

In a similar fashion, one can write a functional equation for coloured non-separable planar maps [84]. It is equivalent to (5.2) via a simple composition argument [15, Section 14]. Writing equations for coloured maps with prescribed face degrees is harder, as the contraction of the root-edge changes the degree of the finite face located to the left of the root-edge. This is however not a serious problem if one counts *proper* colourings of triangulations (the faces of degree 2 that occur can be "smashed" into a single edge), and in 1971, Tutte came up with the following equation, the solution of which kept him busy during the following decade:

$$T(x,y) = xy^2 q(q - 1) + \frac{xz}{yq}T(1,y)T(x,y)$$

$$+ xz\frac{T(x,y) - y^2 T_2(x)}{y} - x^2 yz\frac{T(x,y) - T(1,y)}{x - 1} \quad (5.4)$$

where $T_2(x) = [y^2]T(x,y)$. The series $T(x,y)$ defined by this equation is

$$T(x,y) = \sum_{T} z^{f(T)-1} x^{dv(T)} y^{df(T)} P_T(q, 0), \quad (5.5)$$

where the sum runs over all non-separable near-triangulations (maps in which all finite faces have degree 3). Note that the number of edges and the number of vertices of T can be obtained from $f(T)$ and $df(T)$, using

$$v(T) + f(T) = 2 + e(T) \quad \text{and} \quad 2e(T) = 3(f(T) - 1) + df(T). \quad (5.6)$$

There seems to be no straightforward extension to the Potts generating function[5], and it is not until recently that an equation was obtained for the Potts generating

---

[5]Although another type of functional equation is given in [56, Sec. 3] for the Potts generating function of cubic maps, using matrix integrals.

function $Q(x,y)$ of *quasi-triangulations* [15]. We refer to that paper for the precise definition of this class of maps, which is not so important here. What *is* important is that the series $Q(0,y)$ is the Potts generating function of near-triangulations:

$$Q(0,y) \equiv Q(q,\nu,t,z;0,y) = \frac{1}{q} \sum_{T \in \mathcal{T}} t^{\mathrm{e}(T)} z^{\mathrm{f}(T)-1} y^{\mathrm{df}(T)}\, \mathrm{P}_T(q,\nu).$$

Proposition 5.2 *The Potts generating function of quasi-triangulations satisfies*

$$Q(x,y) = 1 + zt\,\frac{Q(x,y) - 1 - yQ_1(x)}{y} + xzt(Q(x,y)-1) + xyztQ_1(x)Q(x,y)$$

$$+\, yzt(\nu-1)Q(x,y)(2xQ_1(x) + Q_2(x)) + y^2 t\left(q + \frac{\nu-1}{1-xzt\nu}\right)Q(0,y)Q(x,y)$$

$$+\, \frac{yt(\nu-1)}{1-xzt\nu}\,\frac{Q(x,y)-Q(0,y)}{x} \qquad (5.7)$$

where $Q_1(x) = [y]Q(x,y)$ and $Q_2(x) = [y^2]Q(x,y) = \dfrac{(1-2xzt\nu)}{zt\nu}Q_1(x)$.

Tutte's equation (5.4) for non-separable, properly coloured near-triangulations can be recovered from this proposition [15, Section 14]. In Section 5.4, we will use the following (equivalent) equation for the generating function of quasi-triangulations, weighted by their Tutte polynomial:

$$\tilde{Q}(x,y) = 1 + tz\,\frac{\tilde{Q}(x,y) - 1 - y\tilde{Q}_1(x)}{y} + xtz\left(\tilde{Q}(x,y) - 1\right) + xytz\tilde{Q}_1(x)\tilde{Q}(x,y)$$

$$+\, tzy(\nu-1)\tilde{Q}(x,y)\left(2x\tilde{Q}_1(x) + \tilde{Q}_2(x)\right)$$

$$+\, y^2 t\left(\mu + \frac{tx\nu z}{1-x\nu tz}\right)\tilde{Q}(0,y)\tilde{Q}(x,y) + \frac{yt}{1-x\nu tz}\,\frac{\tilde{Q}(x,y)-\tilde{Q}(0,y)}{x}, \qquad (5.8)$$

where $\tilde{Q}_1(x) = [y]\tilde{Q}(x,y)$ and $\tilde{Q}_2(x) = [y^2]\tilde{Q}(x,y) = \dfrac{(1-2xzt\nu)}{zt\nu}\tilde{Q}_1(x)$. We call the specialisation $\tilde{Q}(0,y)$ the *Tutte generating function of near-triangulations*:

$$\tilde{Q}(0,y) \equiv \tilde{Q}(q,\nu,t,z;0,y) = \sum_{T \in \mathcal{T}} t^{\mathrm{e}(T)} z^{\mathrm{f}(T)-1} y^{\mathrm{df}(T)}\, \mathrm{T}_T(\mu,\nu).$$

5.3   A linear case: bipolar orientations of maps

Let G be a connected graph with a root-edge (s,t). A *bipolar orientation* of G is an acyclic orientation of the edges of G such that s is the single source and t the single sink. Such orientations exist if and only if G is non-separable. It is known [70, 77] that the number of bipolar orientations of G is:

$$(-1)^{\mathrm{v}(G)}\frac{\partial \mathrm{P}_G}{\partial q}(1,0).$$

This number is also called the *chromatic invariant* of G [19, p. 355]. This expression implies that the generating function of (non-atomic) planar maps equipped with a bipolar orientation, counted by edges (t), non-root vertices (w), degree of the root-vertex (x) and of the root-face (y) is

$$B(t, w; x, y) = -\frac{\partial}{\partial q}\left(qM(q, 0, t, -w; x, y) - q\right)\Big|_{q=1} = -\frac{\partial M}{\partial q}(1, 0, t, -w; x, y),$$

where $M(x, y)$ is the Potts generating function of planar maps, defined by (5.1) (we have used $M(1, 0, t, w; x, y) = 1$). By differentiation, it is easy to derive from (5.2) an equation with two catalytic variables satisfied by $B(x, y)$. Using again $M(1, 0, t, w; x, y) = 1$, this equation is found to be *linear*:

$$\left(1 + \frac{xytw}{1-x} + \frac{xyt}{1-y}\right)B(x, y) = xy^2wt + \frac{x^2ywt}{1-x}B(1, y) + \frac{xy^2t}{1-y}B(x, 1). \quad (5.9)$$

Similarly, using (5.6), one finds that the generating function of planar near-triangulations equipped with a bipolar orientation, counted by non-root faces (z), degree of the root-vertex (x) and of the root-face (y) is

$$B^{\triangleleft}(z; x, y) = -\frac{\partial T}{\partial q}(0, iz; x, iy),$$

where $i^2 = -1$ and $T(q, z; x, y)$ is Tutte's generating function for coloured non-separable near-triangulations, defined by (5.5). Given that $T(1, z; x, y) = 0$, the equation satisfied by $B^{\triangleleft}$ is again linear:

$$\left(1 - \frac{xz}{y} - \frac{zx^2y}{x-1}\right)B^{\triangleleft}(x, y) = xy^2 - xzyB_2^{\triangleleft}(x) - \frac{zx^2yB^{\triangleleft}(1, y)}{x-1} \quad (5.10)$$

with $B_2^{\triangleleft}(x) = [y^2]B^{\triangleleft}(x, y)$.

In the past few years, much progress has been made in the solution of linear equations with two catalytic variables [27, 26, 32, 30, 89, 90]. It is now understood that a certain group of rational transformations, which leaves invariant the *kernel* of the equation (the coefficient of $B(x, y)$) plays an important role. In particular, when this group is finite, the equation can often be solved in an elementary way, using what is sometimes called the *algebraic version* of the kernel method [27, 30]. This is the case for both (5.9) and (5.10). We detail the solution of (5.10), and explain how to adapt it to solve (5.9).

Proposition 5.3 *The number of bipolar orientations of near-triangulations having $m + 1$ vertices is*

$$\frac{(3m)!}{(4m^2 - 1)m!^2(m + 1)!}.$$

The number of bipolar orientations of near-triangulations having $m + 1$ vertices and a root-face of degree j is

$$\frac{j(j-1)(3m-j-1)!}{m!(m+1)!(m-j+1)!}. \quad (5.11)$$

For $m \geq 2$, the number of bipolar orientations of near-triangulations having $m + 1$ vertices, a root-vertex of degree i and a root-face of degree j is

$$\frac{(i-1)(j-1)(2m-j-2)!(3m-i-j-1)!}{(m-1)!m!(m-j+1)!(2m-i-j+1)!}\left((2j+i-6)m+i+3j-j^2-ij\right).$$

The corresponding generating functions in 1, 2 and 3 variables are D-finite.

The last two formulas are due to Tutte [110, Eqs. (32) and (34)]. He also derived them from the functional equation (5.10), but his proof involved a lot of guessing, while ours is constructive. The first formula in Proposition 5.3 seems to be new.

Proof It will prove convenient to set $x = 1/(1 - u)$. After multiplying (5.10) by $(x - 1)/x^2/y$, the equation we want to solve reads:

$$u\bar{y}\left(1 - u - z(y\bar{u} + \bar{y})\right) B^{\triangleleft}\left(\frac{1}{1 - u}, y\right) = uy - R(u) - S(y), \qquad (5.12)$$

with $\bar{u} = 1/u$, $\bar{y} = 1/y$, $R(u) = zuB_2^{\triangleleft}\left(\frac{1}{1-u}\right)$ and $S(y) = zB^{\triangleleft}(1, y)$. Let $K(u, y) = 1 - u - z(y\bar{u} + \bar{y})$ be the *kernel* of this equation. This kernel is invariant by the transformations:

$$\Phi : (u, y) \mapsto (yz\bar{u}, y) \quad \text{and} \quad \Psi : (u, y) \mapsto (u, u\bar{y}).$$

Both transformations are involutions, and, by applying them iteratively to (u, y), one obtains 6 pairs (u', y') on which $K(\cdot, \cdot)$ takes the same value:

$$(u, y) \overset{\Phi}{\longrightarrow} (yz\bar{u}, y) \overset{\Psi}{\longrightarrow} (yz\bar{u}, z\bar{u}) \overset{\Phi}{\longrightarrow} (z\bar{y}, z\bar{u}) \overset{\Psi}{\longrightarrow} (z\bar{y}, u\bar{y}) \overset{\Phi}{\longrightarrow} (u, u\bar{y}) \overset{\Psi}{\longrightarrow} (u, y). \quad (5.13)$$

For each such pair (u', y'), the corresponding specialisation of (5.12) reads

$$(u'/y') K(u, y) B^{\triangleleft}(1/(1 - u'), y') = u'y' - R(u') - S(y').$$

We form the alternating sum of the 6 equations of this form obtained from the pairs (5.13). The series $R(\cdot)$ and $S(\cdot)$ cancel out, and, after dividing by $K(u, y)$, we obtain:

$$u\bar{y}B^{\triangleleft}\left(\frac{1}{1-u}, y\right) - z\bar{u}B^{\triangleleft}\left(\frac{1}{1-yz\bar{u}}, y\right) + yB^{\triangleleft}\left(\frac{1}{1-yz\bar{u}}, z\bar{u}\right)$$

$$- u\bar{y}B^{\triangleleft}\left(\frac{1}{1-z\bar{y}}, z\bar{u}\right) + z\bar{u}B^{\triangleleft}\left(\frac{1}{1-z\bar{y}}, u\bar{y}\right) - yB^{\triangleleft}\left(\frac{1}{1-u}, u\bar{y}\right)$$

$$= \frac{uy - y^2z\bar{u} + yz^2\bar{u}^2 - z^2\bar{y}\bar{u} + zu\bar{y}^2 - u^2\bar{y}}{1 - u - z(y\bar{u} + \bar{y})}.$$

The above identity holds in the ring of formal power series in z with coefficients in $\mathbb{Q}(u, y)$, which we consider as a sub-ring of Laurent series in u and y. Recall that $B^{\triangleleft}(x, y)$ has coefficients in $xy^2\mathbb{Q}[x, y]$. Hence, in the left-hand side of this identity, the terms with positive exponents in u and y are exactly those of $u\bar{y}B^{\triangleleft}\left(\frac{1}{1-u}, y\right)$. It follows that the latter series is the positive part (in u and y) of the rational function

$$R(z; u, y) := \frac{uy - y^2z\bar{u} + yz^2\bar{u}^2 - z^2\bar{y}\bar{u} + zu\bar{y}^2 - u^2\bar{y}}{1 - u - z(y\bar{u} + \bar{y})}.$$

It remains to perform a coefficient extraction. One first finds:

$$\frac{1}{1 - u - z(\bar{y} + y\bar{u})} = \sum_{n\geq 0}\sum_{a\geq -n}\sum_{b=-n}^{n} z^n u^a y^b \binom{n}{\frac{b+n}{2}}\binom{\frac{b+n}{2} + n + a}{n}$$

where the sum is restricted to triples (n, a, b) such that $n + b$ is even. An elementary calculation then yields the expansion of $R(z; u, y)$, and finally

$$B^{\triangleleft}\left(z; \frac{1}{1-u}, y\right) = \sum_{n \geq 0} \sum_{i \geq 0} \sum_{j=2}^{n+2} z^n u^i y^j \frac{(i+1)(j-1)(i+j)\left(\frac{3n+j}{2}+i-1\right)!}{\left(\frac{n-j}{2}+1\right)!\left(\frac{n+j}{2}+i+1\right)!\left(\frac{n+j}{2}\right)!}, \quad (5.14)$$

where the sum is restricted to triples (n, i, j) such that $n + j$ is even. In particular, the case $u = 0$ shows that the number of bipolar orientations of near-triangulations having n finite faces and a root-face of degree j is

$$\frac{(j-1)j\left(\frac{3n+j}{2}-1\right)!}{\left(\frac{n-j}{2}+1\right)!\left(\frac{n+j}{2}+1\right)!\left(\frac{n+j}{2}\right)!},$$

which coincides with (5.11), given that the number of vertices of such maps is $1 + (n + j)/2$.

The first formula of the proposition is then obtained by summing over j. The third one is easily verified using (5.14). However, it can also be *derived* from (5.14) if one prefers a constructive proof. One proceeds as follows. First, observe that if a rational function $R(u)$ is of the form $P(1/(1-u))$, for some Laurent polynomial P, then $P(x)$ coincides with the expansion of $R(1 - \bar{x})$ as a Laurent series in $\bar{x}$. In particular, if $P(x) \in x\mathbf{Q}[x]$, then $P(x)$ is the positive part in x of the expansion of $R(1 - \bar{x})$ in $\bar{x}$. The coefficient of $z^n y^j$ in $B^{\triangleleft}(z; x, y)$ is precisely in $x\mathbf{Q}[x]$, so that we can apply this extraction procedure to the right-hand side of (5.14). We first express the coefficient of $z^n y^j$ as a rational function of u, using

$$\sum_{i \geq 0} u^i \binom{a+b+i}{a} = \frac{1}{u^b(1-u)^{a+1}} - \sum_{j=0}^{b-1} \frac{1}{u^{b-j}}\binom{a+j}{a}.$$

Then, we set $u = 1 - \bar{x}$, expand in $\bar{x}$ this rational function, and extract the positive part in x. For the above series, this gives:

$$\begin{aligned}
[x^>] \sum_{i \geq 0} u^i \binom{a+b+i}{a} &= [x^>] \frac{x^{a+1}}{(1-\bar{x})^b} - \sum_{j=0}^{b-1} \frac{1}{(1-\bar{x})^{b-j}}\binom{a+j}{a} \\
&= [x^>] \frac{x^{a+1}}{(1-\bar{x})^b} \\
&= \sum_{k=0}^{a} x^{a+1-k}\binom{k+b-1}{k}.
\end{aligned}$$

Combining these two ingredients yields the third formula of the proposition.

Finally, the form of these three formulas, together with the closure properties of D-finite series [81, 82], imply that the associated generating functions are D-finite. $\square$

The same method allows us to solve the linear equation (5.9) obtained for bipolar orientations of general maps. One sets $x = 1 + u$ and $y = 1 + v$. It is also convenient to write

$$B(x, y) = xy^2 tw + x^2 y^2 t^2 w\, G(x, y).$$

The equation satisfied by G reads

$$uv \left(1 - t(1 + \bar{u})(1 + \bar{v})(u + vw)\right) G(1 + u, 1 + v) =$$
$$uv - tu(1 + u)G(1 + u, 1) - twv(1 + v)G(1, 1 + v).$$

The relevant transformations Φ and Ψ are now

$$\Phi : (u, v) \mapsto (\bar{u}wv, v) \quad \text{and} \quad \Psi : (u, v) \mapsto (u, u\bar{v}\bar{w}).$$

Again, they generate a group of order 6. One finally obtains that $G(1 + u, 1 + v)$ is the non-negative part (in u and v) of the following rational function:

$$\frac{(1 - \bar{u}\bar{v})(u\bar{v} - w\bar{u})(\bar{u}v - \bar{v}\bar{w})}{1 - t(1 + \bar{u})(1 + \bar{v})(u + vw)}.$$

A coefficient extraction, combined with Lemma 6 of [26], yields the following results.

Proposition 5.4 *For $1 \le m < n$, the number of bipolar orientations of planar maps having n edges and $m + 1$ vertices is*

$$\frac{2}{(n-1)n^2} \binom{n}{m-1} \binom{n}{m} \binom{n}{m+1}.$$

For $1 \le m < n$ and $2 \le j \le m + 1$, the number of bipolar orientations of planar maps having n edges, $m + 1$ vertices and a root-face of degree j is

$$\frac{j(j-1)}{(n-1)n^2} \binom{n}{m} \binom{n}{m+1} \binom{n-j-1}{m-j+1}.$$

For $n \ge 3$, $1 \le m < n$, $2 \le i \le n - m + 1$ and $2 \le j \le m + 1$, the number of bipolar orientations of planar maps having n edges, $m + 1$ vertices, a root-vertex of degree i and a root-face of degree j is

$$\frac{(i-1)(j-1)}{(n-1)n} \binom{n}{m} \left[\binom{n-j-1}{n-m-2} \binom{n-i-1}{m-2} - \binom{n-j-1}{n-m-1} \binom{n-i-1}{m-1} \right].$$

The associated generating functions are D-finite.

The solution we have sketched is very close to [26, Section 2]. Eq. (5.9) was also solved independently by Baxter [4], but his solution involved some guessing, while the one we have presented here is constructive.

5.4   A quasi-linear case: spanning trees

When $\mu = \nu = 1$, the Tutte polynomial $T_G(\mu, \nu)$ gives the number of spanning trees of G. The equations (5.3) and (5.8) that define the Tutte generating functions of our main two families of planar maps turn out to be much easier to solve in this case.

Consider first general planar maps, and Tutte's equation (5.3). We replace w by wt and z by zt so that t keeps track of the edge number. When $\mu = \nu = 1$, the equation reads:

$$\left(1 - \frac{x^2 ywt}{x-1} - \frac{xy^2 zt}{y-1} - xyzt(x-1)\tilde{M}(x,1) - xywt(y-1)\tilde{M}(1,y)\right)\tilde{M}(x,y) =$$
$$1 - \frac{xyzt}{y-1}\tilde{M}(x,1) - \frac{xywt}{x-1}\tilde{M}(1,y). \quad (5.15)$$

Observe that, up to a factor $(x-1)(y-1)$, the *same* linear combination of $\tilde{M}(x,1)$ and $\tilde{M}(1,y)$ appears on the right- and left-hand sides. This property was observed, but not fully exploited, by Tutte [117]. Bernardi [10] showed that it allows us to solve (5.15) using the standard kernel method usually applied to *linear* equations with two catalytic variables [26]. We thus obtain a new proof of the following result, due to Mullin [91]. Using Mullin's terminology, we say that a map equipped with a distinguished spanning tree is *tree-rooted*.

Proposition 5.5 *The number of tree-rooted planar maps with n edges is*

$$\frac{(2n)!(2n+2)!}{n!(n+1)!^2(n+2)!}.$$

The number of tree-rooted planar maps with $i+1$ vertices and $j+1$ faces is

$$\frac{(2i+2j)!}{i!(i+1)!j!(j+1)!}.$$

The associated generating functions are D-finite.

Proof Set

$$S(u,v) \equiv S(w,z,t;u,v) = \frac{1}{(1-ut)(1-vt)}\tilde{M}\left(w,z,t^2;\frac{1}{1-ut},\frac{1}{1-vt}\right).$$

Eq. (5.15) can be rewritten as

$$\left(1 - t(u+v+w\bar{u}+z\bar{v})\right)S(u,v) = \left(1 - uvt^2 S(u,v)\right)\left(1 - tz\bar{v}S(u,0) - tw\bar{u}S(0,v)\right).$$
$$(5.16)$$

Observe that the (Laurent) polynomial $(1 - t(u+v+w\bar{u}+z\bar{v}))$ is invariant by the transformation $u \mapsto w\bar{u}$. Seen as a polynomial in v, it has two roots. Exactly one of them, denoted $V \equiv V(w,z,t;u)$ is a formal power series in t with coefficients in $\mathbb{Q}[w,z,u,\bar{u}]$, satisfying

$$V = t\left(z + (u+w\bar{u})V + V^2\right). \quad (5.17)$$

In (5.16), specialise v to V. The left-hand side vanishes, and hence the right-hand side vanishes as well. Since its first factor is not zero, there holds

$$tzuS(u,0) + twVS(0,V) = uV.$$

Now replace u by $w\bar{u}$ in (5.16), and specialise again v to V. This gives

$$tzw\bar{u}S(w\bar{u},0) + twVS(0,V) = w\bar{u}V.$$

By taking the difference of the last two equations, we obtain

$$tzuS(u,0) - tzw\bar{u}S(w\bar{u},0) = (u - w\bar{u})V. \tag{5.18}$$

Since $S(u,0)$ is a series in t with coefficients in $\mathbf{Q}[w,z,u]$, this equation implies that $tzuS(u,0)$ is the positive part in u of $(u - w\bar{u})V$. The number $TR(i,j)$ of tree-rooted planar maps having $i+1$ vertices and $j+1$ faces is the coefficient of $w^i z^j t^{i+j}$ in $\tilde{M}(w,z,t;1,1)$, that is, the coefficient of $w^i z^j t^{2i+2j}$ in $S(u,0)$, or the coefficient of $w^i z^{j+1} t^{2i+2j+1} u$ in $tzuS(u,0)$. The Lagrange inversion formula, applied to (5.17), yields

$$[w^i z^j t^n u^{n+1-2i-2j}]V = \frac{(n-1)!}{i!(j-1)!j!(n+1-i-2j)!}. \tag{5.19}$$

Hence

$$\begin{aligned}
TR(i,j) &= [w^i z^{j+1} t^{2i+2j+1} u](tzuS(u,0)) \\
&= [w^i z^{j+1} t^{2i+2j+1} u^0]V - [w^{i-1} z^{j+1} t^{2i+2j+1} u^2]V \quad \text{(by (5.18))},
\end{aligned}$$

which, thanks to (5.19), gives the second result of the proposition. The first one follows by summing over all pairs (i,j) such that $i + j = n$. Alternatively, one can apply the Lagrange inversion formula to the equation satisfied by V when $w = z = 1$, which is $V = t(1 + uV)(1 + \bar{u}V)$. $\qquad\square$

Similarly, we can derive from the functional equation (5.8) defining the Tutte-generating function of quasi-triangulations the number of tree-rooted near-triangulations having a fixed outer degree and number of vertices. This result is also due to Mullin [91], but the proof is new.

Proposition 5.6 *The number of tree-rooted near-triangulations having $i+1$ vertices and a root-face of degree d is*

$$\frac{d}{(i+1)(4i-d)}\binom{3i-d}{i}\binom{4i-d}{i}.$$

The associated generating function is D-finite.

Proof We specialise to $\mu = \nu = z = 1$ the equation (5.8) that defines the Tutte-generating function of quasi-triangulations. We then replace $\tilde{Q}_2(x)$ by its expression in terms of $\tilde{Q}_1$. Again, the same linear combination of $\tilde{Q}_1(x)$ and $\tilde{Q}(0,y)$ occurs in the right- and left-hand sides, and the equation can be rewritten as

$$\left(1 - t\bar{y} - xy(1 - tx) - \frac{ty}{x(1 - tx)}\right)\tilde{Q}(x,y) =$$
$$\left(1 - t\bar{y} - tR_1(x) - \frac{ty}{x(1 - tx)}\tilde{Q}(0,y)\right)\left(1 - xy\tilde{Q}(x,y)\right) \tag{5.20}$$

where $R_1(x) = x + \tilde{Q}_1(x)$. Let us denote $u := x(1 - xt)$. Equivalently, we introduce a new indeterminate u and set

$$x = X(u) := \frac{1 - \sqrt{1 - 4ut}}{2t}.$$

The (Laurent) polynomial $(1 - t\bar{y} - uy - t\bar{u}y)$ occurring in the left-hand side of (5.20) is invariant by the transformation $y \mapsto tu\bar{y}/(t + u^2)$. As a polynomial in u, it has two roots. One of them is a power series in t with constant term 0, satisfying

$$U = t\frac{y + U\bar{y}}{1 - Uy}. \tag{5.21}$$

In (5.20), specialise x to $X(U)$. The left-hand side vanishes, leaving

$$tUR_1(X(U)) + ty\tilde{Q}(0, y) = U(1 - t\bar{y}).$$

If we first replace y by $t\bar{y}U/(t + U^2) = t/(1 - t\bar{y})$ in (5.20) before specialising x to $X(U)$, we obtain instead

$$tUR_1(X(U)) + \frac{t^2}{1 - t\bar{y}}\ \tilde{Q}\left(0, \frac{t}{1 - t\bar{y}}\right) = t\bar{y}U.$$

By taking the difference of the last two equations, one finds:

$$ty\tilde{Q}(0, y) - \frac{t^2}{1 - t\bar{y}}\ \tilde{Q}\left(0, \frac{t}{1 - t\bar{y}}\right)\ =\ U(1 - 2t\bar{y}).$$

Since $\tilde{Q}(0, y)$ is a series in t with coefficients in $\mathbf{Q}[y]$, this equation implies that $ty\tilde{Q}(0, y)$ is the positive part in y of $U(1 - 2t\bar{y})$. The Lagrange inversion formula, applied to (5.21), gives:

$$[t^n y^{3i-n+2}]U = \frac{1}{n}\binom{n}{i+1}\binom{n+i-1}{i}.$$

This yields

$$\begin{aligned}
[t^n y^{3i-n}]\tilde{Q}(0, y)\ &=\ \frac{1}{n+1}\binom{n+1}{i+1}\binom{n+i}{i} - \frac{2}{n}\binom{n}{i+1}\binom{n+i-1}{i} \\
&=\ \frac{3i - n}{(i+1)(n+i)}\binom{n}{i}\binom{n+i}{i},
\end{aligned}$$

which is equivalent to the proposition, as a near-triangulation with n edges and outer degree $3i - n$ has $i + 1$ vertices.   □

5.5   When q is a Beraha number: Algebraicity

We now report on more difficult results obtained recently by Olivier Bernardi and the author [15] by following and adapting Tutte's enumeration of properly q-coloured triangulations [116].

For certain values of q, it is possible to derive from the equation with two catalytic variables defining $M(x, y)$ an equation with a single catalytic variable (namely, y) satisfied by $M(1, y)$. For instance, one can derive from the case $q = 1$ of (5.2) that $M(y) \equiv M(1, y)$ satisfies

$$M(y) = 1 + y^2 t\nu w M(y)^2 + \nu t y \frac{yM(y) - M(1)}{y - 1}.$$

This is only a moderately exciting result, as the latter equation is just the standard functional equation (3.1) obtained by deleting recursively the root-edge in planar maps.

But let us be persistent. When $q = 2$, one can derive from (5.2) that $M(y) \equiv M(1, y)$ satisfies a polynomial equation with one catalytic variable, involving two additional unknown series, namely $M(1)$ and $M'(1)$. This equation is rather big (see [15, Section 12]), and we do not write it here. No combinatorial way to derive it is known at the moment. When $\nu = 0$, the series $M'(1)$ disappears, and one recovers the standard equation (3.3) obtained by deleting recursively the root-edge in bipartite planar maps.

This construction works as soon as $q \neq 0, 4$ is of the form $2 + 2\cos(j\pi/m)$, for integers j and m. These numbers generalise *Beraha's numbers* (obtained for $j = 2$), which occur frequently in connection with chromatic properties of planar graphs [9, 58, 72, 73, 87, 97]. They include the three integer values $q = 1, 2, 3$. Given that the solutions of polynomial equations with one catalytic variable are always algebraic (Theorem 3.1), the following algebraicity result holds [15].

Theorem 5.7 *Let $q \neq 0, 4$ be of the form $2 + 2\cos j\pi/m$ for two integers j and m. Then the series $M(q, \nu, t, w; x, y)$, defined by (5.2), is algebraic over $\mathbf{Q}(q, \nu, t, w, x, y)$.*

A similar method works for quasi-triangulations.

Theorem 5.8 *Let $q \neq 0, 4$ be of the form $2 + 2\cos j\pi/m$ for two integers j and m. Then the series $Q(q, \nu, t, z; x, y)$, defined by (5.7), is algebraic over $\mathbf{Q}(q, \nu, t, z, x, y)$.*

For the two integer values $q = 2$ (the Ising model) and $q = 3$ (the 3-state Potts model), we have applied the procedure described in Section 3.3 to obtain explicit algebraic equations satisfied by $M(q, \nu, t, 1; 1, 1)$ and $Q(q, \nu, t, 1; 1, 1)$. However, when $q = 3$, we could only solve the case $\nu = 0$ (corresponding to proper colourings). The final equations are remarkably simple. We give them here for general planar maps. For triangulations, these equations are not new: the Ising model on triangulations was already solved by several other methods (including bijective ones, see Section 6.3 for details), and properly 3-coloured triangulations are just Eulerian triangulations, as discussed in Section 3.2. With the help of Bruno Salvy, we have also conjectured an algebraic equation of degree 11 for the generating function of properly 3-coloured cubic maps (maps in which all vertices have degree 3). By the duality relation (2.5), this corresponds to the series $Q(q, \nu, t, 1; 1, 1)$ taken at $q = 3$, $\nu = -2$.

Theorem 5.9 *The Potts generating function of planar maps $M(2, \nu, t, w; x, y)$, defined by (5.1) and taken at $q = 2$, is algebraic. The specialisation $M(2, \nu, t, w; 1, 1)$ has degree 8 over $\mathbf{Q}(\nu, t, w)$.*

When $w = 1$, the degree decreases to 6, and the equation admits a rational parametrisation. Let $S \equiv S(t)$ be the unique power series in t with constant term 0 satisfying

$$S = t \, \frac{\left(1 + 3\nu S - 3\nu S^2 - \nu^2 S^3\right)^2}{1 - 2S + 2\nu^2 S^3 - \nu^2 S^4}.$$

Then

$$M(2,\nu,t,1;1,1) = \frac{1 + 3\nu S - 3\nu S^2 - \nu^2 S^3}{(1 - 2S + 2\nu^2 S^3 - \nu^2 S^4)^2} \times$$

$$\left(\nu^3 S^6 + 2\nu^2(1-\nu)S^5 + \nu(1-6\nu)S^4 - \nu(1-5\nu)S^3 + (1+2\nu)S^2 - (3+\nu)S + 1\right).$$

Theorem 5.10 *The Potts generating function of planar maps $M(3,\nu,t,w;x,y)$, defined by (5.1) and taken at $q = 3$, is algebraic.*

The specialisation $M(3,0,t,1;1,1)$ that counts properly three-coloured planar maps by edges, has degree 4 over $\mathbf{Q}(t)$, and admits a rational parametrisation. Let $S \equiv S(t)$ be the unique power series in t with constant term 0 satisfying

$$t = \frac{S(1 - 2S^3)}{(1 + 2S)^3}.$$

Then

$$M(3,0,t,1;1,1) = \frac{(1 + 2S)\left(1 - 2S^2 - 4S^3 - 4S^4\right)}{(1 - 2S^3)^2}.$$

5.6 The general case: differential equations

The culminating, and final point in Tutte's study of properly coloured triangulations was a non-linear differential equation satisfied by their generating function. For the more complicated problem of counting maps weighted by their Potts polynomial, we have come with a *system* of differential equations that defines the corresponding generating function [14]. One compact way to write this system is as follows.

Theorem 5.11 *Let $\beta = \nu - 1$ and*

$$\Delta(t,v) = (q\nu + \beta^2) - q(\nu+1)v + (\beta t(q-4)(wq+\beta) + q)v^2.$$

There exists a unique triple $(A(t,v), B(t,v), C(t,v))$ of polynomials in v with coefficients in $\mathbf{Q}[q,\nu,w][[t]]$, having degree $4, 2$ and 2 respectively in v, such that

$$A(0,v) = (1-v)^2, \qquad A(t,0) = 1,$$
$$B(0,v) = 1 - v, \qquad C(t,0) = w(q + 2\beta) - 1 - \nu,$$

and

$$\frac{1}{C(t,v)} \frac{\partial}{\partial v}\left(\frac{v^4 C(t,v)^2}{A(t,v)\Delta(t,v)^2}\right) = \frac{v^2}{B(t,v)} \frac{\partial}{\partial t}\left(\frac{B(t,v)^2}{A(t,v)\Delta(t,v)^2}\right). \tag{5.22}$$

Let $A_i(t)$ (resp. $B_i(t)$) denote the coefficient of v^i in $A(t,v)$ (resp. $B(t,v)$). Then the Potts generating function of planar maps, $M(1,1) \equiv M(q,\nu,t,w;1,1)$, defined by (5.1), is related to A and B by

$$12\,t^2 w\left(q\nu + \beta^2\right) M(1,1) - A_2(t) + 2B_2(t) - 8t\left(w(q+2\beta) - \nu - 1\right)B_1(t) + B_1(t)^2$$
$$= 4t\left(1 - 3(\beta+2)^2 t + (6(\beta+2)(q+2\beta)t + q + 3\beta)w - 3t(q+2\beta)^2 w^2\right).$$

Comments

1. Let us write

$$A(t,v) = \sum_{j=0}^{4} A_j(t)v^i, \quad B(t,v) = \sum_{j=0}^{2} B_j(t)v^i, \quad C(t,v) = \sum_{j=0}^{2} C_j(t)v^i.$$

The differential equation (5.22) then translates into a system of 9 differential equations (with respect to t) relating the 11 series A_j, B_j, C_j. However, $A_0(t) = A(t,0)$ and $C_0(t) = C(t,0)$ are given explicitly as initial conditions, so that there are really as many unknown series in t as differential equations. Observe moreover that no derivative of the series $C_j(t)$ arise in the system. This is why we only need initial conditions for the series $A_j(t)$ and $B_j(t)$. They are prescribed by the values of $A(0,v)$ and $B(0,v)$.

2. The form of the above result is very close to Tutte's solution of properly coloured planar triangulations, which can be stated as in Theorem 5.12 below. However, Tutte's case is simpler, as it boils down to only 4 differential equations. This explains why Tutte could derive from his system a *single* differential equation for the generating function of properly coloured triangulations. More precisely, it follows from the theorem below that, if $t = z^2$ and $H \equiv H(t) = t^2 T_2(1)$,

$$2q^2(1-q)t + (qt + 10H - 6tH')H'' + q(4-q)(20H - 18tH' + 9t^2H'') = 0. \quad (5.23)$$

So far, we have not been able to derive from Theorem 5.11 a single differential equation for coloured planar maps.

Theorem 5.12 *Let*

$$\Delta(v) = v + 4 - q.$$

There exists a unique pair $(A(z,v), B(z,v))$ of polynomials in v with coefficients in $\mathbf{Q}[q][[z]]$, having degree 3 and 1 respectively in v, such that

$$A(0,v) = 1 + v/4, \quad A(z,0) = 1,$$
$$B(0,v) = 1,$$

and

$$-\frac{4z}{v}\frac{\partial}{\partial v}\left(\frac{v^3}{A(z,v)}\right) = \frac{1}{B(z,v)\Delta(v)}\frac{\partial}{\partial z}\left(\frac{B(z,v)^2}{A(z,v)}\right).$$

Let $A_i(z)$ (resp. $B_i(z)$) denote the coefficient of v^i in $A(z,v)$ (resp. $B(z,v)$). Then the face generating function of properly q-coloured planar near-triangulations having outer-degree 2, denoted $T_2(q, z; 1)$ and defined by (5.4), is related to A and B by

$$20 z^4 (q-4)T_2(q, z; 1)/q - 2 B_1(z)^2 - \left(96 z^2 - 24 z^2 q + 1\right)B_1(z) + 2 A_2(z)$$
$$- 2 z^2 \left(10 - q + 432 z^2 - 216 z^2 q + 27 z^2 q^2\right) = 0.$$

6 Some bijections for coloured planar maps

Certain specialisations of the Potts generating function of planar maps can be determined using a purely bijective approach.

6.1   Bipolar orientations of maps

The numbers that arise in the enumeration of bipolar orientations of planar maps (Proposition 5.4) are known to count other families of objects: *Baxter permutations* (not the same Baxter as in [4]!), pairs of *twin trees*, and certain configurations of *non-intersecting lattice paths*. Several bijections have been established recently between bipolar orientations and these families [20, 57, 64]. Let us mention, however, that the only family that is simple to enumerate is that of non-intersecting lattice paths (via the Lindström-Gessel-Viennot theorem). Hence bijections with this family are the only ones that really provide a self-contained proof of Proposition 5.4.

Regarding bipolar orientations of triangulations (Proposition 5.3), we are currently working on certain bijections with *Young tableaux* of height at most 3, in collaboration with Nicolas Bonichon and Éric Fusy.

6.2   Spanning trees

It is not hard to count in a bijective manner tree-rooted maps with $i + 1$ vertices and $j + 1$ faces (Proposition 5.5). The construction below, which is usually attributed to Lehman and Walsh [118], is actually not far from Mullin's original proof [91]. Starting from a tree-rooted map (M, T), one walks around the tree T in counterclockwise order, starting from the root-edge of M (Figure 18, left), and:

- when an edge e of T is met, one walks along this edge, and writes a when e is met for the first time, $\bar{a}$ otherwise;

- when an edge e not in T is met, one crosses the edge, and writes b when e is met for the first time, $\bar{b}$ otherwise.

This gives a shuffle of two Dyck words[6] u and v, one of length $2i$ on the alphabet $\{a, \bar{a}\}$ (since there are i edges in T), one of length $2j$ on the alphabet $\{b, \bar{b}\}$ (since there are j edges not in the tree). The number of such shuffles is

$$\binom{2i + 2j}{2i} C_i C_j,$$

where $C_i = \binom{2i}{i}/(i+1)$ is the number of Dyck words of length $2i$. The construction is easily seen to be bijective, and this gives the second result of Proposition 5.5.

As already explained, the first result of Proposition 5.5 follows by summing over all i, j such that $i + j = n$. A *direct* bijective proof was only obtained in 2007 by Bernardi [12]. It transforms a tree-rooted map into a pair formed of a plane tree and a non-crossing partition. See [16] for a recent extension to maps of higher genus.

Mullin's original construction [91] decouples the tree-rooted map (M, T) into two objects (Figure 18, right):

- a plane tree with j edges, which is the dual of T and corresponds to the Dyck word v on $\{b, \bar{b}\}$ described above,

---

[6]A Dyck word on the alphabet $\{a, \bar{a}\}$ is a word that contains as many occurrences of a and $\bar{a}$, and such that every prefix contains at least as many a's as $\bar{a}$'s. A shuffle of two Dyck words can be seen as a walk in the first quadrant of $\mathbf{Z}^2$, starting and ending at the origin.

– a plane tree T', which consists of T and of $2j$ half-edges; this tree can be seen as the Dyck word u shuffled with the word c^{2j}.

The vertex degree distribution of M coincides with the degree distribution of T', and Mullin used this property to count tree-rooted maps with prescribed vertex degrees (or dually, with prescribed face degrees, since a map and its dual have the same number of spanning trees). Indeed, it is easy to count trees with a prescribed degree distribution [100, Thm. 5.3.10]. In particular, the number of plane trees with root-degree d, such that n_k non-root vertices have degree k, for $k \geq 1$, and carrying in addition $2j$ half-edges, is

$$T'(d, j, n_1, n_2, \ldots) := \frac{d(2j - 1 + \sum_k n_k)!}{(2j)! \prod_k n_k!},$$

so that the number of tree-rooted maps (M, T) in which the root-vertex has degree d and n_k non-root vertices have degree k, for $k \geq 1$, is

$$\frac{1}{j+1}\binom{2j}{j}T'(d, j, n_1, n_2, \ldots) = \frac{d(2j - 1 + \sum_k n_k)!}{j!(j+1)! \prod_k n_k!},$$

where $j = e(M) - v(M) + 1 = (d + \sum_k (k - 2)n_k)/2$ is the *excess* of M (and also the number of faces, minus 1). In particular, the number of tree-rooted maps (M, T) having a root-vertex of degree d and $2i - d$ non-root vertices of degree 3 is

$$\frac{d(4i - d - 1)!}{i!(i + 1)!(2i - d)!},$$

since such maps have excess i. This is the dual statement of Proposition 5.6.

 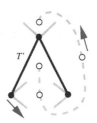

Figure 18: Left: The tour of a tree-rooted map gives an encoding by a shuffle of Dyck words, here $bba a\bar{b}\bar{b}\bar{a}b\bar{b}\bar{a}$. Right: Alternatively, one can decouple a tree-rooted map into the dual plane tree (dashed lines) and a plane tree T' carrying half-edges.

6.3 The Ising model ($q = 2$)

As observed in [34], a simple transformation relates the Potts generating function of maps at $q = 2$ to the enumeration of bipartite maps by vertex degrees.

Proposition 6.1 *Let $B(t, v, w; x)$ be the generating function of planar bipartite maps, counted by edges (t), non-root vertices of degree 2 (variable v), non-root vertices of* *degree $\neq 2$ (variable w), and degree of the root-vertex (x). Let $M(q, \nu, t, w; x, y)$ be the Potts generating function of planar maps, defined by (5.1). Then*

$$M\left(2, tv, \frac{t}{1 - t^2 v^2}, w; x, 1\right) = B(t, v + w, w; x).$$

This identity can be refined by keeping track of the number of non-root vertices of each degree and colour. Let

$$\overline{M}(\nu, t, x_1, x_2, \dots, y_1, y_2, \dots; x) = \sum_M \nu^{m(M)} t^{e(M)} x^{dv(M)} \prod_{i \geq 1} x_i^{v_i^\circ(M)} y_i^{v_i^\bullet(M)},$$

where the sum runs over all 2-coloured maps M rooted at a black vertex, $m(M)$ is the number of monochromatic edges in M, and $v_i^\circ(M)$ (resp. $v_i^\bullet(M)$) is the number of non-root white (resp. black) vertices of degree i. Let

$$\overline{B}(t, x_1, x_2, \dots, y_1, y_2, \dots; x) = \sum_M t^{e(M)} x^{dv(M)} \prod_{i \geq 1} x_i^{v_i^\circ(M)} y_i^{v_i^\bullet(M)},$$

where the sum runs over all bipartite maps, properly bicoloured in such a way the root-vertex is black. Then

$$\overline{M}\left(tv, \frac{t}{1 - t^2 v^2}, x_1, x_2, \dots, y_1, y_2, \dots; x\right) = \overline{B}(t, x_1, v + x_2, x_3, \dots, y_1, v + y_2, y_3, \dots; x).$$

Proof We establish directly the second identity, which implies the first one by specialising each x_i and y_i to w. Take a 2-coloured planar map M, rooted at a black vertex. On each edge, add a (possibly empty) sequence of square vertices of degree 2, in such a way the resulting map is properly bicoloured. An example is shown on Figure 19. Every monochromatic edge receives an odd number of square vertices, while every dichromatic edge receives an even number of these vertices. Each addition of a vertex of degree 2 also results in the addition of an edge. Since M can be recovered from the bipartite map by erasing all square vertices, the identity follows.   □

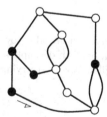
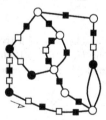

Figure 19: A 2-coloured map and one of the associated bipartite maps.

Recall from Sections 3 and 4 that the enumeration of bipartite maps with prescribed vertex degrees can be addressed via the recursive method (and equations with one catalytic variable) [29], bijections with blossoming trees [34], or bijections with labelled trees [37]. In particular, the last two approaches explain bijectively[7] the algebraicity of the associated generating function, at least when the vertex degrees are bounded.

7 Final comments and questions

We conclude with a number of questions raised by this survey. The first type of question asks what problems have an algebraic solution. Of course, all methods (recursive, bijective, or via matrix integrals...) are welcome to answer them. We then go on with a list of problems that have been solved by a recursive approach, but are still waiting for a purely bijective proof. We also mention questions dealing with asymptotic properties of maps.

7.1 Algebraicity

Theorem 3.2 states several algebraicity results for maps with prescribed face degrees. By comparing the results dealing with general maps to those dealing with Eulerian maps, it appears that our understanding of Eulerian maps with unbounded degrees is probably still incomplete.

Question 7.1 *Let $D_\bullet$ and $D_\circ$ be two subsets of* **N**. *Under what conditions on these sets is the generating function of Eulerian maps such that all black (resp. white) faces have their degree in $D_\bullet$ (resp. $D_\circ$) algebraic?*

This question can in principle be addressed via the equations of [34, 37]. Algebraicity is known to hold when $D_\bullet$ and $D_\circ$ are finite, and when $D_\bullet = \{m\}$ and $D_\circ = m\mathbf{N}$. A natural sub-case that could be addressed first is the following[8].

Question 7.2 *Is the generating function of Eulerian planar maps in which all face degrees are multiples of m algebraic?*

Recall that for $m \geq 3$, these are the maps that admit a cyclic m-colouring (Section 3.2). Algebraicity has already been proved when $m = 2$, that is, for maps that are both Eulerian and bipartite [83, 93].

Eulerian maps are required to have even vertex-degrees. But one could think of other restrictions than parity.

Question 7.3 *Under what condition is the generating function of maps in which both the vertex degrees and the face degrees are constrained algebraic?*

This question seems of course very hard to address. A positive answer is known in at least one case: the generating function of triangulations in which all vertices have degree at least d is algebraic for all d [13, 66].

---

[7]up to minor restrictions imposed at the root of the map

[8]Having raised the question, the author has started to explore it... and come with a positive answer [22]. Algebraicity can be proved either via the equations of [34], or via a bijection with $(m + 1)$-constellations.

7.2 Bijections

Our first question may seem surprising at first sight.

Question 7.4 *Design bijections between families of trees and families of planar maps with unbounded degrees.*

Indeed, it seems that all bijections that can be used to count families of maps with unbounded degrees use a detour via maps with bounded degrees. The simplest example, presented in Section 4, is that of general planar maps: we have first shown that they are in bijection with 4-valent maps (or, dually, quadrangulations), before describing two types of bijections between 4-valent maps and trees. We could actually content ourselves with this situation: after all, isn't a combination of two beautiful bijections twice as beautiful as a single bijection? But there exist problems with an algebraic solution, dealing with maps with unbounded degrees, that have not been solved by a direct bijection so far, like the Ising model on general planar maps (Theorem 5.9), or the hard-particle model on general planar maps [29]. Discovering such bijections could also give an algebra-free proof of the fact that maps in which all degrees are multiples of m are algebraic; the bijection of [35] gives indeed a proof, but requires a bit of algebra. Moreover, this could be a purely bijective way to address the questions raised above on the algebraicity of Eulerian maps in which all face degrees are multiples of m.

Question 7.5 *Design bijections for q-coloured maps.*

This can take several directions:

- find bijections for the special values of q (like $q = 3$) that are known to yield algebraic generating functions (Theorems 5.7 and 5.8);

- find bijections for specialisations of the Potts generating function of maps, like those presented in Section 6 for spanning trees and bipolar orientations;

- finally, one would dream of designing bijections that would establish directly differential equations for coloured maps, starting with the (relatively simple?) case of triangulations (see (5.23)). The author is currently working on an interesting construction of Bouttier *et al.* [39], which allows to count spanning forests on maps and to derive certain differential equations in a simpler way than the recursive approach [23].

Finally, we have discussed in Section 4 two families of bijections, but a third one could exist, as suggested by Bernardi's beautiful construction for loopless triangulations [11].

Question 7.6 *Is the bijection of [11] the tip of some iceberg?*

7.3 Asymptotics of maps

Question 7.7 *What is the asymptotic number of properly q-coloured planar maps having n edges?*

This question has been studied by Odlyzko and Richmond [92] for triangulations, starting from the differential equation (5.23). For $q \in [15/11, 4] \cup [5, \infty)$, they proved that the number of properly q-coloured triangulations with n faces is of the form $\kappa \mu^n n^{-5/2}$. The exponent $-5/2$ is typical in the enumeration of (uncoloured) planar maps.

The asymptotic behaviour of the number of n-edge q-coloured planar maps has been worked out in [15] for $q = 2$ and $q = 3$, using the explicit results of Theorems 5.9 and 5.10. Again, the exponent is $-5/2$. The same question can be asked when a parameter $\nu \neq 0$ weights monochromatic edges. For $q = 2$, the exponent is still $-5/2$, except at the critical value $\nu = (3 + \sqrt{5})/2$, where it becomes $-7/3$. See [75, 21] for similar results on maps of fixed vertex degree.

The proofs of these results use the solutions of the difficult functional equations (5.2) and (5.4). It would be extremely interesting to be able to understand the asymptotic behaviour of these numbers (or the singular behaviour of the associated series) *directly from these equations*. At the moment, we do not known how to do this, even in the case of one catalytic variable.

Question 7.8 *Develop a "singularity analysis" [61] for equations with catalytic variables.*

Finally, the asymptotic geometry of random uncoloured maps has attracted a lot of attention in the past few years [40, 41, 49], and a limit object, *the Brownian map*, has been identified [78, 85, 86]. Similar questions can be addressed for maps equipped with an additional structure.

Question 7.9 *Is there a scaling limit for maps equipped with a spanning tree? a spanning forest? for properly q-coloured maps?*

The final section of [79] suggests a partial, and conjectural answer to this question for maps equipped with certain statistical physics models, including the Ising model. Of course, the first point is to determine how the average distance between two vertices of these maps scales.

Acknowledgements

The author gratefully acknowledges the assistance of Olivier Bernardi in writing this survey.

References

[1] M. Aigner and G. M. Ziegler. *Proofs from The Book*. Springer-Verlag, Berlin, fourth edition, 2010.

[2] C. Banderier, M. Bousquet-Mélou, A. Denise, P. Flajolet, D. Gardy, and D. Gouyou-Beauchamps. Generating functions for generating trees. *Discrete Math.*, 246(1-3):29–55, 2002.

[3] C. Banderier and P. Flajolet. Basic analytic combinatorics of directed lattice paths. *Theoret. Comput. Sci.*, 281(1-2):37–80, 2002.

[4] R. J. Baxter. Dichromatic polynomials and Potts models summed over rooted maps. *Ann. Comb.*, 5(1):17–36, 2001.

[5] E. A. Bender and E. R. Canfield. The asymptotic number of rooted maps on a surface. *J. Combin. Theory Ser. A*, 43(2):244–257, 1986.

[6] E. A. Bender and E. R. Canfield. The number of degree-restricted rooted maps on the sphere. *SIAM J. Discrete Math.*, 7(1):9–15, 1994.

[7] E. A. Bender, E. R. Canfield, and L. B. Richmond. The asymptotic number of rooted maps on a surface. II. Enumeration by vertices and faces. *J. Combin. Theory Ser. A*, 63(2):318–329, 1993.

[8] E. A. Bender and N. C. Wormald. The number of loopless planar maps. *Discrete Math.*, 54(2):235–237, 1985.

[9] S. Beraha, J. Kahane, and N. J. Weiss. Limits of chromatic zeros of some families of maps. *J. Combin. Theory Ser. B*, 28(1):52–65, 1980.

[10] O. Bernardi. Comptage des cartes planaires boisées. Personal communication.

[11] O. Bernardi. Bijective counting of Kreweras walks and loopless triangulations. *J. Combin. Theory Ser. A*, 114(5):931–956, 2007.

[12] O. Bernardi. Bijective counting of tree-rooted maps and shuffles of parenthesis systems. *Electron. J. Combin.*, 14(1):Research Paper 9, 36 pp. (elect.), 2007.

[13] O. Bernardi. On triangulations with high vertex degree. *Ann. Comb.*, 12(1):17–44, 2008.

[14] O. Bernardi and M. Bousquet-Mélou. Counting colored planar maps: differential equations, In preparation.

[15] O. Bernardi and M. Bousquet-Mélou. Counting colored planar maps: algebraicity results. *J. Combin. Theorey Ser. B*, to appear. Arxiv:0909:1695.

[16] O. Bernardi and G. Chapuy. A bijection for covered maps, or a shortcut between Harer-Zagier's and Jackson's formulas. Arxiv:1001.1592, 2010.

[17] O. Bernardi and É. Fusy. A bijection for triangulations, quadrangulations, pentagulations, etc. Arxiv:1007.1292, 2010.

[18] D. Bessis, C. Itzykson, and J. B. Zuber. Quantum field theory techniques in graphical enumeration. *Adv. in Appl. Math.*, 1(2):109–157, 1980.

[19] B. Bollobás. *Modern graph theory*, volume 184 of *Graduate Texts in Mathematics*. Springer-Verlag, New York, 1998.

[20] N. Bonichon, M. Bousquet-Mélou, and É. Fusy. Baxter permutations and plane bipolar orientations. *Sém. Lothar. Combin.*, 61A:Art. B61Ah, 29 pp. (electronic), 2009.

[21] D. V. Boulatov and V. A. Kazakov. The Ising model on a random planar lattice: the structure of the phase transition and the exact critical exponents. *Phys. Lett. B*, 186(3-4):379–384, 1987.

[22] M. Bousquet-Mélou. Algebraicity results for planar Eulerian maps. In preparation.

[23] M. Bousquet-Mélou. Spanning forests on planar maps. In preparation.

[24] M. Bousquet-Mélou. A method for the enumeration of various classes of column-convex polygons. *Discrete Math.*, 154(1-3):1–25, 1996.

[25] M. Bousquet-Mélou. Multi-statistic enumeration of two-stack sortable permutations. *Electron. J. Combin.*, 5(1):Research Paper 21, 12 pp., 1998.

[26] M. Bousquet-Mélou. Four classes of pattern-avoiding permutations under one roof: generating trees with two labels. *Electronic J. Combinatorics*, 9(2):Research Paper 19, 2003.

[27] M. Bousquet-Mélou. Walks in the quarter plane: Kreweras' algebraic model. *Ann. Appl. Probab.*, 15(2):1451–1491, 2005.

[28] M. Bousquet-Mélou and S. Butler. Forest-like permutations. *Ann. Comb.*, 11:335–354, 2007.

[29] M. Bousquet-Mélou and A. Jehanne. Polynomial equations with one catalytic variable, algebraic series and map enumeration. *J. Combin. Theory Ser. B*, 96:623–672, 2006.

[30] M. Bousquet-Mélou and M. Mishna. Walks with small steps in the quarter plane. *Contemp. Math.*, 520:1–40, 2010. ArXiv:0810.4387.

[31] M. Bousquet-Mélou and M. Petkovšek. Linear recurrences with constant coefficients: the multivariate case. *Discrete Math.*, 225(1-3):51–75, 2000.

[32] M. Bousquet-Mélou and M. Petkovšek. Walks confined in a quadrant are not always D-finite. *Theoret. Comput. Sci.*, 307(2):257–276, 2003.

[33] M. Bousquet-Mélou and G. Schaeffer. Enumeration of planar constellations. *Adv. in Appl. Math.*, 24(4):337–368, 2000.

[34] M. Bousquet-Mélou and G. Schaeffer. The degree distribution of bipartite planar maps: applications to the Ising model. In K. Eriksson and S. Linusson, editors, *Formal Power Series and Algebraic Combinatorics*, pages 312–323, Vadstena, Sweden, 2003. ArXiv math.CO/0211070.

[35] J. Bouttier, P. Di Francesco, and E. Guitter. Census of planar maps: from the one-matrix model solution to a combinatorial proof. *Nuclear Phys. B*, 645(3):477–499, 2002. ArXiv:cond-mat/0207682.

[36] J. Bouttier, P. Di Francesco, and E. Guitter. Counting colored random triangulations. *Nucl. Phys. B*, 641:519–532, 2002.

[37] J. Bouttier, P. Di Francesco, and E. Guitter. Planar maps as labeled mobiles. *Electron. J. Combin.*, 11(1):Research Paper 69, 27 pp. (electronic), 2004.

[38] J. Bouttier, P. Di Francesco, and E. Guitter. Combinatorics of bicubic maps with hard particles. *J. Phys. A*, 38(21):4529–4559, 2005.

[39] J. Bouttier, P. Di Francesco, and E. Guitter. Blocked edges on Eulerian maps and mobiles: application to spanning trees, hard particles and the Ising model. *J. Phys. A*, 40(27):7411–7440, 2007.

[40] J. Bouttier and E. Guitter. Statistics in geodesics in large quadrangulations. *J. Phys. A*, 41(14):145001, 30, 2008.

[41] J. Bouttier and E. Guitter. The three-point function of planar quadrangulations. *J. Stat. Mech. Theory Exp.*, 7:P07020, 39, 2008.

[42] E. Brézin, C. Itzykson, G. Parisi, and J. B. Zuber. Planar diagrams. *Comm. Math. Phys.*, 59(1):35–51, 1978.

[43] W. G. Brown. Enumeration of non-separable planar maps. *Canad. J. Math.*, 15:526–545, 1963.

[44] W. G. Brown. Enumeration of triangulations of the disk. *Proc. London Math. Soc. (3)*, 14:746–768, 1964.

[45] W. G. Brown. Enumeration of quadrangular dissections of the disk. *Canad. J. Math.*, 17:302–317, 1965.

[46] W. G. Brown. On the existence of square roots in certain rings of power series. *Math. Ann.*, 158:82–89, 1965.

[47] W. G. Brown and W. T. Tutte. On the enumeration of rooted non-separable planar maps. *Canad. J. Math.*, 16:572–577, 1964.

[48] G. Chapuy, M. Marcus, and G. Schaeffer. A bijection for rooted maps on orientable surfaces. *SIAM J. Discrete Math.*, 23(3):1587–1611, 2009.

[49] P. Chassaing and G. Schaeffer. Random planar lattices and integrated super-Brownian excursion. *Probab. Theory Related Fields*, 128(2):161–212, 2004.

[50] R. Cori and B. Vauquelin. Planar maps are well labeled trees. *Canad. J. Math*, 33:1023–1042, 1981.

[51] J.-M. Daul. *q*-States Potts model on a random planar lattice. ArXiv:hep-th/9502014.

[52] A. de Mier and M. Noy. A solution to the tennis ball problem. *Theoret. Comput. Sci.*, 346(2-3):254–264, 2005.

[53] A. Del Lungo, F. Del Ristoro, and J.-G. Penaud. Left ternary trees and non-separable rooted planar maps. *Theoret. Comput. Sci.*, 233(1-2):201–215, 2000.

[54] P. Di Francesco, B. Eynard, and E. Guitter. Coloring random triangulations. *Nuclear Phys. B*, 516(3):543–587, 1998.

[55] P. Di Francesco, P. Ginsparg, and J. Zinn-Justin. 2D gravity and random matrices. *Phys. Rep.*, 254(1-2), 1995. 133 pp.

[56] B. Eynard and G. Bonnet. The Potts-q random matrix model: loop equations, critical exponents, and rational case. *Phys. Lett. B*, 463(2-4):273–279, 1999.

[57] S. Felsner, É. Fusy, M. Noy, and D. Orden. Bijections for Baxter families and related objects. *J. Combin. Theory Ser. A*, 118(3):993–1020, 2011.

[58] P. Fendley and V. Krushkal. Tutte chromatic identities from the Temperley-Lieb algebra. *Geom. Topol.*, 13(2):709–741, 2009.

[59] S. Feretić and D. Svrtan. On the number of column-convex polyominoes with given perimeter and number of columns. In Barlotti, Delest, and Pinzani, editors, *Proceedings of the 5th Conference on Formal Power Series and Algebraic Combinatorics (Florence, Italy)*, pages 201–214, 1993.

[60] P. Flajolet. Analytic models and ambiguity of context-free languages. *Theoret. Comput. Sci.*, 49(2-3):283–309, 1987.

[61] P. Flajolet and A. Odlyzko. Singularity analysis of generating functions. *SIAM J. Discrete Math.*, 3(2):216–240, 1990.

[62] C. M. Fortuin and P. W. Kasteleyn. On the random cluster model: I. Introduction and relation to other models. *Physica*, 57:536–564, 1972.

[63] É. Fusy, D. Poulalhon, and G. Schaeffer. Dissections and trees, with applications to optimal mesh encoding and to random sampling. *ACM Trans. Algorithms*, 4(2), 2008. Art. 19.

[64] É. Fusy, D. Poulalhon, and G. Schaeffer. Bijective counting of plane bipolar orientations and Schnyder woods. *European J. Combin.*, 30(7):1646–1658, 2009.

[65] Z. Gao. The number of degree restricted maps on general surfaces. *Discrete Math.*, 123(1-3):47–63, 1993.

[66] Z. Gao and N. C. Wormald. Enumeration of rooted cubic planar maps. *Ann. Comb.*, 6(3-4):313–325, 2002.

[67] Z. J. Gao, I. M. Wanless, and N. C. Wormald. Counting 5-connected planar triangulations. *J. Graph Theory*, 38(1):18–35, 2001.

[68] I. P. Goulden and D. M. Jackson. Transitive factorizations in the symmetric group, and combinatorial aspects of singularity theory. *European J. Combin.*, 21(8):1001–1016, 2000.

[69] I. P. Goulden and D. M. Jackson. The KP hierarchy, branched covers, and triangulations. *Adv. Math.*, 219(3):932–951, 2008.

[70] C. Greene and T. Zaslavsky. On the interpretation of Whitney numbers through arrangements of hyperplanes, zonotopes, non-Radon partitions, and orientations of graphs. *Trans. Amer. Math. Soc.*, 280(1):97–126, 1983.

[71] A. Guionnet and É. Maurel-Segala. Combinatorial aspects of matrix models. *ALEA Lat. Am. J. Probab. Math. Stat.*, 1:241–279, 2006.

[72] J. L. Jacobsen, J.-F. Richard, and J. Salas. Complex-temperature phase diagram of Potts and RSOS models. *Nuclear Phys. B*, 743(3):153–206, 2006.

[73] J. L. Jacobsen and J. Salas. Transfer matrices and partition-function zeros for antiferromagnetic Potts models. II. Extended results for square-lattice chromatic polynomial. *J. Statist. Phys.*, 104(3-4):701–723, 2001.

[74] B. Jacquard and G. Schaeffer. A bijective census of nonseparable planar maps. *J. Combin. Theory Ser. A*, 83(1):1–20, 1998.

[75] V. A. Kazakov. Ising model on a dynamical planar random lattice: exact solution. *Phys. Lett. A*, 119(3):140–144, 1986.

[76] D. E. Knuth. *The art of computer programming. Vol. 1: Fundamental algorithms.* Addison-Wesley Publishing Co., Reading, Mass.-London-Don Mills, Ont, 1968.

[77] B. Lass. Orientations acycliques et le polynôme chromatique. *European J. Combin.*, 22(8):1101–1123, 2001.

[78] J.-F. Le Gall. The topological structure of scaling limits of large planar maps. *Invent. Math.*, 169(3):621–670, 2007.

[79] J.-F. Le Gall and G. Miermont. Scaling limits of random planar maps with large faces. *Ann. Probab.*, 39(1):1–69, 2011.

[80] J.-F. Le Gall and F. Paulin. Scaling limits of bipartite planar maps are homeomorphic to the 2-sphere. *Geom. Funct. Anal.*, 18(3):893–918, 2008.

[81] L. Lipshitz. The diagonal of a D-finite power series is D-finite. *J. Algebra*, 113(2):373–378, 1988.

[82] L. Lipshitz. D-finite power series. *J. Algebra*, 122:353–373, 1989.

[83] V. A. Liskovets and T. R. S. Walsh. Enumeration of Eulerian and unicursal planar maps. *Discrete Math.*, 282(1-3):209–221, 2004.

[84] Y. Liu. Chromatic sum equations for rooted planar maps. In *Proceedings of the fifteenth Southeastern conference on combinatorics, graph theory and computing (Baton Rouge, La., 1984)*, volume 45, pages 275–280, 1984.

[85] J.-F. Marckert and G. Miermont. Invariance principles for random bipartite planar maps. *Ann. Probab.*, 35(5):1642–1705, 2007.

[86] J.-F. Marckert and A. Mokkadem. Limit of normalized quadrangulations: the Brownian map. *Ann. Probab.*, 34(6):2144–2202, 2006.

[87] P. P. Martin. The Potts model and the Beraha numbers. *J. Phys. A*, 20(6):L399–L403, 1987.

[88] M. L. Mehta. A method of integration over matrix variables. *Comm. Math. Phys.*, 79(3):327–340, 1981.

[89] M. Mishna. Classifying lattice walks restricted to the quarter plane. *J. Combin. Theory Ser. A*, 116(2):460–477, 2009. ArXiv:math/0611651.

[90] M. Mishna and A. Rechnitzer. Two non-holonomic lattice walks in the quarter plane. *Theoret. Comput. Sci.*, 410(38-40):3616–3630, 2009. ArXiv:math/0701800.

[91] R. C. Mullin. On the enumeration of tree-rooted maps. *Canad. J. Math.*, 19:174–183, 1967.

[92] A. M. Odlyzko and L. B. Richmond. A differential equation arising in chromatic sum theory. In *Proceedings of the fourteenth Southeastern conference on combinatorics, graph theory and computing (Boca Raton, Fla., 1983)*, volume 40, pages 263–275, 1983.

[93] D. Poulalhon and G. Schaeffer. A note on bipartite Eulerian planar maps. Available at http://www.lix.polytechnique.fr/~schaeffe.

[94] D. Poulalhon and G. Schaeffer. A bijection for triangulations of a polygon with interior points and multiple edges. *Theoret. Comput. Sci.*, 307(2):385–401, 2003.

[95] D. Poulalhon and G. Schaeffer. Optimal coding and sampling of triangulations. *Algorithmica*, 46(3-4):505–527, 2006.

[96] H. Prodinger. The kernel method: a collection of examples. *Sém. Lothar. Combin.*, 50:Art. B50f, 19 pp. (electronic), 2003/04.

[97] H. Saleur. Zeroes of chromatic polynomials: a new approach to Beraha conjecture using quantum groups. *Comm. Math. Phys.*, 132(3):657–679, 1990.

[98] G. Schaeffer. Bijective census and random generation of Eulerian planar maps with prescribed vertex degrees. *Electron. J. Combin.*, 4(1):Research Paper 20, 14 pp. (electronic), 1997.

[99] G. Schaeffer. *Conjugaison d'arbres et cartes combinatoires aléatoires*. PhD thesis, Université Bordeaux 1, France, 1998.

[100] R. P. Stanley. *Enumerative combinatorics. Vol. 2*, volume 62 of *Cambridge Studies in Advanced Mathematics*. Cambridge University Press, Cambridge, 1999.

[101] H. N. V. Temperley. Combinatorial problems suggested by the statistical mechanics of domains and of rubber-like molecules. *Phys. Rev. (2)*, 103:1–16, 1956.

[102] W. T. Tutte. A contribution to the theory of chromatic polynomials. *Canadian J. Math.*, 6:80–91, 1954.

[103] W. T. Tutte. A census of planar triangulations. *Canad. J. Math.*, 14:21–38, 1962.

[104] W. T. Tutte. A census of planar maps. *Canad. J. Math.*, 15:249–271, 1963.

[105] W. T. Tutte. On the enumeration of planar maps. *Bull. Amer. Math. Soc.*, 74:64–74, 1968.

[106] W. T. Tutte. Dichromatic sums for rooted planar maps. *Proc. Sympos. Pure Math.*, 19:235–245, 1971.

[107] W. T. Tutte. Chromatic sums for rooted planar triangulations. II. The case $\lambda = \tau + 1$. *Canad. J. Math.*, 25:657–671, 1973.

[108] W. T. Tutte. Chromatic sums for rooted planar triangulations. III. The case $\lambda = 3$. *Canad. J. Math.*, 25:780–790, 1973.

[109] W. T. Tutte. Chromatic sums for rooted planar triangulations. IV. The case $\lambda = \infty$. *Canad. J. Math.*, 25:929–940, 1973.

[110] W. T. Tutte. Chromatic sums for rooted planar triangulations: the cases $\lambda = 1$ and $\lambda = 2$. *Canad. J. Math.*, 25:426–447, 1973.

[111] W. T. Tutte. Chromatic sums for rooted planar triangulations. V. Special equations. *Canad. J. Math.*, 26:893–907, 1974.

[112] W. T. Tutte. On a pair of functional equations of combinatorial interest. *Aequationes Math.*, 17(2-3):121–140, 1978.

[113] W. T. Tutte. Chromatic solutions. *Canad. J. Math.*, 34(3):741–758, 1982.

[114] W. T. Tutte. Chromatic solutions. II. *Canad. J. Math.*, 34(4):952–960, 1982.

[115] W. T. Tutte. Map-colourings and differential equations. In *Progress in graph theory (Waterloo, Ont., 1982)*, pages 477–485. Academic Press, Toronto, ON, 1984.

[116] W. T. Tutte. Chromatic sums revisited. *Aequationes Math.*, 50(1-2):95–134, 1995.

[117] W. T. Tutte. Dichromatic sums revisited. *J. Combin. Theory Ser. B*, 66(2):161–167, 1996.

[118] T. Walsh and A. B. Lehman. Counting rooted maps by genus. II. *J. Combinatorial Theory Ser. B*, 13:122–141, 1972.

[119] T. R. S. Walsh and A. B. Lehman. Counting rooted maps by genus. III: Nonseparable maps. *J. Combinatorial Theory Ser. B*, 18:222–259, 1975.

[120] D. J. A. Welsh and C. Merino. The Potts model and the Tutte polynomial. *J. Math. Phys.*, 41(3):1127–1152, 2000.

[121] D. Zeilberger. A proof of Julian West's conjecture that the number of two-stack-sortable permutations of length n is $2(3n)!/((n+1)!(2n+1)!)$. *Discrete Math.*, 102(1):85–93, 1992.

[122] D. Zeilberger. The umbral transfer-matrix method: I. Foundations. *J. Comb. Theory, Ser. A*, 91:451–463, 2000.

[123] P. Zinn-Justin. The dilute Potts model on random surfaces. *J. Statist. Phys.*, 98(1-2):j10–264, 2000.

CNRS, LaBRI, Université Bordeaux 1
351 cours de la Libération
F-33405 Talence Cedex
France
bousquet@labri.fr

A survey of PPAD-completeness for computing Nash equilibria

Paul W. Goldberg

Abstract

PPAD refers to a class of computational problems for which solutions are guaranteed to exist due to a specific combinatorial principle. The most well-known such problem is that of computing a Nash equilibrium of a game. Other examples include the search for market equilibria, and envy-free allocations in the context of cake-cutting. A problem is said to be complete for **PPAD** if it belongs to **PPAD** and can be shown to constitute one of the hardest computational challenges within that class.

In this paper, I give a relatively informal overview of the proofs used in the **PPAD**-completeness results. The focus is on the mixed Nash equilibria guaranteed to exist by Nash's theorem. I also give an overview of some recent work that uses these ideas to show **PSPACE**-completeness for the computation of specific equilibria found by homotopy methods. I give a brief introduction to related problems of searching for market equilibria.

Acknowledgements

Currently supported by EPSRC Grant EP/G069239/1

1 Total Search Problems

Suppose that you enter a maze without knowing anything in advance about its internal structure. Let us assume that it has only one entrance. To solve the maze, we do not ask to find some central chamber whose existence has been promised by the designer. Instead you need to find either a dead end, or else a place where the path splits, giving you a choice of which way to go; see Figure 1. Observe that this kind of solution is guaranteed to exist, and does not require any kind of promise. This is because, if there are no places where the explorer has a choice, then the interior of the maze (at least, the parts accessible from the entrance) is, topologically, a single path leading to a dead end. If you are asked to find either a dead end or a split of a path, this is informally an example of a (syntactic) *total search problem* — the problem description has been set up so as to guarantee that there exists a solution; the word "total" refers to the fact that *every* problem instance has a solution.

A maze of the sort in Figure 1 could of course be solved in linear time by checking each location. Consider now a more challenging problem defined as follows:

Definition 1.1 CIRCUIT MAZE is a search problem on a $2^n \times 2^n$ grid — the maze is specified using a boolean circuit C that takes as input a bit string of length $O(n)$ that represents the location (coordinates) of a possible wall or barrier between 2 adjacent grid points. C has a single output bit that indicates whether in fact a barrier exists at that location.

A naive search for a grid point that corresponds to a dead end or a split of a path is no longer feasible, so the search problem becomes nontrivial. Notice that not only

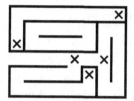
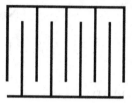

Figure 1: The left-hand maze (with a single entrance on the left) has 5 solutions marked with crosses. Note that while solutions need not be accessible starting from the entrance, at least one will be accessible. The right-hand maze illustrates how there need not be any solutions when there are 2 entrances. In general, an odd number of entrances guarantees the existence of a solution.

must some solution exist, but in addition it is easy to use C to check the validity of a claimed solution.

The problem of finding, or computing, a Nash equilibrium of a game (defined in Section 3) is analogous. Nash's theorem [53] assures us in advance that every game has a Nash equilibrium, and ultimately it works by applying a similar combinatorial principle (explained below) to the one that is being used to assure ourselves that maze problems of the sort defined above must have a solution.

1.1 NP Total Search Problems

Why can we not apply a more standard notion of computational hardness, such as **NP**-hardness, to the problem of Nash equilibrium computation? It turns out to be due precisely to its special status as a *total* search problem, as informally defined above.

Standard **NP**-hard problems do indeed have associated optimisation problems that are total search problems, but they are not **NP** total search problems. Consider for example the (**NP**-complete) travelling salesman problem, commonly denoted TSP. This has an associated optimisation problem, in which we seek a tour of minimal length that visits all the cities. By definition, such a minimal-length tour must exist, so this total search problem is **NP**-hard. But it is not, as far as we know, a member of **NP** — given a solution there is no obvious efficient way to check its optimality.

The complexity class **PPAD** (along with related classes introduced by Papadimitriou [57]) was intended to capture the computational complexity of a relatively small number of problems that seem not to have polynomial-time algorithms, but where there is a mathematical guarantee that every instance has a solution, and furthermore, *given a solution, the validity of that solution may be checked in polynomial time*. Each class of problems (see Section 1.3) has an associated combinatorial principle that guarantees that one has a total search problem.

A simple result due to Megiddo [51] shows that if such a problem is **NP**-complete, then **NP** would have to be equal to co-**NP**— it is proved as follows.

Suppose we have a reduction from any **NP**-hard problem (e.g. SAT) to any **NP** total search problem (e.g. NASH). Thus, from any SAT-instance (a propositional formula) we can efficiently construct a NASH-instance (a game) so that given any solution (Nash equilibrium) to that NASH-instance we can (efficiently) derive an answer to the SAT-instance. That reduction could then be used to construct a nondeterministic algorithm for verifying that an unsatisfiable instance of SAT indeed has *no* solution: Just guess a solution of the NASH-instance, and check that it indeed fails to identify a solution for the SAT-instance.

The existence of such a nondeterministic algorithm for SAT (one that can verify that an unsatisfiable formula is indeed unsatisfiable, hence implying that **NP**=co-**NP**) is an eventuality that is considered by complexity theorists almost as unlikely as **P**=**NP**. We conclude that NASH is very unlikely to be **NP**-complete.

1.2 Reducibility among total search problems

Suppose we have two total search problems X and Y. We say that X is reducible to Y in polynomial time if the following holds. There should be functions f and g both computable in polynomial time, such that given an instance I_X of X, $f(I_X)$ is an instance of Y, and given any solution S to $f(I_X)$, $g(S)$ is a solution to I_X. Thus, if we had a polynomial-time algorithm that solves Y, the reduction would construct a polynomial-time algorithm for X. So, such a reduction shows that Y is "at least as hard" as X.

While reductions in the literature are of the above form, one could use a less restrictive definition, called a Turing reduction, in which problem X reduces to problem Y provided that we we can write down an algorithm that solves X in polynomial time, provided that it has access to an "oracle" for problem Y. As a consequence, if Y does in fact have a polynomial-time algorithm then so does X. However, the reductions used in the literature to date about total search problems, are of the more restricted type.

1.3 PPAD, and some related concepts

PPAD, introduced in [57], stands for "polynomial parity argument on a directed graph". It is defined in terms of a rather artificial-looking problem END OF THE LINE, which is the following:[1]

Definition 1.2 An instance of END OF THE LINE consists of two boolean circuits S and P each of which has n inputs and n outputs, such that $P(0^n) = 0^n \neq S(0^n)$. Find a bit vector x such that $P(S(x)) \neq x$ or $S(P(x)) \neq x \neq 0^n$. [2]

S and P (standing for *successor* and *predecessor*) implicitly define a digraph G on 2^n vertices (bit strings of length n) in which each vertex has indegree and outdegree at most 1. (v, w) is an arc of G (directed from v to w) if and only if $S(v) = w$

---

[1]This is called END-OF-LINE in [13], which is arguably a better name since the "line" is not unique, and there is no requirement that we find the end of any specific line.

[2]Chen et al. [13] define it in terms of a single circuit that essentially combines S and P, that takes an n-bit vector v as input, and outputs $2n$ bits that correspond to $S(v)$ and $P(v)$.

and $P(w) = v$. By construction, 0^n has indegree 0, and either has outdegree 1, or $P(S(0^n)) \neq 0^n$ (in which case 0^n is a solution). Notice that G permits efficient local exploration (the neighbours of any vertex v are easy to compute from v) but non-local properties are opaque. We shall refer to a graph G that is represented in this way as an (S, P)-graph.

The "parity argument on a directed graph" refers to a more general observation: define an "odd" vertex of a graph to be one where the total number of incident edges is an odd number. Then notice that the number of odd vertices must be an even number. Indeed, this observation applies generally to undirected graphs, but we apply it here to (S, P)-graphs, in particular those (S, P)-graphs where vertex 0^n actually has an outgoing arc (so has odd degree). These (S, P)-graphs have an associated total search problem, of finding an alternative odd-degree vertex. One could search for such a vertex by, for example

- checking each bit string in order, looking up its neighbours to see if it's an odd vertex, or

- following a directed path starting at some vertex; an endpoint other than 0^n must be reached,

but since G is exponentially large, these naive approaches will take exponential time in the worst case.

Definition 1.3 A computational problem X belong to the complexity class **PPAD** provided that X reduces to END OF THE LINE in polynomial time. Problem X is **PPAD**-complete provided that X is in **PPAD**, and in addition END OF THE LINE reduces to X in polynomial time.

Thus END OF THE LINE stands in the same relationship to **PPAD**, that CIRCUIT SAT does to **NP** (although **NP** is not actually defined in terms of CIRCUIT SAT, an equivalent definition would say that **NP**-complete problems are those that are polynomial-time equivalent to CIRCUIT SAT, and a member of **NP** is a problem that can be reduced to CIRCUIT SAT).

The above definition of **PPAD** is the one used in [22]; it is noted there that there are many alternative equivalent definitions[3]. Since END OF THE LINE is by construction a total search problem, it follows that members of **PPAD** are necessarily also total search problems.

How hard is "PPAD-complete"?   **PPAD** lies "between **P** and **NP**" in the sense that if **P** were equal to **NP**, then all **PPAD** problems would be polynomial-time solvable, while the assumption that "**PPAD**-complete problems are hard" implies **P** not equal to **NP**. It is a fair criticism of these results that they do not carry as much weight as do **NP**-hardness results, partly for this reason, and partly because there are only a handful of **PPAD**-complete problems, while thousands of problems have been shown to be **NP**-complete.

Why, then, do we take **PPAD**-completeness as evidence that a problem cannot be solved in polynomial time? One argument is that END OF THE LINE is defined

---

[3]The original definition of [57] is in terms of a Turing machine rather than circuits.

in terms of unrestricted circuits, and general boolean circuits seem to be hard to
analyse via polynomial-time algorithms. We should also note the oracle separation
results of [4] in this context.

Related complexity classes   Other combinatorial principles are considered in [57],
that guarantee totality of corresponding search problems. For example, consider the
pigeonhole principle, that given a function $f : X \longrightarrow Y$, if X and Y are finite
and $|X| > |Y|$, there must exist $x, x' \in X$ such that $f(x) = f(x')$. Now define an
associated computational problem:

Definition 1.4 The problem PIGEONHOLE CIRCUIT has as instances, directed boolean
circuits having the same number of inputs and outputs. Let f be the function computed by the circuit. The problem is to identify *either* two distinct bit strings x,
x' with $f(x) = f(x')$, *or* a bit string x with $f(x) = 0$ (where 0 is the all-zeroes bit
string).

Note that by construction, this a total search problem, and it is an **NP** total search
problem since it is computationally easy to check that a given solution is valid. At
the same time, it *seems* to be hard to find a solution, although it is unlikely to be
NP-hard, due to Megiddo's result. The complexity class **PPP** [57] (for "polynomial
pigeonhole principle") is defined as the set of all total search problems that are
reducible to PIGEONHOLE CIRCUIT.

The pigeonhole principle is a generalisation of the parity argument on a directed
graph. To see this, notice that the function $f : X \longrightarrow Y$ can map a vertex v of
an END OF THE LINE graph to the adjacent vertex connected by an arc from v, or
to itself if v has outdegree 0. **PPAD** is a subclass of **PPP**— END OF THE LINE
reduces to PIGEONHOLE CIRCUIT; it would be nice to obtain a reduction the other
way and show equivalence, but that has not been achieved.

If we have an *undirected* graph of degree at most 2 with a known endpoint, then
the search for another endpoint is also a total search problem. The corresponding
complexity class defined in [57] is **PPA**. The CIRCUIT MAZE problem that we considered informally at the start, belongs to **PPA**. As it happens, the problem is likely
to be complete for **PPA**.

2   Sperner's lemma, and an associated computational problem

Sperner's lemma is the following combinatorial result, that can be used to prove
Brouwer's fixed point theorem.

Theorem 2.1 *[64] Let $\{v_0, v_1, \ldots, v_d\}$ be the vertices of a d-simplex S, and suppose
that the interior of S is decomposed into smaller simplices using additional vertices.
Assign each vertex a colour from $\{0, 1, \ldots d\}$ such that v_i gets colour i, and a vertex
on any face of S must get one of the colours of the vertices of that face. Interior vertices may be coloured arbitrarily. Then, this simplicial decomposition must include
a* panchromatic *simplex, i.e. one whose vertices has all distinct colours.*

Proof The proof can be found in many places, so here we just give a sketch for
the 2-dimensional case. Let us define the computational problem SPERNER (dis-

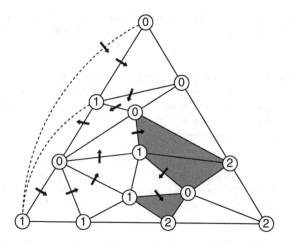

Figure 2: An example to illustrate Sperner's lemma.

The solid lines show the triangle and the simplicisation (triangulation, since we are in 2 dimensions.) The dashed lines add an additional sequence of topological triangles along one side of the triangulation; as a result of those lines, there is only one edge coloured 0-1 when viewed from the exterior.

The arrows are the directed edges of the corresponding graph.

The shaded triangles are trichromatic (End-of-line solutions in the corresponding graph).

cussed in more detail below) to be the problem of exhibiting a trichromatic triangle. Essentially, the proof consists of a reduction from SPERNER to END OF THE LINE!

Choose any two of the colours, say 0 and 1. We begin by adding some further triangles to the triangulation as shown in Figure 2 by the dotted lines: by adding a sequence of triangles that have the original extremal vertex coloured 1, together with two consecutive vertices on the 1-0 edge, we end up with a triangulation that has only one 0/1-coloured edge on the exterior.

Construct a directed graph G whose vertices are triangles of this extended triangulation. Add a directed edge of G between any two triangles that are adjacent and separated by a 0/1-edge; the direction of the edge is so as to cross with 0 on its left and 1 on its right. Consequently, there is a single edge coming into the triangulation from the outside. It is simple to check that in G, trichromatic triangles correspond exactly with degree-1 vertices, solutions to END OF THE LINE. This completes the reduction.   □

To define a challenging computational problem involved with searching for a panchromatic Sperner simplex, we need to work with Sperner triangulations that are represented so compactly that it becomes infeasible to just check every simplex. Hence, we consider exponentially-large simplicial decompositions that satisfy the boundary conditions required for Sperner's lemma.

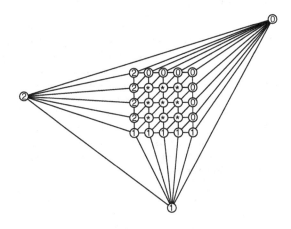

Figure 3: Embedding a triangulated grid inside an instance of Sperner's lemma

Informally, the computational problem SPERNER [57] (with parameter n) takes as input a Sperner triangulation that contains an number of vertices that is exponential in n. The vertices —and their colours— cannot be explicitly listed, since problem instances are supposed to be written down with a syntactic description length that is polynomial in n. Instead, an instance is specified by a circuit C that takes as input the coordinates of a vertex and outputs its colour. This means that the vertices should lie on a regular grid, and C takes as input a bit string that represents the coordinates of a vertex.

Figure 3 shows one way to do this in 2 dimensions (assume that the central grid has exponentially-many points). There is no restriction on how C may colour the vertices labelled $*$; the other labels are fixed boundary conditions to ensure that any trichromatic triangle lies within the grid. A 3-dimensional version would just require some choice of simplicial decomposition of a cube, which would then be applied to each small cube in the corresponding 3-d grid.

2-dimensional SPERNER is known to be **PPAD**-complete [10]; previously it was known from [57] that 3-dimensional SPERNER is **PPAD**-complete. One interesting feature of the **PPAD**-completeness results for Nash equilibrium computation, is that normal-form games do not incorporate generic boolean circuits in an obvious way. Contrast that with SPERNER as defined above, where a problem instance incorporates a generic boolean circuit to determine the colour of a vertex. We continue by explaining Nash equilibrium computation in more detail, and how it is polynomial-time equivalent to SPERNER.

3   Games and Nash Equilibria

A game $\mathcal{G}$ specifies a finite set of $k \geq 2$ players, where each player i has a finite set S_i of (at least two) *actions* or *pure strategies*. Let $S = S_1 \times \ldots \times S_k$ be the set of *pure strategy profiles*, thus an element of S represents a choice made by each player i of a single pure strategy from his set S_i. Finally, given any member of S, $\mathcal{G}$ needs to specify real-valued *utilities* or *payoffs* to each player that result from that pure strategy profile. Let u_s^i denote the payoff to player i that results from $s \in S$.

Informally, a Nash equilibrium is a set of strategies —one for each player— where no player has an "incentive to deviate" i.e. play some alternative strategy. After giving some examples, we provide a precise definition, along with some notation. In general, Nash equilibria do not exist for pure strategies; it is usually necessary to allow the players to randomise over them, as shown in the following examples.

Rock-paper-scissors:

	rock	paper	scissors
rock	$(0,0)$	$(-1,1)$	$(1,-1)$
paper	$(1,-1)$	$(0,0)$	$(-1,1)$
scissors	$(-1,1)$	$(1,-1)$	$(0,0)$

A typical payoff matrix for Stag hunt:

	hunt stag	hunt hare
hunt stag	$(8,8)$	$(0,1)$
hunt hare	$(1,0)$	$(1,1)$

Generalised matching pennies: the row player "wins" (and collects a payment from the column player) whenever both players play the same strategy.

	1	2	$\cdots$	n
1	$(1,-1)$	$(0,0)$	$\cdots$	$(0,0)$
2	$(0,0)$	$(1,-1)$		$(0,0)$
$\vdots$	$\vdots$		$\ddots$	$(0,0)$
n	$(0,0)$	$\cdots$	$(0,0)$	$(1,-1)$

Figure 4: Payoff matrices for example games. Each entry of a matrix contains two numbers: the row player's payoff and the column player's payoff.

Example 3.1 The traditional game of *rock-paper-scissors* fits in to the basic paradigm considered here. The standard payoff matrix shown in Figure 4 awards one point for winning and a penalty of one point for losing. The unique Nash equilibrium has both players randomising uniformly over their strategies, for an expected payoff of zero.

Example 3.2 The *stag hunt* game [60] has two players, each of whom has two actions, corresponding to hunting either a stag, or a hare. In order to catch a stag, both player must cooperate (choose to hunt a stag), but either player can catch a hare on his own. The benefit of hunting a stag is that the payoff (half share of a stag) is presumed to be substantially larger than an entire hare.

A typical choice of payoffs (Figure 4) to reflect this dilemma could award 8 to both players when they both hunt a stag, 0 to a player who hunts a stag while the other player hunts a hare, and 1 to a player who hunts a hare. Using these payoffs, there are 3 Nash equilibria: one where both players hunt a stag, one where both players hunt a hare, and one where each player hunts a stag with probability $\frac{1}{8}$ and a hare with probability $\frac{7}{8}$.

Example 3.3 *Generalised matching pennies* (Figure 4) is a useful example to present here, since it is used as an ingredient of the **PPAD**-completeness proof for NASH.

The game of *matching pennies* is a 2-player game, whose basic version has each player having just two strategies, "heads" and "tails". The row player wins whenever both players make the same choice, otherwise the column player wins. In the original version, a win effectively means that the losing player pays one unit to the winning player; here we modify payoffs so that a win for the column player just results in zero payoffs (he avoids paying the row player). With that modification, generalised matching pennies is the extension to n actions rather than just 2.

There is a unique Nash equilibrium in which both players use the uniform distribution over their pure strategies, and it is easy to see that no other solution is possible.

Comments. The Stag Hunt example shows that there may be multiple equilibria, that some equilibria have more social welfare than others, and that some are more "plausible" that others (thus, it seems reasonable to expect one of the two pure-strategy equilibria to be played, rather than the randomised one). The topic of *equilibrium selection* considers formalisations of the notion of plausibility, and the problem of computing the relevant equilibria. Of course, it is at least as hard to compute one of a subset of equilibria as it is to compute an unrestricted one. For example, it is **NP**-complete to compute equilibria that guarantee one or both players a certain level of payoff, or that satisfy various other properties that are efficiently checkable [18, 35].

If we know the supports of a Nash equilibrium (the strategies played with non-zero probability) then it would be straightforward to compute a Nash equilibrium, since it then reduces to a linear programming problem. Thus (as noted by Papadimitriou in [55]) the search for a Nash equilibrium is —in the 2-player case— essentially a combinatorial problem.

Nash equilibrium and ϵ-Nash equilibrium; definition and notation. A *mixed strategy* for player i is a distribution on S_i, that is, $|S_i|$ nonnegative real numbers summing to 1. We will use x_j^i to denote the probability that player i allocates to strategy $j \in S_i$. If some or all players use mixed strategies, this results in expected payoffs for the players. A *best response* by a player is a strategy (possibly mixed) that maximises that player's expected payoff; the assumption is that players do indeed play to maximise expected payoffs.

Call a set of k mixed strategies x_j^i a *Nash equilibrium* if, for each player i, i's expected payoff $\sum_{s\in S} u_s^i \prod_{r=1}^k x_{s_r}^r$ is maximized over all mixed strategies of i. That is, a Nash equilibrium is a set of mixed strategies from which no player has an incentive to deviate. Let $S_{-i} = S_1 \times \ldots \times S_{i-1} \times S_{i+1} \times \ldots \times S_k$, the set of pure strategy profiles of players other than i. For $s \in S_{-i}$, let $x_s = \prod_{r\neq i; r\in[k]} x_{s_r}^r$ and u_{js}^i be the payoff to i when i plays j and the other players play s. It is well-known (see, e.g., [56]) that the following is an equivalent condition for a set of mixed strategies to be a Nash equilibrium:

$$\forall i, j, j' \sum_{s\in S_{-i}} u_{js}^i x_s > \sum_{s\in S_{-i}} u_{j's}^i x_s \implies x_{j'}^i = 0. \tag{3.1}$$

Also, a set of mixed strategies is an ϵ-Nash equilibrium for some $\epsilon > 0$ if the following holds:

$$\forall i, j, j' \sum_{s\in S_{-i}} u_{js}^i x_s > \sum_{s\in S_{-i}} u_{j's}^i x_s + \epsilon \implies x_{j'}^i = 0. \tag{3.2}$$

The celebrated theorem of Nash [53] states that every game has a Nash equilibrium.

Comments. An ϵ-Nash equilibrium is a weaker notion than a Nash equilibrium; a Nash equilibrium is an ϵ-Nash equilibrium for $\epsilon = 0$, and generally an ϵ-Nash equilibrium is an ϵ'-Nash equilibrium for any $\epsilon' > \epsilon$.

Why focus on approximate equilibria? Approximate equilibria matter from the computational perspective, because for games of more than 2 players, a solution (Nash equilibrium) may be in irrational numbers, even when the utilities u_s^i that specify a game, are themselves rational numbers [53]. The problem of *computing* a Nash equilibrium —as opposed to just knowing that one exists— requires us to specify a format or syntax in which to output the quantities that make up a solution (i.e. the probabilities x_s^i).

In [23] (an expository paper on the results of [22]) we noted an analogy between equilibrium computation and the computation of a root of an odd-degree polynomial f in a single variable. For both problems, there is a guarantee that a solution really exists, and the guarantee is based on the nature of the problem rather than a promise that the given instance has a solution. Furthermore, in each problem we have to deal with the issue of how to represent a solution, since a solution need not necessarily be a rational number. And, in both cases a natural approach is to switch to some notion of approximate solution: instead of searching for x with $f(x) = 0$, search for x with $|f(x)| < \epsilon$, which ensures that x can be written down in a standard syntax.

3.1 Some reductions among equilibrium problems

By way of example, consider the well-known result (that predates the **PPAD**-hardness of NASH; see [55] Chapter 2 for further background) that symmetric 2-player games are as hard to solve as general ones.

Given a $n \times n$ game $\mathcal{G}$, construct a symmetric $2n \times 2n$ game $\mathcal{G}' = f(\mathcal{G})$, such that given any Nash equilibrium of $\mathcal{G}'$ we can efficiently reconstruct a Nash equilibrium of $\mathcal{G}$. To begin, let us assume that all payoffs in $\mathcal{G}$ are positive — the reason why this assumption is fine, is that if there are any negative payoffs, then we can add a sufficiently large constant to all payoffs and obtain a strategically equivalent version (i.e. one that has the same Nash equilibria).

Now suppose we solve the $2n \times 2n$ game $\mathcal{G}' = \begin{pmatrix} 0 & \mathcal{G} \\ \mathcal{G}^T & 0 \end{pmatrix}$, where we assume that entries contain payoffs to both players, and the zeroes represent $n \times n$ matrices of payoffs of zero to both players.

Let p and q denote the probabilities that players 1 and 2 use their first n actions, in some given solution. If we label the rows and columns of the payoff matrix with the probabilities assigned to them by the players, we have

$$
\begin{array}{cc}
 & \begin{array}{cc} q & 1-q \end{array} \\
\begin{array}{c} p \\ 1-p \end{array} & \begin{pmatrix} 0 & \mathcal{G} \\ \mathcal{G}^T & 0 \end{pmatrix}
\end{array}
$$

If $p = q = 1$, both players receive payoff 0, and both have an incentive to change their behaviour, by the assumption that $\mathcal{G}$'s payoffs are all positive (and similarly if $p = q = 0$). So we have $p > 0$ and $1 - q > 0$, or alternatively, $1 - p > 0$ and $q > 0$.

Assume $p > 0$ and $1 - q > 0$ (the analysis for the other case is similar). Let $\{p_1, ..., p_n\}$ be the probabilities used by player 1 for his first n actions, $\{q_1, \ldots q_n\}$ the probabilities for player 2's second n actions.

$$
\begin{array}{cc}
 & \begin{array}{cc} q & (q_1...q_n) \end{array} \\
\begin{array}{c} (p_1,...p_n) \\ 1-p \end{array} & \begin{pmatrix} 0 & \mathcal{G} \\ \mathcal{G}^T & 0 \end{pmatrix}
\end{array}
$$

Note that $p_1 + \ldots + p_n = p$ and $q_1 + \ldots + q_n = 1 - q$. Then $(p_1/p, \ldots, p_n/p)$ and $(q_1/(1-q), \ldots, q_n/(1-q))$ are a Nash equilibrium of $\mathcal{G}$. To see this, consider the diagram; they should form a best response to each other for the top-right part.

This is an example of the kind of reduction defined in Section 1.2. In the context of games, this kind of reduction is called in [1] a Nash homomorphism. Abbott et al. [1] reduce general 2-player games to win-lose 2-player games (where payoffs are 0 or 1). Various other Nash homomorphisms have been derived independently of the work relating NASH to **PPAD**. An important one is Bubelis [8] that reduces (in a more algebraic style) k-player games to 3-player games. The result also highlights a key distinction between k-player games (for $k \geq 3$) and 2-player games; in 2-player games the solutions are rational numbers (provided that the payoff in the games are rational) while for 3 or more players, the solutions may be irrational. The reduction of [8] preserves key algebraic properties of the quantities x_j^i in a solution, such as their degree (i.e the degree of the lowest-degree polynomial with integer coefficients satisfied by x_j^i). Since we have noted that 2-player games have solutions in rational numbers, this kind of reduction could not apply to 2-player games.

3.2 The "in PPAD" result

A proof that Nash equilibrium computation belongs to **PPAD** necessarily incorporates a proof of Nash's theorem itself (for approximate equilibrium), in that a reduction from ϵ-NASH to END OF THE LINE assures us that ϵ-NASH is a total search problem, since it is clear that END OF THE LINE is. To derive Nash's theorem itself, that an exact equilibrium exists, the summary as in [57] is as follows: Consider an infinite sequence of solutions to ϵ-NASH for smaller and smaller ϵ. Since the space of mixed-strategy profiles is compact, these solutions have a limit point, which must be an exact solution.

3.3   The Algebraic Properties of Nash Equilibria

A reader who is interested on the combinatorial aspects of the topic can skip most of this subsection; the main point to note is that since a Nash equilibrium is not necessarily in rational numbers, this motivates the focus on approximate equilibria (i.e. ϵ-Nash equilibria), to ensure that there is a natural syntax in which to output a computed solution to a game. Since any exact equilibrium is an approximate one, any hardness result for approximate equilibria applies automatically to exact ones. The question of polynomial-time computation of approximate equilibria was introduced in [37] in order to finesse the irrationality of exact solutions. However, algorithms for approximate equilibria go back earlier, notably Scarf's algorithm [62].

For positive ϵ, an ϵ-Nash equilibrium need not be at all close to a true Nash equilibrium. Etessami and Yannakakis [31] show that even for exponentially small ϵ, an ϵ-Nash equilibrium may be at variation distance 1 from any true Nash equilibrium, for a 3-player game. Furthermore, they also show that computing an approximate equilibrium that is within variation distance ϵ from a true one (an ϵ-near equilibrium) is square-root-sum hard. This refers to the following well-known computational problem:

Definition 3.4 The *square root sum* problem takes as input two sets of positive integers, say $\{x_1, \ldots, x_n\}$ and $\{y_1, \ldots, y_n\}$. The question is: is the sum of the square roots of the first set, greater than the sum of square roots of the second?

The problem is not known to belong to **NP**, although neither is it known to be hard for any well-known complexity class. In terms of upper-bounding the complexity of this problem, and also for that matter, computing an ϵ-near equilibrium, we may note that the criteria for a set of numbers x^i_j to be an ϵ-near equilibrium can be expressed in the existential theory of real arithmetic, which places the problem in **PSPACE** [58].

The fact that Nash equilibrium probabilities can be irrational numbers (while payoffs are rational numbers) appears in an example in Nash's paper [53]. He describes a simple poker-style game with 3 players and a unique (irrational) Nash equilibrium. We noted earlier that any game with rational payoffs has Nash equilibria probabilities that are algebraic numbers. But even for a 3-player game, if n is the number of actions, the solution may require algebraic numbers of degree exponentially large in n; as noted in [37], the constructions introduced there give us a flexible way to construct, from a given polynomial p, an arithmetic circuit whose fixpoints are the roots of p, and [37] shows how to construct a 4-player game whose equilibria encode those fixpoints. The paper of Bubelis [8] gave an "algebraic" reduction from any k-player game (for $k > 3$) to a corresponding 3-player game, in a way that preserves the algebraic properties of the solutions. Consequently 3-player games have the same feature of 4-player games identified in [37]. The reduction of [8] can in addition be used to show that 3-player games are as computationally hard to solve as k-players, but it does not highlight the expressive power of games as arithmetic circuits. Recently, Feige and Talgam-Cohen [32] give a reduction from (approximate) k-player NASH to 2-player NASH, that does not explicitly proceed via END OF THE LINE.

Is Nash equilibrium computation ever harder than PPAD?   Daskalakis et al. [20] show that most standard classes of concisely-represented games (such as graphical games and polymatrix games) have equilibrium computation problems that reduce to 2-NASH and so belong to **PPAD**. Schoenebeck and Vadhan [63] study the question for "circuit games", a very general concise representation of games where payoffs are given by a boolean circuit. In this context they obtain hardness results, but this is not an **NP** total search problem. The zero-sum version [33] is also hard.

4   Brouwer functions, and discrete Brouwer functions

Brouwer's fixpoint theorem states that every continuous function from a convex compact domain to itself, must have a *fixpoint*, a point x for which $f(x) = x$.

Proving Brouwer's fixpoint theorem using Sperner's lemma   Suppose to begin with that the domain in question is the d-simplex Δ^d. We have continuous $f : \Delta^d \longrightarrow \Delta^d$ and we seek a point $x \in \Delta^d$ with $f(x) = x$. Suppose f is evaluated on a set S of "sample points" in Δ^d, and at each point $x \in S$, consider the value of $f(x) - x$. If the $d + 1$ vertices of Δ^d are given distinct colours $\{0, 1, \ldots, d\}$, we colour-code any $x \in S$ according to the direction of $f(x) - x$: we give it the colour of a vertex which is at least as distant from $f(x)$ as from x. Such a colouring respects the constraints of Sperner's lemma. If S is used as the vertices of a simplicial decomposition, a panchromatic simplex becomes a plausible location for a fixpoint (f is displacing co-located vertices in all different directions). One obtains a fixpoint from the limit of increasingly fine simplicial decompositions.

The result can then be extended to other domains that are topologically equivalent to simplices, such as cubes/cuboids, as used in the constructions we describe here.

In defining an associated computational total search problem "find the point x", we need to identify a syntax or format in which to represent a class of continuous functions. The format should ensure that f can be computed in time polynomial in f's description length (so that solutions x are easily checkable, and the search problem is in **NP**). The class of functions that we define are based on *discrete Brouwer functions*, defined in detail below in Section 4.1.

Discrete Brouwer functions form the bridge between the discrete circuit problem END OF THE LINE and the continuous, numerical problem of computing a Nash equilibrium.

Consider the following obvious fact[4] that if we colour the integers $\{0, \ldots, n\}$ such that

---

[4]pointed out in [13] (Section 5.1)

- each integer is coloured either red or black, and

- 0 is coloured red, and n is coloured black

then there must be two consecutive numbers that have different colours.

Now suppose that n is exponentially large, but for any number x in the range $\{0, \ldots, n\}$ we have an efficient test to identify the colour of x. In that case, we still have an efficient search for consecutive numbers that have opposite colours, namely we can use binary search. If $\lfloor n/2 \rfloor$ is coloured black, we would search for the pair of numbers in the range $\{0, \ldots, \lfloor n/2 \rfloor\}$, otherwise we would search in the range $\{\lfloor n/2 \rfloor, \ldots, n\}$, and so on.

Now consider a 2-dimensional version with pairs of numbers (x, y), $x, y \in \{0, \ldots, n\}$, where pairs of numbers are coloured as follows.

- if $x = 0$, (x, y) is coloured red,

- if $y = 0$ and $x > 0$ then (x, y) is coloured yellow,

- if $x, y > 0$ and either $x = n$ or $y = n$ (or both), then (x, y) is coloured black.

Thus, we have fixed the colouring of the perimeter of the grid of points, but we impose no constraints on how the interior should be coloured. We claim that there exists a square of size 1 that has all 3 colours, but notice that binary search no longer finds such a square efficiently[5]. The fact that such a square exists is due to Sperner's lemma, whose proof envisages following a path that enters the $n \times n$ grid, ending up at a trichromatic square; intuitively, this can be achieved by entering at the point on the perimeter where red and yellow are adjacent, and walking around the outside of the red region. Since there is only one exterior red/yellow interval, we are guaranteed to reach a stage where the red region is no longer bounded by yellow points, but by black points.

4.1 Discrete Brouwer functions

We next give a detailed definition of discrete Brouwer functions (DBFs), together with an associated total search problem (on an exponential-sized domain), that is used in the reductions from END OF THE LINE to the search for a Nash equilibrium of a game. Figure 5 shows how an END OF THE LINE graph is encoded by a DBF.

Definition 4.1 Partition the unit d-dimensional cube K into "cubelets"— 2^{dn} axis aligned cubes of edge length 2^{-n}. A *Brouwer-mapping circuit* (in d dimensions) takes as input a bit string of length dn that represents the coordinates of a cubelet, and outputs its colour, subject to boundary constraints that generalise the 2-D case discussed above. A *Discrete Brouwer function* (DBF) is the function computed by such a circuit, and its syntactic complexity is the size of the circuit that represents it, which we restrict to be polynomial in n.

---

[5]Hirsch et al. [45] give lower bounds for algorithms that search for fixed points based on function evaluation. In this setting, an algorithm that searches for a fixed point of f has "black-box access" to f; f may be computed on query points but the internal structure of f is hidden. They contrast the one-dimension case with the 2-dimensional case by showing that in 2 dimensions, the search for an approximate fixed point of a Lipschitz continuous function in 2D, accurate to p binary digits, requires $\Omega(c^p)$ for some constant c, in contrast to 1D where bisection may be used.

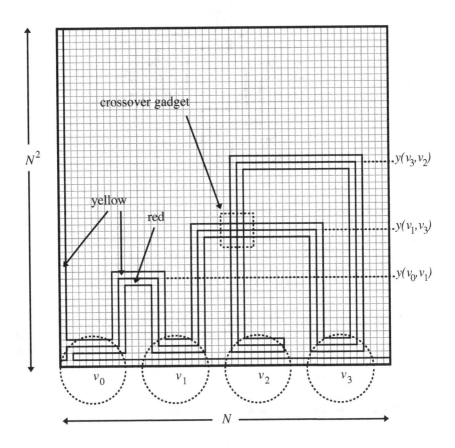

Figure 5: The figure shows how the line graph $(v_0, v_1), (v_1, v_3), (v_3, v_2)$ is encoded as a discrete Brouwer function in two dimensions.

The function colours the grid as a long thin red/yellow strip on a black background. Each vertex v_i is associated with a horizontal line segment close to the x-axis (circled in the diagram). Each edge (v_i, v_j) is associated with a "bridge" from the right-hand side of v_i's line segment to the left-hand side of v_j's.

Each bridge consists of three red/yellow line segments, going up, across and down. The y-coordinate of the horizontal segment $y(i, j)$ should efficiently encode the values i and j (for example, $y(i, j) = i + N.j$ where N is the number of x-values possible). Hence no two bridges can have overlapping horizontal sections. Unfortunately, a horizontal section may need to cross a vertical section of a different bridge, as shown in the diagram. Here a crossover gadget of [10] (Figure 6) may be used.

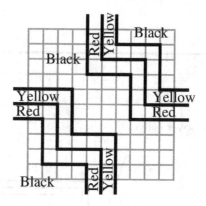

Figure 6: The crossover gadget referred to in Figure 5.

To avoid all three colours meeting in the vicinity of the crossing of two red/yellow path segments: the simple solution is to re-connect the paths in a way that the topological structure of the red-yellow line no longer mimics the structure of the END OF THE LINE graph from which it was derived. The 3-dimensional version, invented in [57] and refined in [22] does not require crossover gadgets.

This gives rise to the total search problem of finding a vertex that belongs to cubelets of all the different colours (a "panchromatic cubelet"). A superficial difference between the presentation of this concept in [22] compared with [13] is that [22] associates each cubelet with a colour while [13] associated each vertex with a colour, and the search is for a cubelet that has vertices of all colours.

5   From Discrete Brouwer functions to Games

The reduction is broken down into two stages: first we take a circuit C representing a DBF and encode it as an arithmetic circuit C' that computes a continuous function from K^d to K^d, where K^d is the d-dimensional unit cube. Then we express C' as a game $\mathcal{G}$ in such a way that given any Nash equilibrium of $\mathcal{G}$, we can efficiently extract the coordinates of a fixpoint of C'.

Figure 7 gives the general idea of how to construct a continuous function f from a DBF, in such a way that fixpoints of the continuous function correspond to solutions of the DBF. The idea is that each colour in the range of the DBF corresponds to a direction of $f(x) - x$. For a colour that lies on the boundary of the discretised cube, we choose a direction away from that boundary, so that f avoids displacing points outside the cube. Indeed, this is where we require the boundary constraints on the colouring of DBFs. Points at the centres of the cubelets of the DBF will be displaced in the direction of that cubelet's colour, while points on or near the boundary of two or more cubelets should get directions that interpolate the individual colour-directions (which is necessary for continuity). A natural choice for these colour-directions results in the following property: *For a proper subset of these*

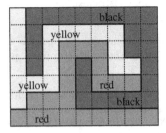

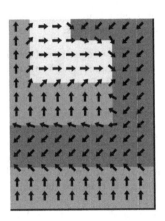

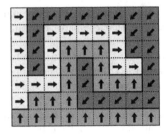

Figure 7: The top left-hand diagram shows a simple discrete Brouwer function in 2 dimensions, and the arrows in the bottom left-hand diagram show the direction of the corresponding continuous Brouwer function f within each small square.

The right-hand diagram is a magnified view of the bottom right-hand section of the continuous Brouwer function, showing how the direction of the function (i.e the direction of $f(x) - x$) can be interpolated on the boundaries of the small squares.

It is always possible to smoothly interpolate between 2 different directions without introducing fixed points of f. But, in the vicinity of the point where all 3 colours meet, there will necessarily be a point x where $f(x) = x$.

directions, their average cannot be zero. On the other hand, a weighted average of all the directions can indeed be zero.

Arithmetic circuits The papers [13, 22] show how to construct arithmetic circuits that compute these kinds of continuous functions. An arithmetic circuit consists of a directed graph each of whose nodes belongs to one of a number of distinct types, such as addition and multiplication. Each node will compute a real number, and the type of a node dictates how that number is obtained. For example, an addition node should have 2 incoming arcs, and it will compute the sum of the values located at the adjacent nodes for those two arcs; it may also have a number of outgoing arcs that allow its value to feed in to other nodes.

A significant difference between the circuits of [22] and those of [13] is that the latter do not allow nodes that compute the *product of two computed quantities* (located at other nodes with incoming edges). They do allow the multiplication of a computed quantity by a constant. The result of this is

- The functions that can be computed are more constrained; however they are still expressive enough to simulate the behaviour of DBF circuits,

- two-player normal-form games can simulate the circuits of [13], while three-player games are required to simulate circuits that can multiply a pair of computed quantities.

The more general circuits introduced in [22] correspond to the algebraic complexity class FIXP of Etessami and Yannakakis [31]; the ones without these product nodes correspond to **PPAD**, the linear version of FIXP, called LINEAR-FIXP in [31].

Interpolating the directions For a point x that is *not* on a cubelet boundary, one can design a polynomial-sized arithmetic circuit that compares the values of its coordinates against various numbers, eventually determining which cubelet it belongs to. Its colour assigned by the original DBF can be computed using a further component of the arithmetic circuit, that simulates the DBF. A further component of the circuit can translate the colour to a direction-vector, which gets added to x to yield $f(x)$.

For x on or close to a cubelet boundary, both papers [13, 22] perform the interpolation as follows. f is computed not just at x, but also at a small cluster of points in the vicinity of x. Then the average value of $f(x')$ is computed, for all points x' in this cluster. [13] show how this can be done in non-constant dimension, which allows a stronger hardness-of-approximation result to be obtained; see below. For constant dimension this clustering trick can be avoided; [38] shows the interpolation can be done more directly, and without the multiplication nodes discussed above.

Encoding approximate solutions, and snake embeddings *Snake embeddings*, invented by Chen et al. [13] allow hardness results for ϵ-approximate solutions where ϵ is inverse polynomial in n. A snake embedding maps a discrete Brouwer function $\mathcal{F}$ to a lower-dimension DBF $\mathcal{F}'$, in such a way that a solution to $\mathcal{F}'$ efficiently encodes a solution to $\mathcal{F}$, and the number of cubelets along the edges decreases by a constant factor. Applying these repeatedly, we obtain a DBF in $\Theta(n)$ dimensions where the cubelets have $O(1)$ edge length. The coordinates of points in the cube can be perturbed by relatively large amounts without leaving a cubelet.

5.1 Graphical Games

Graphical games were introduced in [47, 50] as a succinct means of representing certain games having many players. In a graphical game, each player has an associated vertex of an underlying graph G. The payoff to a player at vertex v is assumed to depend only on the behaviour of himself and his neighbours. Consequently, for a low-degree graph, the payoffs can be described much more concisely than would be allowed by normal form.

Graphical games were used by Daskalakis et al. [22] as an intermediate stage between discrete Brouwer functions, and normal-form games. That is, it was shown in [22] how a discrete Brouwer function can be converted into a graphical game $\mathcal{GG}$ such that any Nash equilibrium of $\mathcal{GG}$ encodes a solution to the discrete Brouwer function. Essentially, the underlying graph of $\mathcal{GG}$ has the same structure as the arithmetic circuit derived from the DBF, and the probabilities in the Nash equilibrium correspond to the quantities computed in the circuit. Chen et al [13] reduce more directly from discrete Brouwer functions to 2-player games.

5.2 From graphical to normal-form games

Recall the game of *generalised matching pennies* (GMP) from Example 3.3. The construction used in [13, 22] starts with a "prototype" zero-sum game consisting of a version of GMP where all strategies have been duplicated, as shown in the following payoff matrix R for the row player (M is some large positive quantity).

$$R = \begin{pmatrix} M & M & 0 & 0 & \cdots & 0 & 0 \\ M & M & 0 & 0 & \cdots & 0 & 0 \\ 0 & 0 & M & M & \cdots & 0 & 0 \\ 0 & 0 & M & M & \cdots & 0 & 0 \\ \vdots & \vdots & \vdots & \vdots & \ddots & \vdots & \vdots \\ 0 & 0 & 0 & 0 & \cdots & M & M \\ 0 & 0 & 0 & 0 & \cdots & M & M \end{pmatrix}$$

Having noted that GMP has a unique equilibrium in which both players randomise uniformly, we next note that in the above game, each player will randomise uniformly over pairs of duplicated strategies. So, each player allocates probability $\frac{1}{n}$ to each such pair. Within each pair, that allocation of $\frac{1}{n}$ may be split arbitrarily; $\mathcal{G}$ imposes no constraints on how to divide it.

The idea next, is to add certain quantities to the payoffs (that are relatively small in comparison with M) so that a player's choice of how to split between a pair of "duplicated" strategies may affect the opponents' choice for other pairs of his strategies. It can be shown that in Nash equilibria of the resulting game,

- the probability allocated to any strategy-pair is not exactly $\frac{1}{n}$ but is fairly close, since the value of M dominates the other payoffs that are introduced,

- the probabilities p and p' allocated to the members of a strategy-pair are used to represent the number $p/(p+p') \in [0, 1]$,

- the numbers thus represented can be made to affect each others' values in the same way that players in a graphical game, having just 2 pure strategies $\{0, 1\}$, affect each others' probabilities of playing 1.

6 Easy and hard classes of games

The **PPAD**-completeness results for normal-form games have led to a line of research addressing the very natural question of what types of games admit polynomial-time algorithms, and which ones are also **PPAD**-hard. In particular, with regard to **PPAD**-hardness, the general aim is to obtain hardness results for games that are more and more syntactically restricted; the first results, for 4 players [21], then 3 players [12, 25], and then 2 players [11] can be seen as the initial chapter in this narrative. In this section, our focus is on results on the frontier of **PPAD**-completeness and membership of **P**. We do not consider games where solutions can be shown to exist by means of a potential function argument (known to be equivalent to *congestion games* [52, 59]).

While normal-form games are the most natural ones to consider from a theoretical perspective, there are many alternative ways to specify a game. It is, after all, unnatural to write down a description of a game in normal form; only for very small games is this feasible.

6.1 Hard equilibrium computation problems

So, one natural direction is to look for **PPAD**-hardness results for more restricted types of games than the ones currently known to be **PPAD**-hard.

Restricted 2-player games We currently know that 2-player games are hard to solve even when they are sparse [13, 14]; the problem SPARSE BIMATRIX denotes the problem of computing a Nash equilibrium for 2-player games having a constant bound on the number of non-zero entries in each row and column. (For example, Theorem 10.1 of [13] show that 2-player games remain **PPAD**-hard when rows and columns have up to 10 non-zero entries.) 2-player games are also **PPAD**-complete to solve when they have 0/1 payoffs [1][6].

Restricted graphical games It is shown in [30] that degree-2 graphical games are solvable in polynomial time, but **PPAD**-complete for graphs with constant pathwidth (it is an open question precisely what pathwidth is required to make the problem **PPAD**-complete).

Ranking games Brandt et al. [5] study ranking games from a computational-complexity viewpoint. Ranking games are proposed as model of various real-world competitions; the outcome of the players' behaviour is a ranking of its participants, thus is maps any pure-strategy profile to a ranking of the players. Let u_r^i be the payoff to player i that results from being ranked r-th, where we assume $u_r^i \geq u_{r+1}^i$ (players prefer to be ranked first to being ranked second, and so on.)

In the 2-player case, any ranking game is strategically equivalent to a zero-sum game. For more than 2 players, one can apply the observation of [54] that any k-player game is essentially equivalent to a $k+1$-player zero-sum game, where the additional player is given payoffs that set the total payoff to zero, but takes no part in the game, in that his actions do not affect the other players' payoffs. Using this, it is easy to reduce 2-player win-lose games to 3-player ranking games.

Polymatrix games A polymatrix game is a multiplayer game in which any player's payoff is the sum of payoffs he obtains from bimatrix games with the other players. A special case of interest is where there is a limit on the number of pure strategies per player. In general, these polymatrix games are **PPAD**-complete [20], which also follows from [22] — specifically Section 6 of [22] presents a **PPAD**-hardness proof

---

[6]The paper of Abbott et al. [1] appeared before the first **PPAD**-hardness results for games, and reduced the search for Nash equilibria of general 2-player games to the search for Nash equilibria in 0/1-payoff 2-player games (i.e. win-lose games).

for 2-NASH by using graphical game gadgets that have the feature that the payoff to a vertex can be expressed as the sum of payoffs of 2-player games with his neighbours. (This extension to 2-NASH applies the observation that it is unnecessary to have a player who computes the product of his neighbours; such a vertex has payoffs that cannot decompose as the sum of bimatrix games.)

6.2 Polynomial-time equilibrium computation problems

In reviewing some of the computational positive results (types of games that have polynomial-time algorithms) we focus on games where the search problem is guaranteed to be total due to being in **PPAD**, as opposed to *potential games* for example, where equilibria can be shown to exist due to a potential function argument.

Restricted win-lose games   The papers [2, 16] identify polynomial-time algorithms for subclasses of the win-lose games shown to be hard by [1]. A win-lose game has an associated bipartite graph whose vertices are the pure strategies of the two players. We add an edge from s to s' if when the players play s and s', the player of s obtains a payoff of 1. Addario-Berry et al. [2] study the special case when this graph is planar, and obtain an algorithm based on a search in the graph for certain combinatorial structures within the graph, that runs in polynomial time for planar graphs. Codenotti et al. [16] study win-lose games in which the number of winning entries in each row or column is at most 2. Their approach also involves searching for certain structures within a corresponding digraph.

Approximate equilibria   The main question is: for what values of $\epsilon > 0$ can one compute ϵ-Nash equilibria in polynomial time? The question has mainly been considered in the 2-player case. Note that [13] established that there is not a fully polynomial-time approximation scheme for computing ϵ-Nash equilibria. However, it is also known that for any constant $\epsilon > 0$, the problem is subexponential [49].

There has not been much progress in the past couple of years on reducing the value of ϵ that is obtainable; the papers cited in [22] represent the state of the art in this respect. Perhaps the simplest non-trivial algorithm for computing approximate Nash equilibria is the following, due to Daskalakis et al. [24]. Figure 8 shows a generalisation of the algorithm of [24], to k players, obtained independently in [6, 40]. These papers also obtain the lower bounds for solutions having constant support. The approximation guarantee is $1 - \frac{1}{k}$; it is noted in [6] that one can do slightly better by using a more sophisticated 2-player algorithm at the midpoint of the procedure.

Ranking games   A restriction of the ranking games mentioned above, to those having "competitiveness-based strategies" were recently proposed in [36] as a subclass of ranking games that seems to have better computational properties, while still being able to capture features of many real-world competitions for rank. In these games, a player's strategies may be ordered in a sequence that reflects how "competitive" they are. If we let $\{a_1, \ldots, a_n\}$ be the actions available to a player, then each a_j has two associated quantities, a cost c_j and a return r_j, where return is a monotonically increasing function of cost. Given any pure-strategy profile, players

---

1. For $i = 1, 2, \ldots, k-1$

   (a) Player i allocates probability $1 - \frac{1}{k+1-i}$ to some arbitrary strategy

2. For $i = k, k-1, \ldots, 1$

   (a) Player i allocates his remaining probability to a best response to the strategy combination played so far.

Figure 8: Approximation algorithm for k-player NASH

are ranked on the returns of their actions and awarded prizes from that ranking. The payoff to a player is the value of the prize he wins, minus the cost of his action. This results in a trade-off between saving on cost, versus spending more with the aim of a larger prize. Notice that, in contrast to unrestricted ranking games [5], this restriction allows games with many players to be written down concisely.

Anonymous games Anonymous games represent a useful way to concisely describe certain games having a large number of players — in an anonymous game, each player has the same set of pure strategies $\{a_1, \ldots, a_k\}$, and the payoff to player i for playing a_j depends only on i and the total number of players to play a_j (but not the identities of those players; hence the phrase "anonymous game"). Daskalakis and Papadimitriou [26] give a polynomial-time approximation scheme for these games, but note that it is an open problem whether an exact equilibrium may be computed in polynomial time.

Polymatrix games Daskalakis and Papadimitriou [27] show that polymatrix games may be solved exactly when the bimatrix games between pairs of players are zero-sum. This result is applied in [36] to a subclass of the games studied there, specifically ranking games where prize values are a *linearly* decreasing function of rank placement.

7 The Complexity of Path-following Algorithms

In this section we report on recent progress [38] that shows an even closer analogy between equilibrium computation and the problem END OF THE LINE. Namely, that the solutions found by certain well-known "path-following" algorithms have the same computational complexity as the search for the solution to END OF THE LINE that is obtained by following the line. Specifically, both are **PSPACE**-complete.

Consider the algorithm for END OF THE LINE that works by simply "following the line" from the given starting-point 0^n until an endpoint is reached. Clearly this takes exponential time in the worst case, but in fact we can say something stronger. Let OEOTL denote the problem "other end of this line", which requires as output, the endpoint reached by following directed edges from the given starting-point. Papadimitriou observed in [57] that the following holds:

Theorem 7.1 *[57]* OEOTL *is* **PSPACE***-complete.*

Notice that a solution to OEOTL is apparently no longer in **NP**, since while we can check that it solves END OF THE LINE, there is no obvious way to efficiently check that it is the correct end-of-line, i.e. the one connected to the given starting-point.

Proof (sketch) We reduce from the problem of computing the final configuration of a polynomial space bounded Turing machine (TM).

The proof uses the fact that polynomial-space-bounded Turing machines can be simulated by polynomial-space-bounded *reversible Turing machines* [19]. By a *con-figuration* of a TM computation we mean a complete description of an intermediate state of computation, including the contents of the tape, and the state and location of the TM on the tape. Reversible TMs have the property that given an interme-diate configuration, one can readily obtain not just the subsequent configuration, but also the previous one; this is done by memorising a carefully-chosen subset of previous configurations that the machine uses in the course of a computation [19].

From there, it is straightforward to simulate a reversible Turing machine using the two circuits S and P of an instance of OEOTL. Each vertex v of an (S, P)-graph encodes a configuration of a linear-space-bounded TM; S computes the subsequent configuration, and P computes the previous configuration. $\square$

In the context of searching for Nash equilibria, a number of algorithms have been proposed that work by following paths is a graph G associated with game $\mathcal{G}$, where G is derived from $\mathcal{G}$ via a reduction to END OF THE LINE. Examples include Scarf's algorithm [62] (for approximate solutions of k-player games) and the Lemke-Howson algorithm [48] (an exact algorithm for 2-player games). These algorithms were proposed as being in practice more computationally efficient than a brute-force approach. Related algorithms include the *linear tracing procedure* and homotopy methods [39, 41, 42, 43, 44] discussed below: in these papers the motivation is *equilibrium selection* — in cases where multiple equilibria exist, we seek a criterion for identifying a "plausible" one.

Homotopy methods for Game Theory In topology, a homotopy refers to a continuous deformation from some geometrical object to another one. Suppose we want to solve game $\mathcal{G}$. Let $\mathcal{G}_0$ denote another game obtained by changing the nu-merical utilities of $\mathcal{G}$ in such a way that there is some "obvious" equilibrium; for example, we could change the payoffs so that each player obtains 1 for playing his first strategy, and 0 for any other strategy, regardless of the behaviour of the other player(s). Then, we can continuously deform $\mathcal{G}_0$ to get back to $\mathcal{G}$. For $t \in [0, 1]$ let $\mathcal{G}_t$ be a game whose payoffs are weighted averages of those in $\mathcal{G}_0$ with those in $\mathcal{G}$: we write this as $\mathcal{G}_t = (1 - t)\mathcal{G}_0 + t\mathcal{G}$. A homotopy method for solving $\mathcal{G}$ involves keeping track of the equilibria of $\mathcal{G}_t$, which should move continuously as t increases from 0 to 1. There should be a continuous path from the known equilibrium of $\mathcal{G}_0$ to some equilibrium of $\mathcal{G}$ (an application of Browder's fixed point theorem [7].) The catch is, that in following this path it may be necessary for t to go down as well as up.

The main theorem of [38] is that

Theorem 7.2 *[38] It is* **PSPACE***-complete to compute any of the equilibria that are obtained by the Lemke-Howson algorithm.*

The Lemke-Howson algorithm is usually presented as a **discrete path-following algorithm**, in which a 2-player game has an associated degree-2 graph G as follows. Vertices of G are mixed strategy profiles that are characterised by the subset of the players' strategies that are either played with probability zero, or are best responses to the other player's mixed strategy. To be a vertex of G, a mixed-strategy profile should satisfy one of the following:

- All pure strategies are labelled, or

- All but one pure strategies are labelled, and one pure strategy is labelled twice, due to being both a best response, and for being played with probability zero.

In the first case, we either have a Nash equilibrium, or else we have assigned probability zero to all strategies. In the second case, it can be shown that there are two "pivot" operations that we may perform, that remove one of the duplicate labels of the doubly-labelled strategy, and add a label to some alternative strategy; see von Stengel [65] for details. These operations correspond to following edges on a degree-2 graph.

Viewed as a **homotopy method** (see Herings and Peeters [44] Section 4.1) the algorithm works as follows. For a $n \times n$ game $\mathcal{G}$ with players 1 and 2, choose any $i \in \{1, 2\}$ and $j \in [n]$ and give player i a large bonus payoff for using strategy a_j^i; the bonus should be large enough to make a_j^i a dominating strategy for player i. Hence, there is a unique Nash equilibrium consisting of pure-strategy a_j^i together with the pure best response to a_j^i. Now, we reduce that bonus to zero, and the equilibrium should change continuously; however, the requirement that we keep track of a continuously changing equilibrium means that in general, the bonus cannot reduce monotonically to zero; it may have to go up as well as down.

The choice of a_j^i at the start of the homotopy corresponds to the initial "dropped label" in the discrete path-following description.

Figure 9: Two views of the Lemke-Howson algorithm

Savani and von Stengel [61] established earlier that the Lemke-Howson does indeed take exponentially many steps in the worst case; this new result, then, says that moreover there are no "short cuts" to the solution obtained by Lemke-Howson, subject only to the very weak assumption that **PSPACE**-hard problems require exponential time in the worst case. (The **PPAD**-hardness of NASH already established that these solutions are hard to find subject to the hardness of **PPAD**, but the hardness of **PPAD** is a stronger assumption.)

Proof (the main ideas) Recall the role of discrete Brouwer functions in the reductions from END OF THE LINE to NASH, as discussed in Section 4. Suppose $\mathcal{G} = \mathcal{G}_1$ is associated with a discrete Brouwer function having the kind of structure indicated in Figure 5; the bottom left-hand corner looks roughly like the top right-hand "$t = 1$" diagram in Figure 10. Suppose that $\mathcal{G}_0$ is associated with a DBF in which

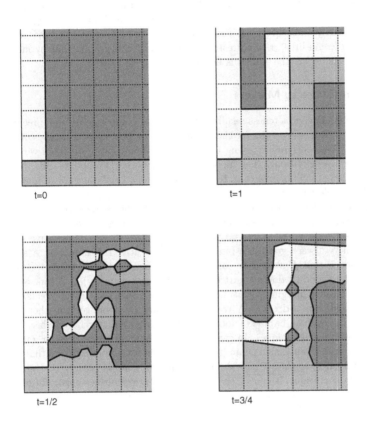

Figure 10: Homotopy for Brouwer function in the plane

the line-encoding structure has been stripped out: the top left "$t = 0$" diagram in Figure 10.

Consider the linear homotopy between the corresponding continuous Brouwer functions $\mathcal{F}_0$ and $\mathcal{F}_1$; $\mathcal{F}_t(x) = (1-t)\mathcal{F}_0(x) + \mathcal{F}_1(x)$. $\mathcal{F}_t$ colours the square according to the direction of $\mathcal{F}_t(x) - x$. Figure 10 (the bottom half) shows how these colourings may evolve for intermediate values of t. For intermediate values of t, in regions where $\mathcal{F}_0 = \mathcal{F}_1$ we have $\mathcal{F}_t(x) = \mathcal{F}_0(x) = \mathcal{F}_1(x)$ for all t. This means that fixpoints of $\mathcal{F}_t$ cannot be located in the region where $\mathcal{F}_1$ gets the colour 0 (or black). So, a continuous path of fixpoints necessarily follows the line of non-black regions of $\mathcal{F}_1$, and necessarily ends up at the end of that particular line. The crossover gadget (Figure 6) would break this correspondence with OEOTL; we fix that by moving to 3-dimensions, where the gadget is not needed. □

8 From Games to Markets

The **PPAD**-completeness results for variants of NASH has led, more recently, to **PPAD**-completeness results for the problems of computing certain types of market equilibria. We continue by giving the general idea and intuition, then we proceed to the formal definitions. For more details on the background to this topic, see Chapters 5,6 of [55]; here we aim to focus on giving a brief overview of more recent **PPAD**-completeness results.

Suppose we have a set G of *goods* and a set T of *traders*. Each trader i in T has a utility function f^i that maps bundles of goods to non-negative real values. Assume the goods are divisible, so f^i is a mapping from vectors of non-negative numbers (the quantity of each good in a bundle) to non-negative real numbers. A standard assumption is that f^i should be continuous, and non-decreasing when restricted to an individual good (it doesn't hurt to receive a bigger quantity), and concave (there is non-increasing marginal utility). Now, suppose that each good j gets a (non-negative real) unit price p_j. A trader with some given budget can then identify (from f^i) the optimal bundle of goods that his budget will purchase (where optimal bundles need not be unique). In general, the aim is to identify prices for goods such that each trader can exchange an initial allocation for an optimal one that has the same total price, so that the total quantity of each good is conserved. Under some fairly mild conditions, it can be shown that such prices always exist. Intuitively, if all traders try to exchange their initial allocations for ones that are optimal, and there is too much demand for good j as a result, then we can fix that problem by raising the price of j.

Definition 8.1 In an *Arrow-Debreu market* [3], each trader has an initial *endowment* consisting of a bundle of goods. Suppose that the prices are set such that the following can happen: each trader exchanges his endowment for an optimal bundle having the same total value, and furthermore, the total quantity of each good is conserved. In that case, we say that these prices allow the market to *clear*. It is shown in [3] that there always exist prices that allow the market to clear.

A *Fisher market* is a special case of an Arrow-Debreu market, in which traders are initially endowed with quantities of money, and there are fixed quantities of goods for sale; in this case the prices should ensure that if each trader buys a bundle of goods that is optimal for his budget, then all goods are sold. This can be seen to be a special case of Arrow-Debreu market, by regarding money as one of the goods in G.

The existence proof of market-clearing prices [3] works by expressing a market satisfying the relevant constraints in terms of an *abstract economy*, a generalisation of the standard notion of a game, and applying Nash's theorem. So, indirectly the proof uses Brouwer's fixed point theorem. This indicates that **PPAD** is the relevant complexity class for studying the hardness of computing a price equilibrium. Results about the computational complexity of finding market-clearing prices are in terms of the types of utility functions of the traders.

Interesting classes of utility functions include the following (where $(x_1, \ldots, x_m)$ denotes a vector of quantities of m goods):

1. Additively separable functions, where a trader's function $f^i(x_1, \ldots, x_m)$ is of the form $f^i(x_1, \ldots, x_m) = \sum_{j=1}^m f_j^i(x_j)$. We still need to specify the structure of the functions f_j^i in order to have a well-defined problem.

2. piecewise linear: in conjunction with additively separable, this would require that each function $f_j^i(x_j)$ be piecewise linear, and the syntactic complexity of such a function would be the total number of pieces of the functions f_j^i. More generally, without the additively separable property, f^i could take the form $f^i(x_1, \ldots, x_m) = \min_{a \in A} f_a^i(x_1, \ldots, x_m)$ where A indexes a finite set of linear functions, and $f_a^i(x_1, \ldots, x_m) = \sum_{j=1}^m \lambda_j^a x_j$ for non-negative coefficients λ_j^a.

3. Leontiev economies: In a *Leontiev economy* we have $f^i(x_1, \ldots, x_m) = \min_{j \in [m]} \lambda_j^i x_j$, a special case of the piecewise linear functions above; note that these functions are not however additively separable.

Polynomial-time algorithms   Linear utility functions take the form $f^i(x_1, \ldots, x_m) = \sum_{j=1}^m \lambda_j^i x_j$ — they are both additively separable and piecewise linear, but not Leontiev; they were initially considered in [34], and are known to be solvable in polynomial time [28, 46] in the case of Fisher markets. For the Arrow-Debreu case, Devanur and Vazirani [29] give a strongly polynomial-time approximation scheme but leave open the problem of finding an exact one in polynomial time.

PPAD-hardness results   The first such result applied to Leontiev economies, for which there is a reduction from 2-player games [17]. Chen et al. [9] show that it is **PPAD**-complete to compute an Arrow-Debreu market equilibrium for the case of additively separable, piecewise linear and concave utility functions. Chen and Teng [15] show that it is **PPAD**-hard to compute Fisher equilibrium prices, from utility functions that are additively separable and piecewise linear concave; this is done by reduction from SPARSE BIMATRIX [14]. Vazirani and Yannakakis [66] show that finding an equilibrium in Fisher markets is **PPAD**-hard in the case of additively-separable, piecewise-linear, concave utility functions. On the positive side, they show that with these utility functions, market equilibria can however be written down with rational numbers.

References

[1] T.G. Abbott, D.M. Kane and P. Valiant, On the complexity of two-player win-lose games, in *46th Symposium on Foundations of Computer Science* IEEE Computer Society, (2005), 113-122.

[2] L. Addario-Berry, N. Olver and A. Vetta, A polynomial time algorithm for finding Nash equilibria in planar win-lose games, *J. Graph Algorithms Appl.* **11(1)** (2007), 309-319.

[3] K.J. Arrow and G. Debreu, Existence of an Equilibrium for a Competitive Economy, *Econometrica* **22(3)** (1954), 265-290.

[4] P. Beame, S. Cook, J. Edmonds, R. Impagliazzo and T. Pitassi, The relative complexity of NP search problems, in *27th ACM Symposium on Theory of Computing* (1995), 303-314.

[5] F. Brandt, F. Fischer, P. Harrenstein and Y. Shoham, Ranking Games, *Artificial Intelligence* **173(3)** (2009), 221-239.

[6] P. Briest, P.W. Goldberg and H. Röglin, Approximate Equilibria in Games with Few Players, *CoRR abs/0804.4524* (2008),

[7] F.E. Browder, On continuity of fixed points under deformations of continuous mappings, *Summa Brasiliensis Math.* **4** (1960), 183-191.

[8] V. Bubelis, On equilibria in finite games, *International Journal of Game Theory* **8** (1979), 65-79.

[9] X. Chen, D. Dai, Y. Du and S.-H. Teng, Settling the Complexity of Arrow-Debreu Equilibria in Markets with Additively Separable Utilities, in *50th Symposium on Foundations of Computer Science* IEEE Computer Society, (2009), 273-282.

[10] X. Chen and X. Deng, On the complexity of 2D discrete fixed point problem, in *33rd International Colloquium on Automata, Languages and Programming* (2006), 489-500.

[11] X. Chen and X. Deng, Settling the complexity of 2-player Nash Equilibrium, in *47th Symposium on Foundations of Computer Science* IEEE Computer Society, (2006), 261-272.

[12] X. Chen and X. Deng, 3-NASH is PPAD-Complete, *Technical report TR05-134, Electronic Colloquium on Computational Complexity* (2005),

[13] X. Chen, X. Deng and S.-H. Teng, Settling the complexity of computing two-player Nash equilibria, *Journal of the ACM* **56(3)** (2009), 1-57.

[14] X. Chen, X. Deng and S.-H. Teng, Sparse Games are Hard, in *2nd Workshop on Internet and Network Economics (WINE)* (2006), 262-273.

[15] X. Chen and S.-H. Teng, Spending Is Not Easier Than Trading: On the Computational Equivalence of Fisher and Arrow-Debreu Equilibria, in *20th International Symposium on Algorithms and Computation* (2009), 647-656.

[16] B. Codenotti, M. Leoncini and G. Resta, Efficient computation of Nash equilibria for very sparse win-lose games, in *14th European Symposium on Algorithms* (2006), 232-243.

[17] B. Codenotti, A. Saberi, K. Varadarajan and Y. Ye, Leontief economies encode nonzero sum two-player games, in *17th Annual ACM-SIAM Symposium on Discrete Algorithms* (2006), 659-667.

[18] V. Conitzer and T. Sandholm, Complexity results about Nash equilibria, in *18th International Joint Conference on Artificial Intelligence* (2003), 765-771.

[19] P. Crescenzi and C.H. Papadimitriou, Reversible Simulation of Space-Bounded Computations, *Theoretical Computer Science* **143(1)** (1995), 159-165.

[20] C. Daskalakis, A. Fabrikant, C.H. Papadimitriou, The Game World Is Flat: The Complexity of Nash Equilibria in Succinct Games, in *33rd International Colloquium on Automata, Languages and Programming* (2006), 513-524.

[21] C. Daskalakis, P.W. Goldberg and C.H. Papadimitriou, The complexity of computing a Nash equilibrium, in *38th ACM Symposium on Theory of Computing* (2006), 71-78.

[22] C. Daskalakis, P.W. Goldberg and C.H. Papadimitriou, The Complexity of Computing a Nash Equilibrium, *SIAM Journal on Computing* **39(1)** (2009), 195-259.

[23] C. Daskalakis, P.W. Goldberg and C.H. Papadimitriou, The Complexity of Computing a Nash Equilibrium, *Communications of the ACM* **52(2)** (2009), 89-97.

[24] C. Daskalakis, A. Mehta and C.H. Papadimitriou, A Note on Approximate Nash Equilibria, *Theoretical Computer Science* **410(17)** (2009), 1581-1588.

[25] C. Daskalakis and C.H. Papadimitriou, Three-player games are hard, *Technical Report TR05-139, Electronic Colloquium on Computational Complexity* (2005), 1-10.

[26] C. Daskalakis and C.H. Papadimitriou, Discretized Multinomial Distributions and Nash Equilibria in Anonymous Games, in *49th Symposium on Foundations of Computer Science* IEEE Computer Society, (2008), 25-34.

[27] C. Daskalakis and C.H. Papadimitriou, On a Network Generalization of the Minmax Theorem, in *36th International Colloquium on Automata, Languages and Programming* (2009), 423-434.

[28] N. Devanur, C.H. Papadimitriou, A. Saberi and V.V. Vazirani, Market Equilibrium via a Primal-Dual-Type Algorithm for a Convex Program, *Journal of the ACM* **55(5)** (2008), 1-18.

80 P.W. Goldberg

[29] N.R. Devanur and V.V. Vazirani, An improved approximation scheme for computing Arrow-Debreu prices for the linear case, in *23rd Conference, Foundations of Software Technology and Theoretical Computer Science* (2003), 149-155.

[30] E. Elkind, L.A. Goldberg and P.W. Goldberg, Nash Equilibria in Graphical Games on Trees Revisited, in *7th ACM Conference on Electronic Commerce* (2006), 100-109.

[31] K. Etessami and M. Yannakakis, On the Complexity of Nash Equilibria and Other Fixed Points, *SIAM Journal on Computing* **39(6)** (2010), 2531-2597.

[32] U. Feige and I. Talgam-Cohen, A Direct Reduction from k-Player to 2-Player Approximate Nash, in *3rd Symposium on Algorithmic Game Theory* Springer LNCS 6386, (2010), 138-149.

[33] L. Fortnow, R, Impagliazzo, V. Kabinets and C. Umans, On the Complexity of Succinct Zero-Sum Games, in *20th IEEE Conference on Computational Complexity* (2005), 323-332.

[34] D. Gale, *Theory of Linear Economic Models*, McGraw Hill, NY (1960).

[35] I. Gilboa and E. Zemel, Nash and correlated equilibria: Some complexity considerations, *Games and Economic Behavior* **1** (1989), 80-93.

[36] L.A. Goldberg, P.W. Goldberg, P. Krysta and C. Ventre, Ranking Games that have Competitiveness-based Strategies, in *11th ACM Conference on Electronic Commerce* (2010), 335-344.

[37] P.W. Goldberg and C.H. Papadimitriou, Reducibility Among Equilibrium Problems, in *38th ACM Symposium on Theory of Computing* (2006), 61-70.

[38] P.W. Goldberg, C.H. Papadimitriou and R. Savani, The Complexity of the Homotopy Method, Equilibrium Selection, and Lemke-Howson Solutions, *Arxiv technical report 1006.5352* (2010), 1-22.

[39] J.C. Harsanyi, The tracing procedure: a Bayesian approach to defining a solution for n-person noncooperative games, *International Journal of Game Theory* **4** (1975), 61-95.

[40] S. Hémon, M. de Rougemont and M. Santha, Approximate Nash Equilibria for Multi-player Games, in *1st Symposium on Algorithmic Game Theory* (2008), 267-278.

[41] P.J-J. Herings, Two simple proofs of the feasibility of the linear tracing procedure, *Economic Theory* **15** (2000), 485-490.

[42] P.J-J. Herings and A. van den Elzen, Computation of the Nash Equilibrium Selected by the Tracing Procedure in N-Person Games, *Games and Economic Behavior* **38** (2002), 89-117.

[43] P.J-J Herings and R.J.A.P. Peeters, A differentiable homotopy to compute Nash equilibria of n-person games, *Economic Theory* **18(1)** (2001), 159-185.

[44] P.J-J. Herings and R. Peeters, Homotopy methods to compute equilibria in game theory, *Economic Theory* **42(1)** (2010), 119-156.

[45] M.D. Hirsch, C.H. Papadimitriou and S. Vavasis, Exponential Lower Bounds for Finding Brouwer Fixed Points, *Journal of Complexity* **5(4)** (1989), 379-416.

[46] K. Jain, A polynomial-time algorithm for computing the Arrow-Debreu market equilibrium for linear utilities, in *45th Symposium on Foundations of Computer Science* IEEE Computer Society, (2004), 286-294.

[47] M. Kearns, M.L. Littman and S. Singh, Graphical models for game theory, in *17th Conference on Uncertainty in Artificial Intelligence* Morgan Kaufmann, (2001), 253-260.

[48] C.E. Lemke and J.T. Howson, Jr., Equilibrium points of bimatrix games, *SIAM J. Appl. Math.* **12(2)** (1964), 413-423.

[49] R. Lipton, V. Markakis and A. Mehta, Playing Large Games using Simple Strategies, in *4th ACM Conference on Electronic Commerce* (2003), 36-41.

[50] M. Littman, M. Kearns and S. Singh, An efficient, exact algorithm for single connected graphical games, in *15th Annual Conference on Neural Information Processing Systems* MIT Press, (2001), 817-823.

[51] N. Megiddo, A note on the complexity of P-matrix LCP and computing an equilibrium, *Res. Rep. RJ6439, IBM Almaden Research Center, San Jose.* (1988), 1-5.

[52] D. Monderer and L. Shapley, Potential Games, *Games and Economic Behavior* **14** (1996), 124-143.

[53] J. Nash, Noncooperative Games, *Annals of Mathematics* **54** (1951), 289-295.

[54] J. von Neumann and O. Morgenstern, *The Theory of Games and Economic Behavior, second ed.*, Princeton University Press, (1947).

[55] N. Nisan, T. Roughgarden, E. Tardos and V.V. Vazirani, *Algorithmic Game Theory*, Cambridge University Press, (2007).

[56] M.J. Osborne and A. Rubinstein, *A Course in Game Theory*, MIT Press, (1994).

[57] C.H. Papadimitriou, On the complexity of the parity argument and other inefficient proofs of existence, *J. Comput. System Sci.* **48** (1994), 498-532.

[58] J. Renegar, A faster PSPACE algorithm for deciding the existential theory of the reals, in *29th Symposium on Foundations of Computer Science* IEEE Computer Society, (1988), 291-295.

[59] R.W. Rosenthal, A Class of Games Possessing Pure-Strategy Nash Equilibria, *International Journal of Game Theory* **2** (1973), 65-67.

[60] J.-J. Rousseau, *Discourse on Inequality*, (1754).

[61] R. Savani and B. von Stengel, Hard-to-Solve Bimatrix Games, *Econometrica* **74(2)** (2006), 397-429.

[62] H. Scarf, The approximation of fixed points of a continuous mapping, *SIAM J. Appl. Math.* **15** (1967), 1328-1343.

[63] G. Schoenebeck and S. Vadhan, The Computational Complexity of Nash Equilibria in Concisely Represented Games, in *7th ACM Conference on Electronic Commerce* (2006), 270-279.

[64] E. Sperner, Neuer Beweis für die Invarianz der Dimensionszahl und des Gebietes, *Abhandlungen aus dem Mathematischen Seminar Universität Hamburg* **6** (1928), 265-272.

[65] B. von Stengel, *Computing equilibria for two-person games. Chapter 45, Handbook of Game Theory, Vol 3*, (eds. R.J. Aumann and S. Hart), North-Holland, Amsterdam (2002). 1723-1759.

[66] V.V. Vazirani and M. Yannakakis, Market Equilibrium under Separable, Piecewise-Linear, Concave Utilities, in *1st Symposium on Innovations in Computer Science* (2010), 156-165.

Dept. of Computer Science, Ashton Building, Ashton Street, Liverpool L69 3BX
P.W.Goldberg@liverpool.ac.uk

Hypergraph Turán problems

Peter Keevash

Abstract

One of the earliest results in Combinatorics is Mantel's theorem from 1907 that the largest triangle-free graph on a given vertex set is complete bipartite. However, a seemingly similar question posed by Turán in 1941 is still open: what is the largest 3-uniform hypergraph on a given vertex set with no tetrahedron? This question can be considered a test case for the general hypergraph Turán problem, where given an r-uniform hypergraph F, we want to determine the maximum number of edges in an r-uniform hypergraph on n vertices that does not contain a copy of F. To date there are very few results on this problem, even asymptotically. However, recent years have seen a revitalisation of this field, via significant developments in the available methods, notably the use of stability (approximate structure) and flag algebras. This article surveys the known results and methods, and discusses some open problems.

Acknowledgements

Research supported in part by ERC grant 239696 and EPSRC grant EP/G056730/1. Thanks to Dan Hefetz, Dhruv Mubayi, Richard Mycroft and Oleg Pikhurko for helpful comments and corrections.

1 Introduction

The *Turán number* $\mathrm{ex}(n, F)$ is the maximum number of edges in an F-free r-graph on n vertices.[1] It is a long-standing open problem in Extremal Combinatorics to develop some understanding of these numbers for general r-graphs F. Ideally, one would like to compute them exactly, but even asymptotic results are currently only known in certain cases. For ordinary graphs ($r = 2$) the picture is fairly complete. The first step was taken by Turán [190], who solved the case when $F = K_t$ is a complete graph on t vertices. The most obvious examples of K_t-free graphs are $(t-1)$-partite graphs. On a given vertex set, the $(t-1)$-partite graph with the most edges is complete and *balanced*, in that the part sizes are as equal as possible (any two sizes differ by at most 1). Turán's theorem is that this construction always gives the largest K_t-free graph on a given vertex set, and furthermore it is unique (up to isomorphism). This result inspired the development of Extremal Graph Theory, which is now a substantial field of research (see [19]). For general graphs F we still do not know how to compute the Turán number exactly, but if we are satisfied with an approximate answer the theory becomes quite simple: it is enough to know the chromatic number of F. Erdős and Stone [62] showed that if $\chi(F) = t$ then $\mathrm{ex}(n, F) \leq \mathrm{ex}(n, K_t) + o(n^2)$. As noted in [58], since $(t-1)$-partite graphs are F-free, this implies that $\mathrm{ex}(n, F) = \mathrm{ex}(n, K_t) + o(n^2)$. When F is not bipartite this gives an asymptotic result for the Turán number. When F is bipartite we can only

---

[1] An *r-graph* (or *r-uniform hypergraph*) G consists of a vertex set and an edge set, each edge being some r-set of vertices. We say G is *F-free* if it does not have a (not necessarily induced) subgraph isomorphic to F.

deduce that $\mathrm{ex}(n, F) = o(n^2)$; in general it is a major open problem to determine even the order of magnitude of Turán numbers for bipartite graphs. However, we will not consider these so-called 'degenerate' problems here.

By contrast with the graph case, there is comparatively little understanding of the hypergraph Turán problem. Having solved the problem for $F = K_t$, Turán [191] posed the natural question of determining $\mathrm{ex}(n, F)$ when $F = K_t^r$ is a complete r-graph on t vertices. To date, no case with $t > r > 2$ of this question has been solved, even asymptotically. Erdős [54] offered \$500 for the solution of any case and \$1000 for a general solution. A comprehensive survey of known bounds on these Turán numbers was given by Sidorenko [180], see also the earlier survey of de Caen [41]; a survey of more general Turán-type problems was given by Füredi [79]. Our focus will be on fixed F and large n, rather than the 'covering design' problems which occur for small n (see [180]). Despite the lack of progress on the Turán problem for complete hypergraphs, there are certain hypergraphs for which the problem has been solved asymptotically, or even exactly, and most of these results have been obtained since the earlier surveys. These special cases may only be scratching the surface of a far more complex general problem, but they are nevertheless interesting for the rich array of different ideas that have been developed for their solutions, ideas that one may hope can be applied or developed to much greater generality. Thus we feel it is most helpful to organise this survey around the methods; we conclude with a summary of the results for easy reference.

The contents by section are as follows: 1: Introduction, 2: Basic arguments, 3: Hypergraph Lagrangians, 4: Link graphs and multigraphs, 5: Stability, 6: Counting, 7: Flag algebras, 8: The remaining exact results, 9: Bounds for complete hypergraphs, 10: The infinitary perspective, 11: Algebraic methods, 12: Probabilistic methods, 13: Further topics, 14: Summary of results.

We use the following notation. Suppose G is an r-graph. We write $V(G)$ for the vertex set of G and $E(G)$ for the edge set of G. We write $v(G) = |V(G)|$ and $e(G) = |E(G)|$. We often identify G with its edge set, so that $|G|$ means $|E(G)|$. For $X \subseteq V(G)$, the induced subhypergraph $G[X]$ has vertex set X and edge set all edges of G that are contained in X. We often abbreviate 'subhypergraph' to 'subgraph'. A k-set is a set of size k. Usually G has n vertices, and asymptotic notations such as $o(1)$ refer to the limit for large n.

2 Basic arguments

We start with a simple but important averaging argument of Katona, Nemetz and Simonovits [101]. Suppose G is an r-graph on n vertices with $\theta\binom{n}{r}$ edges. We say that G has *density* $d(G) = \theta$, as this is the fraction of all possible r-sets that are edges. Now fix any $r \le m < n$ and consider restricting G to subsets of its vertex set of size m. It is easy to check that the average density of these restrictions is also θ. Taking $m = n - 1$, for any fixed r-graph we see that $\binom{n}{r}^{-1}\mathrm{ex}(n, F) \le \binom{n-1}{r}^{-1}\mathrm{ex}(n-1, F)$. Indeed, if $\theta = \binom{n-1}{r}^{-1}\mathrm{ex}(n-1, F)$ then we cannot have an F-free r-graph on n vertices with density more than θ, as the averaging argument would give a restriction to $n - 1$ vertices with at least the same density, contradicting the definition of $\mathrm{ex}(n-1, F)$. Thus the ratios $\binom{n}{r}^{-1}\mathrm{ex}(n, F)$ form a decreasing sequences of real numbers in $[0, 1]$. It follows that they have a limit, which is called the *Turán density*, and denoted $\pi(F)$.

Determining the Turán density is equivalent to obtaining an asymptotic result $\text{ex}(n, F) \sim \pi(F)\binom{n}{r}$, provided that we are in the 'non-degenerate' case when $\pi(F) > 0$. An r-graph F is degenerate if and only if it is r-partite, meaning that the vertices of F can be r-coloured so that every edge has exactly one vertex of each colour. One direction of this implication is clear: if F is not r-partite then the complete r-partite r-graph on n vertices gives a non-zero lower bound for the Turán density. It has about $(n/r)^r$ edges, so we obtain the bound $\pi(F) \geq r!/r^r$. The other direction is a result of Erdős [50]: if F is r-partite then $\pi(F) = 0$, and in fact $\text{ex}(n, F) < n^{r-c}$ for some $c = c(F) > 0$. We note for future reference that this argument shows that there are no Turán densities in the range $(0, r!/r^r)$. We will return to the question of what values may be taken by Turán densities in Section 13.1.

A similar averaging argument establishes the important 'supersaturation' phenomenon discovered by Erdős and Simonovits [59]. Informally, this states that once the density of an r-graph G exceeds the Turán density of F, we not only find a copy of F, but in fact a constant fraction of all $v(F)$-sets from $V(G)$ span a copy of F.

Lemma 2.1 (Supersaturation) *For any r-graph F and $a > 0$ there are $b, n_0 > 0$ so that if G is an r-graph on $n > n_0$ vertices with $e(G) > (\pi(F) + a)\binom{n}{r}$ then G contains at least $b\binom{n}{v(F)}$ copies of F.*

Proof Fix k so that $\text{ex}(k, F) \leq (\pi(F) + a/2)\binom{k}{r}$. There must be at least $\frac{1}{2}a\binom{n}{k}$ k-sets $K \subseteq V(G)$ inducing an r-graph $G[K]$ with $e(G[K]) > (\pi(F) + \frac{1}{2}a)\binom{k}{r}$.[2] Otherwise, we would have $\sum_K e(G[K]) \leq \binom{n}{k}(\pi(F) + \frac{1}{2}a)\binom{k}{r} + \frac{1}{2}a\binom{n}{k}\binom{k}{r} = (\pi(F) + a)\binom{n}{k}\binom{k}{r}$. But we also have $\sum_K e(G[K]) = \binom{n-r}{k-r}e(G) > \binom{n-r}{k-r}(\pi(F) + a)\binom{n}{r} = (\pi(F) + a)\binom{n}{k}\binom{k}{r}$, so this is a contradiction. By choice of k, each of these k-sets contains a copy of F, so the number of copies of F in G is at least $\frac{1}{2}a\binom{n}{k}/\binom{n-v(F)}{k-v(F)} = \frac{1}{2}a\binom{n}{v(F)}/\binom{k}{v(F)}$, i.e. at least $b = \frac{1}{2}a\binom{k}{v(F)}^{-1}$ fraction of all $v(F)$-sets span a copy of F. $\square$

Supersaturation can be used to show that 'blowing up' does not change the Turán density. The *t-blowup* $F(t)$ of F is defined by replacing each vertex x of F by t 'copies' $x^1, \cdots, x^t$ and each edge $x_1 \cdots x_r$ of F by the corresponding complete r-partite r-graph of copies, i.e. all $x_1^{a_1} \cdots x_r^{a_r}$ with $1 \leq a_1, \cdots, a_r \leq t$. Then we have the following result.

Theorem 2.2 (Blowing up) $\pi(F(t)) = \pi(F)$.

First we point out a special case that will be used in the proof. When $F = K_r^r$ consists of a single edge we trivially have $\text{ex}(n, F) = 0$ and so $\pi(F) = 0$. Also $F(t) = K_r^r(t)$ is the complete r-partite r-graph with t vertices in each part. Then the result of Erdős mentioned above gives $\pi(F(t)) = 0$.

Proof By supersaturation, for any $a > 0$ there is $b > 0$ so that if n is large and G is an r-graph on n vertices with $e(G) > (\pi(F) + a)\binom{n}{r}$ then G contains at least

---

[2]In fact, large deviation estimates imply that almost all k-sets $K \subseteq V(G)$ have this property when k is large.

$b\binom{n}{v(F)}$ copies of F. Consider an auxiliary $v(F)$-graph H on the same vertex set as G where edges of H correspond to copies of F in G. For any $T > 0$, if n is large enough we can find a copy K of $K_{v(F)}^{v(F)}(T)$ in H. We colour each edge of K by one of $v(F)!$ colours, corresponding to which of the $v(F)!$ possible orders the vertices of F are mapped to the parts of K. Now a standard result of Ramsey theory implies that for large enough T there is a monochromatic copy of $K_{v(F)}^{v(F)}(t)$, which gives a copy of $F(t)$ in G. □

One application of blowing up is to deduce the Erdős-Stone theorem from Turán's theorem: if $\chi(H) = t$ then H is contained in $K_t(s)$ for some s, so $\pi(H) = \pi(K_t) = \frac{t-2}{t-1}$.

Another useful perspective on blowing up is a formulation in terms of homomorphisms. Given r-graphs F and G we say $f : V(F) \to V(G)$ is a *homomorphism* if it preserves edges, i.e. $f(e) \in E(G)$ for all $e \in E(F)$. Note that f need not be injective; if it is then F is a subgraph of G. We say that G is *F-hom-free* if there is no homomorphism from F to G. Clearly, G is F-hom-free if and only if $G(t)$ is F-free for every t. We can make analogous definitions to the Turán number and density for homomorphic copies of F: we let $\mathrm{ex}_{hom}(n, F)$ be the maximum number of edges in an F-hom-free r-graph on n vertices, and $\pi_{hom}(F) = \lim_{n\to\infty} \binom{n}{r}^{-1}\mathrm{ex}_{hom}(n, F)$. Then blowing up implies that $\pi_{hom}(F) = \pi(F)$.

We can in principle approximate $\pi(F)$ to any desired accuracy by an exhaustive search of small examples. For suppose that $m < n$ and we have found that H is a largest F-hom-free r-graph on m vertices with $\theta\binom{m}{r}$ edges. Then averaging gives $\pi(F) \le \theta$. On the other hand, for any t, the blowup $H(t)$ is an F-free r-graph on tm vertices with $t^r \cdot \theta\binom{m}{r}$ edges, so $\pi(F) \ge \lim_{t\to\infty}\binom{tm}{r}^{-1}t^r\theta\binom{m}{r} = \theta\prod_{i=1}^{r-1}(1 - i/m)$. Thus by examining all r-graphs on m vertices one can approximate $\pi(F)$ to within an error of $O(r^2/m)$. Simple brute force search becomes infeasible even for quite small values of m on very powerful computers. However, more sophisticated search techniques can be much faster, and in some cases they give the best known bounds: see Section 7.

3   Hypergraph Lagrangians

The theory in this section was developed independently by Sidorenko [173] and Frankl and Füredi [69], generalising work of Motzkin and Straus [135] and Zykov [193]. Suppose G is an r-graph on $[n] = \{1, \cdots, n\}$. Recall that the t-blowup $G(t)$ of G is obtained by replacing each vertex by t copies. More generally, we can have different numbers of copies of each vertex: for any vector $t = (t_1, \cdots, t_n)$ we let $G(t)$ be obtained by replacing vertex i with t_i copies, where as before, each edge is replaced by the corresponding complete r-partite r-graph of copies. Then $e(G(t)) = p_G(t) := \sum_{e\in E(G)}\prod_{i\in e} t_i$. Note that $p_G(t)$ is a polynomial where for each edge e of G we have the monomial $\prod_{i\in e} t_i$ in variables corresponding to the vertices of e.

Now suppose that F is an r-graph and G is F-hom-free. We will derive an expression for the best lower bound on $\pi(F)$ that can be obtained from blowups of G. Note that $G(t)$ is an F-free r-graph on $|t| := \sum_{i=1}^{n} t_i$ vertices with density $d(G(t)) =$

$\binom{|t|}{r}^{-1} p_G(t_1, \cdots, t_n)$. Then $\pi(F) \geq \lim_{m \to \infty} d(G(tm)) = r! p_G(t_1/|t|, \cdots, t_n/|t|)$. Thus we want to maximise $p_G(x)$ over the set S of all $x = (x_1, \cdots, x_n)$ with $x_i \geq 0$ for $1 \leq i \leq n$ and $|x| = 1$. (Sometimes S is called the *standard simplex*.) We denote this maximum by $\lambda(G) = \max_{x \in S} p_G(x)$: it is known as the *Lagrangian* of G. Note that the maximum is achieved by some $x \in S$, as S is compact and $p_G(x)$ is continuous. Also, x can be approximated to arbitrary precision by vectors $(t_1/|t|, \cdots, t_n/|t|)$ with integral t_i. We deduce that $\pi(F) \geq b(G) := r! \lambda(G)$, where we refer to $b(G)$ as the *blowup density* of G.[3] We have the following approximate bound for the blowup density by the usual density: $b(G) \geq r! p_G(1/n, \cdots, 1/n) = r! n^{-r} e(G) = d(G) - O(1/n)$. Since $\pi(F)$ is the limit supremum of $d(G)$ over F-hom-free G, we deduce that $\pi(F)$ is also the supremum of $b(G)$ over F-hom-free G.

We say that G is *dense* if every proper subgraph G' satisfies $b(G') < b(G)$. This is equivalent to saying that the maximum of $p_G(x)$ over $x \in S$ is only achieved by vectors x with $x_i > 0$ for $1 \leq i \leq n$, i.e. lying in the interior of S. Then $\pi(F)$ is clearly also the supremum of $b(G)$ over F-hom-free dense G. We say that G *covers pairs* if for every pair of vertices i, j in G there is an edge of G containing both i and j. We claim that if G is dense then G covers pairs. This can be seen from the following simple variational argument. Suppose on the contrary that there is no edge containing both i and j for some pair i, j. Then if we consider $p_G(x)$ with any fixed values for the other variables x_k, $k \neq i, j$ we obtain some linear function $a x_i + b x_j + c$ of x_i and x_j. However, a linear function cannot have an internal strict maximum, so the maximum value of $p_G(x)$ can be achieved with one of x_i or x_j equal to 0. This contradicts the assumption that G is dense, so we deduce that G covers pairs.

We can now derive several results from the theory above. First we recover the results for ordinary graphs ($r = 2$). Note that only complete graphs cover pairs, so only complete graphs can be dense. We have $b(K_t) = 1 - 1/t$, so complete graphs are dense. Suppose that G is a K_t-free graph on n vertices. Then $b(G) = b(G')$ for some dense subgraph G', which must be K_s for some $s < t$. We deduce that $2e(G)/n^2 = 2p_G(1/n, \cdots, 1/n) \leq b(G) = b(G') = 1 - 1/s \leq \frac{t-2}{t-1}$. This gives Turán's theorem in the case when n is divisible by $t - 1$. (This argument is due to Motzkin and Strauss [135].) Also, K_t is F-hom-free if and only if $\chi(F) > t$. Since $\pi(F)$ is the supremum of $b(G)$ for F-hom-free dense G we deduce the Erdős-Stone theorem.

Next we give some hypergraph results. Let H_t^r be the r-graph obtained from the complete graph K_t by extending each edge with a set of $r - 2$ new vertices. More precisely, H_t^r has vertices x_i for $1 \leq i \leq t$ and y_{ij}^k for $1 \leq i < j \leq t$ and $1 \leq k \leq r - 2$ and edges $x_i x_j y_{ij}^1 \cdots y_{ij}^{r-2}$ for $1 \leq i < j \leq t$. We will refer to H_t^r as an *extended complete graph*. (Sometimes 'expanded' is used, but we will use this terminology in a different context later.) Natural examples of H_{t+1}^r-free r-graphs are the blowups $K_t^r(s)$ of the complete r-graph on t vertices. To see that these are H_{t+1}^r-free note that K_t^r is H_{t+1}^r-hom-free, as any map from H_{t+1}^r to K_t^r will map some pair x_i, x_j to the same vertex, so cannot be a homomorphism. On the other

---

[3] Given the simple relationship between $b(G)$ and $\lambda(G)$ it is arguably unnecessary to give them both names. However, the name *Lagrangian* is widely used, so should be mentioned here, whereas *blowup density* is more descriptive and often notationally more convenient.

hand, if G covers pairs and has at least $t+1$ vertices then it cannot be H_{t+1}^r-hom-free: to define a homomorphism $f : V(H_{t+1}^r) \to V(G)$ we arbitrarily choose distinct vertices as $f(x_1), \cdots, f(x_{t+1})$, then for each $1 \le i < j \le t+1$ we fix an edge e_{ij} containing $f(x_i)f(x_j)$ and map y_{ij}^k, $1 \le k \le r-2$ to $e_{ij} \setminus \{f(x_i), f(x_j)\}$. It follows that $\pi(H_{t+1}^r) = b(K_t^r) = r!t^{-r}\binom{t}{r} = \prod_{i=1}^{r-1}(1-i/t)$.

The previous result is due to Mubayi [138], who also gave an exact result for the following family of r-graphs including H_t^r. Let $\mathcal{H}_t^r$ be the set of r-graphs F that have at most $\binom{t}{2}$ edges, and have some set T of size t such that every pair of vertices in T is contained in some edge. We extend our earlier definitions to a family $\mathcal{F}$ of r-graphs in the obvious way: we say G is $\mathcal{F}$-free if it does not contain any F in $\mathcal{F}$, and then we can define $\mathrm{ex}(n, \mathcal{F})$ and $\pi(\mathcal{F})$ as before. Mubayi [138] showed that the unique largest $\mathcal{H}_{t+1}^r$-free r-graph on n vertices is the balanced blowup of K_t^r. This was subsequently refined by Pikhurko [155], who showed that for large n, the unique largest H_{t+1}^r-free r-graph on n vertices is the balanced blowup of K_t^r.

More generally, suppose F is any r-graph that covers pairs. For any $t \ge v(F)$ we define a hypergraph H_t^F as follows. We label the vertices of F as $v_1, \cdots, v_{v(F)}$. We add new vertices $v_{v(F)+1}, \cdots, v_t$. Then for each pair of vertices v_i, v_j not both in F we add another $r-2$ new vertices u_{ij}^k, $1 \le k \le r-2$ and the edge $v_i v_j u_{ij}^1 \ldots u_{ij}^{r-2}$. Thus every pair of vertices in F is contained in an edge of H_t^F (although H_t^F does not cover pairs because of the new vertices u_{ij}^k). As an example, if we take F to be the r-graph with no vertices then $H_t^F = H_t^r$ as defined above. The following theorem generalises Mubayi's density result.

Theorem 3.1 *If F is an r-graph that covers pairs and $t \ge v(F)$ satisfies $\pi(F) \le b(K_t^r) = \prod_{i=1}^{r-1}(1-i/t)$ then $\pi(H_{t+1}^F) = b(K_t^r)$.*

Proof The same argument used for H_{t+1}^r shows that K_t^r is H_{t+1}^F-hom-free, so $\pi(H_{t+1}^F) \ge b(K_t^r)$. For the converse, it suffices to show that any H_{t+1}^F-hom-free dense G satisfies $b(G) \le b(K_t^r)$. This holds by monotonicity if G has at most t vertices, so we can assume G has at least $t+1$ vertices. Now we claim that G is F-hom-free. To see this, note that since F covers pairs, any homomorphism f from F to G is injective, i.e. maps F to a copy of F in G. Then f can be extended to a homomorphism from H_{t+1}^F to G, by the same argument used for H_{t+1}^r. This contradicts our choice of G, so G is F-hom-free. Then $b(G) \le \pi(F) \le b(K_t^r)$. $\square$

The argument of the above theorem is due to Sidorenko [174] where it is given in the special case when F is the 3-graph with 3 edges on 4 vertices and $t = 4$. We will see later (Section 6) that $\pi(F) \le 1/3$. Since $b(K_4^3) = 3/8 > 1/3$ we deduce that in this case $\pi(H_4^F) = 3/8$. Another simple application is to the case when F consists of a single edge. Then $\pi(F) = 0$, so $\pi(H_{t+1}^F) = b(K_t^r)$ for all $t \ge r$. The corresponding exact result for this configuration when n is large is given by Mubayi and Pikhurko [141]; we will call it the *generalised fan*, as they call the case $t = r$ a *fan*. As it has not been explicitly pointed out in the earlier literature, we remark that for every r-graph F that covers pairs, the theorem above gives an infinite family of r-graphs for which we can determine the Turán density.

Sidorenko [174] also applied his method to give the asymptotic result for a construction based on trees that satisfy the Erdős-Sós conjecture. This conjecture (see

[51]) states that if T is a tree on k vertices and G is a graph on n vertices with more than $(k-2)n/2$ edges then G contains T. Although this conjecture is open in general, it is known to hold for many families of trees (e.g. Sidorenko proves it in this case when some vertex is adjacent to at least $(k-2)/2$ leaves, and a proof for large k has been claimed by Ajtai, Komlós, Simonovits and Szemerédi). Suppose that T is a tree satisfying the Erdős-Sós conjecture. Let F be the r-graph obtained from T by adding a set S of $r-2$ new vertices to every edge of T (note that it is the same set for each edge). Let F' be the r-graph obtained from F by adding an edge for each uncovered pair consisting of that pair and $r-2$ new vertices (i.e. $F' = H^F_{v(F)}$). We call F' an *extended tree*. The result is $\pi(F') = b(K^r_{k+r-3})$, provided that $k \geq M_r$, where M_r is a small constant that can be explicitly computed. For example $M_3 = 2$, so when $r = 3$ we have $\pi(F') = b(K^r_k)$ for all $k \geq 2$.

We conclude this section with an application of general optimisation techniques to Turán problems given by Bulò and Pelillo [31]. Suppose that G is a k-graph and consider minimising the polynomial $h(x) = p_{\overline{G}}(x) + a \sum_i x_i^k$ over x in S, where $\overline{G}$ is the complementary k-graph whose edges are the non-edges of G. The intuition for this function is that the first term is minimised when x is supported on a clique of G, whereas the second term is minimised when $x = (1/n, \cdots, 1/n)$, so in combination one might expect a maximum clique to be optimal. It is shown in [31] that this is the case when $0 < a < \frac{1}{k(k-1)}$. One can immediately deduce a bound on the Turán number of K^k_{t+1}. Indeed, the minimum of $h(x)$ is achieved by putting weight $1/t$ on the vertices of a K^k_t, giving value at^{1-k}. On the other hand, substituting $x = (1/n, \cdots, 1/n)$ gives an upper bound of $|\overline{G}|n^{-k} + an^{1-k}$. This gives $|\overline{G}|n^{-k} + an^{1-k} \geq at^{1-k}$, so $\mathrm{ex}(n, K^k_{t+1}) \leq \binom{n}{k} - at(n/t)^k + an$ for any $0 < a < \frac{1}{k(k-1)}$. We will see later that this is not as good as bounds obtained by other methods, but the technique is interesting and perhaps more widely applicable.

4   Link graphs and multigraphs

This section explores the following constructive strategy that can be employed for certain Turán problems. Given an r-graph G and a vertex x of G, the *link* (or *neighbourhood*) $G(x)$ is the $(r-1)$-graph consisting of all $S \subseteq V(G)$ with $|S| = r-1$ and $S \cup \{x\} \in E(G)$. Suppose we are considering the Turán problem for an r-graph F of the following special form: there is some $X \subseteq V(F)$ such that every edge e of F is either contained in X or has exactly one point in X. Then the strategy for finding F is to first find a copy of the subgraph $F[X]$, then extend it to F by consideration of the links of the vertices in X.

Our first example will be to the following question of Katona. Say that an r-graph G is *cancellative* if whenever A, B, C are edges of G with $A \cup C = B \cup C$ we have $A = B$. For example, an (ordinary) graph G is cancellative if and only if it is triangle free. Katona asked for the maximum size of a cancellative 3-graph of n vertices. This was answered as follows by Bollobás [20].

Theorem 4.1 *The unique largest cancellative 3-graph on n vertices is 3-partite.*

It is not hard to see that a 3-partite 3-graph is cancellative. The largest 3-partite 3-graph on n vertices is clearly complete and balanced (meaning as before that the 3 parts are as equal as possible). We denote this 3-graph by $S_3(n)$ and write $s_3(n) = e(S_3(n))$. We will sketch a short proof given by Keevash and Mubayi [106] using link graphs. In this application of the method, the subgraph $F[X]$ described above will just be a single edge $e = xyz$. The links $G(x)$, $G(y)$ and $G(z)$ are pairwise edge-disjoint graphs, for if, say, we had edges xab and yab then $xab \cup xyz = yab \cup xyz$ contradicts G being cancellative. We consider their union U restricted to $V(G) \setminus \{x, y, z\}$ as a 3-edge-coloured graph. Any triangle in U must be 'rainbow' (use all 3 colours), as if, say, ab and ac both have colour x and bc has colour y then $xab \cup bcy = xac \cup bcy$ contradicts G being cancellative.

Proof Suppose G is a cancellative 3-graph on n vertices. For simplicity we just show the inequality $e(G) \leq s_3(n)$, though the uniqueness statement also follows easily. We use induction on n. The result is obvious for $n \leq 4$ so suppose $n \geq 5$. If any triple of vertices is incident to at most $s_3(n) - s_3(n-3)$ edges then we can delete it and apply induction. Thus we can assume that every triple is incident to more than $s_3(n) - s_3(n-3) = t_3(n) - n + 1$ edges, where $t_3(n)$ denotes the number of edges in the balanced complete 3-partite 'Turán graph' on n vertices. Now consider an edge $e = xyz$. Note that there are at most $n - 3$ edges that intersect e in 2 vertices, otherwise there would be some w that forms an edge with 2 pairs of e, but G is cancellative. Since e is incident to at least $t_3(n) - n + 2$ edges (including itself), the number of edges in U is at least $t_3(n) - n + 2 - (n-3) - 1 = t_3(n-3) + 1$. By Turán's theorem U contains a K_4; let its vertex set be $abcd$. This K_4 is 3-edge-coloured in such a way that every triangle is rainbow, which is only possible when it is properly 3-edge-coloured, i.e. each colour is a matching of two edges. Finally we consider the 7-set $S = xyzabcd$. The colouring of $abcd$ implies that every pair of vertices in S is contained in an edge of G, so have disjoint links. But by averaging, the total size of the links of vertices in S is at least $\frac{7}{3}(t_3(n) - n + 2) > \binom{n}{2}$, contradiction. □

A similar argument was applied in [106] to give a new proof of a theorem of Frankl and Füredi [66]. Note that a 3-graph is cancellative if and only if it does not contain either of the following 3-graphs: $F_4 = \{123, 124, 134\}$, $F_5 = \{123, 124, 345\}$. Thus we can write Bollobás' theorem as $ex(n, \{F_4, F_5\}) = s_3(n)$. This was improved in [106] to the 'pure' Turán result $ex(n, F_5) = s_3(n)$ for large n. (This was the first hypergraph Turán theorem.) The original proof required $n \geq 3000$; this was improved to $n \geq 33$ in [106], where the extremal example $S_3(n)$ was also characterised. Very recently, Goldwasser [86] has determined $ex(n, F_5)$ and characterised the extremal examples for all n: $S_3(n)$ is the unique extremal example for $n > 10$, the 'star' (all triples containing some fixed vertex) is the unique extremal example for $n < 10$, and both $S_3(n)$ and the star are extremal for $n = 10$. A new proof of the asymptotic form of the Frankl-Füredi theorem had previously been given by Mubayi and Rödl [143]. That paper applied the link method to obtain several other bounds on Turán densities. They also gave 5 specific 3-graphs each of which has Turán density $3/4$. One of these, denoted $F(3,3)$, is obtained by taking an edge abc, three additional vertices d, e, f, and all edges with one vertex from abc and two from def.

We remark that induction arguments as in the above proof are often very useful for Turán problems. Above it was convenient to consider deleting triples, but usually one considers deleting a single vertex. Then in order to prove the statement $e(G) \leq$

$f(n)$ for an F-free r-graph G on n vertices one can assume that the minimum degree of a vertex in G is more than $f(n) - f(n-1)$. (This argument gives one of the simplest proofs of Turán's theorem.) A caveat is that this induction argument depends on being able to prove a base case, which is not always convenient, as the desired bound may not even be true for small n. Then the following proposition is a more convenient method for obtaining a minimum degree condition. (The proof is to repeatedly delete vertices with degree less than the stated bound: a simple calculation shows that this process terminates and that the final graph has many vertices.)

Proposition 4.2 *For any $\delta, \epsilon > 0$ and $n_0 \geq r \geq 2$ there is n_1 so that any r-graph on $n \geq n_1$ vertices with at least $(\delta + 2\epsilon)\binom{n}{r}$ edges contains an r-graph on $m \geq n_0$ vertices with minimum degree at least $(\delta + \epsilon)\binom{m-1}{r-1}$.*

Our next example using links is by de Caen and Füredi [42], who were the originators of the method. They gave a surprisingly short proof of a conjecture of Sós [183] on the Turán number of the Fano plane. The Fano plane is an ubiquitous object in combinatorics. It is the unique 3-graph on 7 vertices in which every pair of vertices is contained in exactly one edge. It can be constructed by identifying the vertices with the non-zero vectors of length 3 over $\mathbb{F}_2$ (the field with two elements), and the edges with triples $\{x, y, z\}$ with $x + y = z$. It is easy to check that the Fano plane is not bipartite, in that for any partition of its vertex set into two parts, at least one of the parts must contain an edge. Thus a natural construction of a Fano-free 3-graph on n vertices is to take the balanced complete *bipartite* 3-graph: the vertex set has two parts of size $\lfloor n/2 \rfloor$ and $\lceil n/2 \rceil$, and the edges are all triples that intersect both parts. Sós conjectured that this construction gives the exact value for the Turán number of the Fano plane. The following result from [42] verifies this conjecture asymptotically (see Section 5 for the exact result).

Theorem 4.3 $ex(n, Fano) \sim \frac{3}{4}\binom{n}{3}$.

As in the previous example, the construction starts with a single edge $e = xyz$. We combine the links to create a 3-edge-coloured *link multigraph* $L = G(x) + G(y) + G(z)$, where $+$ denotes multiset union; note that unlike the previous example the links need not be edge-disjoint. To find a Fano plane we need to find the same object which appeared in the previous proof: a properly 3-edge-coloured K_4. (Since an edge may have more than one colour, this means we can select a colour for each edge to obtain the required colouring.) To prove an asymptotic result we can assume that G has at least $(3/4 + 2\epsilon)\binom{n}{3}$ edges for some small $\epsilon > 0$, and then that G has minimum degree at least $(3/4 + \epsilon)\binom{n}{2}$ by Proposition 4.2. However, the argument now seems to get stuck at the point of using this lower bound on the links to find a properly 3-edge-coloured K_4.

The key idea is to instead start with a copy of K_4^3, with the intention of using one of its edges as the edge e above. Since G has edge density more than $3/4$, averaging shows that it has a 4-set $wxyz$ of density more than $3/4$, i.e. spanning K_4^3 (we will see better bounds later on the Turán density of K_4^3). Now we consider the 4-edge-coloured link multigraph L, obtained by restricting $G(w) + G(x) + G(y) + G(z)$ to $V(G) \setminus \{w, x, y, z\}$. It suffices to find a properly 3-edge-coloured K_4. This can be achieved by temporarily forgetting the colours of the edges and just counting multiplicities. Thus we consider L as a multigraph with edge multiplicities at most 4 and at least $(3+4\epsilon)\binom{n}{2}$ edges. Such a multigraph must have a 4-set $abcd$ that spans at least 21 edges in L: this is a special case of a theorem of Füredi and Kündgen [80]. Finally we put the colours back. We may consider the bipartite graph B in which one part B_1 is the 4-set $wyxz$, the other part B_2 is the 3 matchings of size 2 formed by $abcd$, and edges in B correspond to edges of G in the obvious way, e.g. we join w to $\{ab, cd\}$ if wab and wcd are edges. Since $L[abcd]$ is at most 3 edges from being complete, the same is true of B. This implies (e.g. using Hall's theorem) that B has a matching that covers B_2. This gives the proper 3-edge-colouring of $abcd$ required to prove the theorem.

5   Stability

Many extremal problems have the property that there is a unique extremal example, and moreover any construction of close to maximum size is structurally close to this extremal example. For example, in the Turán problem for the complete graph K_t, Turán's theorem determines $\text{ex}(n, K_t)$ and describes the unique extremal example as the balanced complete $(t-1)$-partite graph on n vertices. More structural information is given by the Erdős-Simonovits Stability Theorem [182], which may be informally stated as saying that any K_t-free graph G on n vertices with $e(G) \sim \text{ex}(n, K_t)$ is structurally close to the extremal example. More precisely, we have the following statement.

Theorem 5.1 *For any $\epsilon > 0$ there is $\delta > 0$ such that if G is a K_t-free graph with at least $(1-\delta)\text{ex}(n, K_t)$ edges then there is a partition of the vertices of G as $V_1 \cup \cdots \cup V_{t-1}$ with $\sum_i e(V_i) < \epsilon n^2$.*

As well as being an interesting property of extremal problems, this phenomenon gives rise to a surprisingly useful tool for proving exact results. This stability method has two stages. First one proves a stability theorem, that any construction of close to maximum size is structurally close to the conjectured extremal example. Armed with this, we can consider any supposed better construction as being obtained from the extremal example by introducing a small number of imperfections into the structure. The second stage is to analyse any possible imperfection and show that it must lead to a suboptimal configuration, so in fact the conjectured extremal example must be optimal.

This approach can be traced back to work of Erdős and Simonovits in the 60's in extremal graph theory (see [182]). More recently it was applied independently by Keevash and Sudakov [110] and by Füredi and Simonovits [84] to prove the conjecture of Sós mentioned above in an exact form: for large n the unique largest Fano-free 3-graph on n vertices is the balanced complete bipartite 3-graph. Since then it has been applied to many problems in hypergraph Turán theory and more broadly in combinatorics as whole. We will discuss the other Turán applications later; we refer the reader to [107] for an application in extremal set theory and some further references using the method.

To understand how the method works in more detail, it is helpful to consider the 'baby' case of the Turán problem for the 5-cycle C_5. This is not hard to handle by other means, but is sufficiently simple to illustrate the method without too many technicalities. Since $\chi(C_5) = 3$, the Erdős-Stone theorem gives $ex(n, C_5) \sim ex(n, K_3) = \lfloor n^2/4 \rfloor$. In fact $ex(n, C_5) = \lfloor n^2/4 \rfloor$ for $n \geq 5$. We will sketch a proof of this equality for large n. This is a special case of a theorem of Simonovits, that if F is a graph with $\chi(F) = t$ and $\chi(F \setminus e) < t$ for some edge e of F then $ex(n, F) = ex(n, K_t)$ for large n. (We say that such graphs F are *critical*: examples are cliques and odd cycles.) The first step is the following stability result (we state it informally, the precise statement is similar to that in Theorem 5.1).

Lemma 5.2 *Suppose G is a C_5-free graph on n vertices with $e(G) \sim n^2/4$. Then G is approximately complete bipartite.*

Proof (Sketch.) First we claim that we can assume G has minimum degree $\delta(G) \sim n/2$. This is a similar statement to that in Proposition 4.2, although we cannot apply that result, as we cannot remove too many vertices if we want to obtain the structure of G. The solution is to use the same vertex deletion argument and use the bound from the Erdős-Stone theorem to control the number of vertices deleted. The calculation is as follows, for some small $\delta > 0$. If $e(G) > (1 - \delta)n^2/4$ then we can delete at most $\delta^{1/2}n$ vertices of proportional degree less than $1/2 - \delta^{1/2}$, otherwise we arrive at a C_5-free graph G' on $n' = (1 - \delta^{1/2})n$ vertices with $e(G') >$

$$e(G) - \sum_{i=(1-\delta^{1/2})n+1}^{n}(1/2-\delta^{1/2})i > (1-\delta)n^2/4 - \delta^{1/2}n^2/2 + \delta^{1/2}\left(\binom{n+1}{2} - \binom{n'+1}{2}\right) =$$

$n'^2/4 - \delta n^2/2 + \delta(1 - \delta^{1/2}/2)n^2 - O(n) > (1 + \delta)n'^2/4$, contradiction.

Thus we can assume that $\delta(G) \sim n/2$. Next we choose a 4-cycle $abcd$ in G. These are plentiful, as G has edge density about $1/2$, whereas the 4-cycle is bipartite, so has zero Turán density. Now we note that the neighbourhoods $N(a)$ and $N(b)$ cannot share a vertex x other than c or d, otherwise $axbcd$ is a 5-cycle. Furthermore each of these neighbourhoods does not contain a path of length 3, e.g. if $wxyz$ is a path of length 3 in $N(a)$ then $awxyz$ is a 5-cycle. Thus each is very sparse, e.g. $N(a)$ cannot have average degree at least 6, as it is not hard to show that it would then have a subgraph of minimum degree at least 3, and so a path of length 3. Thus we have found two disjoint sets of size about $n/2$ containing only $O(n)$ edges. $\square$

The second step is to refine the approximate structure and deduce an exact result.

Theorem 5.3 $ex(n, C_5) = \lfloor n^2/4 \rfloor$ *for large n.*

Proof (Sketch.) Suppose G is a maximum size C_5-free graph on n vertices. We claim that we can assume G has minimum degree $\delta(G) \geq \lfloor n/2 \rfloor$. For suppose we have proved the result under this assumption for all $n \geq n_0$. Then suppose that n is much larger then n_0 and repeatedly delete vertices while the minimum degree condition fails. A similar calculation to that in the lemma shows that this process terminates with a C_5-free graph G' on $n' \geq n_0$ vertices with $\delta(G') \geq \lfloor n'/2 \rfloor$, and moreover if any vertices were deleted we have $e(G') > \lfloor n'^2/4 \rfloor$, contradiction. Thus we can assume $\delta(G) \geq \lfloor n/2 \rfloor$.

By the lemma G is approximately complete bipartite. Consider a bipartition $V(G) = A \cup B$ that is optimal, in that $e(A) + e(B)$ is minimised. Then $e(A) + e(B) <$

ϵn^2, for some small $\epsilon > 0$. Also, A and B each have size about $n/2$, say $(1/2 \pm \epsilon^{1/2})n$, otherwise $e(G) < |A||B| + \epsilon n^2 < n^2/4$, contradicting G being maximum size. Write $d_A(x) = |N(x) \cap A|$ and $d_B(x) = |N(x) \cap B|$ for any vertex x. Note that for any $a \in A$ we have $d_A(a) \le d_B(a)$, otherwise we could improve the partition by moving a to B. Similarly, $d_B(b) \le d_A(b)$ for any $b \in B$.

Next we claim that the 'bad degrees' must be small, e.g. that $d_A(a) < cn$ for all $a \in A$ where $c = 2\epsilon^{1/2}$. For suppose this fails for some a. Then $N(a) \cap A$ and $N(a) \cap B$ both have size at least cn. Moreover they span a bipartite graph with no path of length 3, so only $O(n)$ edges. This gives $(cn)^2 - O(n) > e(A) + e(B)$ 'missing edges' between A and B, so $e(G) < |A||B| \le n^2/4$, contradiction.

Finally we claim that there are no 'bad edges', i.e. that (A, B) gives a bipartition of G. For suppose that aa' is an edge in A. Then $|N_B(a) \cap N_B(a')| > d(a) - cn + d(a') - cn - |B| > (1/2 - 5\epsilon^{1/2})n$. But there is no path $ba''b'$ with $b, b' \in B' = N_B(a) \cap N_B(a')$ and $a'' \in A' = A \setminus \{a, a'\}$, so A' and B' span a bipartite graph with only $O(n)$ edges - a very emphatic contradiction! $\qquad\square$

The above proof illustrates a template that is followed by many (but not all) applications of the stability method. In outline, the steps are: (i) G has high minimum degree, (ii) G has approximately correct structure, so consider an optimal partition, (iii) the bad degrees are small, (iv) there are no bad edges. For example, the deduction in [110] of the exact result for the Fano plane from the stability result follows this pattern. It is instructive to note that as well as the link multigraph method of the previous section, considerable use is made of an additional property of the Fano plane: there is a vertex whose deletion leaves a 3-partite 3-graph (any vertex has this property). It is an intriguing problem to understand what properties of an r-graph F make it amenable to either of the two steps of the stability method, i.e. whether a stability result holds, and whether it can be used to deduce an exact result.

A variant form of the stability approach is to prove a statement analogous to the following theorem of Andrásfai, Erdős and Sós [7]: any triangle-free graph G on n vertices with minimum degree $\delta(G) > 2n/5$ is bipartite. The approach taken in [84] to the exact result for the Fano plane is to prove the following statement: if $\delta > 0$, n is large, and G is a Fano-free 3-graph on n vertices with $\delta(G) > (3/4 - \delta)\binom{n}{2}$ then G is bipartite. One might think that this is a stronger type of statement than the stability result, but in fact it is equivalent in difficulty: it follows by exactly the same proof as that of the refinement argument sketched above (the second stage of the stability method). It would be interesting, and probably rather more difficult, to determine the smallest minimum degree for which this statement holds. For example, in the Andrásfai-Erdős-Sós theorem the bound $2n/5$ is tight, as shown by the blowup of a 5-cycle; what is the analogous 'second-best' construction for the Fano plane? (Or indeed, for other hypergraph Turán results...?)

We conclude this section by mentioning a nice application of the stability method to showing the 'non-principality' of Turán densities. If $\mathcal{F}$ is a set of graphs then it is clear that $\pi(\mathcal{F}) = \min_{F \in \mathcal{F}} \pi(F)$; one can say that the Turán density for a set of graphs is 'principal', in that it is determined by just one of its elements. However, Balogh [10] showed that this is not the case for hypergraphs, confirming a conjecture of Mubayi and Rödl [143]. Mubayi and Pikhurko [142] showed that even a set of two hypergraphs may not be principal. Let F denote the Fano plane and let F' be the 'cone' of K_t, i.e. the 3-graph on $\{0, \cdots, t\}$ with edges $0ij$ for $1 \le i < j \le t$. Note that F' is not contained in the blowup of K_t^3, so $\pi(F') \to 1$ as $t \to \infty$. Thus we can choose t so that $\min\{\pi(F), \pi(F')\} = \pi(F) = 3/4$. Now we claim that $\pi(\{F, F'\}) < 3/4$. To see this, consider a large Fano-free 3-graph G with edge density $3/4 - o(1)$. Then G is approximately a complete bipartite 3-graph. But the complete bipartite 3-graph contains many copies of F', and these cannot all be destroyed by the approximation, so G contains a copy of F', as required.

6 Counting

The arguments in this section deduce bounds on the edge density of an F-free r-graph G from counts of various small subgraphs. The averaging argument discussed in Section 2 gives the essence of this idea, but this uses F-freeness only to say that r-graphs containing F have a count of zero in G. However, there are various methods that can extract information about other counts. Arguments using the Cauchy-Schwartz inequality or non-negativity of squares play an important role here. We will illustrate this in the case when F is the (unique) 3-graph on 4 vertices with 3 edges (we called this F_4 in Section 4). First consider what can be obtained from a basic averaging argument. If G is an F-free 3-graph then every 4-set of G has at most 2 edges, i.e. density at most $1/2$. It follows that the density of G is at most $1/2$.

For an improvement, consider the sum $S = \sum_{xy} \binom{d(x,y)}{2}$, where the sum is over unordered pairs of vertices xy and $d(x,y)$ denotes the number of edges containing xy. Note that S counts the number of unordered pairs ab such that axy and bxy are edges. Since we are assuming that every 4-set has at most 2 edges, this is exactly the number of 4-sets $abxy$ with 2 edges. On the other hand, the Cauchy-Schwartz inequality (more precisely, convexity of the function $f(t) = \binom{t}{2}$) gives $S \ge \binom{n}{2}\binom{d}{2}$, where d is the average value of $d(x,y)$. We have $\binom{n}{2}d = \sum_{xy} d(x,y) = 3e(G) = 3\theta\binom{n}{3}$, where θ denotes the edge density of G, so $S \ge \frac{1}{4}\theta^2 n^4 + O(n^3)$. For an upper bound on S, we can double-count pairs (T, e) where T is a 4-set with 2 edges and e is an edge of T. This gives $2S \le (n-3)e(G)$, so $S \le \frac{1}{12}\theta n^4 + O(n^3)$. Combining the bounds on S we obtain $\theta \le 1/3 + O(1/n)$ as a bound on the density of G.

It is remarkable that the Turán problem is still open for such a seemingly simple 3-graph – we do not even know its Turán density! For many years the above argument (due to de Caen [38]) gave the best known upper bound. More recently it has been improved in a series of papers [134, 136, 186, 144, 158]; the current best bound is 0.2871 due to Baber and Talbot [9]. The best known lower bound is $\pi(F) \ge 2/7 = 0.2857\cdots$, due to Frankl and Füredi [67]. It is worth noting their unusual iterative construction. For most Turán problems, the known or conjectured optimal construction is obtained by dividing the vertex set into a fixed number of parts and defining edges by consideration only of intersection sizes with the parts. However, this construction is obtained by first blowing up a particular 3-graph on 6 vertices, then iteratively substituting copies of the construction inside each of the 6 parts. The 3-graph in question has the 10 edges 123, 234, 345, 451, 512, 136, 246, 356, 416, 526. This has blowup density $3! \frac{10}{6^3} = \frac{5}{18}$, so the overall density is given by the geometric series $\frac{5}{18} \sum_{i \ge 0}(1/36)^i = \frac{2}{7}$. If this is indeed the optimal construction, as conjectured in [136], then its complicated nature gives some indication as to why it has so far eluded proof. It is also an intriguing example from the point of view of the finiteness questions discussed in [123, 156] (see Section 7).

Our next example of a counting argument, due to Sidorenko [175], gives a bound on the Turán density that depends only on the number of edges. If F is an r-graph with f edges then the bound is $\pi(F) \leq \frac{f-2}{f-1}$. Before proving this we set up some notation. Suppose that G is an r-graph on n vertices. Label the vertices of F as $\{v_1, \cdots, v_t\}$ for some t and the edges as $\{e_1, \cdots, e_f\}$. Let $x = (x_1, \cdots, x_t)$ denote an arbitrary t-set of vertices in G. We think of the map $\pi_x : v_i \mapsto x_i$ as a potential embedding of F in G. Write $e_i(x)$ for the indicator function that is 1 if $\pi_x(e_i)$ is an edge of G or 0 otherwise. Then $F(x) = \prod_{i=1}^{f} e_i(x)$ is 1 if π_x is a homomorphism from F to G or 0 otherwise. Now suppose that G is F-free. Then $F(x)$ can only be non-zero if $x_i = x_j$ for some $i \neq j$. If we choose x randomly this has probability $O(1/n)$, so $\mathbb{E}F(x) = O(1/n)$.

Now comes a trick that can only be described as pulling a rabbit out of a hat. We claim that the following inequality holds pointwise: $F(x) \geq e_1(x) + \sum_{i=2}^{f} e_1(x)(e_i(x) - 1)$. To see this, note that since $F(x)$ and all the $e_i(x)$ are $\{0,1\}$-valued, the only case where the inequality is not obvious is when $F(x) = 0$ and $e_1(x) = 1$. But then $F(x) = 0$ implies that $e_i(x) = 0$ for some i, and then the term $e_1(x)(e_i(x) - 1)$ gives a -1 to cancel $e_1(x)$, so the inequality holds. Taking expected values gives $\mathbb{E}F(x) \geq \sum_{i=2}^{f} \mathbb{E}e_1(x)e_i(x) - (f-2)\mathbb{E}e_1(x)$. Now $\mathbb{E}e_1(x) = \theta + O(1/n)$, where θ is the edge density of G. Also, the Cauchy-Schwartz inequality implies that $\mathbb{E}e_1(x)e_i(x) \geq \theta^2 + O(1/n)$ for each i (this is similar to the lower bound on S in the previous example). We deduce that $(f-1)\theta^2 - (f-2)\theta \leq O(1/n)$, i.e. $\theta \leq \frac{f-2}{f-1} + O(1/n)$, as required.

It is an interesting question to determine the extent to which this bound can be improved. When $f = 3$ no improvement is possible for general r, as the bound of $1/2$ is achieved by the triangle for $r = 2$, or the 'expanded triangle' for even r (see [111]). For $r = f = 3$ there are only two non-degenerate cases to consider, namely the 3-graphs F_4 and F_5 discussed in Section 4. The Turán number of F_5 is given by the complete 3-partite 3-graph, which has density $2/9$. Then from the discussion of F_4 above we see that when $r = f = 3$ the bound of $1/2$ is quite far from optimal. We do not know if $1/2$ can be achieved when $f = 3$ and $r \geq 5$ is odd. A refinement of Sidorenko's argument given in [102] shows that the bound $\pi(F) \leq \frac{f-2}{f-1}$ can be improved if r is fixed and f is large. One might think that the worst case is when F is a complete r-graph, which would suggest an improvement to $\pi(F) \leq 1 - \Omega(f^{-(r-1)/r})$ (see [102] for further discussion). Another general bound obtained by the Local Lemma will be discussed later in the section on probabilistic methods.

The above two examples show that the 'right' counting argument can be surprisingly effective, but they do not give much indication as to how one can find this argument. For a general Turán problem there are so many potential inequalities that might be useful that one needs a systematic approach to understand their capabilities. The most significant steps in this direction have been taken by Razborov [158], who has obtained many of the sharpest known bounds on Turán densities using his theory of flag algebras [156]. We will describe this in Section 7, but first we note that earlier steps in this direction appear in the work of de Caen [39] and Sidorenko [172].

We briefly describe the quadratic form method in [172], as it has some additional features that are not exploited by other techniques. Suppose G is an r-graph on $[n]$. Let y be the vector in $\mathbf{R}^{\binom{n}{r-1}}$, where co-ordinates correspond to $(r-1)$-sets of vertices, and the entry for a given $(r-1)$-set is its degree, i.e. the number of edges containing it. Let u be the all-1 vector. Writing $(\cdot,\cdot)$ for the standard inner product, we have $(u,u) = \binom{n}{r-1}$, $(y,u) = r|G|$ and $(y,y) = r|G|+p$, where p is the number of ordered pairs of edges that intersect in $r-1$ vertices. Let $\overline{G} = \{[n]\setminus e : e \in E(G)\}$ be the complementary $(n-r)$-graph and define $\overline{y}, \overline{u}, \overline{p}$ similarly. Note that the degree of an $(n-r-1)$-set in $\overline{G}$ is equal to the number of edges of G in the complementary $(r+1)$-set. Note also that $\overline{p} = p$. Let Q_i be the number of $(r+1)$-sets that contain at least i edges. Thus $\overline{y}$ has $Q_i - Q_{i+1}$ co-ordinates equal to i. For any $t \geq 1$ we have the inequality $(\overline{y} - (r+1)\overline{u}, \overline{y} - t\overline{u}) \leq \sum_{i=0}^{t}(r+1-i)(t-i)(Q_i - Q_{i+1})$. Then come some algebraic manipulations which we will just summarise: (i) use summation by parts, (ii) substitute $Q_0 = (\overline{u},\overline{u}) = \binom{n}{r+1}$, (iii) rewrite in terms of y and u. The resulting inequality is $(y,y)+((1+t/r)(n-r)+1)(y,u)+\sum_{i=1}^{t}(t+r+2-2i)Q_i \leq 0$.

Now we come to the crux of the method, which is an inequality for (y,u) for general vectors y, u satisfying a quadratic inequality as above. The inequality is that if $(y,y) - 2a(y,u) + b(u,u) \leq 0$ then $(y,u) \geq (a - \sqrt{a^2-b})(u,u)$. To see this add $-(y-su, y-su) \leq 0$ to the first inequality, which gives $2(s-a)(y,u) \leq (s^2-b)(u,u)$, so $(y,u) \geq \frac{b-s^2}{2(a-s)}(u,u)$ for $s \leq a$; the optimal choice is $s = a - \sqrt{a^2-b}$ (note that the first inequality implies $a^2 - b \geq 0$). For example, let us apply the inequality to 3-graphs in which every 4-set spans at least t edges, where $t \in \{1,2,3\}$ (it turns out that we should choose the same t above). Then $Q_i = \binom{n}{4}$ for $0 \leq i \leq t$. We apply the general inequality with $a = \frac{1}{2}((1+t/3)(n-3)+1) \sim \frac{1}{2}(1+t/3)n$ and $b = \sum_{i=1}^{t}(t+5-2i)\binom{n}{4}\binom{n}{2}^{-1} \sim \frac{1}{3}tn^2$. Then $s/n \sim (1+t/3)/2 - \sqrt{(1+t/3)^2/4 - t/3} = t/3$, so $|G| = \frac{1}{3}(y,u) \geq \frac{tn}{9}\binom{n}{2} \sim \frac{t}{3}\binom{n}{3}$. In particular we have re-proved the earlier example (in complementary form). An interesting additional feature of this method is that one can obtain a small improvement by exploiting integrality of the vectors y and u. Instead of adding the inequality $-(y-su, y-su) \leq 0$ above, one can use $-(y-\lfloor s\rfloor u, y - (\lfloor s\rfloor+1)u) \leq 0$. This does not alter the asymptotic bound, but can be used to improve Turán bounds for small n, which in turn can be used in other asymptotic arguments.

7 Flag algebras

A systematic approach to counting arguments is provided by the theory of flag algebras. This is abstract and difficult to grasp in full generality, but for many applications it can be boiled down to a form that is quite simple to describe. Our discussion here will be mostly based on the nice exposition given in [9]. The starting point is the following description of the averaging bound. Given r-graphs H and G we write $i_H(G)$ for the 'induced density' of H in G. This is defined as the probability that a random $v(H)$-set in $V(G)$ induces a subgraph isomorphic to H; we think of H as fixed and G as large. For example, if H is a single edge then $i_H(G) = d(G)$ is the edge density of G. Fix some $\ell \geq r$, and let $\mathcal{G}_\ell$ denote the set of all r-graphs on ℓ vertices (up to isomorphism). Then we can write $d(G) = \sum_{H \in \mathcal{G}_\ell} i_H(G)d(H)$. Now suppose F is an r-graph and let $\mathcal{F}_\ell$ denote the set of all F-free r-graphs on ℓ vertices.

If G is F-free then $i_H(G) = 0$ for $H \in \mathcal{G}_\ell \setminus \mathcal{F}_\ell$, so $d(G) = \sum_{H \in \mathcal{F}_\ell} i_H(G)d(H)$. In particular we have the averaging bound $d(G) \leq \max_{H \in \mathcal{F}_\ell} d(H)$, but this is generally rather weak. The idea of the method is to generate further inequalities on the densities $i_H(G)$ that improve this bound. If we have an inequality $\sum_{H \in \mathcal{F}_\ell} c_H i_H(G) \geq 0$ then we have $d(G) \leq \sum_{H \in \mathcal{F}_\ell} i_H(G)(d(H) + c_H) \leq \max_{H \in \mathcal{F}_\ell}(d(H) + c_H)$, which may be an improvement if some coefficients c_H are negative.

These inequalities can be generated by arguments similar to that used on the sum $S = \sum_{xy} \binom{d(x,y)}{2}$ in Section 6 when F is the 3-graph with 4 vertices and 3 edges. In this context, one should view that argument as generating an inequality for a 4-vertex configuration (two edges) from two 3-vertex configurations (edges) that overlap in two points. We will use this as a running example to illustrate the flag algebra definitions. In general, we will consider overlapping several pairs of r-graphs along a common labelled subgraph. To formalise this, we define a *type* σ to consist of an F-free r-graph on k vertices together with a bijective labelling function $\theta : [k] \to V(\sigma)$ for some $k \geq 0$ (if $k < r$ then σ has no edges, and if $k = 0$ it has no vertices). Then we define a σ-*flag* to be an F-free r-graph H containing an induced copy of σ, labelled by θ. In our example we take $\sigma = xy$ to be a 3-graph with 2 vertices and no edges, labelled as $\theta(1) = x$ and $\theta(2) = y$. Then we take $H = e^\sigma$ to be a single edge xyz, with the same labelling $\theta(1) = x$ and $\theta(2) = y$.

Next we define induced densities for flags. Let Φ be the set of all injective maps $\phi : [k] \to V(G)$. For any fixed $\phi \in \Phi$ we define $i_{H,\phi}(G)$ as the probability that a random $v(H)$-set S in $V(G)$ containing the image of ϕ induces a σ-flag isomorphic to H. Note that $i_{H,\phi}(G)$ can only be non-zero if $\phi([k])$ induces a copy of σ that can be identified with σ in such a way that $\phi = \theta$. If this holds, then $i_{H,\phi}(G)$ is the probability that $G[S]$ induces a copy of the underlying r-graph of H consistent with the identification of $\phi([k])$ with σ. We can relate flag densities to normal densities by averaging over $\phi \in \Phi$; we have $\mathbb{E}_\phi i_{H,\phi}(G) = p_\sigma(H)i_H(G)$, where we also let H denote the underlying r-graph of the σ-flag H, and $p_\sigma(H)$ is the probability that a random injective map $\phi : [k] \to V(H)$ gives a copy of the type σ. In our example, for any $u, v \in V(G)$ we consider the function ϕ defined by $\phi(1) = u$ and $\phi(2) = v$; we denote this function by uv. Then $i_{e^\sigma,uv}(G)$ is the probability that a random vertex $w \in V(G) \setminus \{u, v\}$ forms an edge with uv, i.e. $i_{e^\sigma,uv}(G) = \frac{d(u,v)}{n-2}$. Then we have $\mathbb{E}_{uv} i_{e^\sigma,uv}(G) = \frac{1}{n(n-1)} \sum_u \sum_{v \neq u} \frac{d(u,v)}{n-2} = \frac{6e(G)}{n(n-1)(n-2)} = d(G) = i_e(G)$; note that $p_\sigma(e) = 1$.

Given two σ-flags H and H' we have the approximation $i_{H,\phi}(G)i_{H',\phi}(G) = i_{H,H',\phi}(G) + o(1)$, where we define $i_{H,H',\phi}(G)$ to be the probability that when we independently choose a random $v(H)$-set S and a random $v(H')$-set S' in $V(G)$ subject to $S \cap S' = \phi([k])$ we have $G[S] \cong H$ and $G[S'] \cong H$ as σ-flags. Here the $o(1)$ term tends to zero as $v(G) \to \infty$: this expresses the fact that random embeddings of H and H' are typically disjoint outside of $\phi([k])$. Note that we can compute $i_{H,H',\phi}(G)$ by choosing a random ℓ-set L containing $S \cup S'$ for some $\ell \geq v(H) + v(H') - k$ and conditioning on the σ-flag J induced by L. Writing $\mathcal{F}_\ell^\sigma$ for the set of σ-flags on ℓ vertices we have $i_{H,H',\phi}(G) = \sum_{J \in \mathcal{F}_\ell^\sigma} i_{H,H',\phi}(J)i_{J,\phi}(G)$. Thus we can express $i_{H,H',\phi}(G)$ as a linear combination of flag densities $i_{J,\phi}(G)$, where the coefficients $i_{H,H',\phi}(J)$ are given by a finite computation.

In our running example we consider $H = H' = e^\sigma$. Then $i_{e^\sigma,e^\sigma,uv}(G)$ is the probability that a random pair w, w' of vertices in $V(G) \setminus \{u, v\}$ each form an edge with uv, i.e. $i_{e^\sigma, e^\sigma, uv}(G) = \binom{d(u,v)}{2} \binom{n-2}{2}^{-1}$. We can also compute this by conditioning on $J = G[L]$ for a random 4-set L containing uv. Let J denote the σ-flag on 4 vertices where 2 vertices are labelled x and y, and there are 2 edges, both containing x and y. Then $i_{e^\sigma, e^\sigma, uv}(G) = i_{J, uv}(G)$; note that this uses our particular choice of F, otherwise we would have additional terms corresponding to the 3-edge and 4-edge configurations. Averaging over uv we get $\frac{1}{n(n-1)} \sum_u \sum_{v \neq u} \binom{d(u,v)}{2} \binom{n-2}{2}^{-1} = p_\sigma(J) i_J(G) = \frac{1}{6} i_J(G)$, i.e. $i_J(G) = \binom{n}{4}^{-1} \sum_{uv} \binom{d(u,v)}{2}$, similarly to the calculation of $S = \sum_{uv} \binom{d(u,v)}{2}$ in Section 6.

Now we consider several σ-flags $F_1, \cdots, F_m$, assign them some real coefficients $a_1, \cdots, a_m$ and consider the inequality $(\sum_{i=1}^m a_i i_{F_i, \phi}(G))^2 \geq 0$. Expanding the square and using the identity for overlapping flags we obtain $\sum_{i,j=1}^m a_i a_j i_{F_i, F_j, \phi}(G) \geq o(1)$. Then we convert this into an inequality for densities of subgraphs (rather than flags) by averaging over $\phi \in \Phi$. Provided that we have chosen $\ell \geq v(F_i) + v(F_j) - k$ we can compute the average $\mathbb{E}_\phi i_{F_i, F_j, \phi}(G)$ by choosing a random ℓ-set L and conditioning on the subgraph $H \in \mathcal{F}_\ell$ induced by L. Let Φ_H be the set of all injective maps $\phi : [k] \to V(H)$. Then $\mathbb{E}_\phi i_{F_i, F_j, \phi}(G) = \sum_{H \in \mathcal{F}_\ell} b_{ij}(H) i_H(G)$, where $b_{ij}(H) = \mathbb{E}_{\phi \in \Phi_H} i_{F_i, F_j, \phi}(H)$. In any application H and $F_1, \cdots, F_m$ are fixed small r-graphs, so these coefficients $b_{ij}(H)$ can be easily computed. Finally we have an inequality of the required form: $\sum_{H \in \mathcal{F}_\ell} c_H i_H(G) \geq o(1)$, where $c_H = \sum_{i,j=1}^m a_i a_j b_{ij}(H)$.

Continuing the previous example, let $\sigma = xy$, let F_0 be the σ-flag on 3 vertices with no edges, taken with coefficient $a_0 = -1$, and let $F_1 = e^\sigma$ be the σ-flag on 3 vertices with one edge, taken with coefficient $a_1 = 2$. There are three 3-graphs in $\mathcal{F}^4$, which we label H_0, H_1 and H_2 according to the number of edges. The coefficients $b_{ij}(H_k)$ may be computed as follows: $b_{00}(H_0) = 1$, $b_{00}(H_1) = 1/2$, $b_{00}(H_2) = 1/6$, $b_{01}(H_0) = 0$, $b_{01}(H_1) = 1/4$, $b_{01}(H_2) = 1/3$, $b_{11}(H_0) = 0$, $b_{11}(H_1) = 0$, $b_{11}(H_2) = 1/6$. Then we obtain the coefficients $c_{H_0} = 1$, $c_{H_1} = -1/2$, $c_{H_2} = -1/2$, so the inequality $i_{H_0}(G) - i_{H_1}(G)/2 - i_{H_2}(G)/2 \geq o(1)$. We also have the averaging identity $d(G) = i_{H_1}(G)/4 + i_{H_2}(G)/2$. Adding the inequality $o(1) \leq i_{H_0}(G)/3 - i_{H_1}(G)/6 - i_{H_2}(G)/6$ gives $d(G) \leq i_{H_0}(G)/3 + i_{H_1}(G)/12 + i_{H_2}(G)/3 \leq 1/3 + o(1)$, so we recover the previous bound.

The reader may now be thinking that this is more obscure than the earlier argument and we are no nearer to a systematic approach! Indeed, there is no obvious way to choose σ and the σ-flags $F_1, \cdots, F_m$; in results so far these have been obtained by guesswork and computer experimentation. However, once these have been fixed an optimal inequality can be determined by solving a semidefinite program. (We refer the reader to [122] for background information on semidefinite programming that we will use below.) We first remark that once we have chosen σ and the σ-flags $F_1, \cdots, F_m$ we can sum several inequalities of the form $(\sum_{i=1}^m a_i i_{F_i, \phi}(G))^2 \geq 0$. Equivalently, we can fix a positive semidefinite m by m matrix $Q = (q_{ij})_{i,j=1}^m$ and use the inequality $\sum_{i,j=1}^m q_{ij} i_{F_i, \phi}(G) i_{F_j, \phi}(G) \geq 0$. Averaging over ϕ gives $\sum_{H \in \mathcal{F}_\ell} c_H(Q) i_H(G) \geq o(1)$, where $c_H(Q) = \sum_{i,j=1}^m q_{ij} b_{ij}(H)$. We can write $c_H(Q) = Q \cdot B(H)$, where $B(H)$ is the matrix $(b_{ij}(H))_{i,j=1}^m$ (note that it is symmetric) and $X \cdot Y = tr(XY)$ denotes inner product of symmetric matrices. We obtain a bound $\pi(F) \leq V + o(1)$, where V is the solution of the following optimisation problem in the variables $\{q_{ij}\}_{i,j=1}^m$ (we write $Q \geq 0$ to mean that Q is positive semidefinite):

$$V = \inf_{Q \geq 0} \max_{H \in \mathcal{F}_\ell} d(H) + c_H(Q).$$

This formulation is amenable to solution by computer, at least when ℓ is small, so that $\mathcal{F}_\ell$ is not too large. Razborov [158] has applied this method to re-prove many results on Turán densities that were obtained by other methods, and to obtain the sharpest known bounds for several Turán problems that are still open. At this point we recall the question of Turán mentioned earlier: what is the largest 3-graph on n vertices with no tetrahedron? Turán proposed the following construction: form a balanced partition of the n vertices into sets V_0, V_1, V_2, and take the edges to be those triples that either have one vertex in each part, or have two vertices in V_i and one vertex in V_{i+1} for some i, where $V_3 := V_0$. One can check that this construction gives a lower bound $\pi(K_4^3) \geq 5/9$. Until recently, the best known upper bound was $\pi(K_4^3) \leq \frac{3+\sqrt{17}}{12} = 0.593592\cdots$, given by Chung and Lu [32]. Razborov [158] announced computations that suggest the bound $\pi(K_4^3) \leq 0.561666$. These computations were also verified in [9], so the bound is probably correct. Here we should stress that, unlike some computer-aided mathematical arguments where there is potential to doubt whether they are really 'proofs', the flag algebra computations described here can in principle be presented in a form that can be verified (very tediously) by hand. However, there is not much point in doing this for a bound that is very unlikely to be tight, so this bound is likely to remain unrigorous!

The main result of [158] is the following asymptotic result for the tetrahedron problem under an additional assumption: if G is a 3-graph on n vertices in which no 4-set of vertices spans exactly 1 edge or exactly 4 edges then $e(G) \leq (5/9 + o(1))\binom{n}{3}$. This is given by an explicit flag computation that can be verified (laboriously) by hand.[4] This was subsequently refined by Pikhurko [154] using the stability method to give an exact result: for large n, the Turán construction for no tetrahedron gives the unique largest 3-graph on n vertices in which no 4-set of vertices spans exactly 1 edge or exactly 4 edges. It is interesting to contrast the uniqueness and stability of this restricted problem with the full tetrahedron problem, where there are exponentially many constructions that achieve the best known bound, see [26, 117, 64, 78].

We briefly describe the elegant Fon-der-Flaass construction [64] here. Suppose that Γ is an oriented graph with no induced oriented 4-cycle. Let G be the 3-graph on the same vertex set in which abc is an edge if it induces a subgraph of Γ which either has an isolated vertex or a vertex of outdegree 2. It is not hard to see that any 4-set contains at least one edge of G, so the complement of G is K_4^3-free. The picture below (from [64]) illustrates a suitable class of orientations of a complete tripartite graph. The parts A, B, C are represented by line segments. They are partitioned into subparts, represented by subsegments, e.g. A is partitioned into parts labelled by $\ell_1, \ell_2, \cdots$. The direction of edges is represented by an arrow in the appropriate rhombus. A weakened form of Turán's conjecture raised in [64] is whether all constructions of this type have density at least $4/9$. Some progress towards this was recently made by Razborov [159].

---

[4]It is easy to verify that the matrices used in the argument are positive definite without the floating point computations referred to in [158]; for example one can just verify that all symmetric minors have positive determinant (symmetric Gaussian elimination is a more efficient method).

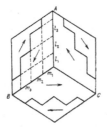

It is natural to ask whether all hypergraph Turán problems can be solved using flag algebras (at least in principle, given enough computation). A related general question posed by Lovász [123] is whether every linear inequality $\sum_H c_H i_H(G) \geq 0$ valid for all graphs G can be expressed as a finite sum of squares. Razborov [156] posed a similar question in terms of a certain 'Cauchy-Schwartz' calculus. Both these questions were recently answered in the negative by Hatami and Norine [92]. Furthermore, they proved that the problem of determining the validity of a linear inequality is undecidable. Their proof is a reduction to Matiyasevich's solution to Hilbert's tenth problem, which shows undecidability of the problem of determining whether a multivariate polynomial with integer coefficients is non-negative for every assignment of integers to its variables. There does not seem to be a direct consequence of their results for deciding inequalities of the form $a\pi(F) \leq b$ for r-graphs F and integers a and b, but perhaps this problem may also be 'difficult', or even undecidable.

8 The remaining exact results

Exact results for hypergraph Turán numbers are so rare that we can finish off a description of the known results in this section. By exact results we mean that the Turán number is determined for large n (it would be of course be nice to know it for all n, but then this section would already be finished!) We have mentioned earlier the exact results for F_5 (which implies that for cancellative 3-graphs), the Fano plane, extended complete graphs and generalised fans. Sidorenko [175, 176] and Frankl [65] considered the Turán problem for the following $2k$-graph which we call the *expanded triangle* C_3^{2k}. The vertex set is $K_1 \cup K_2 \cup K_3$ where K_1, K_2 and K_3 are disjoint k-sets, and there are three edges $K_1 \cup K_2$, $K_1 \cup K_3$ and $K_2 \cup K_3$. Thus the expanded triangle is obtained from a graph triangle by expanding each vertex into a k-set. Suppose that G is a $2k$-graph on n vertices with no expanded triangle. There is a natural auxiliary graph J on k-sets of vertices, where we join two k-sets in J if their union is an edge of G. Then J is triangle-free, and applying Mantel's theorem gives the bound $\pi(C_3^{2k}) \leq 1/2$. For a construction, consider a partition of n vertices into two roughly equal parts, and take the edges to be all $2k$-sets that intersect each part in an odd number of vertices: we call this *complete oddly bipartite*. To see that this does not contain the expanded triangle, consider an attempted embedding and look at the intersection sizes of the k-sets K_1, K_2 and K_3 with one of the parts. Some two of these have the same parity, so combine to form an edge with an even intersection with this part. This gives a lower bound that matches the upper bound asymptotically, so $\pi(C_3^{2k}) = 1/2$.

Keevash and Sudakov [111] proved an exact result, confirming a conjecture of Frankl, that for large n, the unique largest $2k$-graph on n vertices with no expanded triangle is complete oddly bipartite. One should note that the number of edges is this construction is maximised by a partition that is slightly unbalanced (by an amount of order $\sqrt{n}$). Finding the optimal partition sizes is in fact an open problem, equivalent to finding the minima of binary Krawtchouk polynomials. Nevertheless, known bounds on this problem are sufficient to allow an application of the stability method, with the conclusion that the optimal construction is complete oddly bipartite, even if we do not know the exact part sizes. Sidorenko also considered the *expanded clique* C_r^{2k}, obtained by expanding each vertex of K_r into a k-set. Applying Turán's theorem to the auxiliary graph J on k-sets gives $\pi(C_r^{2k}) \leq \frac{r-2}{r-1}$. On the other hand, Sidorenko only gave an asymptotic matching lower bound in the case when r is of the form $2^p + 1$. The construction is to partition n vertices into 2^p parts, labelled by the vector space $\mathbb{F}_2^p$, and take edges to be all $2k$-sets whose labels have a non-zero sum. This is C_r^{2k}-free, as for any r k-sets, the labels of some two will have the same sum (by the pigeonhole principle), so form an edge whose labels sum to zero. This raised the question of what happens for r not of this form. One might think that a combinatorial problem of this nature will not depend on a number theoretic condition, so there ought to be other constructions. However, we showed in [111] that this is not the case, in a somewhat different application of the stability method. We studied structural properties of a putative C_r^4-free 4-graph with density close to $\frac{r-2}{r-1}$ and showed that they give rise to certain special proper edge-colourings of K_{r-1}. It then turns out that these special edge-colourings have a natural $\mathbb{F}_2$ vector space structure on the set of colours, so we get a contradiction unless r is of the form $2^p + 1$.

It would be interesting to give better bounds for other r. We present here a new construction showing that $\pi(C_4^4) \geq 9/14 = 0.6428\cdots$; this is not far from the upper bound of $2/3$ (which we know is not sharp). Partition a set V of n vertices into sets A, B, C, where A is further partitioned as $A = A_1 \cup A_2$. We will optimise the sizes of these sets later. We say that $S \subseteq V$ has *type ijk* if $|S \cap A| = i$, $|S \cap B| = j$, $|S \cap A| = k$. Let G be the 4-graph in which 4-tuples of the following types are edges: (i) all permutations of 310, (ii) 121 and 112 (but not 211), (iii) 400 with 3 vertices in one A_i and 1 vertex in the other. We claim that G is C_4^4-free. To see this, suppose for a contradiction that we can choose 4 pairs of vertices such that any pair of these pairs forms an edge of G. We can naturally label each pair as AA, BB, CC, AB, AC or BC. Since every pair of pairs forms an edge, we see that each label apart from AA can occur at most once, and at most one of AA, BB, CC can occur. Now consider cases according to how many times AA occurs. If AA does not occur, then BB and CC can account for at most one pair, so the other 3 pairs must be AB, AC, BC; however, AB and AC do not form an edge. If AA occurs once or twice then BB, CC and BC cannot occur, so we must have AB and AC; however, again these do not form an edge. If AA occurs at least 3 times then some two AA's do not form an edge, as the restriction of G to A is C_3^4-free. In all cases we have a contradiction, so G is C_4^4-free. Computations show that the optimal set sizes are $|A_1| = |A_2| = (7 - \sqrt{21})/28$ and $|B| = |C| = (7 + \sqrt{21})/28$, and that then G has density $9/14$. Since this relatively simple construction gives a density quite close to the easy upper bound, we conjecture that it is optimal.

The question of improving the auxiliary graph bound described above gives rise to the following 'coloured Turán problem' that seems to be of independent interest. (A similar problem is also discussed in Section 13.4.) Suppose H is a C_4^4-free 4-graph on n vertices with $\alpha\binom{n}{4}$. Let J be the K_4-free auxiliary graph on pairs: J has $N = \binom{n}{2}$ vertices and $\sim \alpha\binom{N}{2}$ edges. If α is close to $2/3$ then for a 'typical' triangle xyz in J the common neighbourhoods N_{xy}, N_{xz}, N_{yz} partition most of $V(J)$ into 3 independent sets. These must account for almost all of the missing edges. Going back to the original set of n vertices, we can interpret N_{xy}, N_{xz}, N_{yz} as a 3-edge-colouring of (most of) the $\binom{n}{2}$ pairs, such that any choice of a pair of disjoint pairs of the same colour gives a non-edge of H. Counting then implies that almost all non-edges are properly 3-edge-coloured. It is not hard to see that this is impossible, but the question is to quantify the extent to which this property is violated. For example, what is the minimum number of 4-cycles in which which precisely one pair of opposite edges has both edges of the same colour?

Now we will return to cancellative r-graphs. Bollobás conjectured that the natural generalisation of his theorem for cancellative 3-graphs should hold, namely that the largest cancellative r-graph on n vertices should be r-partite. This was proved by Sidorenko [173] in the case $r = 4$. Note that there are four configurations that are forbidden in a cancellative 4-graph, there is the 4-graph expanded triangle mentioned above, and another three which are formed by taking two edges as $abcx$, $abcy$ and a third edge that contains xy and intersects abc in either 0, 1 or 2 vertices. Sidorenko showed that even just forbidding these last three configurations but allowing the expanded triangle one obtains the same result, that the largest such 4-graph is 4-partite. This was further refined by Pikhurko [152] who showed that it is enough to forbid just one configuration, the *generalised triangle* $\{abcx, abcy, uvxy\}$: for large n, the largest 4-graph with no generalised triangle is 4-partite.

Sidorenko's argument is an instructive application of hypergraph Lagrangians. We will sketch the proof that if G is a cancellative 4-graph then $e(G) \le (n/4)^4$, which is tight when n is divisible by 4. Since $e(G)/n^4 = p_G(1/n, \cdots, 1/n) \le \lambda(G)$ it suffices to prove that the Lagrangian $\lambda(G)$ is at most 4^{-4}. We can assume that G covers pairs; then it follows that any two vertices in G have disjoint link 3-graphs (or we get one of Sidorenko's three forbidden configurations). Recall that $\lambda(G) = \max_{x \in S} p_G(x)$ where S is the set of all $x = (x_1, \cdots, x_n)$ with $x_i \ge 0$ for all i and $\sum_i x_i = 1$. Suppose that the maximum occurs at some x with $x_i > 0$ for $1 \le i \le m$ and $x_i = 0$ for $i > m$ (without loss of generality). We can discard all $i > m$ and then regard the maximum as being at an interior point of the corresponding region S_m defined for the vector $x = (x_1, \cdots, x_m)$. Next comes an ingredient from the theory of optimisation we have not yet mentioned: the gradient of $p_G(x)$ is normal to the constraint plane $\sum_i x_i = 1$, i.e. the partial derivatives $\partial_i p_G(x)$, $1 \le i \le m$ are all equal to some constant c. We can compute it by $c = c \sum_i x_i = \sum x_i \partial_i p_G(x) = r p_G(x) = r\lambda(G)$. Also, since vertices have disjoint links, any monomial $x_a x_b x_c$ occurs at most once in $\sum_i \partial_i p_G(x)$, so $mr\lambda(G) = \sum_i \partial_i p_G(x) \le \max_{x \in S} \sum_{1 \le a < b < c < m} x_a x_b x_c = m^{-3}\binom{m}{3}$. This gives the required bound if $m = 4$ or $m \ge 6$, and we cannot have $m = 5$ if G covers pairs, so we are done.

Pikhurko's proof is an ingenious combination of Sidorenko's argument with the stability method. And what happens for larger r? Shearer [170] showed that the Bollobás conjecture is false for $r \ge 10$. The intermediate values are still open.

Suppose now that we alter the general problem and just forbid the configurations analogous to those in Sidorenko's result; thus we consider r-graphs such that there do not exist two edges that share an $(r-1)$-set T and a third edge containing the two vertices not in T. When $r = 5$ and $r = 6$ this problem was solved by Frankl and Füredi [70]: the extremal constructions are the blowups of the small Witt designs, the $(11,5,4)$-design for $r = 5$ and the $(12,6,5)$-design for $r = 6$. They obtain the bounds $e(G) \leq \frac{6}{11^4}n^5$ when $r = 5$, with equality only when $11|n$, and $e(G) \leq \frac{11}{12^5}n^6$ when $r = 6$, with equality only when $12|n$. (The exact result for all large n remains open.) The proofs involve some intricate computations with hypergraph Lagrangians. They make the following appealing conjecture that would have greatly simplified some of these computations were it known. Consider the problem of maximising the Lagrangian $\lambda(G)$ for r-graphs G on m edges. Is the maximum attained when G is an initial segment of the colexicographic order?

Some further exact results can be grouped under the general umbrella of 'books'. The r-book with p pages is the r-graph obtained by taking $p \leq r$ edges that share a common $(r-1)$-set T, and one more edge that is disjoint from T and contains the vertices not in T. For example, the generalised triangle of Pikhurko's result mentioned above is the 4-book with 2 pages. Füredi, Pikhurko and Simonovits [83] gave an exact result for the 4-book with 3 pages: for large n the unique extremal 4-graph is obtained by a balanced partition into two parts and taking edges as all 4-sets with 2 vertices in each part. Next consider r-graphs that do not have an r-book with r pages. A nice reformulation of this property is to say that such r-graphs G have *independent neighbourhoods*: for any $(r-1)$-set T, the neighbourhood $N(T) = \{x \in V(G) : T \cup \{x\} \in E(G)\}$ does not contain any edges of G. Füredi, Pikhurko and Simonovits [82] gave an exact result for 3-graphs with independent neighbourhoods: for large n, the unique extremal 3-graph is obtained by taking a partition into two parts A, B and taking edges as all 3-sets with 2 vertices in A and 1 vertex in B (the optimal class sizes are $|A| = 2n/3$, $|B| = n/3$ when n is divisible by 3). Füredi, Mubayi and Pikhurko [81] gave an exact result for 4-graphs with independent neighbourhoods: for large n, the unique extremal 4-graph is complete oddly bipartite (the same construction as for the expanded triangle). There is a conjecture in [81] for general r that the largest r-graphs with independent neighbourhoods are obtained by a partition into two parts A, B and taking edges as all r-sets that intersect B in an odd number of vertices, but are not contained in B. The results mentioned above confirm this for $r = 3$ and $r = 4$. However, the conjecture was disproved for $r \geq 7$ by Bohman, Frieze, Mubayi and Pikhurko [17]. The conjecture would have implied that r-graphs with independent neighbourhoods have edge density at most $1/2$. In fact, the construction in [17]. shows that the maximum edge density is roughly $1 - \frac{2\log r}{r}$, which approaches 1 for large r. The authors of [17] believe that the conjecture is true for $r = 5$ and $r = 6$.

There is one more exact result (to the best of this author's knowledge). We ask the expert readers to take note, as it seems to be have been overlooked in earlier bibliographies on this subject. The motivating problem is the Turán problem for K_5^3, where Turán conjectured that the complete bipartite 3-graph gives the extremal construction. This was disproved for $n = 9$ by Surányi (the affine plane over $\mathbb{F}_3$) and for all odd $n \geq 9$ by Kostochka and Sidorenko (see construction 5 in [180]). However, they did not disprove the asymptotic conjecture, so it may be that $\pi(K_5^3) = 3/4$.

Zhou [192] obtained an exact result when one forbids a larger class of 3-graphs that includes K_5^3. Say that two vertices x, y in a 3-graph G are *t-connected* if there are vertices a, b, c such that every triple with 2 vertices from abc and 1 from xy is an edge. Say that xyz is a *t-triple* if xyz is an edge and each pair in xyz is *t*-connected. For example, K_5^3 is a *t*-triple. The result of [192] is that the unique largest 3-graph on n vertices with no *t*-triple is complete bipartite. Note that the 3-graph $F(3,3)$ mentioned earlier is an example of a *t*-triple, so the result of [143] strengthens the asymptotic form of Zhou's result (but not the exact result).

9   Bounds for complete hypergraphs

We return now to the original question of Turán, concerning the Turán numbers for the complete hypergraphs K_t^r. None of the Turán densities $\pi(K_t^r)$ with $t > r > 2$ has yet been determined, so here we have a more modest goal of giving reasonable bounds. Most of these can be found in an excellent survey of Sidorenko [180], to which we refer the reader for full details. We will not reproduce this here, but instead outline the ideas behind the main bounds, summarise the other bounds, and also mention a few more recent developments. For the purpose of this section it is convenient to change to the 'complementary' notation that was preferred by many early writers on Turán numbers. They define the 'Turán number' $T(n, k, r)$ to be the minimum number of edges in an r-graph G on n vertices such that any subset of k vertices contains at least one edge of G. Note that G has this property if and only if the 'complementary' r-graph of r-sets that are not edges of G is K_k^r-free; thus $T(n, k, r) + \mathrm{ex}(n, K_k^r) = \binom{n}{r}$. They also define the density $t(k, r) = \lim_{n \to \infty} \binom{n}{r}^{-1} T(n, k, r)$; thus $t(k, r) + \pi(K_k^r) = 1$.

We start with the lower bound on $t(k, r)$, which is equivalent to an upper bound on $\pi(K_k^r)$. The trivial averaging argument gives $t(k, r) \geq \binom{k}{r}^{-1}$. In general, the best known bound is $t(k, r) \geq \binom{k-1}{r-1}^{-1}$, due to de Caen [38]. This follows from his exact bound of $T(n, k, r) \geq \frac{n-k+1}{n-r+1} \binom{k-1}{r-1}^{-1} \binom{n}{r}$. This in turn is deduced from a hypergraph generalisation of a theorem of Moon and Moser that relates the number of cliques of various sizes in a graph. Suppose that G is an r-graph on n vertices and let N_k be the number of copies of K_k^r in G. Then the inequality is

$$N_{k+1} \geq \frac{k^2 N_k}{(k-r+1)(k+1)} \left(\frac{N_k}{N_{k-1}} - \frac{(r-1)(n-k)+k}{k^2} \right), \qquad (9.1)$$

provided that $N_{k-1} \neq 0$. Given this inequality, the bound on $T(n, k, r)$ follows from some involved calculations; the main step is to show by induction on k that $N_k \geq N_{k-1} \frac{r^2 \binom{k}{r}}{k^2 \binom{n}{r-1}} (e(G) - F(n, k, r))$, where $F(n, k, r) = (r^{-1}(n-r+1) - \binom{k-1}{r-1}^{-1}(n-k+1)) \binom{n}{r-1}$. Inequality (9.1) is proved by the following double counting argument. Let P be the number of pairs (S, T) where S and T are each sets of k vertices, such that S spans a K_k^r, T does not span a K_k^r, and $|S \cap T| = k-1$. For an upper bound on P, we enumerate the N_{k-1} copies of K_{k-1}^r and let a_i be the number of K_k^r's containing the ith copy. Since $\sum_{i=1}^{N_{k-1}} a_i = k N_k$ we have $P = \sum_{i=1}^{N_{k-1}} a_i (n - k + 1 - a_i) \leq (n - k + 1) k N_k - N_{k-1}^{-1} k^2 N_k^2$. For a lower bound, we enumerate the copies of K_k^r as $B_1, \cdots, B_{N_k}$, and let b_i be the number of K_{k+1}^r's containing the ith copy. For each

B_j, there are $n - k - b_j$ ways to choose $x \notin B_j$ such that $B_j \cup \{x\}$ does not span a K_{k+1}^r. Given such an x, there is some $C \subseteq B_j$ of size $k - 1$ such that $C \cup \{x\}$ is not an edge. Then for each $y \in B_j \setminus C$ the pair $(B_j, B_j \cup x \setminus y)$ is counted by P. This gives $P \geq \sum_{j=1}^{N_k}(n - k - b_j)(k - r - 1) = (k - r - 1)((n - k)N_k - (k+1)N_{k+1})$. Combining with the lower bound and rearranging gives the required inequality.

Next we consider the upper bound on $t(k, r)$, which is equivalent to a lower bound on $\pi(K_k^r)$. The best general construction is due to Sidorenko [171]; it implies the bound $t(k, r) \leq \left(\frac{r-1}{k-1}\right)^{r-1}$. For comparison with the lower bound, note that $\left(\frac{r-1}{k-1}\right)^{r-1}\binom{k-1}{r-1} = \prod_{i=1}^{r-1}\frac{k-i}{k-1}\frac{r-1}{r-i}$; if k is large compared to r, the ratio of the bounds is roughly $(r - 1)^{r-1}(r - 1)!^{-1}$, which is exponential in r, but independent of k. To explain the construction, we will rephrase it here using the following simple fact.

The lorry driver puzzle. A lorry driver needs to follow a certain closed route. There are several petrol stations along the route, and the total amount of fuel in these stations is sufficient for the route. Show that there is some starting point from which the route can be completed.

The construction is to divide n vertices into $k-1$ roughly equal parts $A_1, \ldots, A_{k-1}$, and say a set B of size r is an edge of G if there is some j such that $\sum_{i=1}^{s}|B \cap A_{j+i}| \geq s + 1$ for each $1 \leq s \leq r - 1$ (where $A_i := A_{i-k+1}$ if $i > k - 1$). To interpret this in the lorry driver framework, consider any set K of size k, imagine that each element of K represents a unit of fuel, and that it takes $\frac{k}{k-1}$ units of fuel to drive from A_i to A_{i+1}. Then K contains enough fuel for a complete circuit, so the lorry driver puzzle tells us that there is some starting point from which a complete circuit is possible. (For completeness we now give the solution to the puzzle. Imagine that the driver starts with enough fuel to drive around the route and consider the journey starting from an arbitrary point, in which she still picks up all the fuel at any station, even though she doesn't need it. Then the point at which the fuel reserves are lowest during this route can be used as a starting point for another route which satisfies the requirements.) Let B be the set of the first r elements of K that are encountered on this circuit (breaking ties arbitrarily). Since $r \geq (r - 1)\frac{k}{k-1}$, the lorry can advance distance $r - 1$ using just the fuel from B. This implies that B is an edge, as $\lceil s\frac{k}{k-1}\rceil = s + 1$ for $1 \leq s \leq r - 1$. Thus any set of size k contains an edge, as required.

It is not obvious how to estimate the number of edges in the construction without tedious calculations, so we will give a simple combinatorial argument here. It is convenient to count edges together with an order of the vertices in each edge, thus counting each edge $r!$ times. We can form an ordered edge $B = x_1 \ldots x_r$ using the following three steps: (i) choose the starting index j, (ii) assign each x_ℓ to one of the parts A_{j+i}, $1 \leq i \leq r - 1$, (iii) choose a vertex for each x_ℓ within its assigned part. Clearly there are $k - 1$ choices in step (i) and $\left(\frac{n}{k-1}\right)^r + O(n^{r-1})$ choices in step (iii). In step (ii) there are $(r - 1)^r$ ways to assign the parts if we ignore the required inequalities on the intersection sizes (i.e. that there should be enough fuel for the lorry). Now we claim that given any assignment, there is exactly one cyclic permutation that satisfies the required inequalities. More precisely, if we assign b_i of the x_ℓ's to A_{j+i} for $1 \leq i \leq r - 1$, then there is exactly one c with $1 \leq c \leq r - 1$ such that the shifted sequence $b_i' = b_{c+i}$ (where $b_i := b_{i-r+1}$ for $i > j + r - 1$) satisfies

$\sum_{i=1}^{s} b_i' \geq s + 1$ for each $1 \leq s \leq r - 1$. To see this consider a lorry that makes a circuit of A_{j+i}, $1 \leq i \leq r - 1$, where as before each of the x_ℓ's is a unit of fuel, but now it takes one unit of fuel to advance from A_i to A_{i+1}, and the lorry is required to always have one spare unit of fuel. Clearly a valid starting point for the lorry is equivalent to a shifted sequence satisfying the required inequalities. As in the solution to the original puzzle, we imagine that the driver starts with enough fuel to drive around the route and consider the journey starting from an arbitrary point. Then the point at which the fuel reserves are lowest during this route is a starting point for a route where there is always one spare unit of fuel. Furthermore, this is the unique point at which the fuel reserves are lowest, and so it gives the unique cyclic permutation satisfying the required inequalities. We deduce that there are $(r - 1)^{r-1}$ valid assignments in step (ii). Putting everything together, the number of edges is $r!^{-1} \cdot (k - 1) \cdot (r - 1)^{r-1} \cdot (1 + O(1/n)) \left(\frac{n}{k-1}\right)^r \sim \left(\frac{r-1}{k-1}\right)^{r-1} \binom{n}{r}$.

Having discussed the general case in detail, we now summarise some better bounds that have been found in specific cases. One natural case to focus on is $t(r + 1, r) = 1 - \pi(K_{r+1}^r)$. For large r, a construction of Sidorenko [181] gives the best known upper bound, which is $t(r+1, r) \leq (1 + o(1)) \frac{\log r}{2r}$. Other known bounds are effective for small r; these are $t(r + 1, r) \leq \frac{1+2\ln r}{r}$ by Kim and Roush [114] and $t(2s+1, 2s) \leq 1/4 + 2^{-2s}$ by de Caen, Kreher and Wiseman [43]. On the other hand, the known lower bounds are very close to the bound $t(r+1, r) \geq 1/r$ discussed above in the general case. Improvements to the second order term were given by Chung and Lu [32], who showed that $t(r + 1, r) \geq \frac{1}{r} + \frac{1}{r(r+3)} + O(r^{-3})$ when r is odd, and by Lu and Zhao [129], who obtained some improvements when r is even, the best of which is $t(r+1, r) \geq \frac{1}{r} + \frac{1}{2r^3} + O(r^{-4})$ when r is of the form $6k + 4$. Thus the known upper and lower bounds are separated by a factor of $(1/2 + o(1)) \log r$. As a first step towards closing this gap, de Caen [41] conjectured that $r \cdot t(r + 1, r) \to \infty$ as $r \to \infty$.

We have already discussed the known bounds for K_4^3 in Section 7. For K_5^4 the following nice construction was given by Giraud [85]. Suppose M is an m by m matrix with entries equal to 1 or 0. We define a 4-graph G on $n = 2m$ vertices corresponding to the rows and columns of M. Any 4-set of rows or 4-set of columns is an edge. Also, any 4-set of 2 rows and 2 columns inducing a 2 by 2 submatrix with even sum is an edge. We claim that any 5-set of vertices of G contains an edge. This is clear if we have at least 4 rows or at least 4 columns, so suppose without loss of generality that we have 3 rows and 2 columns. Then in the induced 3 by 2 submatrix we can choose 2 rows whose sums have the same parity, i.e. a 2 by 2 submatrix with even sum, which is an edge. To count edges in G, first note that we have $2\binom{m}{4}$ from 4-sets of rows and 4-sets of columns. Also, for any pair i, j of columns, we can divide the rows into two classes O_{ij} and E_{ij} according to whether the entries in columns i and j have odd or even sum. Then the number of 2 by 2 submatrices using columns i and j with even sum is $\binom{|O_{ij}|}{2} + \binom{|E_{ij}|}{2} \geq 2\binom{m/2}{2}$. Furthermore, for some values of m there is a construction that achieves equality for every pair i, j: take a Hadamard matrix, i.e. a matrix with ± 1 entries in which every pair of columns is orthogonal, then replace the -1 entries by 0.

This shows that $t(5, 4) \leq \lim_{m \to \infty} \binom{2m}{4}^{-1} \left(2\binom{m}{4} + 2\binom{m/2}{2}\binom{m}{2}\right) = 5/16$; equivalently $\pi(K_5^4) \geq 11/16 = 0.6875$. Sidorenko [180] conjectured that equality holds.

Markström [131] gave an upper bound $\pi(K_5^4) \leq \frac{1753}{2380} = 0.73655\cdots$. This was achieved by an extensive computer search to find all extremal 4-graphs for $n \leq 16$. Based on this evidence, he made the stronger conjecture that this construction (modified according to divisibility conditions) is always optimal for $n \geq 12$. Markström [133] has also compiled a web archive of small constructions for various hypergraph Turán problems. For K_5^3 the Turán numbers were computed for $n \leq 13$ by Boyer, Kreher, Radziszowski and Sidorenko [25]. The collinear triples of points of the projective plane of order 3 form the unique 3-graph on 13 vertices such that every 5-set contains at least one edge. It follows that $t(5,3) \geq 52/\binom{13}{3} = 2/11$, i.e. $\pi(K_5^3) \leq 9/11$. As we mentioned in Section 8, Turán conjectured that $\pi(K_5^3) = 3/4$, corresponding to the complete bipartite 3-graph.

More generally, Turán conjectured that $\pi(K_{t+1}^3) = 1 - (2/t)^2$. Together with Mubayi we found the following family of examples establishing the lower bound (previously unpublished). It is convenient to work in the complementary setting; thus we describe 3-graphs of density $(2/t)^2$ such that every $(t+1)$-set contains at least one edge. Let D be any directed graph on $\{1,\ldots,t\}$ that is the vertex-disjoint union of directed cycles (we allow cycles of length 2, but not loops). Let $V_1,\cdots,V_t$ be a balanced partition of a set V of n vertices. Let G be the 3-graph on V where the edges consist of all triples that are either contained within some V_i or have 2 points in V_i and 1 point in V_j, for every directed edge (i,j) of D. Then G has $t\binom{n/t}{3} + t\binom{n/t}{2}n/t \sim (2/t)^2\binom{n}{3}$ edges. Also, if $S \subseteq V$ does not contain any edge of G, then S has at most 2 points in each part, and whenever it has 2 points in a part it is disjoint from the next part on the corresponding cycle, so we must have $|S| \leq t$. Thus G has the required properties.

10 The infinitary perspective

A new perspective on extremal problems can be obtained by stepping outside of the world of hypergraphs on finite vertex sets, and viewing them as approximations to an appropriate 'limit object'. This often leads to more elegant formulations of results from the finite world, after one has put in the necessary theoretical ground work to make sense of the 'limit'. That alone may justify this perspective for those of a theoretical bent, though others will ask whether it can solve problems not amenable to finite methods. Since the theory itself is quite a recent development, it is probably too soon to answer this latter question, other than to say that elegant reformulations usually lead to progress in mathematics.

We will approach the subject by first returning to flag algebras (see Section 7), which we will now describe in the theoretical framework of [156]. Recall that the aim when applying flag algebras to Turán problems was to generate a 'useful' inequality of the form $\sum_H c_H i_H(G) \geq 0$, valid for any F-free r-graph G. We can package the coefficients c_H as a 'formal sum' $\sum_H c_H H$ in $\mathbf{R}\mathcal{F}$, by which we mean the real vector space of formal finite linear combinations of F-free r-graphs. We can think of any F-free r-graph G as acting on $\mathbf{R}\mathcal{F}$ via the map $\sum_H c_H H \mapsto \sum_H c_H i_H(G)$; we will identify this map with G. Our goal will be to understand F-free r-graphs purely as appropriate maps on formal finite linear combinations. First we note that certain elements always evaluate to zero, so they should be factored out. If H is an F-free r-graph and $\ell \geq v(H)$ then $i_H(G) = \sum_{J \in \mathcal{F}_\ell} i_H(J)i_J(G)$, so the linear combination $H - \sum_{J \in \mathcal{F}_\ell} i_H(J)J$ is mapped to zero by G. Let $\mathcal{K}$ be the subspace generated by all such combinations in the kernel, and let $\mathcal{A} = \mathbf{R}\mathcal{F}/\mathcal{K}$ be the quotient space. Now we make $\mathcal{A}$ into an algebra by defining a multiplication operator: we let $HH' = \sum_{J \in \mathcal{F}_{v(H)+v(H')}} i_{H,H'}(J)J$, where $i_{H,H'}(J)$ is the probability that when $V(J)$ is randomly partitioned as $S \cup S'$ with $|S| = v(H)$ and $|S'| = v(H')$ we have $J[S] \cong H$ and $J[S'] \cong H'$. (One needs to prove that this is well-defined.) Then we have $G(HH') = G(H)G(H') + o(1)$ when $v(G)$ is large, so the map G is an 'approximate homomorphism' from $\mathcal{A}$ to $\mathbf{R}$. One final property to bear in mind is that $G(H) := i_H(G)$ is always non-negative. (A similar construction gives rise to an algebra $\mathcal{A}^\sigma = \mathbf{R}\mathcal{F}^\sigma/\mathcal{K}^\sigma$ for any type σ; we have just described the case when $|\sigma| = 0$ for simplicity.)

Now we come to the point of the above discussion: it gives an approximate characterisation of F-free r-graphs, in the following sense. Given an r-graph G, we can identify the map $G : \mathbf{R}\mathcal{F} \to \mathbf{R}$ defined above with the vector $(i_H(G))_{H \in \mathcal{F}} \in [0,1]^{\mathcal{F}}$; we will also identify this vector with G. The space $[0,1]^{\mathcal{F}}$ is compact in the product topology, so any sequence of r-graphs contains a convergent subsequence. Let Φ be the set of homomorphisms ϕ from $\mathcal{A}$ to $\mathbf{R}$ such that $\phi(H) \geq 0$ for every H in $\mathcal{F}$. The following key result is Theorem 3.3 in [156]: for any convergent sequence of F-free r-graphs, the limit is in Φ; conversely, any element of Φ is the limit of some sequence of F-free r-graphs. (For simplicity, we have stated this result just for F-free r-graphs, but there is a much more general form that applies to flags in theories.) This result establishes a correspondence between the finite world inequalities $\sum_H c_H i_H(G) \geq o(1)$ for r-graphs G (which we were interested in above) and inequalities $\phi(\sum_H c_H H) \geq 0$ for ϕ in Φ in the infinitary world. In particular, the Turán density $\pi(F) = \limsup_{G \in \mathcal{F}} d(G)$ can be rewritten as $\pi(F) = \max_{\phi \in \Phi} \phi(e)$. Note the maximum value $\pi(F)$ is achieved by some extremal homomorphism $\phi \in \Phi$ (this is because Φ is compact and $\phi \mapsto \phi(e)$ is continuous). This permits 'differential methods' (see section 4.3 of [156]), i.e. deriving inequalities from the fact that any small perturbations of ϕ must reduce $\phi(e)$, which are potentially very powerful. For example, perturbation with respect to a single vertex is analogous to the deletion argument in Proposition 4.2, but general perturbations do not have any obvious analogue in the finite setting.

Graph limits were first studied by Borgs, Chayes, Lovász, Sós and Vesztergombi (2003 unpublished and [24]) and by Lovász and Szegedy [126]. A substantial theory has been developed since then, of which we will only describe a couple of ingredients here: a convenient description of limit objects and the equivalence of various notions of convergence. The starting point is a very similar notion of convergence to that used by [156]. Let $t_H(G)$ denote the *homomorphism density* of H in G, defined as the probability that a random map from V_H to V_G is a homomorphism. Say that a sequence $G_1, G_2, \ldots$ is left-convergent if $t_H(G_i)$ converges for every H. (It is not hard to see via inclusion-exclusion that using induced densities is equivalent.) The limit objects can be described as *graphons*, which are symmetric measurable[5] functions $W : [0,1]^2 \to [0,1]$. For such a function we can define the homomorphism

---

[5]We will assume in this discussion that the reader is familiar with the basics of measure theory. A careful exposition for the combinatorial reader that fills in much of this background is given in [153]. Note also that we are using a more restricted definition of 'graphon' than the original definition given in [126].

density of H in W as $t_H(W) = \int \prod_{ij \in E(H)} W(x_i, x_j)dx$, where $x = (x_1, \cdots, x_{v(H)})$
and the integral is over $[0,1]^{v(H)}$. We can recover $t_H(G)$ as a case of this definition
by defining a graphon W_G as a step function based on the adjacency matrix of G.
Label $V(G)$ by $[n]$, partition $[0,1]^2$ into n^2 squares of side $1/n$, and set W_G equal
to 1 on subsquare (i,j) if ij is an edge, otherwise 0. Then $t_H(G) = t_H(W_G)$. The
main result of [126] is that for any left-convergent sequence (G_i) there is a graphon
W such that $t_H(G_i) \to t_H(W)$ for every graph H, and conversely, any graphon W
can be obtained in this way from a left-convergent sequence (G_i). This gives an
intuitive picture of a graph limit as an 'infinite adjacency matrix'. A more formal
justification of this intuition is given by the W-random graph $G(n, W)$. This is a
random graph on $[n]$ defined by choosing independent $X_1, \cdots, X_n$ uniformly in $[0,1]$
and connecting vertices i and j with probability $W(X_i, X_j)$. Corollary 2.6 of [126]
shows that $G(n, W)$ converges to W with probability 1.

An alternative description of convergence is given by the cut-norm on graphons,
defined by $\|W\|_\square = \sup_{S,T \subseteq [0,1]} \left| \int_{S \times T} W(x,y)\, dx\, dy \right|$. First we need to take account
of the lack of uniqueness in graphons. Suppose $\phi : [0,1] \to [0,1]$ is a measure-
preserving bijection and define W^ϕ by $W^\phi(x,y) = W(\phi(x), \phi(y))$. Then W^ϕ is
equivalent to W in the sense that $t_H(W^\phi) = t_H(W)$ for any H. We define the
cut-distance between two graphons as $\delta_\square(W, W') = \inf_\phi \|W^\phi - W'\|_\square$, where the
infimum is over all measure-preserving bijections. Then another equivalent definition
of convergence for (G_i), given in [24], is to say that $\delta_\square(W_{G_i}, W) \to 0$ for some
graphon W. Furthermore, W is essentially unique, in that if $G_n \to W$ and $G_n \to W'$
then $\delta_\square(W, W') = 0$. The equivalence classes $[W] = \{W' : \delta_\square(W, W') = 0\}$ are
named *graphits* by Pikhurko [153], in a paper that introduces an analytic approach
to stability theorems. Theorem 15 in [153] contains a characterisation of stability
that can be informally stated as follows: an extremal graph problem is stable if and
only if there is a unique graphit that can be obtained by limits of approximately
extremal graphs. (We say that an r-graph F is *stable* if for any $\epsilon > 0$ there is $\delta > 0$
and n_0 such that for any two F-free r-graphs G and G' on $n > n_0$ vertices, each
having at least $(\pi(F) - \delta)\binom{n}{r}$ edges, we can obtain G' from G by adding or deleting
at most ϵn^r edges.) The analytic proof of the Erdős-Simonovits stability theorem
given in [153] is much more complicated than a straightforward approach, but may
point the way to other stability results that cannot be obtained by simpler methods.

The limit theory above also has close connections with the theory of regularity
for graphs and hypergraphs, which are explored by Lovász and Szegedy [127]. We
start with the Szemerédi's Regularity Lemma, which is a fundamental tool in modern
graph theory. Our discussion here will be rather brief; for an extensive survey we
refer the reader to [116]. Roughly speaking, the regularity lemma allows any graph
G to be approximated by a weighted graph R, in which the size of R depends only
the desired accuracy of approximation, but is independent of the size of G. A precise
statement of the lemma (in its simplest form) is as follows: for any $\epsilon > 0$, there is
a number $m = m(\epsilon)$, such that for any number n, and any graph G on n vertices,
there is a partition of $V(G)$ as $V_1 \cup \cdots \cup V_r$ for some $r \leq m$, so that all but at most
ϵn^2 pairs of vertices belong to induced bipartite subgraphs $G_{ij} := G(V_i, V_j)$ that
are 'ϵ-regular'. We have not yet defined 'ϵ-regular': this is a notion that captures
the idea that the bipartite subgraph $G(V_i, V_j)$ looks like a random bipartite graph.

The formal definition is as follows. Suppose G is a bipartite graph with parts A and B. The density of G is $d(G) = \frac{|E(G)|}{|A||B|}$. We say that G is ϵ-regular if for any $A' \subseteq A$, $B' \subseteq B$ with $|A'| > \epsilon|A|$, $|B'| > \epsilon|B|$, writing G' for the bipartite subgraph of G induced by A' and B' we have $d(G') = d(G) \pm \epsilon$. Note that the definition fits well with the randomness heuristic: standard large deviation estimates imply that a random bipartite graph is ϵ-regular with high probability.

After applying Szemerédi's Regularity Lemma, we can use the resulting partition to define an approximation of G. This is the *reduced graph* R, defined on the vertex set $[r] = \{1, \cdots, r\}$. The vertices of R correspond to the parts $V_1, \cdots, V_r$ (which are also known as *clusters*). The edges of R correspond to pairs of clusters that induce bipartite subgraphs that look random and are sufficiently dense: we fix a 'density parameter' d, and include an edge ij in R with weight $d_{ij} := d(G_{ij})$ whenever G_{ij} is ϵ-regular with $d_{ij} \geq d$. A key property of this approximation of G by R is that it satisfies a 'counting lemma', whereby the number of copies of any fixed graph in G can be accurately predicted by the weighted number of copies of this graph in R. For example, we have the following Triangle Counting Lemma. Suppose $1 \leq i, j, k \leq r$ and write $T_{ijk}(G)$ for the set of triangles in G with one vertex in each of V_i, V_j and V_k. Write $d_{ijk} = \frac{|T_{ijk}(G)|}{|V_i||V_j||V_k|}$ for the corresponding 'triangle density', i.e. the proportion of all triples with one vertex in each of V_i, V_j and V_k that are triangles. Suppose $0 < \epsilon < 1/2$ and G_{ij}, G_{ik} and G_{jk} are ϵ-regular. Then $d_{ijk} = d_{ij}d_{ik}d_{jk} \pm 20\epsilon$. Note that this corresponds well to the randomness intuition: if the graphs were indeed random, with each edge being independently selected with probability equal to the corresponding density, then the probability of any particular triple uvw being a triangle would be the product of the probabilities for each of its three pairs uv, uw, vw being edges. Furthermore, there is a Counting Lemma for general subgraphs along similar lines, which starts to indicate the connection with the notion of convergence discussed above using subgraph densities.

The following weaker form of the regularity lemma was obtained by Frieze and Kannan [77]. Given a partition P of $V(G)$ as $V_1 \cup \cdots \cup V_r$, for $S, T \subseteq V(G)$ we write $e_P(S, T) = \sum_{i,j=1}^{r} d_{ij}|V_i \cap S||V_j \cap T|$. Note that $e_P(S, T)$ is the expected value of $e_G(S, T)$ (the number of edges of between S and T) if each G_{ij} were a random bipartite graph of density d_{ij}. The result of [77] is that there is a partition P into $r \leq 2^{2/\epsilon^2}$ classes such that $|e_G(S, T) - e_P(S, T)| \leq \epsilon n^2$ for all $S, T \subseteq V(G)$. This is rather weaker than the regularity lemma, as it has replaced a uniformity condition holding locally for most pairs of classes by a global uniformity condition. The compensation is that the number of classes needed is much smaller, only an exponential function, as opposed to the tower bound that is necessary in the regularity lemma (see [87]). The weak regularity lemma can be reformulated in analytic language as follows (Lemma 3.1 of [127]): for any graphon W and $\epsilon > 0$ there is a graphon W' that is a step function with at most $2^{2/\epsilon^2}$ steps such that $\|W - W'\|_\square \leq \epsilon$. The full regularity lemma, indeed even a stronger form due to Alon, Fischer, Krivelevich and Szegedy [4], can also be obtained from an analytic form. The key fact is that graphits form a compact metric space with the distance $\delta_\square$ defined above (Theorem 5.1 of [127]). This implies the following (Lemma 5.2 of [127]): let $h(\epsilon, r) > 0$ be an arbitrary fixed function; then for any ϵ there is $m = m(\epsilon)$ such that any graphon W can be written as $W = U + A + B$, where U is a step function with $r \leq m$ steps, $\|A\|_\square \leq h(\epsilon, r)$

and $\|B\|_1 \leq \epsilon$. Informally, this says that one can change W by a small function B to obtain a function $U + A$ which corresponds to an extremely regular partition. Regular approximation results of this type were first obtained in [162, 187].

Regularity theory for hypergraphs is much more complicated, so we will only make a few comments here and refer the reader to the references for more information. The theory was first developed independently in different ways by Rödl et al. [148, 163, 162] and Gowers [89]. Alternative perspectives and refinements were given in [8, 94, 187, 188]. The analytic theory discussed above is generalised to hypergraphs by Elek and Szegedy [46, 47] as follows. For any sequence of r-graphs $G_1, G_2, \ldots$ which is convergent in the sense that $t_H(G_i)$ converges for every r-graph H, there is a limit object W, called a *hypergraphon*, such that $t_H(G_i) \to t_H(W)$ for every r-graph H. Hypergraphons are functions of $2^r - 1$ variables, corresponding to the non-empty subsets of $[r]$, that are symmetric under permutations of $[r]$. The need for $2^r - 1$ variables (actually $2^r - 2$, as $[r]$ is unnecessary) reflects the fact that a complete theory of hypergraph regularity needs to consider the simplicial r-complex generated by an r-graph, and regularise k-sets with respect to $(k-1)$-sets for each $2 \leq k \leq r$ (see Section 5 of [88] for further explanation of this point).

In the case of 3-graphs we consider a function $W(x_1, x_2, x_3, x_{12}, x_{13}, x_{23})$ from $[0,1]^6$ to $[0,1]$ that is symmetric under permutations of 123. Given a fixed 3-graph H, we can define the homomorphism density of H in W similarly to above by $t_H(W) = \int \prod_{e \in H} W_e dx$, where $x = (x_1, \cdots, x_{v(H)})$ and the integral is over $[0,1]^{v(H)}$ as before, and W_e is evaluated according to some fixed labelling $e = e_1 e_2 e_3$ by $W_e = W(x_{e_1}, x_{e_2}, x_{e_3}, x_{e_1 e_2}, x_{e_1 e_3}, x_{e_2 e_3})$. Similarly to the graph case, any 3-graph G can be realised by a hypergraphon W_G that is a step function (which need only depend on the first 3 co-ordinates). Some intuition for hypergraphons can be obtained by consideration of the W-random 3-graph $G(n, W)$. This can be defined as a random 3-graph on $[n]$ by choosing independent random variables X_i, $1 \leq i \leq n$ and X_{ij}, $1 \leq i < j \leq n$ uniformly in $[0,1]$ and including the edge ijk for $i < j < k$ with probability $W(X_i, X_j, X_k, X_{ij}, X_{ik}, X_{jk})$. Theorem 12 of [47] shows that $G(n, W)$ converges to W with probability 1. If we approximate W by a step function then this gives us the following informal picture of a regularity partition of a 3-graph G: a piece of the partition is obtained by taking three classes V_i, V_j, V_k, then three 'random-like' bipartite graphs $V_{ij} \subseteq V_i \times V_j$, $V_{ik} \subseteq V_i \times V_k$, $V_{jk} \subseteq V_j \times V_k$, and then a 'random-like' subset of the triangles formed by V_{ij}, V_{ik} and V_{jk}. Further generalisations of the theory from graphs to hypergraphs given in [47] are the equivalence of various definitions of convergence (Theorem 14), and a formulation of regularity as compactness (Theorem 4).

We conclude this section with a concrete situation where hypergraph regularity theory gives some insight into Turán problems. This is via the removal lemma, a straightforward consequence of hypergraph regularity theory that can be easily stated as follows. For any $b > 0$ and r-graph F there is $a > 0$, so that if G is a r-graph on n vertices with fewer than $an^{v(F)}$ copies of F, then one can delete at most bn^r edges from G to obtain an F-free r-graph. This was used by Pikhurko [155, Lemma 4] to show that if the Turán problem for F is stable then so is the Turán problem for any blowup $F(t)$. A sketch of the proof is as follows. Suppose that $0 < 1/n \ll a \ll b \ll c$, and G_1 and G_2 are $F(t)$-free r-graphs on n vertices, each having at least $(\pi(F(t)) - b)\binom{n}{r}$ edges. Since G_1 and G_2 are $F(t)$-free, supersaturation implies that they each have at most $an^{v(F)}$ copies of F. The removal lemma implies that one can delete at most bn^r edges from G_1 and G_2 to obtain F-free r-graphs G'_1 and G'_2, each having at least $(\pi(F) - 2b)\binom{n}{r}$ edges (recall that $\pi(F(t)) = \pi(F)$). Now by stability of F we can obtain G'_2 from G'_1 by adding or deleting at most cn^r edges. Thus we can obtain G_2 from G_1 by adding or deleting at most $2cn^r$ edges, so $F(t)$ is stable. In particular, this enables the application of the stability method in [155] to the extended complete graph H_t^r (see Section 3); stability of H_t^r follows from stability of $\mathcal{H}_t^r$, which was proved by Mubayi [138].

11 Algebraic methods

Kalai [98] proposed the following conjecture generalising Turán's tetrahedron problem. Suppose that G is a 3-graph on $[n] = \{1, \cdots, n\}$ such that every 4-set of vertices spans at least one edge (thus G is the complement of a K_4^3-free 3-graph). Fix $s \geq 1$ and consider the following matrix $M_s(G)$. The rows are indexed by edges of G. The columns are divided into s blocks, each of which contains $\binom{n}{2}$ columns indexed by all pairs of vertices. The entry in row e and column uv in block i is $\pm x_{i,w}$ if $e = uvw$ for some w or is 0 otherwise, where $\{x_{i,w} : 1 \leq i \leq s, w \in V(G)\}$ are indeterminate variables, and the sign is positive if w lies between u and v, otherwise negative. Let $r_s(G)$ be the rank of $M_s(G)$. The conjecture is that

$$r_s(G) \geq \sum_{i=1}^{s} \left(\binom{n-2i}{2} - \binom{i}{2} \right) = s\binom{n}{2} - 2\binom{s+1}{2}n + 3\binom{s+2}{3}.$$

Note that the sum in the conjecture is maximised when $s = \lfloor n/3 \rfloor$, and the value obtained is the number of edges in (the complementary form of) Turán's conjecture.

What is the motivation for this conjecture? The definition of $M_s(G)$ is reminiscent of the incidence matrices, which have seen many applications in Combinatorics (see [75]). The (pair) *incidence matrix* for G has rows indexed by edges of G, columns by pairs of vertices, and the entry in row e and column uv is 1 if $e = uvw$ for some w or is 0 otherwise. Thus $M_s(G)$ is obtained by concatenating s copies of the incidence matrix and replacing the 1's by certain weights. In the case $s = 1$, we can set all the variables $x_w := x_{1,w}$ equal to 1 without changing the rank: to see this note that the variables cancel if we multiply each column uv by $x_u x_v$ and divide each row uvw by $x_u x_v x_w$. Thus we obtain the *signed incidence matrix*, which is obtained from the incidence matrix by attaching signs according to the order of u, v, w as above.

To understand signed incidence matrices it is helpful to start with graphs. Suppose G is a graph on $[n]$. Then the signed incidence matrix has rows indexed by edges of G, columns indexed by $[n]$, and a row ij with $i < j$ has -1 in column i, 1 in column j, and 0 in the other columns. Note that the set of rows corresponding to a cycle in G can be signed so that the sum is zero, so is linearly dependent. Conversely, it is not hard to show by induction that a set of rows corresponding to an acyclic subgraph is linearly independent. Another way to say this is that the signed incidence matrix is a linear representation of the cycle matroid of G. (We refer the reader to [149] for an introduction to Matroid Theory.) Thus the maximum rank is $n - 1$, with equality if and only if G is connected.

Kalai [97] developed a 'hyperconnectivity' theory for graphs using generalised signed incidence matrices. Similarly to the 3-graph case, when G is a graph, we define the matrix $M_s(G)$, which has rows indexed by $E(G)$, s blocks of columns each indexed by $[n]$, and a row uv with $u < v$ has $x_{i,v}$ in column u of block i, $-x_{i,u}$ in column v of block i, and is 0 otherwise. The resulting matrix is then considered to be a linear representation of the s-hyperconnectivity matroid. The maximum possible rank is $sn - \binom{s+1}{2}$. Any G achieving this maximum is called s-hyperconnected. One result of [97] is that any s-hyperconnected graph is s-connected, in the usual sense that deleting any $s - 1$ vertices leaves a connected graph. Another is that K_{s+2} is a circuit, i.e. a minimally dependent set in the matroid, which leads us to an interesting digression on saturation problems.

The *saturation problem* for H is to determine $s(n, H)$, defined as the minimum number of edges in a maximal H-free graph on n vertices. Thus we want an H-free graph G such that adding any new edge to G creates a copy of H, and G has as few edges as possible. Suppose that G is K_{s+2}-saturated. Then for any pair $uv \notin E(G)$ there is a copy of K_{s+2} in $G \cup uv$, which is a circuit, so uv is in the span of G. It follows that G spans the entire s-hyperconnectivity matroid. In particular, G has at least $sn - \binom{s+1}{2}$ edges. This bound is tight, as may be seen from the example $K_s + E_{n-s}$, i.e. a clique of size s completely joined to an independent set of size $n - s$. More generally, the same argument applies to any G that has the weaker property that there is a sequence $G = G_0, G_1, \ldots, G_t = K_n$, where each G_{i+1} is obtained from G_i by adding an edge that creates a copy of K_{s+2} in G_{i+1} that was not present in G_i. Thus Kalai showed that such G also must have at least $sn - \binom{s+1}{2}$ edges, giving a new proof of a conjecture of Bollobás [19, Exercise 6.17]. See the recent survey [63] for more information on saturation problems.

Now we return to consider the meaning of the signed incidence matrix for 3-graphs. First we give another interpretation for graphs. We can think of the signed incidence matrix for K_n as a linear map from $\mathbb{F}^{\binom{n}{2}}$ to $\mathbb{F}^n$, for some field $\mathbb{F}$, acting on row vectors from the right. Then an edge uv with $u < v$ is mapped to the vector $v - u$, where we are identifying edges and vertices with their corresponding basis vectors. Geometrically, this is a *boundary* operation: we think of the line segment from u to v as having boundary points u and v, with the sign indicating the order. Similarly, we can think of the signed incidence matrix for K_n^3 as a linear map from $\mathbb{F}^{\binom{n}{3}}$ to $\mathbb{F}^{\binom{n}{2}}$, where an edge uvw with $u < v < w$ is mapped to $-vw + uw - uv$. It is convenient to write $vu = -uv$. Then we can think of the boundary operation as taking a 2-dimensional triangle uvw to its bounding cycle, oriented cyclically as wv, vu, uw. This cycle has 'no boundary', in that if we apply the boundary operation to $wv + vu + uw$ we get $v - w + w - u + u - v = 0$. In general, an oriented cycle has no boundary, which conforms to the geometric picture of it as a closed loop. The cycles generate the *cycle space*, which is the subspace of $\mathbb{F}^{\binom{n}{3}}$ of vectors that have no boundary, i.e. are mapped to zero by the signed incidence matrix.

Now consider a 3-graph G on $[n]$. We can create a simplicial complex C which has G as its two-dimensional faces, and the complete graph K_n as its one-dimensional faces; i.e. we take the 1-skeleton of the n-simplex and glue in triangles according to the edges of G. We interpret the rows of the signed incidence matrix of G as the boundary cycles of the triangles. These generate the *boundary space* of G, which is a subspace of the cycle space of K_n. The quotient space is the first *homology* space $H_1(C)$: it is a measure of the number of 1-dimensional 'holes' in the complex C. A lower bound on $r_1(G)$ is equivalent to an upper bound on the first Betti number $\beta_1(C) = \dim H_1(C)$, so the case $s = 1$ of Kalai's conjecture can be rephrased as saying that $\beta_1(C) \leq n - 2$: this was proved by Kalai (unpublished). He also established the corresponding algebraic generalisation of Turán's theorem on complete graphs.

Kalai also introduced a procedure of *algebraic shifting*, which is an intriguing and potentially powerful tool for a variety of combinatorial problems. In general, 'shifting' or 'compression' refers to a commonly employed technique in extremal set theory, where a problem for general families is reduced to the problem for an initial segment in some order; e.g. most proofs of the Kruskal-Katona theorem have this flavour. We refer the reader to [99] for a survey; here we just give a very brief taste of the operation and its properties. Suppose G is a k-graph on $[n]$ and $X = (x_{ij})_{i,j=1}^n$ is a matrix of indeterminates. Let $X^{\wedge k}$ be the $\binom{n}{k}$ by $\binom{n}{k}$ matrix indexed by k-sets, in which the (S,T)-entry is the determinant of the k by k submatrix of X corresponding to S and T. Let $M(G)$ be the submatrix of $X^{\wedge k}$ formed by the rows corresponding to edges of G. Now construct a basis for the column space of $M(G)$ by the greedy algorithm, at each step choosing the first column not in the span of those chosen previously. The k-sets indexing the chosen columns give the *(exterior) shifted family* $\Delta(G)$. This rather obscure process has some remarkable properties. Björner and Kalai [16] showed that it preserves the face numbers and Betti numbers of any simplicial complex. Even for a graph G, the presence of certain edges in $\Delta(G)$ encodes non-trivial information. For example, 23 appears iff G has a cycle, 45 appears iff G is non-planar, and dn appears iff G is d-hyperconnected. It seems computationally hard to compute $\Delta(G)$, although the randomised algorithm of substituting random constants for the variables and using Gaussian elimination is very likely to give the correct result. Potential applications are discussed in Section 6 of [99], but they still are yet to be realised!

Another application of homological methods was given by Csakany and Kahn [37]. A *d-simplex* is a collection of $d+1$ sets with empty intersection, every d of which have nonempty intersection. A few examples serve to illustrate that many common extremal problems have a forbidden configuration that is a simplex: the Erdős-Ko-Rado theorem [57] forbids 2 disjoint sets, which is a 1-simplex; the Ruzsa-Szemerédi $(6,3)$-theorem [165] forbids the *special triangle* $\{123, 345, 561\}$, which is a 2-simplex; the Turán tetrahedron problem forbids the 3-simplex K_4^3. Chvátal [33] posed the problem of determining the largest r-graph on n vertices with no d-simplex (Erdős [52] had posed the triangle problem earlier). He conjectured that when $r \geq d+1 \geq 2$ and $n > r(d+1)/d$ the maximum number of edges is $\binom{n-1}{r-1}$, with equality only for a star (all sets containing some fixed vertex). The known cases are $r = d + 1$ (Chvátal [33]), fixed r, d and large n (Frankl and Füredi [68]), $d = 2$ (Mubayi and Verstraëte [145]), and $\Omega(n) < r < n/2 - O(1)$ (Keevash and Mubayi [107]).

Csakany and Kahn gave new proofs of Chvátal's result (and also a similar result of Frankl and Füredi on the special triangle). They work with homology over the field $\mathbb{F}_2$, which has the advantage that there is no need to worry about signs $(+1 = -1)$, so boundary maps are given by incidence matrices. They note that a star G is acyclic, meaning that the boundary map is injective on the space generated by the edges of G. Furthermore, for any acyclic G the size of G is equal to the dimension of its boundary space, which is at most $\binom{n-1}{r-1}$, as this is the dimension of the boundary space of the complete r-graph K_n^r. Thus it suffices to consider the case when G has a non-trivial cycle space. Next they show that all minimal cycles in G are copies of K_{r+1}^r, and that no edge can overlap a K_{r+1}^r in precisely $r-1$ points. Thus each cycle substantially reduces the dimension of the boundary space for the acyclic part of G, and (omitting some substantial details) the result follows after some rank computations.

We mention one final application of algebraic methods with a different flavour. Suppose G is a graph on $[n]$. Assign variables $x_1, \cdots, x_n$ to the vertices and consider the polynomial $f_G(x) = \prod_{ij \in E(G)} (x_i - x_j)$. Thus f_G vanishes iff $x_i = x_j$ for some edge ij. Note that G has independence number $\alpha(G) < k$ iff f_G belongs to the ideal I of polynomials in $\mathbf{Z}[x]$ that vanish on any assignment x with at least k equal variables. Li and Li [121] showed that I is generated by the polynomials $f_G(x)$ for graphs G that are a disjoint union of $k-1$ cliques, and moreover the sizes of the cliques may be taken as equal as possible. One can also show that the degree of any polynomial in I is at least the degree of the generators. Applying this to f_G for any G with $\alpha(G) < k$, the resulting lower bound on the number of edges gives a proof of Turán's theorem (in complementary form). It would be interesting to obtain similar generalisations for other Turán problems.

12   Probabilistic methods

While probabilistic methods are generally very powerful in Combinatorics, they seem to be less effective for Turán problems, perhaps because the extremal constructions tend to be quite orderly. Some exceptions to this are random constructions for the tetrahedron codegree problem (see Section 13.2) and the bipartite link problem (see Section 13.6). For certain bipartite Turán problems the best known constructions are random, although these are in cases where the upper bound is quite far from the lower bound, so it is by no means an indication that the best construction is random. Consider the Turán problem for the complete bipartite graph $K_{r,r}$. Kővári, Sós and Turán [105] obtained the upper bound $\mathrm{ex}(n, K_{r,r}) = O(n^{2-1/r})$. A simple probabilistic lower bound due to Erdős and Spencer [61] is obtained by taking the random graph $G_{n,p}$ and deleting an edge from each copy of $K_{r,r}$. Then the expected number of edges has order $\Theta(pn^2) - \Theta(p^{r^2}n^{2r})$, so choosing $p = \Theta(n^{-2/(r+1)})$ gives a lower bound of order $\Omega(n^{2-2/(r+1)})$. Recently, Bohman and Keevash [18] obtained a small improvement to $\Omega(n^{2-2/(r+1)}(\log n)^{1/(r^2-1)})$ from the analysis of the H-free process. However, there is still a polynomial gap between the bounds.

Lu and Székely [128] applied the Lovász Local Lemma to Turán problems (among others). The general framework is as follows. Suppose that $A_1, \cdots, A_n$ are 'bad' events. A graph G on $[n]$ is a *negative dependency* graph for the events if $\mathbb{P}(A_i \mid \cap_{j \in S} \overline{A_j}) \leq \mathbb{P}(A_i)$ for any i and S such that there are no edges from i to S and $\mathbb{P}(\cap_{j \in S} \overline{A_j}) > 0$. The general form of the local lemma states that if there are $x_1, \cdots, x_n \in [0,1)$ such that $\mathbb{P}(A_i) \leq x_i \prod_{j:ij \in E(G)} (1-x_j)$ for all i then $\mathbb{P}(\cap_{i=1}^n \overline{A_i}) \geq \prod_{i=1}^n (1-x_i) > 0$, i.e. there is a positive probability that none of the bad events occur. This is applied to give the following packing result for hypergraphs. Suppose that H_1 and H_2 are r-graphs such that H_i has m_i edges and every edge of H_i intersects at most d_i other edges of H_i, for $i = 1, 2$. Suppose that $n \geq \max\{v(H_1), v(H_2)\}$ and

$(d_1 + 1)m_2 + (d_2 + 1)m_1 \leq \frac{1}{e}\binom{n}{r}$, where e is the base of natural logarithms. Then there are edge-disjoint embeddings of H_1 and H_2 on the same set of n vertices. This result is in turn used to deduce several results, including the following Turán bound. Suppose that F is an r-graph such that every edge intersects at most d other edges. Then $\pi(F) \leq 1 - \frac{1}{e(d+1)}$. This may be compared with the result of Sidorenko mentioned above (Section 6) that bounds $\pi(F)$ in terms of the number of edges in F. In many cases the bound in terms of d is an improvement, but the appearance of e makes it seem very unlikely that it is ever tight!

13 Further topics

This section gives a brief taste of a few areas of research closely related to the Turán problem. It is necessarily incomplete, both in the selection of topics and in the choice of references for each topic. The topics by subsection are 13.1: Jumps, 13.2: Minimum degree problems, 13.3: Different host graphs, 13.4: Coloured Turán problems, 13.5: The speed of properties, 13.6: Local sparsity, 13.7: Counting subgraphs.

13.1 Jumps

Informally speaking, 'jumps' refer to the phenomenon that r-graphs of a certain density are often forced to have large subgraphs with a larger density. For example, the Erdős-Stone theorem implies that a large graph of density bigger than $1 - 1/t$ contains blowups $K_{t+1}(m)$ of K_{t+1}, so has large subgraphs of density more than $1 - 1/(t + 1)$. Another example is the result of Erdős mentioned earlier (Section 2) that a large r-graph of positive density contains complete r-partite r-graphs $K_r^r(m)$, so has large subgraphs of density more than $r!/r^r$. Formally, the density d is a *jump* for r-graphs if there is some $c > 0$ such that for any $\epsilon > 0$ and $m \geq r$ there is n_0 sufficiently large such that any r-graph on n vertices with density at least $d + \epsilon$ has a subgraph on m vertices with density at least $d + c$. For example, every $d \in [0, 1)$ is a jump for graphs, and every $d \in [0, r!/r^r)$ is a jump for r-graphs. Deciding whether $r!/r^r$ is a jump for r-graphs is a long-standing open problem of Erdős [53]. In fact, Erdős made the stronger conjecture that every $d \in [0, 1)$ is a jump for r-graphs, but this was disproved by Frankl and Rödl [72]. The distribution of jumps and non-jumps is not at all understood, and very few specific examples are known. Further examples of non-jumps are given by Frankl, Peng, Rödl and Talbot [71] and Peng, e.g. [150]. On the positive side, Baber and Talbot [9] recently applied flag algebras to show that every $d \in [0.2299, 0.2316)$ is a jump for 3-graphs.

We give a brief sketch of the Frankl-Rödl method, as applied in [71] to prove that $5/9$ is not a jump for 3-graphs. One uses the following reformulation: d is a jump for r-graphs if and only if there is a finite family $\mathcal{F}$ of r-graphs with Turán density $\pi(\mathcal{F}) \leq d$ and blowup density $b(F) > d$ for all $F \in \mathcal{F}$. Suppose for a contradiction that $5/9$ is a jump for 3-graphs. Choose $\mathcal{F}$ with $\pi(\mathcal{F}) \leq 5/9$ and $b(F) > 5/9$ for all $F \in \mathcal{F}$. Let t be large and G be the Turán construction with parts of size t. Now the idea is to add $O(t^2)$ random edges inside each part, obtaining G^* such that $b(G^*) > 5/9$, but $b(H) \leq 5/9$ for any small subgraph H of G^* (here we are omitting a lot of the proof). Since $b(G^*) > 5/9$ and $\pi(\mathcal{F}) \leq 5/9$, a sufficiently large blowup

$G^*(m)$ must contain some $F \in \mathcal{F}$. We can write $F \subseteq H(m)$ for some small subgraph H of G^*. But then $b(F) \leq b(H(m)) = b(H) \leq 5/9$ contradicts the choice of $\mathcal{F}$, so $5/9$ is not a jump for 3-graphs.

13.2 Minimum degree problems

Turán problems concern the maximum number of edges in an F-free r-graph, but it is also natural to ask about the maximum possible minimum degree. More precisely, there is a minimum s-degree parameter $\delta_s(G)$ for each $0 \leq s \leq r-1$, defined as the minimum over all sets S of s vertices of the number of edges containing S. Then we can define a generalised Turán number $\mathrm{ex}_s(n, F)$ as the largest value of $\delta_s(G)$ attained by an F-free r-graph G on n vertices. Note that $\delta_0(G) = e(G)$, so $\mathrm{ex}_0(n, F) = \mathrm{ex}(n, F)$ is the usual Turán number. We can also define generalised Turán densities $\pi_s(F) = \lim_{s\to\infty} \mathrm{ex}_s(n, F)\binom{n-s}{r-s}^{-1}$ (it is non-trivial to show that the limit exists). A simple averaging argument shows that $\pi_i(F) \geq \pi_j(F)$ when $i \leq j$. The vertex deletion method in Proposition 4.2 shows that $\pi_1(F) = \pi_0(F) = \pi(F)$, so minimum 1-degree problems are not essentially different to Turán problems. However, in general we obtain a rich source of new problems, and it is not apparent how they relate to each other. The case $s = r-1$ was introduced by Mubayi and Zhao [146] under the name of *codegree density*. Their main result is that for $r \geq 3$ there are no jumps for codegree problems. In particular, the set of codegree densities is dense in $[0, 1)$. Moreover, they conjecture that any $d \in [0, 1)$ is the codegree density of some family.

As for Turán problems, there are few known results for codegree problems, even asymptotically. The tetrahedron K_4^3 is again one of the first interesting examples. Here the asymptotically best known construction is to take a random tournament on $[n]$ and say that a triple ijk with $i < j < k$ is an edge if i has one edge coming in and one edge coming out. This shows that the codegree density of the tetrahedron is at least $1/2$. For an upper bound, nothing better is known than the bounds for the usual Turán density, which are also upper bounds on the codegree density by averaging. One known result is for the Fano plane, where Mubayi [137] showed that the codegree density is $1/2$. Turán and codegree problems for other projective geometries were considered in [103, 113, 104]. An exact codegree result for the Fano plane was obtained by Keevash [104]: if G is a Fano-free 3-graph on n vertices, where n is large, and $\delta_2(G) \geq n/2$, then n is even and G is a balanced complete bipartite 3-graph. The argument used a new 'quasirandom counting lemma' for regularity theory, which extends the usual counting lemma by not only counting copies of a particular subgraph, but also showing that these copies are evenly distributed. Even for graphs, this quasirandom counting lemma has consequences that are not immediately obvious; for example, given a tripartite graph G in which each bipartite graph is dense and ϵ-regular (for some small ϵ), for any choice of dense graphs H_1, H_2, H_3 inside the parts V_1, V_2, V_3 of G, there are many copies of $K_3(2)$ in G in which the pairs inside each part are edges of the graphs H_1, H_2, H_3. Results of this type are potentially very powerful in assembling hypergraphs from smaller pieces.

Minimum degree conditions also lead to the study of spanning configurations. Here we look for conditions on a hypergraph G on n vertices that guarantee a particular subgraph F that also has n vertices. The prototype is Dirac's theorem

[44] that every graph on $n \geq 3$ vertices with minimum degree at least $n/2$ contains a Hamilton cycle. Other classical minimum degree result for graphs is the Hajnal-Szemerédi theorem [91] that minimum degree $(1 - 1/t)n$ gives a perfect packing by copies of K_t (when t divides n). A generalisation by Kómlos, Sarközy and Szemerédi [115] states that the same minimum degree even gives the $(t-1)$th power of a Hamilton cycle (when n is large). Another generalisation by Kühn and Osthus [120] determines the threshold for packing an arbitrary graph H up to an additive constant (the precise statement is technical). An example result for hypergraphs is a theorem of Rödl, Ruciński and Szemerédi [160] that any r-graph on n vertices with minimum codegree $(1 + o(1))n/2$ has a 'tight' Hamilton cycle, i.e. a cyclic ordering of the vertices such that every consecutive r-set is an edge. We refer the reader to the surveys [119] for graphs and [161] for hypergraphs.

13.3   Different host graphs

A range of new problems open up when we consider additional properties for Turán problems, besides that of not containing some forbidden r-graph. For any host r-graph H and fixed r-graph F, we may define $ex(H, F)$ as the maximum number of edges in an F-free subgraph of H. Thus the usual Turán number $ex(n, F) = ex(K_n^r, F)$ is the case when H is complete. We stick to graphs $(r = 2)$ for simplicity.

In principle one can consider any graph H, but some host graphs seem particular natural. An important case is when H is given by a random model, e.g. the Erdős-Rényi random graph $G_{n,p}$. This is motivated by considerations of *resilience* of properties. Here we consider some property of $G_{n,p}$ (i.e. a property that holds with high probability) and ask how resilient it is when some edges are deleted. For example, if p is not 'too small', $G_{n,p}$ not only has a triangle, but one even needs to delete asymptotically half of the edges to destroy all triangles. Equivalently, any triangle-free subgraph of $G_{n,p}$ has asymptotically at most half of its edges. This is tight, as any graph has a bipartite subgraph that contains at least half of its edges. To clarify what 'too small' means, note that if the number of triangles is much smaller than the number of edges then such a result will not hold, as one can delete one edge from each triangle with negligible effect. This suggests $p = n^{-1/2}$ as a threshold for the problem, which is indeed the case (this follows from a result of Frankl and Rödl [74]). There is a large literature on generalising this result, which we do not have space to go into here. A comprehensive generalisation to many extremal problems was recently given independently by Schacht [169] and Conlon and Gowers [36]. Among these results is Turán's theorem for random graphs, that when p is not too small the largest K_{t+1}-free subgraph of $G_{n,p}$ has asymptotically $1 - 1/t$ of its edges; again the threshold for p is the value for which the number of K_t's is comparable with the number of edges. Similar results apply for hypergraph Turán problems, and to certain extremal problems from number theory, such as Szemerédi's theorem on arithmetic progressions. Another direction of research is that started by Sudakov and Vu [185] on *local resilience*. Here the question is how many edges one needs to delete from each vertex to destroy a certain property of $G_{n,p}$. This is a better question for global properties such as Hamiltonicity, which one can destroy by deleting all edges at one vertex: this is not a significant global change, but a huge local change.

The above only concerns the case when the host graph H is random. Mubayi and Talbot [144] consider Turán problems with colouring conditions, which could also be viewed from the perspective of a constrained host graph. Say that an r-graph G is *strongly t-colourable* if there is a t-colouring of its vertices such that no edge has more than one vertex of the same colour. (They call this 't-partite', but we use this term differently in this paper; our use of 't-partite' is equivalent to their use of 't-colourable'.) Their main result (in our language) is that the asymptotic maximum density of an F-free r-graph on n vertices that is strongly t-colourable is equal to the maximum blowup density $b(G)$ over all hom-F-free r-graphs G on t vertices. For example, the maximum density in a strongly 4-colourable K_4^3-free 3-graph is 8/27; this is achieved by a construction with 4 parts of sizes $n/3, 2n/9, 2n/9, 2n/9$, with edges equal to all triples with one vertex in the large part and the other two vertices in two different smaller parts. Chromatic Turán problems were considered earlier by Talbot [186] as a tool for obtaining bounds on Turán density of the 3-graph on 4 vertices with 3 edges. (These chromatic bounds were subsequently improved by Markström and Talbot [132].) Here the problem is to estimate the maximum density of an F-free r-graph on n vertices that is t-partite. Mubayi and Talbot solved this problem for the extended complete graph, in the sense that they have a procedure for computing the maximum density, which is in principle finite, although not practical except in small cases. They conjecture that the natural example is optimal, but can only prove this for $r = 2$ and $r = 3$. One example of their result shows that chromatic Turán densities can be irrational: the maximum density of a bipartite K_6^3-free 3-graph is $(13\sqrt{13} - 35)/27 \approx 0.4397$, achieved by blowing up $K_5^3 - e$.

Another case which has received a lot of attention is when $H = Q_n$ is the graph of the n-cube, i.e. $V(H)$ consists of all subsets of $[n]$ and edges join sets that differ in precisely one element. Erdős [55] posed the problem of determining the maximum proportion of edges in a C_4-free subgraph of Q_n. Noting that any consecutive levels of the cube span a C_4-free subgraphs, a lower bound of 1/2 is obtained by taking the union of the subgraphs spanned by levels $2i$ and $2i + 1$ for $0 \le i < n/2$. The best known upper bound is approximately 0.6226, due to Thomason and Wagner [189]. We will not attempt to survey the literature on these problems, but refer the reader to Conlon [34] for a simpler proof of many of the known results and several references. We draw the reader's attention to the problem of deciding whether $ex(Q_n, C_{10})$ has the same order of magnitude as $e(Q_n)$; this is the only unsolved instance of this problem for cycles (the answer is 'yes' for C_4 and C_6; 'no' for C_8 and longer cycles).

We also remark that even 'vertex Turán problems' in the cube seem to be hard. For example, what is the smallest constant a_d such that there is a set of $\sim a_d 2^n$ vertices in the n-cube that hits every subcube of dimension d? This problem was introduced by Alon, Krech and Szabó [5], who showed $\frac{\log d}{2^{d+2}} \le a_d \le \frac{1}{d+1}$; there is a surprisingly large gap between the upper and lower bounds! A variant on this problem introduced by Johnson and Talbot [96] is to find a particular subset $F \subseteq V(Q_d)$: what is the large constant λ_F such that there exists $S \subseteq V(Q_n)$ of size $|S| \sim \lambda_F 2^n$ such that there is no subgraph embedding $i : Q_d \to Q_n$ with $i(F) \subseteq S$? In particular, they conjecture that $\lambda_F = 0$ for $|F| \le \binom{d}{d/2}$ (it is not hard to see that this can be false for larger F). Bukh (personal communication) observed that this conjecture is equivalent to the following hypergraph Turán problem. For $r \ge s > t$ we define the following r-graph $S^r K_s^t$, which may be regarded as a 'suspension' of K_s^t. The vertex set of $S^r K_s^t$ is the union of disjoint sets S of size s and A of size $r - t$. The edges consist of all r-tuples containing A. The conjecture is that $\lim_{r\to\infty} \pi(S^r K_s^t) = 0$ for any s, t. Even the case $s = 4$ and $t = 2$ is currently open! The case $s = 3$ and $t = 2$ is straightforward: $S^r K_3^2$ just consists of (any) 3 edges on a set of $r + 1$ vertices, so $\pi(S^r K_3^2) \leq 2/(r + 1)$. However, it is an interesting problem to determine the order of magnitude of $\pi(S^r K_3^2)$ for large r: Alon (communication via Bukh) gave a lower bound of order $(\log r)/r^2$. In general, given the apparent difficulty of determining Turán densities exactly, it seems that such problems involving additional limits may be a fruitful avenue for developing the theory.

13.4   Coloured Turán problems

There are a variety of generalisations of the Turán problem that allow additional structures, such as directed edges, multiple edges, or coloured edges. Even for graphs this leads to rich theories and several unsolved problems. Brown, Erdős and Simonovits initiated this field with a series of papers on problems for digraphs and multigraphs. For multigraph problems, we fix some positive integer q and consider multigraphs with no loops and edge multiplicity at most q. Then given a family $\mathcal{F}$ of multigraphs, we want to determine $\mathrm{ex}(n, \mathcal{F})$, defined as the maximum number of edges in a multigraph not containing any F in $\mathcal{F}$. A further generalisation allows directions on the edges; for simplicity we ignore this here. In the case $q = 1$ this is the usual Turán problem. For $q = 2$ (or digraphs), it is shown in [27] that any extremal problem has a blowup construction that is asymptotically optimal. Here a *blowup* is defined by taking some symmetric $t \times t$ matrix A whose entries are integers between 0 and q, dividing a vertex set into t parts, and putting a_{ij} edges between any pair of vertices u, v with u in part i and v in part j (we may have $i = j$).

Similarly to the usual Turán problem, one can define the *blowup density* $b(A)$ which is the density achieved by this construction: formally $b(A)$ is the maximum value of $x^t A x$ over the standard simplex S of all $x = (x_1, \cdots, x_n)$ with $x_i \geq 0$ for $1 \leq i \leq n$ and $\sum x_i = 1$. Say that such a matrix is *dense* if any proper principal submatrix has lower blowup density. It is shown in [28] that for any dense matrix A there is a finite family $\mathcal{F}$ such that A is the unique matrix whose blowup gives asymptotically optimal constructions of $\mathcal{F}$-free multigraphs. Furthermore, for $q = 2$ (or digraphs), in [29] they describe an algorithm that determines all optimal matrices for a given family (the algorithm is finite, but not practical). Simpler proofs of these results were given by Sidorenko [178], who also showed that analogous statements do not hold for $q > 2$, thus disproving a conjecture of Brown, Erdős and Simonovits. Brown, Erdős and Simonovits also conjectured that all densities are jumps (as for graphs), but this was disproved by Rödl and Sidorenko [164] for $q \geq 4$. The conjecture is true for $q = 2$, but is open for $q = 3$.

Another coloured variant on many problems of extremal set theory, including Turán problems, was introduced by Hilton [93] and later by Keevash, Saks, Sudakov and Verstraëte [109]. Given a list of set systems, which we think of as colours, we call another set system *multicoloured* if for each of its sets we can choose one of the colours it belongs to in such a way that each set gets a different colour. Given an integer k and some forbidden configurations, the multicoloured extremal problem is to choose k colours with total size as large as possible subject to containing no multicoloured forbidden configuration. Let f be the number of sets in the forbidden configuration. One possible extremal construction for this problem is to take $f - 1$ colours to consist of all possible sets, and the other colours to be empty. Another construction is to take all k colours to be equal to some fixed family that is of maximum size subject to not containing a forbidden configuration. In [109] we solved the multicolour version of Turán's theorem, by showing that one of these two constructions is always optimal. In other words, if $G_1, \cdots, G_k$ are graphs on the same set of n vertices for which there is no multicoloured K_t, then $\sum_{i=1}^{k} e(G_i)$ is maximised either by taking $\binom{t}{2} - 1$ complete graphs and the rest empty graphs, or by taking all k graphs equal to some fixed Turán graph $T_{t-1}(n)$. Simple calculations show that there is a threshold value k_c so that the first option holds for $k < k_c$ and the second option holds for $k \geq k_c$. This proved a conjecture of Hilton [93] (although we were not aware of this paper at the time of writing). It would be interesting to understand which other extremal problems exhibit this phenomenon of having only two extremal constructions. It is not universal, as shown by an example in [109], but it does hold for several other classical problems of extremal set theory, as shown in [21] and [112].

A related problem posed by Diwan and Mubayi (unpublished) concerns the minimum size of a colour, rather than the total size of colour. Specifically, for any n and a fixed graph F with edges coloured red or blue, they ask for the threshold m such that, given any red graph and blue graph on the same set of n vertices each with more than m edges, one can find a copy of F with the specified colouring. They pose a conjecture when F is a coloured clique, and prove certain cases of their conjecture. Their proof uses a stronger result which replaces the minimum size of a colour by a weighted linear combination of the colours. Such problems have been recently studied in a much more general context by Marchant and Thomason [130], who gave applications to the probability of hereditary graph properties (see Section 13.5).

We conclude this subsection with another coloured generalisation studied by Keevash, Mubayi, Sudakov and Verstraëte [108]. For a fixed graph H, we ask for the maximum number of edges in a properly edge-coloured graph on n vertices which does not contain a rainbow H, i.e. a copy of H all of whose edges have different colours. This maximum is denoted $\text{ex}^*(n, H)$, and we refer to it as the *rainbow Turán number* of H. For any non-bipartite graph H we showed that $\text{ex}^*(n, H) \sim \text{ex}(n, H)$, and for large n we have $\text{ex}^*(n, H) = \text{ex}(n, H)$ when H is critical (e.g. a clique or an odd cycle). Bipartite graphs H are a source of many open problems. The case when $H = C_{2k}$ is an even cycle is particularly interesting because of its connection to additive number theory. We conjecture that $\text{ex}^*(n, C_{2k}) = O(n^{1+1/k})$, which would generalise a result of Ruzsa on B_k^*-sets in abelian groups. (We proved it for $k = 2$ and $k = 3$.) More generally, there is considerable scope to investigate number theoretic consequences of extremal results on coloured graphs, as applied to Cayley graphs.

13.5 The speed of properties

Suppose $\mathcal{P}$ is a graph property, i.e. a set of graphs that is closed under isomorphism. We consider properties $\mathcal{P}$ that are *hereditary*, meaning that they are closed under taking induced subgraphs, or even *monotone*, meaning that they are closed under taking arbitrary subgraphs. A monotone property can be characterised as the set of $\mathcal{F}$-free graphs, for some (possibly infinite) family $\mathcal{F}$. Similarly, a hereditary property can be characterised as the set of *induced-$\mathcal{F}$-free* graphs, for some $\mathcal{F}$, i.e. graphs with no copy of any F in $\mathcal{F}$ as an induced subgraph. The *speed* $s(n)$ of $\mathcal{P}$ is the number of labelled graphs in $\mathcal{P}$ on $[n]$. There is a large literature on the speed of properties, too large to adequately cite here, so we refer the reader to [3] as a recent paper with many references.

Consider the problem of counting F-free graphs on $[n]$, for some fixed graph F. By taking all subgraphs of any fixed F-free graph of maximum size $\mathrm{ex}(n, F)$ we obtain at least $2^{\mathrm{ex}(n,F)}$ distinct F-free graphs. In fact, this is essentially tight for non-bipartite graphs F, as Erdős, Frankl and Rödl [56] showed an upper bound of $2^{(1+o(1))\mathrm{ex}(n,F)}$. (The case when F is bipartite is another story, see [15] for some recent results.) The corresponding generalisation to hereditary properties was proved by Alekseev [1] and by Bollobás and Thomason [22]. They showed that the speed of $\mathcal{P}$ is $2^{(1-1/r+o(1))n^2/2}$, where r is a certain parameter of $\mathcal{P}$ known as the 'colouring number' (informally, it is the maximum number of parts in a partite construction for graphs in $\mathcal{P}$, where each part is complete or empty, and the graph is otherwise arbitrary). These results have been refined to give more precise error terms and even a description of the structure of almost all graphs in a hereditary property. For monotone properties the results are due to Balogh, Bollobás and Simonovits [11, 12], and for hereditary properties to Alon, Balogh, Bollobás and Morris [3]. Bollobás and Thomason [23] studied a generalisation in which a property is measured by its probability of occurring in the random graph $G(n, p)$ (thus the speed corresponds to $p = 1/2$). This generalised problem exhibits extra complexities, analysed by Marchant and Thomason [130] (see Section 13.4).

It is natural to pose the same questions for hypergraph properties. Dotson and Nagle [45] and Ishigami [95] showed that the speed of a hereditary r-graph property $\mathcal{P}$ is $2^{\mathrm{ex}(n,\mathcal{P})+o(n^r)}$. Here $\mathrm{ex}(n, \mathcal{P})$ is the maximum size of an r-graph G on $[n]$ on n vertices such that there exists an r-graph H on $[n]$ that is edge-disjoint from G such that $H \cup G' \in \mathcal{P}$ for every subgraph G' of G. (In the case when $\mathcal{P}$ is monotone this is the usual Turán number, i.e. the maximum size of an r-graph in $\mathcal{P}$.) In principle this is analogous to the Alexeev-Bollobás-Thomason result, but we lack a concrete description of $\mathrm{ex}(n, \mathcal{P})$ analogous to the colouring number (even in the monotone case, which is the point of this survey!) In the case of the Fano plane a refined result was obtained by Person and Schacht [151], who showed that almost every Fano-free 3-graph on n vertices is bipartite. One might expect similar results to hold for other Turán problems where we know uniqueness and stability of the extremal construction. This has been established by Balogh and Mubayi [13, 14] for cancellative 3-graphs and 3-graphs with independent neighbourhoods.

13.6 Local sparsity

Brown, Erdős and Sós [30] generalised the hypergraph Turán problem by asking
for the maximum number of edges in an r-graph satisfying a 'local sparsity' condition
that bounds the number of edges in any set of a given size. Write $\mathrm{ex}^r(n, v, e)$ for
the maximum number of edges in an r-graph on n vertices such that no set of v
vertices spans at least e edges. For example $\mathrm{ex}^3(n, 4, 4) = \mathrm{ex}(n, K_4^3)$. A result of
[30] is $\mathrm{ex}^r(n, e(r - k) + k, e) = \Theta(n^k)$ for any $1 \leq k \leq r$; the upper bound follows by
noting that any k-set belongs to at most $e - 1$ edges, and the lower bound by taking
a random r-graph of small constant density and deleting all edges in $e(r - k) + k$-sets
with at least e edges. They described the case $r = 3$, $v = 6$, $e = 3$ as 'the most
interesting question we were unable to answer'. This was addressed by the celebrated
'(6,3)-theorem' of Ruzsa and Szémeredi [165] that $n^{2-o(1)} < \mathrm{ex}^3(n, 6, 3) < o(n^2)$. It
would be very interesting to tighten these bounds: this is connected with regularity
theory (see Section 10) and bounds for Roth's theorem (see [90, 166].) Further
results on the general problem are $n^{2-o(1)} < \mathrm{ex}^r(n, 3(r - 1), 3) < o(n^2)$ in [56],
$\mathrm{ex}^r(n, e(r - k) + k + \lfloor\log_2(e)\rfloor, e) < o(n^k)$ in [167], $\mathrm{ex}^r(n, 4(r - k) + k + 1, 4) < o(n^k)$
for $k \geq 3$ in [168], and $n^{k-o(1)} < \mathrm{ex}^r(n, 3(r - k) + k + 1, 3) < o(n^k)$ in [6]. An
interesting open problem is to determine whether $\mathrm{ex}^3(n, 7, 4)$ is $o(n^2)$.

A weighted generalisation of this problem is to determine the largest total weight
$\mathrm{ex}_{\mathbf{Z}}(n, k, r)$ that can be obtained by assigning integer weights to the edges of a graph
on n vertices such that any set of k vertices spans a subgraph of weight at most r.
(We stick to graphs for simplicity.) Note that negative weights are allowed, but for
comparison with multigraph problems one can also consider the analogous quantity
$\mathrm{ex}_{\mathbf{N}}(n, k, r)$ in which weights have to be non-negative. We remarked earlier that
the example $\mathrm{ex}_{\mathbf{N}}(n, 4, 20) \sim 3\binom{n}{2}$ was crucial in determining the Turán density of
the Fano plane. In general, Füredi and Kündgen [80] have determined $\mathrm{ex}_{\mathbf{Z}}(n, k, r)$
asymptotically for all k and r, but there remain several interesting open problems,
such as determining exact values and extremal constructions, and obtaining similar
results for $\mathrm{ex}_{\mathbf{N}}(n, k, r)$.

Another generalisation is to specify exactly what numbers of edges are allowed
in any set of a given size. In Section 7 we discussed the problem for 3-graphs in
which every 4-set spans 0, 2 or 3 edges. In Section 6 we mentioned the lower bound
$\pi(F) \geq 2/7$ given by Frankl and Füredi [67] when F is the 3-graph with 4 vertices
and 3 edges. The main result of [67] is an exact result for 3-graphs in any 4-set spans
0 or 2 edges. In fact they classify all such 3-graphs: they are either obtained by
(i) blowing up the 3-graph on 6 vertices described in Section 6, or (ii) by placing n
points on a circle and taking the edges as all triples that form a triangle containing
the centre (assume that the centre is not on the line joining any pair). It is easy to
check that the blowup construction (i) is larger for $n \geq 6$. A related problem is a
conjecture of Erdős and Sós [60] that any 3-graph with bipartite links has density
at most $1/4$. Construction (ii) is an example that would be tight for this conjecture.
Another example is to take a random tournament and take the edges to be all
triples that induce cyclic triangles. In Section 9 we mentioned the improvements
on $\pi(K_{r+1}^r)$ given by Lu and Zhao [129]. These were based on a structural result
for r-graphs in which every $(r + 1)$-set contains 0 or r edges, answering a question
of de Caen [41]: if $r = 2$ then G is a complete bipartite graph, and if $r \geq 3$ and

Hypergraph Turán problems

125

$n > r(p-1)$, where p is the smallest prime factor of $r-1$, then G is either the empty graph or a star (all r-sets containing some fixed vertex).

13.7 Counting subgraphs

A further generalisation of the Turán problem is to look not only for the threshold at which a particular r-graph F appears, but how many copies of F are guaranteed by a given number of edges. Even the most basic case counting triangles in graphs is a difficult problem that was open for many years. The following asymptotic solution was recently given by Razborov [157] using flag algebras. Among graphs on n vertices with edge density between $1 - 1/t$ and $1 - 1/(t+1)$, the asymptotic minimum number of triangles is achieved by a complete $(t+1)$-partite graph in which t parts are of equal size and larger than the remaining part. (One can give an explicit formula in terms of the edge density, but the resulting expression is rather unwieldy.) It is conjectured that the same example minimises the number of copies of K_s for any s. Nikiforov [147] established this for $s = 4$, and also re-proved Razborov's result for $s = 3$ by different means. In general, Bollobás (see [19, Chapter 6]) showed a lower bound that is equal to the conjecture for densities of the form $1 - 1/t$, and a linear function on each interval $[1 - 1/t, 1 - 1/(t+1)]$. Very recently, Reiher announced an asymptotic solution to the full conjecture.

The problem takes on a different flavour when one considers graphs where the number of edges exceeds the Turán number, but is asymptotically the same. For example, Rademacher (unpublished) extended Mantel's result by showing that a graph on n vertices with $n^2/4 + 1$ edges contains at least $\lfloor n/2 \rfloor$ triangles (which is tight). This was extended by Erdős [49] and then by Lovász and Simonovits [125], who showed that if $q < n/2$ then $n^2/4 + q$ edges guarantee at least $q\lfloor n/2 \rfloor$ triangles. Mubayi [139, 140] has extended these results in several directions. For a critical graph F he showed that there is $\delta > 0$ such that if n is large and $1 \le q < \delta n$ then any graph on n vertices with $\mathrm{ex}(n, F) + q$ edges contains at least $qc(n, F)$ copies of F. Here $c(n, F)$ is the minimum number of copies of F created by adding a single edge to the Turán graph, which is easy to compute for any particular example, although a general formula is complicated. The bound is sharp up to an error of $O(qc(n, F)/n)$. For hypergraphs he obtains similar results in many cases where uniqueness and stability of the extremal example is known.

For bipartite graphs F there is an old conjecture of Sidorenko [177] that random graphs achieve the minimum number of copies of F. A precise formulation may be given in terms of homomorphisms. Recall that the homomorphism density $t_F(G)$ is the probability that a random map from $V(F)$ to $V(G)$ is a homomorphism. Then the conjecture is that $t_F(G) \ge d(G)^{e(F)}$, where $d(G) = t_e(G)$ is the edge density of G. This may be viewed as a correlation inequality for the events that edges of F are embedded as edges of G. It also has an equivalent analytic formulation as $t_F(W) \ge t_e(W)^{e(F)}$ for any graphon W, which is worth noting as integrals similar to $t_F(W)$ appear in other contexts (e.g. Feynmann integrals in quantum field theory). Sidorenko was a pioneer of the analytic approach, and surveyed many of his results in [179]. Recent partial results on the Sidorenko conjecture include a local form by Lovász [124] and an approximate form by Conlon, Fox and Sudakov [35]. Note that examples in [177] show that the natural hypergraph generalisation of the conjecture

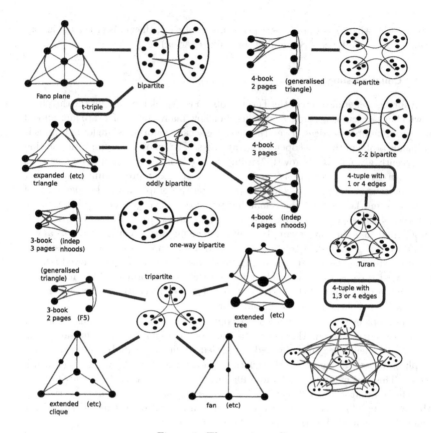

Figure 1: The exact results

is false.

In the other direction, one may ask to maximise the number of copies of a fixed r-graph F in an r-graph G, given the number of edges and vertices in G. We start with the case when $F = K_t^r$ is a clique. This turns out not to depend on the number of vertices in G. For example, when $e(G) = \binom{m}{r}$ the extremal example is K_m^r, which has $\binom{m}{t}$ copies of K_t^r. In general the extremal example is determined by the Kruskal-Katona theorem [118, 100]. Results for general graphs were obtained by Alon [2] and for hypergraphs by Friedgut and Kahn [76]. Here we do not expect precise answers, but just seek the order of magnitude. The result of [76] is that the maximum number of copies of an r-graph F in an r-graph G with e edges has order $e^{\alpha^*(F)}$, where $\alpha^*(F)$ is the fractional independence number of F.

Going back to cliques in graphs, Sós and Straus [184] proved the following (generalisation of a) conjecture of Erdős [48]. Suppose G is a graph and let N_t denote the number of K_t's in G. If $N_{k+1} = 0$ (i.e. G is K_{k+1}-free) and $t \geq 0$ then $N_{t+1} \leq \binom{k}{t+1}\binom{k}{t}^{-(t+1)/t} N_t^{(t+1)/t}$. Note that equality holds if G is a blowup of K_k. Repeated application gives a bound for the number of K_t's in a K_{k+1}-free graph in terms of the number of edges N_2: we have $N_t \leq \binom{k}{t}\binom{k}{2}^{-t/2} N_2^{t/2}$. The proof uses a far-reaching generalisation of the Lagrangian method considered in Section 3. In fact, it is hard to appreciate the scope of the method in the generality presented in [184], and it may well have applications in other contexts yet to be discovered. The idea is to assign a variable x_T to each K_t in G and consider the polynomial $f_G(x) = \sum_S \prod_{T \subseteq S} x_T$ in the variables $x = (x_T)$, where the sum is over all K_{t+1}'s S in G. Let λ be the maximum value of $f_G(x)$ over all x with every $x_T \geq 0$ and $\sum_T x_T^t = 1$ (note the power). A general transfer lemma in [184] implies that a maximising x can be chosen with the property that the vertices incident to variables of positive weight induce a complete subgraph. This implies that x is supported on the K_t's contained in some clique, which has size at most k, since $N_{k+1} = 0$. The maximum is achieved with equal weights $\binom{k}{t}^{-1/t}$, which gives $\lambda = \binom{k}{t+1}\binom{k}{t}^{-(t+1)/t}$. On the other hand, setting every x_T equal to $N_t^{-1/t}$ is a valid assignment, and gives a lower bound $\lambda \geq N_{t+1} N_t^{-(t+1)/t}$, so the result follows.

14 Summary of results

This survey has been organised by methods, so for easy reference we summarise the results here. The exact results are illustrated in Figure 1 (some infinite families are indicated by a representative example). A list follows: F_5 [66] (generalising cancellative 3-graphs [20]), Fano plane [84, 110], expanded triangle [111], generalised 4-graph triangle = 4-book with 2 pages [152] (generalising cancellative 4-graphs [173]), 4-book with 3 pages [83], 3-graphs with independent neighbourhoods [82], 4-graphs with independent neighbourhoods = 4-book with 4 pages [81], extended complete graphs [155] (refining [138]), generalised fans [141], extended trees [174], 3-graph 4-sets with 1, 3 or 4 edges [67] 3-graph 4-sets with 1 or 4 edges [154] (refining [158]), 3-graph t-triples [192]. Besides these, there is an 'almost exact' result for generalised 5-graph and 6-graph triangles [70], and asymptotic results (i.e. exact Turán densities) for expanded cliques [176] and 5 3-graphs related to $F(3,3)$ [143]. Many further asymptotic results follow from Theorem 3.1.

References

[1] V.E. Alekseev, On the entropy values of hereditary classes of graphs, *Discrete Math. Appl.* **3** (1993), 191–199.

[2] N. Alon, On the number of subgraphs of prescribed type of graphs with a given number of edges, *Israel J. Math.* **38** (1981), 116–130.

[3] N. Alon, J. Balogh, B. Bollobás and R. Morris, The structure of almost all graphs in a hereditary property, arXiv:0905.1942, 2009.

[4] N. Alon, E. Fischer, M. Krivelevich, and M. Szegedy, Efficient testing of large graphs, *Combinatorica* **20** (2000), 451–476.

[5] N. Alon, A. Krech and T. Szabó, Turán's theorem in the hypercube, *SIAM J. Disc. Math.* **21** (2007), 66–72.

[6] N. Alon and A. Shapira, On an Extremal Hypergraph Problem of Brown, Erdős and Sós, *Combinatorica* **26** (2006), 627–645.

[7] B. Andrásfai, P. Erdős and V. T. Sós, On the connection between chromatic number, maximal clique and minimal degree of a graph. *Disc. Math.* **8** (1974), 205–218.

[8] T. Austin, On exchangeable random variables and the statistics of large graphs and hypergraphs, *Probability Surveys* **5** (2008), 80–145.

[9] R. Baber and J. Talbot, Hypergraphs do jump, to appear in *Combin. Probab. Comput.*

[10] J. Balogh, The Turán density of triple systems is not principal, *J. Combin. Theory Ser. A* **100** (2002), 176–180.

[11] J. Balogh, B. Bollobás and M. Simonovits, On the number of graphs without forbidden subgraph, *J. Combin. Theory Ser. B* **91** (2004), 1–24.

[12] J. Balogh, B. Bollobás and M. Simonovits, The typical structure of graphs without given excluded subgraphs, to appear in *Random Structures Algorithms*.

[13] J. Balogh and D. Mubayi, Almost all triple systems with independent neighborhoods are semi-partite, arXiv:1002.1925, 2010.

[14] J. Balogh and D. Mubayi, Almost all cancellative triple systems are tripartite, arXiv:0910.2941, 2009.

[15] J. Balogh and W. Samotij, The number of K_{st}-free graphs, to appear in *J. London Math. Soc.*

[16] A. Björner and G. Kalai, An extended Euler-Poincaré theorem, *Acta Math.* **161** (1988), 279–303.

[17] T. Bohman, A. Frieze, D. Mubayi and O. Pikhurko, Hypergraphs with independent neighborhoods, to appear in *Combinatorica*.

[18] T. Bohman and P. Keevash, The early evolution of the H-free process, *Invent. Math.* **181** (2010), 291–336.

[19] B. Bollobás, *Extremal graph theory*, Reprint of the 1978 original, Dover Publications, Inc., Mineola, NY, 2004.

[20] B. Bollobás, Three-graphs without two triples whose symmetric difference is contained in a third, *Disc. Math.* **8** (1974), 21–24.

[21] B. Bollobás, P. Keevash and B. Sudakov, Multicoloured extremal problems, *J. Combin. Theory Ser. A* **107** (2004), 295–312.

[22] B. Bollobás and A. Thomason, Projections of bodies and hereditary properties of hypergraphs, *Bull. London Math. Soc.* **27** (1995) 417–424.

[23] B. Bollobás and A. Thomason, The structure of hereditary properties and colourings of random graphs *Combinatorica* **20** (2000), 173–202.

[24] C. Borgs, J. Chayes, L. Lovász, V. T. Sós and K. Vesztergombi, Convergent sequences of dense graphs I: Subgraph frequences, metric properties and testing, *Adv. Math.* **219** (2008), 1801—1851.

[25] D. E. Boyer, D. L. Kreher, S. Radziszowski and A. Sidorenko, On $(n, 5, 3)$-Turán systems, *Ars Combin.* **37** (1994), 13–31.

[26] W. G. Brown, On an open problem of Paul Turán concerning 3-graphs, *Studies in Pure Mathematics*, 91–93, Basel-Boston: Birkhauser, 1983.

[27] W. G. Brown, P. Erdős and M. Simonovits, Extremal problems for directed graphs, *J. Combin. Theory Ser. B* **15** (1973), 77–93.

[28] W. G. Brown, P. Erdős and M. Simonovits, Inverse extremal digraph problems, Finite and infinite sets, Vol. I, II (Eger, 1981), 119–156, *Colloq. Math. Soc. János Bolyai* **37**, North-Holland, Amsterdam, 1984.

[29] W. G. Brown, P. Erdős and M. Simonovits, Algorithmic solution of extremal digraph problems, *Trans. Amer. Math. Soc.* **292** (1985), 421–449.

[30] W. G. Brown, P. Erdős and V. T. Sós, Some extremal problems on r-graphs, *New directions in the theory of graphs* (Proc. Third Ann Arbor Conf., Univ. Michigan, Ann Arbor, Mich., 1971), 53–63, Academic Press, New York, 1973.

[31] S. R. Buló and M. Pelillo, A generalization of the Motzkin-Straus theorem to hypergraphs, *Optim Lett* **3** (2009), 287–295.

[32] F. Chung and L. Lu, An upper bound for the Turán number $t_3(n, 4)$, *J. Combin. Theory Ser. A* **87** (1999), 381–389.

[33] V. Chvátal, An extremal set-intersection theorem, *J. London Math. Soc.* **9** (1974/75), 355–359.

[34] D. Conlon, An extremal theorem in the hypercube, arXiv:1005.0582, 2010.

[35] D. Conlon, J. Fox and B. Sudakov, An approximate version of Sidorenko's conjecture, to appear in *Geom. Funct. Anal.*

[36] D. Conlon and W. T. Gowers, Combinatorial theorems in sparse random sets, arXiv:1011.4310, 2011.

[37] R. Csákány and J. Kahn, A homological approach to two problems on finite sets, *J. Algebraic Combin.* **9** (1999), 141–149.

[38] D. de Caen, Extension of a theorem of Moon and Moser on complete subgraphs. *Ars Combin.* **16** (1983), 5–10.

[39] D. de Caen, Linear constraints related to Turán's problem, Proceedings of the fourteenth Southeastern conference on combinatorics, graph theory and computing (Boca Raton, Fla., 1983), *Congr. Numer.* **39** (1983), 291–303.

[40] D. de Caen, On upper bounds for 3-graphs without tetrahedra, *Congressus Numer.* **62** (1988), 193–202.

[41] D. de Caen, The current status of Turán's problem on hypergraphs, *Extremal Problems for Finite Sets*, Visegrád, 1991, Bolyai Soc. Math. Stud., Vol. 3, pp. 187–197, János Bolyai Math. Soc., Budapest, 1994.

[42] D. de Caen and Z. Füredi, The maximum size of 3-uniform hypergraphs not containing a Fano plane, *J. Combin. Theory Ser. B* **78** (2000), 274–276.

[43] D. de Caen, D.L. Kreher and J. Wiseman, On constructive upper bounds for the Turán numbers T(n,2r+1,2r), Nineteenth Southeastern Conference on Combinatorics, Graph Theory, and Computing (Baton Rouge, LA, 1988), Congr. Numer. 65 (1988) 277–280.

[44] G.A. Dirac, Some theorems on abstract graphs, *Proc. Lond. Math. Soc.* **2** (1952), 69–81.

[45] R. Dotson and B. Nagle, Hereditary properties of hypergraphs, *J. Combin. Theory Ser. B* **99** (2009), 460–473.

[46] G. Elek and B. Szegedy, Limits of Hypergraphs, Removal and Regularity Lemmas. A Non-standard Approach, arXiv:0705.2179, 2007.

[47] G. Elek and B. Szegedy, A measure-theoretic approach to the theory of dense hypergraphs, arXiv:0810.4062, 2008.

[48] P. Erdős, On the number of complete subgraphs contained in certain graphs, *Magyar Tud. Akad. Mat. Kutató Int. Közl.* **7** (1962), 459–464.

[49] P. Erdős, On a theorem of Rademacher-Turán, *Illinois J. Math* **6** (1962), 122–127.

[50] P. Erdős, On extremal problems of graphs and generalized graphs, Israel J. Math. 2 (1964) 183–190.

[51] P. Erdős, Extremal problems in graph theory, Theory of Graphs and its Applications (Proc. Sympos. Smolenice, 1963), pp. 29–36, Publ. House Czech. Acad. Sci., Prague, 1964.

[52] P. Erdős, Topics in combinatorial analysis, *Proc. Second Louisiana Conf. on Comb., Graph Theory and Computing*, 2–20, Louisiana State Univ., Baton Rouge 1971.

[53] P. Erdős, On some extremal problems on r-graphs, *Disc. Math.* 1 (1971), 1–6.

[54] P. Erdős, On the combinatorial problems which I would most like to see solved, *Combinatorica* 1 (1981), 25–42.

[55] P. Erdős, Some problems in graph theory, combinatorial analysis and combinatorial number theory, Graph Theory and Combinatorics, Academic Press (1984), 1–17.

[56] P. Erdős, P. Frankl and V. Rödl, The asymptotic number of graphs not containing a fixed subgraph and a problem for hypergraphs having no exponent, *Graphs Combin.* 2 (1986), 113–121.

[57] P. Erdős, H. Ko and R. Rado, Intersection theorems for systems of finite sets, *Quart. J. Math. Oxford Ser.* 12 (1961) 313–320.

[58] P. Erdős and M. Simonovits, A limit theorem in graph theory, *Studia Sci. Math. Hung. Acad.* 1, 51–57, (1966).

[59] P. Erdős and M. Simonovits, Supersaturated graphs and hypergraphs, *Combinatorica* 3 (1982), 181–192.

[60] P. Erdős and V. T. Sós, On Ramsey-Turán type theorems for hypergraphs, *Combinatorica* 2 (1982), 289–295.

[61] P. Erdős and J. H. Spencer, *Probabilistic Methods in Combinatorics*, Academic Press, 1974.

[62] P. Erdős and A.H. Stone, On the structure of linear graphs, Bull. Am. Math. Soc. 52 1087–1091, (1946).

[63] J. Faudree, R. Faudree and J. R. Schmitt, A survey of minimum saturated graphs and hypergraphs, preprint.

[64] D. G. Fon-Der-Flaass, A method for constructing (3, 4)-graphs, *Math. Zeitschrift* 44 (1988), 546–550.

[65] P. Frankl, Asymptotic solution of a Turán-type problem, *Graphs and Combinatorics* 6 (1990), 223–227.

[66] P. Frankl and Z. Füredi, A new generalization of the Erdős-Ko-Rado theorem, *Combinatorica* 3 (1983), 341–349.

[67] P. Frankl and Z. Füredi, An exact result for 3-graphs, *Disc. Math.* 50 (1984), 323–328.

[68] P. Frankl and Z. Füredi, Exact solution of some Turán-type problems, *J. Combin. Theory Ser. A.* **45** (1987), 226–262.

[69] P. Frankl and Z. Füredi, Extremal problems and the Lagrange function of hypergraphs, *Bulletin Institute Math. Academia Sinica* **16** (1988), 305–313.

[70] P. Frankl and Z. Füredi, Extremal problems whose solutions are the blowups of the small Witt-designs, *J. Combin. Theory Ser. A* **52** (1989), 129–147.

[71] P. Frankl, Y. Peng, V. Rödl and J. Talbot, A note on the jumping constant conjecture of Erdős, *J. Combin. Theory Ser. B* **97** (2007), 204–216.

[72] P. Frankl and V. Rödl, Hypergraphs do not jump, *Combinatorica* **4** (1984), 149–159.

[73] P. Frankl and V. Rödl, Lower bounds for Turán's problem, *Graphs Combin.* **1** (1985), 213–216.

[74] P. Frankl and V. Rödl, Large triangle-free subgraphs in graphs without K_4, *Graphs Combin.* **2** (1986), 135–144.

[75] P. Frankl and R. M. Wilson, Intersection theorems with geometric consequences, *Combinatorica* **1** (1981), 357–368.

[76] E. Friedgut and J. Kahn, On the number of copies of one hypergraph in another, *Israel J. Math.* **105** (1998), 251–256.

[77] A. Frieze and R. Kannan, Quick approximation to matrices and applications, *Combinatorica* **19** (1999), 175–220.

[78] A. Frohmader, More constructions for Turán's (3,4)-conjecture, *Electronic J. Combin.* **15** (2008), R137.

[79] Z. Füredi, Turán type problems, *Surveys in combinatorics*, London Math. Soc. Lecture Note Ser. 166, Cambridge Univ. Press, Cambridge, 1991, 253–300.

[80] Z. Füredi and A. Kündgen, Turán problems for integer-weighted graphs, *J. Graph Theory* **40** (2002), 195–225.

[81] Z. Füredi, D. Mubayi and O. Pikhurko, Quadruple systems with independent neighborhoods, *J. Comb. Theory Ser. A* **115** (2008), 1552–1560.

[82] Z. Füredi, O. Pikhurko and M. Simonovits, On triple systems with independent neighborhoods, *Combin. Probab. Comput.* **14** (2005), 795–813.

[83] Z. Füredi, O. Pikhurko and M. Simonovits, 4-books of three pages, *J. Comb. Theory Ser. A* **113** (2006), 882–891.

[84] Z. Füredi and M. Simonovits, Triple systems not containing a Fano configuration, *Combin. Probab. Comput.* **14** (2005), 467–484.

[85] G. R. Giraud, Remarques sur deux problèmes extrémaux, *Disc. Math.* **84** (1990), 319–321.

[86] J. Goldwasser, On the Turán Number of {123, 124, 345}, manuscript.

[87] W. T. Gowers, Lower bounds of tower type for Szemerédi's uniformity lemma, *Geom. Funct. Anal.* **7** (1997), 322–337.

[88] W. T. Gowers, Quasirandomness, counting and regularity for 3-uniform hypergraphs, *Combin. Probab. Comput.* **15** (2006), 143–184.

[89] W. T. Gowers, Hypergraph regularity and the multidimensional Szemerédi theorem, *Ann. of Math.* **166** (2007), 897–946.

[90] B. J. Green and J. Wolf, A note on Elkin's improvement of Behrend's construction, arXiv:0810.0732, 2010.

[91] A. Hajnal and E. Szemerédi, Proof of a conjecture of Erdős, *Combinatorial Theory and Its Applications*, Amsterdam, Netherlands: North-Holland, pp. 601-623, 1970.

[92] H. Hatami and Serguei Norine, Undecidability of linear inequalities in graph homomorphism densities, to appear in *J. AMS*, arXiv:1005.2382, 2010.

[93] A. J. W. Hilton, On ordered set systems and some conjectures related to the Erdős-Ko-Rado Theorem and Turán's Theorem, *Mathematika* **28** (1981), 54–66.

[94] Y. Ishigami, A simple regularization of hypergraphs, arXiv:math/0612838, 2006.

[95] Y. Ishigami, The number of hypergraphs and colored hypergraphs with hereditary properties, arXiv:0712.0425, 2007.

[96] J. R. Johnson and J. Talbot, Vertex Turán problems in the hypercube, *J. Combin. Theory Ser. A* **117** (2010), 454–465.

[97] G. Kalai, Hyperconnectivity of graphs, *Graphs Combin.* **1** (1985), 65–79.

[98] G. Kalai, A new approach to Turán's conjecture, *Graphs Combin.* **1** (1985), 107–109.

[99] G. Kalai, Algebraic shifting, Computational commutative algebra and combinatorics (Osaka, 1999), 121–163, *Adv. Stud. Pure Math.* **33**, Math. Soc. Japan, Tokyo, 2002.

[100] G. Katona, A theorem of finite sets, *Theory of graphs* (Proc. Colloq., Tihany, 1966), 187–207, Academic Press, New York, 1968.

[101] G. Katona, T. Nemetz and M. Simonovits, On a problem of Turán in the theory of graphs, *Mat. Lapok* **15** (1964), 228–238.

[102] P. Keevash, The Turán problem for hypergraphs of fixed size, *Electron. J. Combin.* **12** (2005), N11.

[103] P. Keevash, The Turán problem for projective geometries, *J. Combin. Theory Ser. A* **111** (2005), 289–309.

[104] P. Keevash, A hypergraph regularity method for generalised Turán problems, *Random Structures Algorithms* **34** (2009), 123–164.

[105] T. Kövari, V. T. Sós and P. Turán, On a problem of K. Zarankiewicz, *Colloquium Math.* **3** (1954), 50–57.

[106] P. Keevash and D. Mubayi, Stability theorems for cancellative hypergraphs, *J. Combin. Theory Ser. B* **92** (2004), 163–175.

[107] P. Keevash and D. Mubayi, Set systems without a simplex or a cluster, *Combinatorica* **30** (2010), 175–200.

[108] P. Keevash, D. Mubayi, B. Sudakov and J. Verstraëte, Rainbow Turán problems, *Combin. Probab. Comput.* **16** (2007), 109–126.

[109] P. Keevash, M. Saks, B. Sudakov and J. Verstraëte, Multicolour Turan problems, *Adv. in Applied Math.* **33** (2004), 238–262.

[110] P. Keevash and B. Sudakov, The Turán number of the Fano plane, *Combinatorica* **25** (2005), 561–574.

[111] P. Keevash and B. Sudakov, On a hypergraph Turán problem of Frankl, *Combinatorica* **25** (2005), 673–706.

[112] P. Keevash and B. Sudakov, Set systems with restricted cross-intersections and the minimum rank of inclusion matrices, *SIAM J. Discrete Math.* **18** (2005), 713–727.

[113] P. Keevash and Y. Zhao, Codegree problems for projective geometries, *J. Combin. Theory Ser. B* **97** (2007), 919–928.

[114] K. Kim and F. Roush, On a problem of Turán, Studies in pure mathematics, 423–425, Birkhäuser, Basel-Boston, Mass. 1983.

[115] J. Komlós, G. N. Sárközy and E. Szemerédi, Proof of the Seymour conjecture for large graphs, *Ann. Comb.* **2** (1998), 43–60.

[116] J. Komlós and M. Simonovits, Szemerédi's regularity lemma and its applications in graph theory, Combinatorics, Paul Erdős is eighty, Vol. 2 (Keszthely, 1993), 295–352, *Bolyai Soc. Math. Stud.* **2** János Bolyai Math. Soc., Budapest, 1996.

[117] A. Kostochka, A class of constructions for Turán's (3,4) problem, *Combinatorica* **2** (1982), 187–192.

[118] J. B. Kruskal, The number of simplices in a complex, *Mathematical optimization techniques*, 251–278 Univ. of California Press, Berkeley, Calif, 1963.

[119] D. Kühn and D. Osthus, Embedding large subgraphs into dense graphs, *Surveys in Combinatorics*, Cambridge University Press, 2009, 137–167.

[120] D. Kühn and D. Osthus, The minimum degree threshold for perfect graph packings, *Combinatorica* **29** (2009), 65–107.

[121] R.S. Li and W.W. Li, Independence numbers of graphs and generators of ideals, *Combinatorica* **1** (1981), 55–61.

[122] L. Lovász, Semidefinite programs and combinatorial optimization, *Recent advances in algorithms and combinatorics*, CMS Books Math. vol. 11, 137–194, Springer, New York, 2003.

[123] L. Lovász, Graph homomorphisms: Open problems, manuscript, 2008.

[124] L. Lovász, Subgraph densities in signed graphons and the local Sidorenko conjecture, arXiv:1004.3026, 2010.

[125] L. Lovász and M. Simonovits, On the number of complete subgraphs of a graph. II, Studies in pure mathematics, 459–495, Birkhäuser, Basel, 1983.

[126] L. Lovász and B. Szegedy, Limits of dense graph sequences, *J. Combin. Theory Ser. B* **96** (2006), 933–957.

[127] L. Lovász and B. Szegedy, Szemerédi's lemma for the analyst, *Geom. Funct. Anal.* **17** (2007), 252–270.

[128] L. Lu and L. Székely, Using Lovász Local Lemma in the space of random injections, *Electronic J. Combin.* **14** (2007), R63.

[129] L. Lu and Y. Zhao, An exact result for hypergraphs and upper bounds for the Turán density of K_{r+1}^r, *SIAM J. Disc. Math.* **23** (2009), 1324–1334.

[130] E. Marchant and A. Thomason, Extremal graphs and multigraphs with two weighted colours, *Fete of Combinatorics and Computer Science*, Bolyai Society Mathematical Studies, Vol. 20, 2010.

[131] K. Markström, Extremal hypergraphs and bounds for the Turán density of the 4-uniform K_5 *Disc. Math.* **309** (2009), 5231–5234.

[132] K. Markström and J. Talbot, On the density of 2-colorable 3-graphs in which any four points span at most two edges, *J. Combin. Des.* **18** (2010), 105–114.

[133] K. Markström, A web archive of Turán graphs, URL: http://abel.math.umu.se/klasm/Data/hypergraphs/turanhypergraphs.html.

[134] U. Matthias, Hypergraphen ohne vollstände r-partite Teilgraphen, PhD thesis, Heidelberg, 1994.

[135] T. S. Motzkin and E. G. Straus, Maxima for graphs and a new proof of a theorem of Turán, *Canad. J. Math.* **17** (1965), 533–540.

[136] D. Mubayi, On hypergraphs with every four points spanning at most two triples, *Electronic J. Combin.* **10** (2003), N10.

[137] D. Mubayi, The co-degree density of the Fano plane, *J. Combin. Theory Ser. B* **95** (2005), 333–337.

[138] D. Mubayi, A hypergraph extension of Turán's theorem, *J. Combin. Theory Ser. B* **96** (2006), 122–134.

[139] D. Mubayi, Counting substructures I: color critical graphs, to appear in *Adv. Math.*

[140] D. Mubayi, Counting substructures II: hypergraphs, preprint.

[141] D. Mubayi and O. Pikhurko, A new generalization of Mantel's theorem to k-graphs, *J. Combin Theory Ser. B* **97** (2007), 669–678.

[142] D. Mubayi and O. Pikhurko, Constructions of non-principal families in extremal hypergraph theory, *Disc. Math.* **308** (2008), 4430–4434.

[143] D. Mubayi and V. Rödl, On the Turán number of triple systems, *J. Combin. Theory Ser. A* **100** (2002), 136–152.

[144] D. Mubayi and J. Talbot, Extremal problems for t-partite and t-colorable hypergraphs, *Electronic J. Combin.* **15** (2008), R26.

[145] D. Mubayi and J. Verstraëte, Proof of a conjecture of Erdős on triangles in set-systems, *Combinatorica* **25** (2005), 599–614.

[146] D. Mubayi and Y. Zhao, Co-degree density of hypergraphs, *J. Combin. Theory Ser. A* **114** (2007), 1118–1132.

[147] V. Nikiforov, The number of cliques in graphs of given order and size, to appear in *Trans. AMS.*

[148] B. Nagle, V. Rödl and M. Schacht, The counting lemma for regular k-uniform hypergraphs, *Random Structures Algorithms* **28** (2006), 113–179.

[149] J. Oxley, *Matroid Theory*, Oxford University Press, 1992.

[150] Y. Peng, Non-jumping numbers for 4-uniform hypergraphs, *Graphs Combin.* **23** (2007), 97–110.

[151] Y. Person and M. Schacht, Almost all hypergraphs without Fano planes are bipartite, *Proc. Twentieth Annual ACM-SIAM Symposium on Discrete Algorithms (SODA 09)*, 217–226.

[152] O. Pikhurko, An exact Turán result for the generalized triangle, *Combinatorica* **28** (2008) 187–208.

[153] O. Pikhurko, An analytic approach to stability, *Disc. Math.* **310** (2010), 2951–2964.

[154] O. Pikhurko, The minimum size of 3-graphs without a 4-set spanning no or exactly three edges, preprint.

[155] O. Pikhurko, Exact computation of the hypergraph Turán function for expanded complete 2-graphs, accepted by *J. Combin Theory Ser. B*, publication suspended for an indefinite time, see http://www.math.cmu.edu/ pikhurko/Copyright.html

[156] A. A. Razborov, Flag Algebras, J. Symbolic Logic **72** (2007), 1239–1282,

[157] A. A. Razborov, On the Minimal Density of Triangles in Graphs, *Combin. Probab. Comput.* **17** (2008), 603–618.

[158] A. A. Razborov, On 3-hypergraphs with forbidden 4-vertex configurations, *SIAM J. Disc. Math.* **24** (2010), 946–963.

[159] A. A. Razborov, On the Fon-der-Flaass interpretation of extremal examples for Turán's (3,4)-problem, arXiv:1008.4707, 2010.

[160] V. Rödl, A. Ruciński and E. Szemerédi, An approximate Dirac-type theorem for *k*-uniform hypergraphs, *Combinatorica* **28** (2008), 229–260.

[161] V. Rödl and A. Ruciński, Dirac-type questions for hypergraphs – a survey (or more problems for Endre to solve), An Irregular Mind (Szemerédi is 70), *Bolyai Soc. Math. Studies* **21** (2010).

[162] V. Rödl and M. Schacht, Regular partitions of hypergraphs: regularity lemmas, *Combin. Probab. Comput.* **16** (2007), 833–885.

[163] V. Rödl and J. Skokan, Regularity lemma for uniform hypergraphs, *Random Structures Algorithms* **25** (2004), 1–42.

[164] V. Rödl and A. Sidorenko, On the jumping constant conjecture for multigraphs, *J. Combin. Theory Ser. A* **69** (1995), 347–357.

[165] I. Z. Ruzsa and E. Szemerédi, Triple systems with no six points carrying three triangles, Combinatorics (Proc. Fifth Hungarian Colloq., Keszthely, 1976), Vol. II, pp. 939–945, *Colloq. Math. Soc. János Bolyai* **18**, North-Holland, Amsterdam-New York, 1978.

[166] T. Sanders, On certain other sets of integers, arXiv:1007.5444, 2010.

[167] G. N. Sárközy and S. M. Selkow, An extension of the Ruzsa-Szemerédi Theorem, *Combinatorica* **25** (2005), 77–84.

[168] G. N. Sárközy and S. M. Selkow, On a Turán-type hypergraph problem of Brown, Erdos and T. Sós, *Disc. Math.* **297** (2005), 190–195.

[169] M. Schacht, Extremal results for random discrete structures, preprint.

[170] J. B. Shearer, A new construction for cancellative families of sets, *Electron. J. Combin.* **3** (1996), R15.

[171] A. F. Sidorenko, Systems of sets that have the T-property, *Moscow University Mathematics Bulletin* **36** (1981), 22–26.

[172] A. F. Sidorenko, Method of quadratic forms in the Turán combinatorial problem, *Moscow University Mathematics Bulletin* **37** (1982), 3–6.

[173] A. F. Sidorenko, The maximal number of edges in a homogeneous hypergraph containing no prohibited subgraphs, *Math Notes* **41** (1987), 247–259. Translated from *Mat. Zametki*.

[174] A. F. Sidorenko, Asymptotic solution for a new class of forbidden r-graphs. *Combinatorica* **9** (1989), 207–215.

[175] A. F. Sidorenko, Extremal combinatorial problems in spaces with continuous measure (in Russian), *Issledovanie Operatchiy i ASU* **34** (1989), 34–40.

[176] A. Sidorenko, Asymptotic solution of the Turán problem for some hypergraphs, *Graphs Combin.* **8** (1992), 199–201.

[177] A. Sidorenko, A correlation inequality for bipartite graphs, *Graphs Combin.* **9** (1993), 201–204.

[178] A. Sidorenko, Boundedness of optimal matrices in extremal multigraph and digraph problems, *Combinatorica* **13** (1993), 109–120.

[179] A. Sidorenko, An analytic approach to extremal problems for graphs and hypergraphs, *Extremal problems for finite sets* (Visegrád, 1991), 423–455, Bolyai Soc. Math. Stud. 3, János Bolyai Math. Soc., Budapest, 1994.

[180] A. Sidorenko, What we know and what we do not know about Turán numbers, *Graphs and Combinatorics* **11** (1995), 179–199.

[181] A. Sidorenko, Upper bounds for Turán numbers, *J. Combin. Theory Ser. A* **77** (1997), 134–147.

[182] M. Simonovits, A method for solving extremal problems in graph theory, stability problems, *Theory of Graphs (Proc. Colloq. Tihany, 1966)*, Academic Press, New York, and Akad. Kiadó, Budapest, 1968, 279–319.

[183] V. Sós, Remarks on the connection of graph theory, finite geometry and block designs, *Teorie Combinatorie*, Tomo II, Accad. Naz. Lincei, Rome, 1976, 223–233.

[184] V. T. Sós and E. G. Straus, Extremals of functions on graphs with applications to graphs and hypergraphs. *J. Combin. Theory Ser. B* **32** (1982), 246–257.

[185] B. Sudakov and V. Vu, Local resilience of graphs, *Random Structures Algorithms* **33** (2008), 409–433.

[186] J. Talbot, Chromatic Turán problems and a new upper bound for the Turán density of K_4^-, *European J. Combin.* **28** (2007), 2125–2142.

[187] T. Tao, A variant of the hypergraph removal lemma, *J. Combin. Theory Ser. A* **113** (2006), 1257–1280.

[188] T. Tao, A correspondence principle between (hyper)graph theory and probability theory, and the (hyper)graph removal lemma, *J. Anal. Math.* **103** (2007), 1–45.

[189] A. Thomason and P. Wagner, Bounding the size of square-free subgraphs of the hypercube, *Disc. Math.* **309** (2009), 1730–1735.

[190] P. Turán, On an extremal problem in graph theory (in Hungarian), *Mat. Fiz. Lapok* **48** (1941), 436–452.

[191] P. Turán, Research problem, *Közl MTA Mat. Kutató Int.* **6** (1961), 417–423.

[192] B. Zhou, A Turán-type problem on 3-graphs, *Ars Combin.* **31** (1991), 177–181.

[193] A. A. Zykov, On some properties of linear complexes, *Mat. Sbornik (N. S.)* **24** (66) (1949) 163–188.

School of Mathematical Sciences,
Queen Mary, University of London,
Mile End Road,
London E1 4NS,
UK.
p.keevash@qmul.ac.uk

Some new results in extremal graph theory

Vladimir Nikiforov

Abstract

In recent years several classical results in extremal graph theory have been improved in a uniform way and their proofs have been simplified and streamlined. These results include a new Erdős-Stone-Bollobás theorem, several stability theorems, several saturation results and bounds for the number of graphs with large forbidden subgraphs.

Another recent trend is the expansion of spectral extremal graph theory, in which extremal properties of graphs are studied by means of eigenvalues of various matrices. One particular achievement in this area is the casting of the central results above in spectral terms, often with additional enhancement. In addition, new, specific spectral results were found that have no conventional analogs.

All of the above material is scattered throughout various journals, and since it may be of some interest, the purpose of this survey is to present the best of these results in a uniform, structured setting, together with some discussions of the underpinning ideas.

Contents

1   Introduction

The purpose of this survey is to give a systematic account of two recent lines of research in extremal graph theory. The first one, developed in [14],[15],[16],[63, 68], improves a number of classical results grouped around the theorem of Turán. The main progress is along the following three guidelines: replacing fixed parameters by variable ones; giving explicit conditions for the validity of the statements; developing and using tools of general scope. Among the results obtained are a new Erdős-Stone-Bollobás theorem (see Section 2.2), several stability theorems (see Section 2.4), several saturation results, and bounds for the number of graphs without given large subgraphs.

The second line of research, developed in [13],[69, 82], can be called *spectral extremal graph theory*, where connections are sought between graph properties and the eigenvalues of certain matrices associated with graphs. As a result of this research, much of classical extremal graph theory has been translated into spectral statements, and this translation has also brought enhancement. Among the results obtained are spectral forms of the Turán theorem and the Erdős-Stone-Bollobás theorem, several stability theorems, along with new bounds for the Zarankiewicz problem (What is the maximum number of edges in a graph with no $K_{s,t}$?).

In the course of this work a few tools were developed, which help to cast systematically some classical results and their proofs into spectral form. The use of this machinery is best exhibited in [66], where we gave a new stability theorem and also its spectral analog - Theorems 2.19 and 3.10 below. As an illustration, in Section 5 we outline the proofs of these two results.

We believe that ultimately the spectral approach to extremal graph theory will turn out to be more fruitful than the conventional one, albeit it is also more difficult, and is still underdeveloped. Indeed, most statements in conventional terms can be cast and proved in spectral terms, but in addition to that, there are a lot of specific spectral results (say, Theorem 3.22) with no conceivable conventional setting.

The rest of the survey is organized as follows. To keep the beginning straightforward, the bulk of the necessary notation and the basic facts have been shifted to Section 6, although some definitions are given also where appropriate. Section 2 covers the conventional, nonspectral problems, while Section 3 presents the spectral results. In Section 4, we have collected some basic and more widely applicable statements, which we have found useful on more than one occasion. Finally Section 5 presents some proof techniques for illustration, and in fact these are the only proofs in this survey.

2   New results on classical extremal graph problems

In extremal graph theory one investigates how graph properties depend on the value of various graph parameters. In a sense almost all of graph theory deals with extremal problems, but there is a bundle of results grouped around Turán's theorem [89], that undoubtedly constitutes the core of extremal graph theory. To state this celebrated theorem, which has stimulated researchers for more than six decades, recall that for $n \geq r \geq 2$, the Turán graph $T_r(n)$ is the complete r-partite graph of order n whose class sizes differ by at most one. We let $t_r(n) = e(T_r(n))$.

Theorem 2.1 *If G is a graph of order n, with no complete subgraph of order $r + 1$, then $e(G) \leq t_r(n)$ with equality holding only when $G = T_r(n)$.*

Here is a more popular, but slightly weaker version, which we shall call the *concise Turán theorem:*

If G is a graph of order n, with $e(G) > (1 - 1/r) n^2/2$, then G contains a complete subgraph of order $r + 1$.

No doubt, Turán's theorem is a nice combinatorial statement and it is not too difficult to prove as well. However, its external simplicity is incomparable with its real importance, since this theorem is a cornerstone on which rest much more general statements about graphs. Thus, in this survey, we shall meet the Turán graph $T_r(n)$ and the numbers $t_r(n)$ on numerous occasions.

2.1   The extremal problems that are studied

Among the many questions motivated by Turán's theorem, the ones that we will discuss in Section 2 fall into the following three broad classes:

(1) Which subgraphs are present in a graph G of order n whenever $e(G) > t_r(n)$ and n is sufficiently large?

As we shall see, here the range of $e(G) - t_r(n)$ determines different problems: when $e(G) - t_r(n) = o(n)$ we have *saturation problems*, and when $e(G) - t_r(n) = o(n^2)$, we have *Erdős-Stone type problems.*

Other questions that we will be interested in give rise to the so called *stability problems*, concerning near-maximal graphs without forbidden subgraphs.

(2) Suppose that H_n is a graph which is present in any graph G of order n whenever $e(G) > t_r(n)$, but H_n is not a subgraph of the Turán graph $T_r(n)$. We can ask the following questions:

- What can be the structure of an H_n-free graph G of order n if $e(G) > t_r(n) - f(n)$, where $f(n) \geq 0$ and $f(n) = o(n^2)$?

- What can be the structure of an H_n-free graph G of order n, with minimum degree $\delta(G) > (1 - c)\delta(t_r(n))$ for some sufficiently small $c > 0$?

Obviously these two general questions have lots of variations, many of which are intensively studied due to their applicability in other extremal problems.

Finally, recall that a long series of results deals with the number of graphs having some monotone or hereditary properties. Here we will discuss a similar and natural question which, however, goes beyond this paradigm:

(3) Let $\{H_n\}$ be a sequence of graphs with $v(H_n) = o(\log n)$. How many H_n-free graphs of order n are there?

2.2 Erdős-Stone type problems

We write $K_r(s_1, ..., s_r)$ for the complete r-partite graph with class sizes $s_1, ..., s_r$, and set for short

$$K_r(p) = K_r(p, ..., p) \quad \text{and} \quad K_r(p; q) = K_r(p, ..., p, q).$$

Let us recall the fundamental theorem of Erdős and Stone [42].

Theorem 2.2 *For all $c > 0$ and natural r, p, there is an integer $n_0(p, r, c)$ such that if G is a graph of order $n > n_0(p, r, c)$ and $e(G) \geq (1 - 1/r + c)n^2/2$, then G contains a $K_{r+1}(p)$.*

Noting that $t_r(n) \approx (1 - 1/r)n^2/2$, we see the close relation of Theorem 2.2 to Turán's theorem. In fact, Theorem 2.2 answers a fairly general question: *what is the maximum number of edges $e(n, H)$ in a graph of order n that does not contain a fixed $(r + 1)$-chromatic subgraph H?* Theorem 2.2 immediately implies that $e(n, H) \leq (1 - 1/r + o(1))n^2/2$. On the other hand, $T_r(n)$ contains no $(r + 1)$-chromatic subgraphs, and so, $e(n, H) = (1 - 1/r + o(1))n^2/2$.

Write $g(n, r, c)$ for the maximal p such that every graph G of order n with

$$e(G) \geq (1 - 1/r + c)n^2/2$$

contains a $K_{r+1}(p)$. For almost 30 years the order of magnitude of $g(n, r, c)$ remained unknown; it was established first by Bollobás and Erdős in [8], as given below. This simplest quantitative form of the Erdős-Stone theorem we call the *Erdős-Stone-Bollobás theorem.*

Theorem 2.3 *There are constants $c_1, c_2 > 0$ such that*

$$c_1 \log n \leq g(n, r, c) \leq c_2 \log n.$$

Subsequently the function $g(n, r, c)$ was determined with great precision in [9], [21], [10], [52], to name a few milestones. However, since Szemerédi's Regularity Lemma is a standard tool in this research, the results are confined to fixed c, and n extremely large.

To overcome these restrictions, in [63], we proposed a different approach, based on the expectation that the presence of many copies of a given subgraph H must imply the existence of large blow-ups of H. As a by-product, this approach gave results in other directions as well, which otherwise do not seem too close to the Erdős-Stone theory; two such topics are outlined in 2.2.2 and 2.2.3.

2.2.1 Refining the Erdős-Stone-Bollobás theorem The general idea above is substantiated for cliques in the following two theorems, given in [63].

Theorem 2.4 *Let $r \geq 2$, let c and n be such that*

$$0 < c < 1/r! \quad and \quad n \geq \exp\left(c^{-r}\right),$$

and let G be a graph of order n. If $k_r(G) > cn^r$, then G contains a $K_r(s;t)$ with $s = \lfloor c^r \log n \rfloor$ and $t > n^{1-c^{r-1}}$.

In a nutshell, Theorem 2.4 says that if a graph contains many r-cliques, then it has large complete r-partite subgraphs. Hence, to obtain Theorem 2.3, all we need to prove is that the hypothesis of the Erdős-Stone theorem implies the existence of sufficiently many r-cliques. This implication is fairly standard, and so we obtain the following explicit version of the Erdős-Stone-Bollobás theorem.

Theorem 2.5 *Let $r \geq 2$, let c and n be such that*

$$0 < c < 1 \quad and \quad n \geq \exp\left((r^r/c)^{r+1}\right),$$

and let G be a graph of order n. If $e(G) \geq (1 - 1/r + c)n^2$, then G contains a $K_r(s;t)$ with

$$s = \left\lfloor (c/r^r)^{r+1} \log n \right\rfloor \quad and \quad t > n^{1-(c/r^r)^r}.$$

In the two theorems above, we would like to emphasize the three principles outlined in the introduction: first, the fundamental parameter c may depend on n, e.g., letting $c = 1/\log\log n$, the conclusion is meaningful for sufficiently large n; note that this fact can be verified precisely because the conditions for validity are stated explicitly. Also, the proof of these theorems relies on more basic statements of wider applicability - Lemma 4.1 and Lemma 4.2.

Another observation about this setup is the peculiarity of the graphs $K_r(s;t)$ in the conclusions of the above theorems: if the statement holds for some c, then it holds also for all positive $c' < c$ as long as n is large enough. That is to say, when n increases, in addition to the graphs $K_r(s;t)$ guaranteed by the theorems, we can find other, larger and more lopsided graphs $K_r(s';t')$ with $s' < s$ and $t' > t$. This same observation can be made on numerous other occasions below, and usually we shall omit it to avoid repetition.

Let us note that Theorem 2.3 implies also the following assertion, which strengthens the observation of Erdős and Simonovits [43]:

Theorem 2.6 *Let $r \geq 3$ and let $F_1, F_2, \ldots$ be $(r+1)$-chromatic graphs satisfying $v(F_n) = o(\log n)$. Then*

$$\max\{e(G) : G \in \mathcal{G}(n) \ and \ F_n \nsubseteq G\} = \frac{r-1}{2r}n^2 + o(n^2).$$

Thus Theorem 2.6 solves asymptotically the Turán problem for families of forbidden subgraphs whose order grows not too fast with n. Moreover, the condition $v(F_n) = o(\log n)$ can be sharpened further using the bounds given by Ishigami in [52].

2.2.2 Graphs with many copies of a given subgraph In this subsection we shall apply the basic idea above to arbitrary subgraphs of graphs, including induced ones.

Let us first define a *blow-up* of a graph H: given a graph H of order r and positive integers $k_1, \ldots, k_r$, we write $H(k_1, \ldots, k_r)$ for the graph obtained by replacing each vertex $u \in V(H)$ with a set V_u of size k_u and each edge $uv \in E(H)$ with a complete bipartite graph with vertex classes V_u and V_v.

We are interested in the following generalization of Theorem 2.4: *Suppose that a graph G of order n contains cn^r copies of a given subgraph H on r vertices. How large a "blow-up" of H must G contain?*

The following theorem from [64] is an analog of Theorem 2.4 for arbitrary subgraphs.

Theorem 2.7 *Let $r \geq 2$, let c and n be such that*

$$0 < c < 1/r! \quad and \quad n \geq \exp\left(c^{r^2}\right),$$

and let H be a graph of order r. If $G \in \mathcal{G}(n)$ and G contains more than cn^r copies of H, then G contains an $H(s, \ldots s, t)$ with $s = \left\lfloor c^{r^2} \log n \right\rfloor$ and $t > n^{1-c^{r-1}}$.

A similar theorem is conceivable for induced subgraphs, but note the obvious bump: the complete graph K_n has $\Theta(n^2)$ edges, i.e. K_2's, but contains no *induced* 4-cycle, i.e. $K_2(2)$. To come up with a meaningful statement, we need the following more flexible version of a blow-up:

*We say that a graph F **is of type** $H(k_1, \ldots, k_r)$, if F is obtained from $H(k_1, \ldots, k_r)$ by adding some (possibly zero) edges within the sets V_u, $u \in V(H)$.*

This definition in hand, we can state the induced graph version of Theorem 2.7, also from [64].

Theorem 2.8 *Let $r \geq 2$, let c and n be such that*

$$0 < c < 1/r! \quad and \quad n \geq \exp\left(c^{r^2}\right),$$

and let H be a graph of order r. If $G \in \mathcal{G}(n)$ and G contains more than cn^r induced copies of H, then G contains an induced subgraph of type $H(s, \ldots s, t)$, where $s = \left\lfloor c^{r^2} \log n \right\rfloor$ and $t > n^{1-c^{r-1}}$.

For constant c, the above theorems give the correct order of magnitude of the subgraphs of type $H(s, \ldots s, t)$, namely, $\log n$ for s and $n^{1-o(c)}$ for t. When c depends on n, the best bounds on s and t are apparently unknown.

2.2.3 Complete r-partite subgraphs of dense r-graphs In this subsection *graph* stands for *r-uniform hypergraph* for some fixed $r \geq 3$. We use again $K_r(s_1, \ldots, s_r)$ to denote the complete r-partite r-graph with class sizes $s_1, \ldots, s_r$.

In the spirit of the previous topics, it is natural to ask: *Suppose that a graph G of order n contains cn^r edges. How large a subgraph $K_r(s)$ must G contain?* As shown by Erdős and Stone [42] and Erdős [32], $s \geq a(\log n)^{1/(r-1)}$ for some $a = a(c) > 0$, independent of n.

In [65] this fundamental result was extended in three directions: c may depend on n, the complete r-partite subgraph may have vertex classes of variable size, and the graph G is taken to be an r-partite r-graph with equal classes. The last setup is obviously more general than just taking r-graphs.

The following three theorems are given in [65].

Theorem 2.9 *Let* $r \geq 3$, *let* c *and* n *be such that*

$$0 < c \leq r^{-3} \quad and \quad n \geq \exp\left(1/c^{r-1}\right),$$

and let the positive integers $s_1, \ldots, s_{r-1}$ *satisfy* $s_1 s_2 \cdots s_{r-1} \leq c^{r-1} \log n$. *Then every graph with* n *vertices and at least* $cn^r/r!$ *edges contains a* $K_r(s_1, \ldots, s_{r-1}, t)$ *with* $t > n^{1-c^{r-2}}$.

Instead of this theorem it is easier and more effective to prove a more general one for r-partite r-graphs.

Theorem 2.10 *Let* $r \geq 3$, *let* c *and* n *be such that*

$$0 < c \leq r^{-3} \quad and \quad n \geq \exp\left(1/c^{r-1}\right),$$

and let the positive integers $s_1, \ldots, s_{r-1}$ *satisfy* $s_1 s_2 \cdots s_{r-1} \leq c^{r-1} \log n$. *Let* $U_1, \ldots, U_r$ *be sets of size* n *and* $E \subset U_1 \times \cdots \times U_r$ *satisfy* $|E| \geq cn^r$. *Then there exist* $V_1 \subset U_1, \cdots, V_r \subset U_r$ *satisfying* $V_1 \times \cdots \times V_r \subset E$ *and*

$$|V_1| = s_1, \cdots, |V_{r-1}| = s_{r-1}, \quad |V_r| > n^{1-c^{r-2}}.$$

In turn, Theorem 2.10 is deduced from a counting result about r-partite r-graphs, which generalizes the double counting argument of Kővári, Sós and Turán for bipartite graphs [57].

Theorem 2.11 *Let* $r \geq 2$ *and let* c *and* n *be such that*

$$2^r \exp\left(-\frac{1}{r}(\log n)^{1/r}\right) \leq c \leq 1.$$

Let G *be an* r-partite r-graph with parts $U_1, \ldots, U_r$ *of size* n, *and with edge set* $E \subset U_1 \times \cdots \times U_r$ *satisfying* $|E| \geq cn^r$. *If the positive integers* $s_1, s_2, \ldots, s_r$ *satisfy* $s_1 s_2 \cdots s_r \leq \log n$, *then* G *contains at least*

$$\left(\frac{c}{2^r}\right)^{rs_1 \cdots s_r} \binom{n}{s_1} \cdots \binom{n}{s_r}.$$

complete r-partite subgraphs with precisely s_i vertices in U_i for every $i = 1, \ldots, r$.

Following Erdős [32] and taking a random r-graph G of order n and density $1-\varepsilon$, a straightforward calculation shows that with probability tending to 1, G does not contain a $K_r(s, \ldots, s)$ for $s > A(\log n)^{1/(r-1)}$, where $A = A(\varepsilon)$ is independent of n. That is to say, Theorems 2.9 and 2.10 are essentially tight.

2.3 Saturation problems

Saturation problems concern the type of subgraphs one necessarily finds in graphs of order n, with $t_r(n) + o(n^2)$ edges. Among all possible saturation problems we will consider only the most important case: which subgraphs necessarily occur in graphs of order n and size $t_r(n) + 1$? Turán's theorem says that such graphs contain a K_{r+1}, but one notes that they contain much larger supergraphs of K_{r+1}.

Our first theorem completes an unfinished investigation started by Erdős in 1963, in [31]. We also present several results related to joints - a class of important subgraphs, whose study was also initiated by Erdős.

2.3.1 Unavoidable subgraphs of graphs in $G(n, t_r(n) + 1)$ Let $s_1 \geq 2$, and write $K_r^+(s_1, s_2, ..., s_r)$ for the graph obtained from $K_r(s_1, s_2, ..., s_r)$ by adding an edge to the first part. For short, we also set

$$K_r^+(p) = K_r^+(p, ..., p) \quad \text{and} \quad K_r^+(p; q) = K_r^+(p, ..., p, q).$$

In [31] Erdős gave the following result:

Theorem 2.12 *For every $\varepsilon > 0$, there exist $c = c(\varepsilon) > 0$ and $n_0(\varepsilon)$ such that if G is a graph of order $n > n_0(\varepsilon)$ and $e(G) > \lfloor n^2/4 \rfloor$, then G contains a*

$$K_2^+\left(\lfloor c\log n\rfloor, \lceil n^{1-\varepsilon}\rceil\right).$$

For some time there was no generalization of this result for $K_r^+(s; t)$ until Erdős and Simonovits [41] came up with a similar assertion valid for all $r \geq 2$.

Theorem 2.13 *Let $r \geq 2$, $q \geq 1$, and let n be sufficiently large. If G is a graph of order n with $t_r(n) + 1$ edges, then G contains a $K_r^+(q)$.*

In a sense Theorem 2.13 is best possible as any graph H that necessarily occurs in all sufficiently large graphs $G \in G(n, t_r(n) + 1)$ can be imbedded in $K_r^+(q)$ for q sufficiently large. To see this, just add an edge to the Turán graph $T_r(n)$ and note that all $(r+1)$-partite subgraphs of this graph are edge-critical with respect to the chromatic number. However, Theorem 2.12 suggests that stronger statements are possible, and indeed, in [67], we extended both Theorems 2.12 and 2.13 to the following one.

Theorem 2.14 *Let $r \geq 2$, let c and n be such that*

$$0 < c \leq r^{-(r+7)(r+1)} \quad \text{and} \quad n \geq e^{2/c},$$

and let G be a graph of order n. If $e(G) > t_r(n)$, then G contains a

$$K_r^+\left(\lfloor c\log n\rfloor; \lceil n^{1-\sqrt{c}}\rceil\right).$$

As usual, in Theorem 2.14 c may depend on n within the given confine. Note also that if the conclusion holds for some c, it holds also for positive $c' < c$, provided n is sufficiently large. This implies Erdős's Theorem 2.12.

2.3.2 Joints and books Erdős [35] proved that if $r \geq 2$ and $n > n_0(r)$, every graph $G = G(n, t_r(n) + 1)$ has an edge that is contained in at least $n^{r-1}/(10r)^{6r}$ cliques of order $(r + 1)$. This fundamental fact seems so important, that in [14] we found it necessary to give the following definition:

*An r-**joint** of size t is a collection of t distinct r-cliques sharing an edge.*

Note that two r-cliques of an r-joint may share up to $r - 1$ vertices and that for $r > 3$ there may be many non-isomorphic r-joints of the same size. We shall write $js_r(G)$ for the maximum size of an r-joint in a graph G; in particular, if $2 \leq r \leq n$ and r divides n, then $js_r(T_r(n)) = \left(\frac{n}{r}\right)^{r-2}$.

In this notation, the above result of Erdős reads: *if $r \geq 2$, $n > n_0(r)$, and $G \in \mathcal{G}(n, t_r(n) + 1)$, then*

$$js_{r+1}(G) \geq \frac{n^{r-1}}{(10r)^{6r}}. \tag{2.1}$$

In fact, the study of $js_3(G)$, also known as the *booksize* of G, was initiated by Erdős even earlier, in [30], and was subsequently generalized in [34] and [35]; it seems that he foresaw the importance of joints when he restated his general result in 1995, in [36]. A quintessential result concerning joints is the "triangle removal lemma" of Ruzsa and Szemerédi [87], which can be stated as a lower bound on the booksize $js_3(G)$ when G is a graph of a particular kind.

In fact joints help to obtain several of the results mentioned in this survey, e.g., the general stability Theorem 2.19 and its spectral version, Theorem 3.10. Later, we shall give also spectral conditions for the existence of large joints, in Theorem 3.8.

In [14], Bollobás and the author enhanced the bound of Erdős (2.1) to the following explicit one.

Theorem 2.15 *Let $r \geq 2$, $n > r^8$, and let G be a graph of order n. If $e(G) \geq t_r(n)$, then*

$$js_{r+1}(G) > \frac{n^{r-1}}{r^{r+5}}$$

unless $G = T_r(n)$.

In [16] an analogous theorem is given in the case when G has many r-cliques, rather than edges. More precisely, letting $k_r(G)$ stand for the number of r-cliques of a graph G, we have

Theorem 2.16 *Let $r \geq 2$, $n > r^8$, and let G be a graph of order n. If $k_s(G) \geq k_s(T_r(n))$ for some s, $(2 \leq s \leq r)$, then*

$$js_{r+1}(G) > \frac{n^{r-1}}{r^{2r+12}}$$

unless $G = T_r(n)$.

Note that Theorems 2.15 and 2.16 cannot be improved too much, as shown by the graph G obtained by adding an edge to $T_r(n)$: we have $k_s(G) \geq k_s(T_r(n))$ but $js_{r+1}(G) \leq \lceil n/r \rceil^{r-1}$. However, the best bound in Theorem 2.15 is known only for 3-joints. Usually a 3-joint of size t is called a *book of size t*. Edwards [28], and independently Khadžiivanov and Nikiforov [56] proved the following theorem.

Theorem 2.17 *If G is a graph of order n with $e(G) > \lfloor n^2/4 \rfloor$, then it contains a book of size greater than $n/6$.*

This theorem is best possible in view of the following graph. Let $n = 6k$. Partition $[n]$ into 6 sets A_{11}, A_{12}, A_{13}, A_{21}, A_{22}, A_{23} with $|A_{11}| = |A_{12}| = |A_{13}| = k - 1$ and $|A_{21}| = |A_{22}| = |A_{23}| = k + 1$. For $1 \leq j < k \leq 3$ join every vertex of A_{ij} to every vertex of A_{ik} and for $j = 1, 2, 3$ join every vertex of A_{1j} to every vertex of A_{2j}. The resulting graph has size $> \lfloor n^2/4 \rfloor + 1$ and its booksize is $k + 1 = n/6 + 1$.

A more recent presentation of these results can be found in [12].

2.4 Stability problems

This subsection has three parts. First we sharpen the classical stability theorem of Erdős [33],[34] and Simonovits [88], which gives information about the structure of graphs without fixed forbidden subgraphs and whose size is close to the maximum possible. Second, we give several specific stability theorems for specific forbidden subgraphs, where stronger conclusions are possible. Lacking a better term, we call such cases *strong stability*.

Finally, we discuss the structure of K_r-free graphs of large minimum degree. This is a rich area with many results and a long history. It is not customary to consider it in the context of stability problems, but we believe this is the general category where this area belongs, since most of its statements can be phrased so that large minimum degree of a K_r-free graph implies a certain structure.

2.4.1 A general stability theorem Let F be a fixed $(r + 1)$-partite graph F and G be a graph of order n. The theorem of Erdős and Stone implies that if $\varepsilon > 0$ and $e(G) > (1 - 1/r + \varepsilon) n^2/2$, then G contains F, when n is sufficiently large. On the other hand, $T_r(n)$ is r-partite and therefore does not contain F, although

$$e(T_r(n)) = t_r(n) \approx (1 - 1/r) n^2/2.$$

Erdős and Simonovits [33],[34],[88] noticed that if a graph G of order n contains no copy of F and has close to $(1 - 1/r) n^2/2$ edges, then G is similar to $T_r(n)$.

Theorem 2.18 *Let $r \geq 2$ and let F be a fixed $(r + 1)$-partite graph. For every $\delta > 0$, there is an $\varepsilon > 0$ such that if G is a graph of order n with $e(G) > (1 - 1/r - \varepsilon) n^2/2$, then either G contains F or G differs from $T_r(n)$ in fewer than δn^2 edges.*

A closer inspection of this statement reveals that ε depends both on δ and on F. To investigate this dependence, we simplify the picture by assuming that F is a complete $(r + 1)$-graph. Moreover, radically departing from the setup of fixed F, we assume that $F = K_{r+1} \left(\lfloor c \log n \rfloor ; \lceil n^{1-\sqrt{c}} \rceil \right)$ for some $c > 0$. Note that for a given n the single real parameter c characterizes F completely. It turns out with this selection of F we still can get an enhancement of Theorem 2.18, as proved in [66].

Theorem 2.19 *Let $r \geq 2$, let c, ε and n be such that*

$$0 < c < r^{-3(r+14)(r+1)}, \quad 0 < \varepsilon < r^{-24}, \quad n > e^{1/c},$$

and let G be a graph of order n. If $e(G) > (1 - 1/r - \varepsilon) n^2/2$, then one of the following statements holds:

(a) G contains a $K_{r+1}\left(\lfloor c \log n \rfloor ; \left\lceil n^{1-\sqrt{c}} \right\rceil\right)$;

(b) G differs from $T_r(n)$ in fewer than $\left(\varepsilon^{1/3} + c^{1/(3r+3)}\right) n^2$ edges.

Note that, as usual, c may depend on n. A natural question is how tight Theorem 2.19 is. The complete answer seems difficult since two parameters, ε and c, are involved. First, the factor $\left(\varepsilon^{1/3} + c^{1/(3r+3)}\right)$ in condition (b) is far from the best one, but is simple. However for fixed c condition (a) is best possible up to a constant factor. Indeed, let $\alpha > 0$ be sufficiently small. A randomly chosen graph of order n with $(1 - \alpha) n^2/2$ edges contains no $K_2(\lfloor c' \log n \rfloor, \lfloor c' \log n \rfloor)$ and differs from $T_r(n)$ in more that $c'' n^2$ edges for some positive c' and c'', independent of n.

2.4.2 Strong stability For certain forbidden graphs condition (ii) of Theorem 2.19 can be strengthened. Such particular stability theorems can be of interest in applications, e.g., Ramsey problems. We start with a theorem in [84], which gives a particular stability condition for K_{r+1}-free graphs.

Theorem 2.20 Let $r \geq 2$ and $0 < \varepsilon \leq 2^{-10} r^{-6}$, and let G be a K_{r+1}-free graph of order n. If $e(G) > (1 - 1/r - \varepsilon) n^2/2$, then G contains an induced r-partite graph H of order at least $(1 - 2\sqrt[3]{\varepsilon}) n$ and with minimum degree $\delta(H) \geq (1 - 1/r - 4\sqrt[3]{\varepsilon}) n$.

Note that the stability condition in this theorem is stronger than condition (b) of Theorem 2.19. Indeed, the classes of H are almost equal, it is almost complete, and contains almost all vertices of G. This type of conclusion is the purpose of the three theorems below. In the first two of them the premise "K_{r+1}-free" will be further weakened; but Theorem 2.20 is still of interest, because it is proved for all conceivable n.

The following two theorems have been proved in [67] and [14].

Theorem 2.21 Let $r \geq 2$, let c, ε and n be such that

$$0 < c < r^{-(r+7)(r+1)}/2, \quad 0 < \varepsilon < r^{-8}/8, \quad n > e^{2/c},$$

and let G be a graph of order n. If $e(G) > (1 - 1/r - \varepsilon) n^2/2$, then one of the following statements holds:

(a) G contains a $K_r^+\left(\lfloor c \log n \rfloor ; \left\lceil n^{1-2\sqrt{c}} \right\rceil\right)$;

(b) G contains an induced r-partite subgraph H of order at least $(1 - \sqrt{2\varepsilon}) n$, with minimum degree

$$\delta(H) > \left(1 - 1/r - 2\sqrt{2\varepsilon}\right) n.$$

Theorem 2.22 Let $r \geq 2$, let c and n be such that

$$r \geq 2, \quad 0 < \varepsilon < r^{-8}/32, \quad n > r^8,$$

and let G be a graph of order n. If $e(G) > (1 - 1/r - \varepsilon) n^2/2$, then one of the following statements holds:

(a) $j s_{r+1}(G) > \left(1 - 1/r^3\right) n^{r-1}/r^{r+5}$;

(b) G contains an induced r-partite subgraph H of order at least $(1 - 4\sqrt{\varepsilon})\,n$, with minimum degree

$$\delta\,(H) > \left(1 - 1/r - 6\sqrt{\varepsilon}\right)n.$$

As one can expect, the analogous statement for books is quite close to the best possible [12].

Theorem 2.23 Let $0 < \varepsilon < 10^{-5}$ and let G be a graph of order n. If $e\,(G) > (1/4 - \varepsilon)\,n^2$, then either G contains a book of size at least $(1/6 - 2\sqrt[3]{\varepsilon})\,n$ or G contains an induced bipartite graph H of order at least $(1 - \sqrt[3]{\varepsilon})\,n$ and with minimal degree $\delta\,(H) \geq (1/2 - 4\sqrt[3]{\varepsilon})\,n$.

2.4.3 K_r-free graphs with large minimum degree   A famous theorem of Andrásfai, Erdős and Sós [1] shows that if $r \geq 2$ and G is a K_{r+1}-free graph of order n and with minimum degree satisfying

$$\delta\,(G) > \left(1 - \frac{3}{3r - 1}\right)n, \tag{2.2}$$

then G is r-partite. They also gave an example showing that equality in (2.2) is not sufficient to get the same conclusion.

In particular, for $r = 2$ this statement says that every triangle-free graph of order n with minimum degree $\delta\,(G) > 2n/5$ is bipartite. On the other hand, Hajnal [41] constructed a triangle-free graph of order n with arbitrary large chromatic number and with minimum degree $\delta\,(G) > (1/3 - \varepsilon)\,n$. In view of Hajnal's example, Erdős and Simonovits [41] conjectured that all K_3-free graphs of order n with $\delta\,(G) > n/3$ are 3-chromatic. However, this conjecture was disproved by Häggkvist [49], who described for every $k \geq 1$ a $10k$-regular, 4-chromatic, triangle-free graph of order $29k$. The example of Häggkvist is based on the Mycielski graph M_3, also known as the Grötzsch graph, which is a 4-chromatic triangle-free graph of order 11. To construct M_3, let $v_1, \ldots, v_5$ be the vertices of a 5-cycle and choose 6 other vertices $u_1, \ldots, u_6$. Join u_i to the neighbors of v_i for all $i = 1, \ldots, 5$, and finally join u_6 to $u_1, \ldots, u_5$.

Other graphs that are crucial in these questions are the triangle-free, 3-chromatic Andrásfai graphs $A_1, A_2, \ldots$, first described in [3]: set $A_1 = K_2$ and for every $i \geq 2$ let A_i be the complement of the $(i - 1)$-th power of the cycle C_{3i-1}.

To state the next structural theorems we need the following definition: a graph G is said to be **homomorphic** to a graph H, if there exists a map $f : V\,(G) \to V\,(H)$ such that $uv \in E\,(G)$ implies that $f\,(u)\,f\,(v) \in E\,(H)$.

In [53], Jin generalized the case $r = 2$ of the theorem of Andrásfai, Erdős and Sós and a result of Häggkvist from [49] in the following theorem.

Theorem 2.24 Let $1 \leq k \leq 9$, and let G be a triangle-free graph of order n. If

$$\delta\,(G) > \frac{k + 1}{3k + 2}n,$$

then G is homomorphic to A_k.

Note that this result is tight: taking the graph A_{k+1}, and blowing it up by a factor t, we obtain a triangle-free graph G of order $n = (3k + 2)t$ vertices, with $\delta(G) = (k + 1)n/(3k + 2)$, which is not homomorphic to A_k.

Note also that all graphs satisfying the premises of Theorem 2.24 are 3-chromatic. Addressing this last issue, Jin [54], and Chen, Jin and Koh [22] gave a finer characterization of all K_3-free graphs with $\delta > n/3$.

Theorem 2.25 *Let G be a triangle-free graph of order n, with $\delta(G) > n/3$. If $\chi(G) \geq 4$, then $M_3 \subset G$. If $\chi(G) = 3$ and*

$$\delta(G) > \frac{k+1}{3k+2}n,$$

then G is homomorphic to A_k.

Finally, Brandt and Thomassé [20] gave the following ultimate result.

Theorem 2.26 *Let G be a triangle-free graph of order n. If $\delta(G) > n/3$, then $\chi(G) \leq 4$.*

It is natural to ask the same questions for K_r-free graphs with large minimum degree. Contrary to expectation, the answers are by far easier. First, the graphs of Andrásfai, Hajnal and Häggkvist are easily generalized by joining them with appropriately chosen complete $(r - 3)$-partite graphs.

In particular, for every ε there exists a K_{r+1}-free graph of order n with

$$\delta(G) > \left(1 - \frac{2}{2r - 1} - \varepsilon\right)n$$

and arbitrary large chromatic number, provided n is sufficiently large.

Hence, the main question is: *how large $\chi(G)$ can be when G is a K_{r+1}-free graph of order n with $\delta(G) > (1 - 2/(2r - 1))n$.* The answer is:

Theorem 2.27 *Let $r \geq 2$ and let G be a K_{r+1}-free graph of order n. If*

$$\delta(G) > \left(1 - \frac{2}{2r - 1}\right)n,$$

then $\chi(G) \leq r + 2$.

This theorem leaves only two cases to investigate, viz., $\chi(G) = r+1$ and $\chi(G) = r + 2$. As one can expect, when $\delta(G)$ is sufficiently large, we have $\chi(G) = r + 1$. The precise statement extends Theorem 2.24 as follows.

Theorem 2.28 *Let $r \geq 2$, $1 \leq k \leq 9$, and let G be a K_{r+1}-free graph of order n. If*

$$\delta(G) > \left(1 - \frac{2k - 1}{(2k - 1)r - k + 1}\right)n$$

then G is homomorphic to $A_k + K_{r-2}$.

As a corollary, under the premises of Theorem 2.28, we find that $\chi(G) \leq r + 1$. Also Theorem 2.28 is best possible in the following sense: for every k and n, there exists an $(r + 1)$-chromatic K_{r+1}-free G of order n with

$$\delta(G) \geq \left(1 - \frac{2k - 1}{(2k - 1)r - k + 1}\right) n - 1$$

that is not homomorphic to $A_k + K_{r-2}$.

Using the example of Häggkvist, we construct for every n an $(r + 2)$-chromatic, K_{r+1}-free graph G with

$$\delta(G) \geq \left(1 - \frac{19}{19r - 9}\right) n - 1,$$

which shows that the conclusion of Theorem 2.28 does not necessarily hold for $k \geq 10$.

To give some further structural information, we extend Theorem 2.26 as follows.

Theorem 2.29 *Let $r \geq 2$ and G be a K_{r+1}-free graph of order n with*

$$\delta(G) > \left(1 - \frac{2}{2r - 1}\right) n.$$

If $\chi(G) \geq r + 2$, then $M_3 + K_{r-2} \subset G$. If $\chi(G) \leq r + 1$ and

$$\delta(G) > \left(1 - \frac{2k - 1}{(2k - 1)r - k + 1}\right) n$$

then G is homomorphic to $A_k + K_{r-2}$.

This result is best possible in view of the examples described prior to Theorem 2.29.

We deduce the proofs of Theorems 2.27, 2.28 and 2.29 by induction on r from Theorems 2.26, 2.24 and 2.25 respectively. The induction step, carried out uniformly in all the three proofs, is based on the crucial Lemma 4.5. This lemma can be applied immediately to extend other results about triangle-free graphs.

The new results in this subsection, together with Lemma 4.5 have been published in [68]. Since the first version of that paper was made public, the author learned that similar research has been undertaken independently by W. Goddard and J. Lyle [48].

2.5 The number of graphs with large forbidden subgraphs

An intriguing question is how many graphs with given properties there are. For certain natural properties such as "G is K_r-free" or "G has no induced graph isomorphic to H" satisfactory answers have been obtained. Thus, given a graph H, write $\mathcal{P}_n(H)$ for the set of all labelled graphs of order n not containing H. A classical result of Erdős, Kleitman and Rothschild [38] states that

$$\log_2 |\mathcal{P}_n(K_{r+1})| = (1 - 1/r + o(1)) \binom{n}{2}. \tag{2.3}$$

Ten years later, Erdős, Frankl and Rödl [37] showed that the conclusion in (2.3) remains valid if K_{r+1} is replaced by an arbitrary fixed $(r + 1)$-chromatic graph H.

In fact, as shown in [15], the conclusion in (2.3) also remains valid if K_{r+1} is replaced by a sequence of forbidden graphs whose order grows with n. Until recently such results seemed to be out of reach; however, the framework laid out in [63] and [64] has opened new possibilities. Here is the theorem that directly generalizes the Erdős-Frankl-Rödl result.

Theorem 2.30 *Given $r \geq 2$ and $0 < \varepsilon \leq 1/2$, there exists $\delta = \delta(\varepsilon) > 0$ such that for n sufficiently large,*

$$(1 - 1/r)\binom{n}{2} \leq \log_2 \left|\mathcal{P}_n\left(K_{r+1}\left(\lfloor \delta \log n \rfloor ; \lceil n^{1-\sqrt{\delta}} \rceil\right)\right)\right| \leq (1 - 1/r + \varepsilon)\binom{n}{2}.$$
(2.4)

Note that the real contribution of Theorem 2.30 is the upper bound in (2.4) since the lower bound follows by counting the labelled spanning subgraphs of the Turán graph $T_r(n)$. Let us mention that the proof of Theorem 2.30 does not use Szemerédi's Regularity Lemma, which is a standard tool for such questions.

Similar statements can be proved for forbidden induced subgraphs, where the role of the chromatic number is played by the *coloring number* χ_c of a graph property, introduced first in [17], and defined below.

Let $0 \leq s \leq r$ be integers and let $\mathcal{H}(r, s)$ be the class of graphs whose vertex sets can be partitioned into s cliques and $r - s$ independent sets. Given a graph property $\mathcal{P}$, the coloring number $\chi_c(\mathcal{P})$ is defined as

$$\chi_c(H) = \max\{r : \mathcal{H}(r, s) \subset \mathcal{P} \text{ for some } s \in [r]\}$$

Also, given a graph H, let us write $\mathcal{P}_n^*(H)$ for the set of graphs of order n not containing H as an induced subgraph. Alexeev [2] and, independently Bollobás and Thomason [17],[18] proved that the exact analog of the result of Erdős, Frankl and Rödl holds:

If H is a fixed graph and $r = \chi_c(\mathcal{P}_n^*(H))$, then

$$\log_2 |\mathcal{P}_n^*(H)| = (1 - 1/r + o(1))\binom{n}{2}.$$
(2.5)

This result also can be extended by replacing H with a sequence of forbidden graphs whose order grows with n. To this end, recall the definition of a graph *of type $H(k_1, \ldots, k_h)$*, where H is a fixed labelled graph of order h and $k_1, \ldots, k_h$ are positive integers (Subsection 2.2.2). Informally, a graph of type $H(k_1, \ldots, k_h)$ is obtained by first "blowing-up" H to $H(k_1, \ldots, k_h)$ and then adding (possibly zero) edges to the vertex classes of the "blow-up" but keeping intact the edges across vertex classes.

Now, given a labelled graph H and positive integers p and q, let $\mathcal{P}_n(H; p, q)$ be the set of labelled graphs of order n that contain no induced subgraph of type $H(p, \ldots, p, q)$.

Here is the result for forbidden induced subgraphs, also from [15].

Theorem 2.31 *Let H be a labelled graph and let $r = \chi_c\left(\mathcal{P}_n^*(H)\right)$. For every $\varepsilon > 0$, there is $\delta = \delta(\varepsilon) > 0$ such that for n sufficiently large*

$$(1 - 1/r)\binom{n}{2} \leq \log_2 \left| \mathcal{P}_n\left(H; \lfloor \delta \log n \rfloor, \left\lceil n^{1-\sqrt{\delta}}\right\rceil \right)\right| \leq (1 - 1/r + \varepsilon)\binom{n}{2}. \quad (2.6)$$

In a sense Theorems 2.30 and 2.31 are almost best possible, in view of the following simple observation, which can be proved by considering the random graph $G_{n,p}$ with $p \to 1$.

Given $r \geq 2$ and $\varepsilon > 0$, there exists $C > 0$ such that the number S_n of labelled graphs which do not contain $K_2\left(\lceil C \log n\rceil, \lceil C \log n\rceil\right)$ satisfies $S_n \geq (1 - \varepsilon)\,2^{\binom{n}{2}}$.

3   Spectral extremal graph theory

Generally speaking, spectral graph theory investigates graphs using the spectra of various matrices associated with graphs, such as the adjacency matrix. For an introduction to this topic we refer the reader to [23].

Given a graph G with vertex set $\{v_1, \ldots, v_n\}$, the adjacency matrix of G is a matrix $A = [a_{ij}]$ of size n given by

$$a_{ij} = \begin{cases} 1, & \text{if } (v_i, v_j) \in E(G); \\ 0, & \text{otherwise.} \end{cases}$$

Note that A is symmetric and nonnegative, and much is known about the spectra of such matrices. For instance, the eigenvalues of A are real numbers, which we shall denote by $\mu_1(G), \ldots, \mu_n(G)$, indexed in non-increasing order. The value $\mu(G) = \mu_1(G)$ is called the spectral radius of G and has maximum absolute value among all eigenvalues.

Another matrix that we shall use is the Laplacian matrix L, defined as $D(G) - A(G)$, where $D(G)$ is the diagonal matrix of the row-sums of A, i.e., the degrees of G. The eigenvalues of the Laplacian are denoted by $\lambda_1(G), \ldots, \lambda_n(G)$, indexed in non-decreasing order.

A third matrix associated with graphs is the Q-matrix or the "signless Laplacian", defined as $Q = D + A$. The eigenvalues of Q are denoted by $q_1(G), \ldots, q_n(G)$, indexed in non-increasing order. The Q-matrix has received a lot of attention in recent years, see, e.g., [24],[25] and [26]. The Laplacian and the Q-matrix are positive semi-definite matrices, and $\lambda_1(G) = 0$.

3.1   The spectral problems that are studied

How large can be the spectral radius $\mu(G)$ when G is a K_r-free graph of order n? Such questions come easily to the mind when one studies extremal graph problems. In fact, with any extremal problem of the type *"What is the maximum number of edges in a graph G of order n with property $\mathcal{P}$?"* goes a spectral analog: *"What is the maximum spectral radius of a graph G of order n with property $\mathcal{P}$?"* This is not merely a superficial analogy since if we have a solution of the spectral problem, then by the fundamental inequality

$$\mu(G) \geq 2e(G)/n, \quad (3.1)$$

we immediately obtain an upper bound on $e(G)$ as well. The use of this implication is illustrated on several occasions below; in particular, for the Zarankiewicz problem we obtain the sharpest bounds on $e(G)$ known so far.

On the other hand, inequality (3.1) suggests a way to conjecture spectral results by taking known nonspectral extremal statements that involve the average degree of a graph and replacing the average degree by $\mu(G)$. More often than not, the resulting statement is correct and even stronger, but of course it needs its own proof. To create suitable proof tools we painstakingly built several technical but rather flexible statements such as Theorem 4.8 and Lemma 4.6.

This smooth machinery is sufficient to prove spectral analogs of most of the extremal problems discussed in Section 2 and of several others as well. Among these results are: various forms of Turán's theorem, the Erdős-Stone-Bollobás theorem, conditions for large joints and for odd cycles; a general stability theorem and several strong stability theorems, an asymptotic solution of the general extremal problem for non-bipartite forbidden subgraphs, the Zarankiewicz problem, sufficient conditions for paths and cycles, sufficient conditions for Hamilton paths and cycles.

Despite these successful translations, more can be expected from spectral extremal graph theory, which seems inherently richer than the conventional one. Indeed, we give also a fair number of spectral results that have no conventional analog, for example, results involving the smallest eigenvalue of the adjacency matrix or the spectral radius of the Laplacian matrix.

3.2 Spectral forms of the Turán theorem

In 1986, Wilf [90] showed that if G is a graph of order n with clique number $\omega(G) = \omega$, then

$$\mu(G) \le (1 - 1/\omega)n. \tag{3.2}$$

Note first that in view of the inequality $\mu(G) \ge 2e(G)/n$, (3.2) implies the concise Turán theorem:

$$e(G) \le (1 - 1/\omega)n^2/2. \tag{3.3}$$

However, inequality (3.2) opens many other new possibilities. Indeed, if we combine (3.2) with other lower bounds on $\mu(G)$, e.g., with

$$\mu^2(G) \ge \frac{1}{n} \sum_{u \in V(G)} d^2(u),$$

we obtain other forms of (3.3). An infinite class of similar lower bounds is given in [70].

Below we sharpen inequality (3.2) in two ways.

A concise spectral Turán theorem In 1970 Nosal [85] showed that every triangle-free graph G satisfies $\mu(G) \le \sqrt{e(G)}$. This result was extended in [69] and [75] in the following theorem, conjectured by Edwards and Elphick in [29]:

Theorem 3.1 *If G is a graph of order n and $\omega(G) = \omega$, then*

$$\mu^2(G) \le 2(1 - 1/\omega)e(G). \tag{3.4}$$

If G has no isolated vertices, then equality is possible if and only if one of the following conditions holds:
(a) $\omega = 2$ and G is a complete bipartite graph;
(b) $\omega \geq 3$ and G is a complete regular ω-partite graph.

In view of (3.3), we see that

$$\mu^2(G) \leq 2(1 - 1/\omega)\, m \leq 2(1 - 1/\omega)(1 - 1/\omega)\, n^2/2 = ((1 - 1/\omega)\, n)^2,$$

and so (3.4) implies (3.2).

As shown in [69], inequality (3.4) follows from the celebrated result of Motzkin and Straus [61]:

Let G be a graph of order n with cliques number $\omega(G) = \omega$. If $(x_1, \ldots, x_n)$ is a vector with nonnegative entries, then

$$\sum_{uv \in E(G)} x_u x_v \leq \frac{\omega - 1}{2\omega} \left(\sum_{u \in V(G)} x_u \right)^2. \tag{3.5}$$

On the other hand, this result follows in turn from the concise Turán theorem, as shown in [71]. The implications

$$(3.4) \implies (3.3) \implies \mathrm{MS} \implies (3.4)$$

justify regarding inequality (3.4) as a flexible spectral form of the concise Turán theorem.

Next we extend Theorem 3.1 in a somewhat unexpected direction. Recall that, a k-walk in a graph G is a sequence of vertices $v_1, \ldots, v_k$ of G such that v_i is adjacent to v_{i+1} for $i = 1, \ldots, k - 1$; write $w_k(G)$ for the number of k-walks in G. Observing that $2e(G) = w_2(G)$, we see that the following theorem, given in [70] , extends inequality (3.4).

Theorem 3.2 *If $r \geq 1$ and G is a graph with clique number $\omega(G) = \omega$, then*

$$\mu^r(G) \leq (1 - 1/\omega)\, w_r(G). \tag{3.6}$$

Suppose that G has no isolated vertices and equality holds for some $r \geq 1$.
(i) If $r = 1$, then G is a regular complete ω-partite graph.
(ii) If $r \geq 2$ and $\omega > 2$, then G is a regular complete ω-partite graph.
(iii) If $r \geq 2$ and $\omega = 2$, then G is a complete bipartite graph, and if r is odd, then G is regular.

It is somewhat surprising that for $r \geq 2$ the number of vertices of G is not relevant in this theorem.

A precise spectral Turán theorem Since Wilf's inequality (3.2) becomes equality only when ω divides n, one can expect that some fine tuning is still possible. Indeed, in [72] we sharpened inequality (3.2), bringing it the closest possible to the Turán theorem.

Theorem 3.3 *If G is a graph of order n with no complete subgraph of order $r + 1$, then $\mu(G) \leq \mu(T_r(n))$. Equality holds if and only if $G = T_r(n)$.*

Here is an equivalent, shorter form of this statement: *If $G \in \mathcal{G}(n)$ and $\omega(G) = \omega$, then $\mu(G) < \mu(T_\omega(n))$ unless $G = T_\omega(n)$.*

Note also that $\mu(T_2(n)) = \sqrt{\lfloor n^2/4 \rfloor}$; for $\omega \geq 3$ there is also a closed expression for $\mu(T_\omega(n))$, but it is somewhat cumbersome.

Spectral radius and independence number   One wonders if there is a theorem about the independence number $\alpha(G)$, similar to the Turán theorem. One obvious answer is obtained by restating the concise Turán theorem in complementary terms

$$2e(G) \geq n^2/\alpha(G) - n,$$

which immediately implies that $\mu(G) \geq n/\alpha(G) - 1$ as well. Note that here the spectral statement follows from the conventional one. However, a proof by induction on α gives the following sharper result.

Theorem 3.4 *If $G \in G(n)$ and $\alpha(G) = \alpha$, then $\mu(G) \geq \lceil n/\alpha \rceil - 1$.*

In a different direction, for connected graphs and some special values of α, more specific results have been proved in [91].

Also, by the well-known inequality $q_1(G) \geq 2\mu_1(G)$, Theorem 3.4 proves Conjecture 27 from [50].

3.3   A spectral Erdős-Stone-Bollobás theorem

Having seen various spectral forms of the Turán theorem, one can expect that many other results that surround it can be cast in spectral form as well; and this is indeed the case. The following theorem, given in [78], is the spectral analog of the Erdős-Stone-Bollobás theorem, more precisely, of Theorem 2.5.

Theorem 3.5 *Let $r \geq 3$, let c and n be such that*

$$0 < c < 1/r!, \quad n \geq \exp\left((r^r/c)^r\right),$$

and let G be a graph of order n. If

$$\mu(G) \geq (1 - 1/(r-1) + c)n, \tag{3.7}$$

then G contains a $K_r(s;t)$ with $s \geq \lfloor (c/r^r)^r \log n \rfloor$ and $t > n^{1-c^{r-1}}$.

Let us emphasize that the functionality of Theorem 2.5 is entirely preserved: in particular, c may depend on n, e.g., letting $c = 1/\log\log n$, the conclusion is meaningful for sufficiently large n.

Since $\mu(G) \geq 2e(G)/v(G)$, Theorem 3.5, in fact, implies Theorem 2.5. Other lower bounds on $\mu(G)$, such as those given in [70], imply other new versions of this theorem.

Suppose that c is a sufficiently small constant. Choosing randomly a graph G of order n with $\lceil (1 - 1/(r-1) + 2c)n^2/2 \rceil$ edges, we have $\mu(G) \geq (1 - 1/(r-1) + c)n$, but G contains no $K_2 \left(\lfloor C \log \rfloor, \lfloor C \log n \rfloor \right)$ for some $C > 0$, independent of n. Hence, for constant c, Theorem 3.5 is best possible up to a constant factor.

We close this topic with a consequence of Theorem 3.5, given in [78], that solves asymptotically the following general extremal problem: *Given a family $\mathcal{F}$ of non-bipartite forbidden subgraphs, what is the maximum spectral radius of a graph of order n containing no member of $\mathcal{F}$.*

Theorem 3.6 *Let $r \geq 3$ and let $F_1, F_2, \ldots$ be r-partite graphs satisfying $v(F_n) = o(\log n)$. Then*

$$\max \{ \mu(G) : G \in \mathcal{G}(n) \text{ and } F_n \not\subseteq G \} = \left(1 - \frac{1}{r-1} \right) n + o(n). \qquad (3.8)$$

It is likely that in the setup of Theorem 3.6 the condition $v(F_n) = o(\log n)$ can be sharpened.

3.4   Saturation problems

The precise spectral Turán theorem implies that if G is a graph of order n with $\mu(G) > \mu(T_r(n))$, then G contains a K_{r+1}. Since this setting is analogous to the case when $e(G) > t_r(n)$, one would expect much larger supergraphs of K_{r+1}. In fact, as we shall see, all results from Subsection 2.3 have their spectral analogs. In addition, it is not difficult to show that if G is a graph of order n, then the inequality $e(G) > e(T_r(n))$ implies the inequality $\mu(G) > \mu(T_r(n))$. Therefore, the spectral theorems below imply the corresponding nonspectral extremal results, albeit with somewhat narrower ranges of the parameters.

We start with the spectral analog of Theorem 2.14, given in [76].

Theorem 3.7 *Let $r \geq 2$, let c and n be such that*

$$0 < c \leq r^{-(2r+9)(r+1)}, \quad n \geq \exp(2/c),$$

and let G be a graph of order n. If $\mu(G) > \mu(T_r(n))$, then G contains a

$$K_r^+ \left(\lfloor c \log n \rfloor ; \left\lceil n^{1-\sqrt{c}} \right\rceil \right).$$

Theorem 3.7 is essentially best possible since for every $\varepsilon > 0$, choosing randomly a graph G of order n with $e(G) = \left\lceil (1 - \varepsilon) n^2/2 \right\rceil$, we see that $\mu(G) > (1 - \varepsilon) n$, but G contains no $K_2 \left(\lfloor c \log n \rfloor \right)$ for some $c > 0$, independent of n.

The theorem corresponding to Theorem 2.15 is given in [76]. We state it here in a somewhat refined form.

Theorem 3.8 *Let $r \geq 2$, $n > r^{15}$, and let G be a graph of order n. If $\mu(G) \geq \mu(T_r(n))$, then*

$$js_{r+1}(G) > n^{r-1}/r^{2r+4}$$

unless $G = T_r(n)$.

Theorem 3.8 and its stability complement Theorem 3.12 are crucial in the proof of several other spectral extremal results.

It is easy to see the Turán graph $T_2(n)$ contains no odd cycles and that $\mu(T_2(n)) = \sqrt{\lfloor n^2/4 \rfloor}$. Hence the following theorem gives a sharp spectral condition for the existence of odd cycles.

Theorem 3.9 *Let G be a graph of sufficiently large order n. If $\mu(G) > \sqrt{\lfloor n^2/4 \rfloor}$, then G contains a cycle of length t for every $t \leq n/320$.*

This theorem, given in [79], is motivated by the following result of Bollobás ([6], p. 150): *if G is a graph of order n with $e(G) > \lfloor n^2/4 \rfloor$, then G contains a cycle of length t for every $t = 3, \ldots, \lceil n/2 \rceil$.*

3.5 Stability problems

We shall show that most stability results from Subsection 2.4 have their spectral analogs. However, we could not find spectral analogs of the stability problems that involve minimum degree (Subsection 2.4.3).

We first state a general spectral stability result, and then two stronger versions for specific graphs. We give here only Theorems 3.11 and 3.12 since they are important for various applications, but our machinery helped to deduce many others of somewhat lesser importance, and they can be found in [76] and [79].

The following analog of Theorem 2.19 was given in [76].

Theorem 3.10 *Let $r \geq 2$, let c, ε and n be such that*

$$0 < c < r^{-8(r+21)(r+1)}, \quad 0 < \varepsilon < 2^{-36}r^{-24}, \quad n > \exp(1/c),$$

and let G be a graph of order n. If $\mu(G) > (1 - 1/r - \varepsilon)n$, then one of the following statements holds:

(a) G contains a $K_{r+1}\left(\lfloor c\log \rfloor; \lceil n^{1-\sqrt{c}} \rceil\right)$;

(b) G differs from $T_r(n)$ in fewer than $\left(\varepsilon^{1/4} + c^{1/(8r+8)}\right) n^2$ edges.

The proofs of Theorem 2.19 and 3.10, given in [76] illustrate the isomorphism between the sets of tools developed for the spectral and nonspectral problems. The texts of the two proofs are almost identical, while the differences come from the use of different tools. We refer the reader to Section 5 for more details.

The next two theorems are crucial for several applications. The first one, proved in [13], is a spectral equivalent of Theorem 2.20.

Theorem 3.11 *Let $r \geq 2$ and $0 \leq \varepsilon \leq 2^{-10}r^{-6}$, and let G be a K_{r+1}-free graph of order n. If*

$$\mu(G) \geq (1 - 1/r - \varepsilon)n, \tag{3.9}$$

then G contains an induced r-partite graph H of order at least $\left(1 - 3\alpha^{1/3}\right)n$ and minimum degree

$$\delta(H) > \left(1 - 1/r - 6\varepsilon^{1/3}\right)n.$$

Finally, we have a spectral stability theorem for large joints, proved in [76].

Theorem 3.12 *Let $r \geq 2$, let ε and n be such that*

$$0 < \varepsilon < 2^{-10}r^{-6}, \quad n \geq r^{20},$$

and let G be a graph of order n. If $\mu(G) > (1 - 1/r - \varepsilon)n$, then G satisfies one of the conditions:

(a) $js_{r+1}(G) > n^{r-1}/r^{2r+5}$;

(b) *G contains an induced r-partite subgraph H of order at least $\left(1 - 4\varepsilon^{1/3}\right)n$ with minimum degree $\delta(H) > \left(1 - 1/r - 7\varepsilon^{1/3}\right)n$.*

3.6   The Zarankiewicz problem

What is the maximum spectral radius of a graph of order n with no $K_{s,t}$? This is a spectral version of the famous Zarankiewicz problem: *what is the maximum number of edges in a graph of order n with no $K_{s,t}$?* Except for few cases, no complete solution to either of these problems is known. For instance, Babai and Guiduli [5] have shown that

$$\mu \leq \left((s-1)^{1/t} + o(1)\right)n^{1-1/t}.$$

Using a different method, in [82] we improved this result as follows:

Theorem 3.13 *Let $s \geq t \geq 2$, and let G be a $K_{s,t}$-free graph of order n.*
(i) *If $t = 2$, then*

$$\mu(G) \leq 1/2 + \sqrt{(s-1)(n-1) + 1/4}. \tag{3.10}$$

(ii) *If $t \geq 3$, then*

$$\mu(G) \leq (s-t+1)^{1/t}n^{1-1/t} + (t-1)n^{1-2/t} + t - 2. \tag{3.11}$$

On the other hand, in view of the inequality $2e(G) \leq \mu(G)n$, we see that if G is a $K_{s,t}$-free graph of order n, then

$$e(G) \leq \frac{1}{2}(s-t+1)^{1/t}n^{2-1/t} + \frac{1}{2}(t-1)n^{2-2/t} + \frac{1}{2}(t-2)n. \tag{3.12}$$

This is a slight improvement of a result of Füredi [46] and this seems the best known bound on $e(G)$ so far.

For some values of s and t the bounds given by (3.10) and (3.11) are tight as we now demonstrate.

The case $t = 2$ For $s = t = 2$ inequality (3.10) shows that every $K_{2,2}$-free graph G of order n satisfies

$$\mu(G) \leq 1/2 + \sqrt{n - 3/4}.$$

This bound is tight because equality holds for the friendship graph (a collection of triangles sharing a single common vertex).

Also, Erdős-Renyi [40] showed that if q is a prime power, the polarity graph ER_q is a $K_{2,2}$-free graph of order $n = q^2 + q + 1$ and $q(q+1)^2/2$ edges. Thus, its spectral radius $\mu(ER_q)$ satisfies

$$\mu(ER_q) \geq \frac{q^3 + 2q^2 + q}{q^2 + q + 1} > q + 1 - \frac{1}{q} = 1/2 + \sqrt{n - 3/4} - \frac{1}{\sqrt{n} - 1},$$

which is also close to the upper bound.

For $s > 2$, equality in (3.10) is attained when G is a strongly regular graph in which every two vertices have exactly $s - 1$ common neighbors. There are examples of strongly regular graphs of this type; here is a small selection from Gordon Royle's webpage:

s	n	$\mu(G)$
3	45	12
4	96	20
5	175	30
6	36	15

We are not aware whether there are infinitely many strongly regular graphs in which every two vertices have the same number of common neighbors. However, Füredi [47] has shown that for any n there exists a $K_{s,2}$-free graph G_n of order n such that

$$e(G_n) \geq \frac{1}{2} n \sqrt{sn} + O\left(n^{4/3}\right),$$

and so,

$$\mu(G_n) \geq \sqrt{sn} + O\left(n^{1/3}\right);$$

thus (3.10) is tight up to low order terms.

The case $s = t = 3$ The bound (3.11) implies that if G is a $K_{3,3}$-free graph of order n, then

$$\mu(G) \leq n^{2/3} + 2n^{1/3} + 1.$$

On the other hand, a construction due to Alon, Rònyai and Szabò [4] implies that for all $n = q^3 - q^2$, where q is a prime power, there exists a $K_{3,3}$-free graph G_n of order n with

$$\mu(G_n) \geq n^{2/3} + \frac{2}{3} n^{1/3} + C$$

for some constant $C > 0$. Thus, the bound (3.11) is asymptotically tight for $s = t = 3$. The same conclusion can be obtained from Brown's construction of $K_{3,3}$-free graphs [19].

The general case   As proved in [4], there exists $c > 0$ such that for all $t \geq 2$ and $s \geq (t - 1)! + 1$, there is a $K_{s,t}$-free graph G_n of order n with

$$e(G_n) \geq \frac{1}{2} n^{2-1/t} + O\left(n^{2-1/t-c}\right).$$

Hence, for such s and t we have

$$\mu(G) \geq n^{1-1/t} + O\left(n^{1-1/t-c}\right);$$

thus, the bound (3.11) and also the bound of Babai and Guiduli give the correct order of the main term.

3.7 Paths and cycles

We give now some results about the maximum spectral radius of graphs of order n without paths or cycles of specified length. Writing C_k and P_k for the cycle and path of order k, let us define the functions

$$f_l(n) = \max\{\mu(G) : G \in \mathcal{G}(n) \text{ and } C_l \nsubseteq G\};$$
$$g_l(n) = \max\{\mu(G) : G \in \mathcal{G}(n) \text{ and } C_l \nsubseteq G, \text{ and } C_{l+1} \nsubseteq G\};$$
$$h_l(n) = \max\{\mu(G) : G \in \mathcal{G}(n) \text{ and } P_l \nsubseteq G\}.$$

For these functions we shall show below some exact expressions or at least good asymptotics. It should be noted that except for $f_l(n)$ when l is odd, these questions are quite different from their nonspectral analogs.

The lower bounds on $f_{2l}(n)$, $g_l(n)$ and $h_l(n)$ are given by two families of graphs, which for sufficiently large n give the exact values of $h_l(n)$, and perhaps also of $f_{2l}(n)$ and $g_l(n)$.

Suppose that $1 \le k < n$.

(1) Let $S_{n,k}$ be the graph obtained by joining every vertex of a complete graph of order k to every vertex of an independent set of order $n - k$, that is, $S_{n,k} = K_k \vee \overline{K}_{n-k}$;

(2) Let $S_{n,k}^+$ be the graph obtained by adding one edge within the independent set of $S_{n,k}$.

Note that $P_{l+1} \nsubseteq S_{n,k}$ and $C_l \nsubseteq S_{n,k}$ for $l \ge 2k + 1$. Likewise, $P_{l+1} \nsubseteq S_{n,k}$ and $C_l \nsubseteq S_{n,k}$ for $l \ge 2k + 2$.

Therefore,

$$h_{2k}(n) \ge \mu(S_{n,k}) = (k-1)/2 + \sqrt{kn - (3k^2 + 2k - 1)/4},$$
$$h_{2k+1}(n) \ge \mu\left(S_{n,k}^+\right) = (k-1)/2 + \sqrt{kn - (3k^2 + 2k - 1)/4} + 1/n + O\left(n^{-3/2}\right),$$
$$g_{2k}(n) \ge \mu(S_{n,k}) = (k-1)/2 + \sqrt{kn - (3k^2 + 2k - 1)/4},$$
$$g_{2k+1}(n) \ge \mu\left(S_{n,k}^+\right) = (k-1)/2 + \sqrt{kn - (3k^2 + 2k - 1)/4} + 1/n + O\left(n^{-3/2}\right),$$
$$f_{2k+2}(n) \ge \mu\left(S_{n,k}^+\right) = (k-1)/2 + \sqrt{kn - (3k^2 + 2k - 1)/4} + 1/n + O\left(n^{-3/2}\right).$$

Below we shall give also rather close upper bounds for these functions.

Forbidden odd cycle   In view of Theorem 3.9, we find that if l is odd and $n > 320l$, then

$$f_l(n) = \sqrt{\lfloor n^2/4 \rfloor}.$$

The smallest ratio n/l for which this equation is still valid is not known.

Clearly, for odd l we have $f_l(n) \sim n/2$, which is in sharp contrast to the value of $f_l(n)$ for even l.

Forbidden cycle C_4   The value of $f_4(n)$ is essentially determined in [72]:

Let G be a graph of order n with $\mu(G) = \mu$. If $C_4 \nsubseteq G$, then

$$\mu^2 - \mu \le n - 1. \tag{3.13}$$

Equality holds if and only if every two vertices of G have exactly one common neighbor, i.e., when G is the friendship graph.

An easy calculation implies that

$$f_4(n) = 1/2 + \sqrt{n - 3/4} + O(1/n),$$

where for odd n the $O(1/n)$ term is zero. Finding the precise value of $f_4(n)$ for even n is an open problem.

Here is a considerably more involved bound on the spectral radius of a C_4-free graph of given size, given in [77].

Theorem 3.14 *Let $m \geq 9$ and G be a graph with m edges. If $\mu(G) > \sqrt{m}$, then G has a 4-cycle.*

This theorem is tight, for all stars are C_4-free graphs with $\mu(G) = \sqrt{m}$. Also, let $S_{n,1}$ be the star of order n with an edge within its independent set. The graph $S_{n,1}$ is C_4-free and has n edges, but $\mu(G) > \sqrt{n}$ for $4 \leq n \leq 8$, while $\mu(S_{9,1}) = 3$.

Forbidden cycle C_{2k} The inequality (3.13) can be generalized for arbitrary even cycles as follows: *if $C_{2k+2} \nsubseteq G$, then*

$$\mu^2 - (k-1)\mu \leq k(n-1).$$

In fact, a slightly stronger assertion was proved in [81].

Theorem 3.15 *Let $k \geq 1$ and G be a graph of order n. If*

$$\mu(G) > k/2 + \sqrt{kn + (k^2 - 4k)/4},$$

then $C_{2l+2} \subset G$ for every $l = 1, \ldots, k$.

In view of the graph $S_{n,k}^+$, Theorem 3.15 implies that

$$(k-1)/2 + \sqrt{kn} + o(n) \leq f_{2k+2}(n) \leq k/2 + \sqrt{kn} + o(n). \qquad (3.14)$$

The exact value of $f_{2k+2}(n)$ is not known for $k \geq 2$, and finding this value seems a challenge. Nevertheless, the precision of (3.14) is somewhat surprising, given that the asymptotics of the maximum number of edges in C_{2k+2}-free graphs of order n is not known for $k \geq 2$.

Forbidden pair of cycles $\{C_{2k}, C_{2k+1}\}$ Let us consider now the function $g_l(n)$. To begin with, Favaron, Mahéo, and Saclé [45] showed that if a graph G of order n contains neither C_3 nor C_4, then $\mu(G) \leq \sqrt{n-1}$. Since the star of order n has no cycles and its spectral radius is $\sqrt{n-1}$, we see that

$$g_3(n) = \sqrt{n-1}.$$

We do not know the exact value of $g_l(n)$ for $l > 3$, but we have the following theorem from [81].

Theorem 3.16 *Let $k \geq 1$ and G be a graph of order n. If*

$$\mu(G) > (k-1)/2 + \sqrt{kn + (k+1)^2/4},$$

then $C_{2k+1} \subset G$ or $C_{2k+2} \subset G$.

Theorem 3.16, together with the graphs $S_{n,k}$ and $S_{n,k}^+$, gives

$$(k-1)/2 + \sqrt{kn} + o(n) \leq g_{2k+1}(n) \leq k/2 + \sqrt{kn} + o(n),$$
$$g_{2k}(n) = (k-1)/2 + \sqrt{kn} + \Theta\left(n^{-1/2}\right).$$

Forbidden path P_k   The function $h_k(n)$ is completely known for large n. As proved in [81]:

Theorem 3.17 *Let $k \geq 1$, $n \geq 2^{4k}$ and let G be a graph of order n.*
(i) If $\mu(G) \geq \mu(S_{n,k})$, then G contains a P_{2k+2} unless $G = S_{n,k}$.
(ii) If $\mu(G) \geq \mu\left(S_{n,k}^+\right)$, then G contains a P_{2k+3} unless $G = S_{n,k}^+$.

Theorem 3.17, together with the graphs $S_{n,k}$ and $S_{n,k}^+$, implies that for every $k \geq 1$ and $n \geq 2^{4k}$, we have

$$h_{2k}(n) = \mu(S_{n,k}) = (k-1)/2 + \sqrt{kn - (3k^2 + 2k - 1)/4},$$
$$h_{2k+1}(n) = \mu\left(S_{n,k}^+\right)$$
$$= (k-1)/2 + \sqrt{kn - (3k^2 + 2k - 1)/4} + 1/n + O\left(n^{-3/2}\right).$$

3.8   Hamilton paths and cycles

In [86], Ore found the following sufficient condition for the existence of Hamilton paths and cycles.

Theorem 3.18 *Let G be a graph of order n. If*

$$e(G) \geq \binom{n-1}{2} \tag{3.15}$$

then G contains a Hamiltonian path unless $G = K_{n-1} + K_1$. If the inequality (3.15) is strict, then G contains a Hamiltonian cycle unless $G = K_{n-1} + e$.

In the line above and further, $K_{n-1} + e$ denotes the complete graph K_{n-1} with a pendent edge.

Recently, Fiedler and Nikiforov [44] deduced a spectral version of Ore's result.

Theorem 3.19 *Let G be a graph of order n. If*

$$\mu(G) \geq n - 2, \tag{3.16}$$

then G contains a Hamiltonian path unless $G = K_{n-1} + K_1$. If the inequality (3.16) is strict, then G contains a Hamiltonian cycle unless $G = K_{n-1} + e$.

A subtler spectral condition for Hamiltonicity was obtained using the spectral radius of the complement of a graph.

Theorem 3.20 *Let G be a graph of order n and $\mu\left(\overline{G}\right)$ be the spectral radius of its complement. If*

$$\mu\left(\overline{G}\right) \leq \sqrt{n-1},$$

then G contains a Hamiltonian path unless $G = K_{n-1} + K_1$. If

$$\mu\left(\overline{G}\right) \leq \sqrt{n-2},$$

then G contains a Hamiltonian cycle unless $G = K_{n-1} + e$.

Zhou [92], adopting the same technique, proved a similar result for the signless Laplacian, which has been subsequently refined in [83] to the following one:

Theorem 3.21 *Let G be a graph of order n and $q\left(\overline{G}\right)$ be the spectral radius of the Q-matrix of its complement.*
(i) If

$$q\left(\overline{G}\right) \leq n, \tag{3.17}$$

then G contains a Hamiltonian path unless G is the union of two disjoint complete graphs or n is even and $G = K_{n/2-1,n/2+1}$.
(ii) If

$$q\left(\overline{G}\right) \leq n-1, \tag{3.18}$$

then G contains a Hamiltonian cycle unless G is a union of two complete graphs with a single common vertex or n is odd and $G = K_{\lfloor n/2 \rfloor, \lceil n/2 \rceil}$.

Note that if the inequality in (3.17) or (3.18) is strict, then the corresponding conclusion holds with no exception. Also, as it turns out, Theorem 3.21 considerably strengthens the classical degree conditions for Hamiltonicity by Ore [86].

3.9 Clique number and eigenvalues

If a triangle-free graph is sufficiently dense, then it contains large independent sets and the modulus of its smallest eigenvalue cannot be very small. A more general statement of such type has been proved in [11] for graphs of bounded clique number. Somewhat later, the following explicit dependence was found in [70].

Theorem 3.22 *If $G \in \mathcal{G}\left(n, m\right)$ and $\omega\left(G\right) = \omega$, then*

$$\mu_n\left(G\right) < -\frac{2}{\omega}\left(\frac{2m}{n^2}\right)^{\omega} n. \tag{3.19}$$

Inequality (3.19) captures pretty well the situation in dense graphs, that is, if G is a dense graph with $\mu_n\left(G\right) = O\left(n^{1-c}\right)$ for some $c \in \left(0, 1/2\right)$, then G contains cliques of order $\Omega\left(\log n\right)$.

Moreover, as shown in [70], inequality (3.19) is tight up to a constant factor for several classes of sparse graphs, but complete investigation of this issue seems difficult.

In [75], inequality (3.19) was used to derive a lower bound on $\alpha\left(G\right)$, thus giving other cases of tightness.

Theorem 3.23 *Let* $G \in \mathcal{G}(n, m)$, $d = 2m/n$, *and* $\tau = \left|\mu_n\left(\overline{G}\right)\right|$. *If* $d \geq 2$, *then*

$$\alpha(G) > \left(\frac{n}{d+1} - 1\right)\left(\log\frac{d+1}{\tau} - \log\log(d+1)\right).$$

Inequality (3.19) is concise, but it is difficult to use because the right-hand side is exponential in $\omega(G)$. The following two somewhat simpler bounds, given in [74], stem from Turán's theorem and some inequalities that will be given in the next subsection.

Theorem 3.24 *Let* $G \in \mathcal{G}(n, m)$, $d = 2m/n$, *and* $\omega(G) = \omega$. *Then*

$$\omega \geq 1 + \frac{dn}{(n-d)(d - \mu_n(G))}.$$

Equality holds if and only if G *is a complete regular* ω-*partite graph.*

Similar inequalities [72] exist also for the Laplacian eigenvalues.

Theorem 3.25 *Let* $G = \mathcal{G}(n, m)$, $d = 2m/n$ *and* $\omega(G) = \omega$. *Then*

$$\omega \geq 1 + \frac{dn}{\lambda_n(G)(n-d)},$$

with equality holding if and only if G *is a regular complete* ω-*partite graph.*
Also,

$$\alpha(G) \geq 1 + \frac{(n-1-d)n}{(n-\lambda_2(G))(1+d)},$$

with equality holding if and only if G *is the union of* $\alpha(G)$ *disjoint cliques of equal order.*

Note that both bounds in the last theorem imply the concise Turán theorem.

3.10 Number of cliques and eigenvalues

It turns out that the numbers of various cliques of a graph are closely related to its most important eigenvalues. Bollobás and Nikiforov [13] proved the following chain of inequalities, which were useful on several occasions.

Theorem 3.26 *Let* G *be a graph with* $\omega(G) = \omega \geq 2$ *and* $\mu(G) = \mu$. *For every* $r = 2, \ldots, \omega$,

$$\mu^{r+1} \leq (r+1)k_{r+1}(G) + \sum_{s=2}^{r}(s-1)k_s(G)\mu^{r+1-s}.$$

Observe that, with $r = \omega - 1$, Theorem 3.26 gives the following inequality from [69]; it has been applied to obtain a two line proof of the spectral precise Turán theorem in [72].

Theorem 3.27 *If* G *is a graph with* $\omega(G) = \omega \geq 2$ *and* $\mu(G) = \mu$, *then*

$$\mu^{\omega} \leq k_2(G)\mu^{\omega-2} + 2k_3(G)\mu^{\omega-3} + \cdots + (\omega-1)k_{\omega}(G).$$

Another important consequence of Theorem 3.26, also in [13], gives a lower bound on the number of cliques of any order as stated below.

Theorem 3.28 *If $r \geq 2$ and $G \in \mathcal{G}(n)$, then*

$$k_{r+1}(G) \geq \left(\frac{\mu(G)}{n} - 1 + \frac{1}{r}\right) \frac{r(r-1)}{r+1} \left(\frac{n}{r}\right)^{r+1}.$$

The remaining two theorems of this subsection are given in [74] and have multiple uses. The first one relates the numbers of triangles, edges and vertices of a graph with the smallest eigenvalue of its adjacency matrix.

Theorem 3.29 *If $G \in \mathcal{G}(n,m)$, then*

$$\mu_n(G) \leq \frac{3n^3 k_3(G) - 4m^3}{nm(n^2 - 2m)} \qquad (3.20)$$

with equality if and only if G is a regular complete multipartite graph.

Inequality (3.20) should be regarded as a multifaceted relation that can be used for different purposes. By way of illustration, let us restate it as a lower bound on $k_3(G)$, getting

$$k_3(G) \geq \frac{\mu_n(G)\left(nm(n^2 - 2m)\right) + 4m^3}{3n^3}, \qquad (3.21)$$

with equality holding for regular complete multipartite graphs. However, for all dense quasi-random graphs we have $\mu_n(G) = o(n)$ and $3k_3(G) = 4(1 + o(1))m^3/n^2$. This implies that

$$\frac{4m^3}{3n^3} + o(1)\frac{m^3}{n^3} = k_3(G) \geq o(1)mn + \frac{4m^3}{3n^3},$$

and we reach the somewhat paradoxical conclusion that inequality (3.21) is tight up to low order additive terms for almost all graphs, since almost all graphs are dense and quasi-random.

Statements similar to Theorem 3.29 have been obtained in [74] for the largest Laplacian eigenvalue $\lambda_n(G)$ as well.

Theorem 3.30 *If $G \in \mathcal{G}(n,m)$, then*

$$6nk_3(G) \geq (n + \lambda_n(G)) \sum_{u \in V(G)} d^2(u) - 2nm\lambda_n(G)$$

with equality if and only if G is a complete multipartite graph, and

$$\lambda_n(G) \geq \frac{2m^2 - 3nk_3(G)}{m(n^2 - 2m)}n,$$

with equality if and only if G is a regular complete multipartite graph.

3.11 Chromatic number

Let G be a graph of order n. One of the best known results in spectral graph theory is the inequality of A.J. Hoffman [51]

$$\chi(G) \geq 1 + \frac{\mu_1(G)}{-\mu_n(G)}, \qquad (3.22)$$

However, it seems that there is a lot more to find in this area. Indeed, in [73] we proved the following alternative bound.

Theorem 3.31 *For every graphs of order n,*

$$\chi(G) \geq 1 + \frac{\mu_1(G)}{\lambda_n(G) - \mu_1(G)}. \qquad (3.23)$$

Equality holds if and only if every two color classes of G induce a regular bipartite graph of degree $|\mu_n(G)|$.

When G is obtained from K_n by deleting an edge, inequality (3.23) gives $\chi(G) = n - 1$, while (3.22) gives only $\chi(G) \geq n/2 + 2$. By contrast, for a sufficiently large wheel $W_{1,n}$, i.e., a vertex joined to all vertices of a cycle of length n, (3.23) gives $\chi \geq 2$, while (3.22) gives $\chi \geq 3$.

However, such comparisons are not too informative since, in [73], both (3.23) and (3.22) have been deduced from the same matrix theorem.

4   Some useful tools

In this section we present some results that we have found useful on multiple occasions. The selection and the arrangement of these results does not follow any particular pattern.

We start with an inequality stated by Moon and Moser in [60]; it seems that Khadžiivanov and Nikiforov [55] were the first to publish its complete proof, see also [58], Problem 11.8. The inequality has been used in many questions, say in the proof of Theorem 2.5.

Lemma 4.1 *Let $1 \leq s < t < n$, and let G be a graph of order n, with $k_t(G) > 0$. Then*

$$\frac{(t+1)\, k_{t+1}(G)}{t k_t(G)} - \frac{n}{t} \geq \frac{(s+1)\, k_{s+1}(G)}{s k_s(G)} - \frac{n}{s}. \qquad (4.1)$$

The following two simple lemmas were used to obtain a number of results in Section 2. The first one was proved in [63], and the second one in [67].

Lemma 4.2 *Let $r \geq 2$, let c, n, m, s be such that*

$$0 < c \leq 1/2, \quad n \geq \exp\left(c^{-r}\right), \quad s = \lfloor c^r \log n \rfloor \leq (c/2)\, m + 1,$$

and let G be a bipartite graph with parts A and B of size m and n. If $e(G) \geq cmn$, then G contains a $K_2(s,t)$ with parts $S \subset A$ and $T \subset B$ such that $|S| = s$ and $|T| = t > n^{1-c^{r-1}}$.

Lemma 4.3 *Let α, c, n, m be such that*

$$0 < \alpha \leq 1, \quad 1 \leq c\log n \leq \alpha m/2 + 1,$$

and let G be a bipartite graph with parts A and B of size m and n. If $e(G) \geq \alpha mn$, then G contains a $K_2(s,t)$ with parts $S \subset A$ and $T \subset B$ such that $|S| = \lfloor c\log n \rfloor$ and $|T| = t > n^{1-c\log\alpha/2}$.

The following lemma, given in [78], strengthens a classical condition for the existence of paths given by Erdős and Gallai [39]. It has been used to obtain results about forbidden cycles and elsewhere.

Lemma 4.4 *Suppose that $k \geq 1$ and let the vertices of a graph G be partitioned into two sets U and W.*
(i) If

$$2e(U) + e(U, W) > (2k - 2)|U| + k|W|,$$

then there exists a path of order $2k$ or $2k + 1$ with both ends in U.
(ii) If

$$2e(U) + e(U, W) > (2k - 1)|U| + k|W|,$$

then there exists a path of order $2k + 1$ with both ends in U.

The following lemma from [68] was used to prove Theorems 2.27, 2.28 and 2.29, but may be used to carry over other stability results from triangle-free graphs to K_r-free graphs for $r > 3$.

Lemma 4.5 *Let $r \geq 3$ and let G be a maximal K_{r+1}-free graph of order n. If*

$$\delta(G) > \left(1 - \frac{2}{2r - 1}\right)n,$$

then G has a vertex u such that the vertices not joined to u are independent.

The following lemma, given in [79], bounds the minimal entry of eigenvectors to the spectral radius of the adjacency matrix. This can be useful in various situations, e.g., in conjunction with Lemma 4.7 from [81] and Theorem 4.8 it can be used to prove upper bounds on $\mu(G)$ by induction. Both lemmas have been used to prove several results in Section 3.

Lemma 4.6 *Let G be a graph of order n with minimum degree $\delta(G) = \delta$ and $\mu(G) = \mu$. If $(x_1, \ldots, x_n)$ is a unit eigenvector to μ, then*

$$\min\{x_1, \ldots, x_n\} \leq \sqrt{\frac{\delta}{\mu^2 + \delta n - \delta^2}}.$$

Lemma 4.7 *Let G be a graph of order n and let $(x_1, \ldots, x_n)$ be a unit eigenvector to $\mu(G)$. If u is a vertex satisfying $x_u = \min\{x_1, \ldots, x_n\}$, then*

$$\mu(G - u) \geq \mu(G)\frac{1 - 2x_u^2}{1 - x_u^2}.$$

The theorem below, given in [79], has been used to prove the spectral analog of several nonspectral results.

Theorem 4.8 *Let* α, β, γ, K *and* n *be such that*

$$0 < 4\alpha \leq 1, \quad 0 < 2\beta \leq 1, \quad 1/2 - \alpha/4 \leq \gamma < 1, \quad K \geq 0, \quad n \geq (42K + 4)/\alpha^2\beta,$$

and let G *be a graph of order* n. *If*

$$\mu(G) > \gamma n - K/n \quad and \quad \delta(G) \leq (\gamma - \alpha) n,$$

then there exists an induced subgraph $H \subset G$ *with* $|H| \geq (1 - \beta) n$, *satisfying one of the following conditions:*

(a) $\mu(H) > \gamma(1 + \beta\alpha/2)|H|$;
(b) $\mu(H) > \gamma|H|$ *and* $\delta(H) > (\gamma - \alpha)|H|$.

The abundance of parameters in Theorem 4.8 may obstruct its understanding. In summary, the theorem can be applied when one has to prove that if $\mu(G)$ is sufficiently large then G contains some subgraph F. If $\delta(G)$ is not large enough, by tossing away not too many low degree vertices, one gets a graph H in which either both $\mu(H)$ and $\delta(H)$ are large enough or $\mu(H)$ is considerably above the expected average. Most likely, either of these properties will help to find a copy of F in H. The many parameters ensure greater flexibility.

In [11], using interlacing, Bollobás and Nikiforov gave the following inequality, which has been used to prove several results involving the minimum eigenvalue of the adjacency matrix, e.g., Theorem 3.22.

Theorem 4.9 *If* $G \in \mathcal{G}(n, m)$, *then for every partition* $V(G) = V_1 \cup V_2$,

$$\mu_n(G) \leq \frac{2e(V_1)}{|V_1|} + \frac{2e(V_2)}{|V_2|} - \frac{2m}{n}.$$

Note that this inequality is analogous to the well-known inequality for the Laplacian (see Mohar, [59]):

$$\lambda_n(G) \geq \frac{e(V_1, V_2)}{|V_1||V_2|} n,$$

and in fact for regular graphs both inequalities are identical.

5 Illustration proofs

The purpose of this section is to illustrate the use of the tools developed for translating nonspectral into spectral results. To this end we shall sketch the proofs of Theorems 2.19 and 3.10.

The structure of both proofs is identical. In both proofs we shall use Theorem 2.4 from Section 2.2. The main difference comes from the fact that in the proof of Theorem 2.19 we use Theorem 2.22 while in the proof of Theorems 3.10 we use the analogous spectral result Theorem 3.12.

Proof of Theorem 2.19 Let G be a graph of order n with $e(G) > (1 - 1/r - \varepsilon) n^2/2$. Define the procedure $\mathcal{P}$ as follows:

While $js_{r+1}(G) > n^{r-1}/r^{r+6}$ **do**
   Select an edge contained in $\lceil n^{r-1}/r^{r+6}\rceil$ *cliques of order* $r+1$ *and remove it from* G.

Set for short $\theta = c^{1/(r+1)}r^{r+6}$ and assume first that $\mathcal{P}$ removes at least $\lceil\theta n^2\rceil$ edges before stopping. Then

$$k_{r+1}(G) \geq \theta n^{r-1}/r^{r+6} = c^{1/(r+1)}n^{r+1},$$

and Theorem 2.4 implies that

$$K_{r+1}\left(\lfloor c\ln n\rfloor,\ldots,\lfloor c\ln n\rfloor,\left\lceil n^{1-\sqrt{c}}\right\rceil\right) \subset G.$$

Thus, in this case condition *(a)* holds, completing the proof.

Assume therefore that $\mathcal{P}$ removes fewer than $\lceil\theta n^2\rceil$ edges before stopping. Writing G' for the resulting graph, we see that

$$e(G') > e(G) - \theta n^2 > (1 - 1/r - \varepsilon - \theta)n^2/2$$

and $js_{r+1}(G') < n^{r-1}/r^{r+6}$. Here we want to apply Theorem 2.22 and so we check for its prerequisites. First, from $\log n \geq 1/c \geq r^{3(r+14)(r+1)}$ we easily get $n > r^8$. Also,

$$\varepsilon + \theta < r^{-8}/8.$$

Now, Theorem 2.22 implies that G' contains an induced r-partite subgraph G_0 satisfying

$$|G_0| \geq \left(1 - \sqrt{2(\varepsilon+\theta)}\right)n \text{ and } \delta(G_0) > \left(1 - 1/r - 2\sqrt{2(\varepsilon+\theta)}\right)n.$$

By routine calculations we find that G differs from $T_r(n)$ in fewer than

$$\left(\theta + (2r^2 - r)\sqrt{2(\varepsilon+\theta)}\right)n^2 < \left(\varepsilon^{1/3} + c^{1/(3r+3)}\right)n^2$$

edges, and condition *(b)* follows, completing the proof of Theorem 2.19.   □

Proof of Theorem 3.10 Let G be a graph of order n with $\mu(G) > (1 - 1/r - \varepsilon)n$. Define the procedure $\mathcal{P}$ as follows:

While $js_{r+1}(G) > n^{r-1}/r^{2r+5}$ **do**
   Select an edge contained in $\lceil n^{r-1}/r^{2r+5}\rceil$ *cliques of order* $r+1$ *and remove it from* G.

Set for short $\theta = c^{1/(r+1)}r^{2r+5}$ and assume first that $\mathcal{P}$ removes at least $\lceil\theta n^2\rceil$ edges before stopping. Then

$$k_{r+1}(G) \geq \theta n^{r-1}/r^{2r+5} = c^{1/(r+1)}n^{r+1},$$

and Theorem 2.4 implies that

$$K_{r+1}\left(\lfloor c\ln n\rfloor,\ldots,\lfloor c\ln n\rfloor,\left\lceil n^{1-\sqrt{c}}\right\rceil\right) \subset G.$$

Thus, in this case condition *(a)* holds, completing the proof.

Assume now that $\mathcal{P}$ removes fewer than $\lceil \theta n^2 \rceil$ edges before stopping. Write G' for the resulting graph; we obviously have $js_{r+1}(G') \leq n^{r-1}/r^{2r+5}$. Letting $\mu(X)$ be the largest eigenvalue of a Hermitian matrix X, recall Weyl's inequality

$$\mu(B) \geq \mu(A) - \mu(A - B),$$

holding for any Hermitian matrices A and B. Also, recall that $\mu(H) \leq \sqrt{2e(H)}$ for any graph H. Applying these results to the graphs G and G', we find that

$$\mu(G') \geq \mu(G) - \sqrt{2\theta}n \geq \left(1 - 1/r - \varepsilon - \sqrt{2\theta}\right)n.$$

Here we want to apply Theorem 3.12 and so we check for its prerequisites. First, from $\log n \geq 1/c \geq r^{8(r+21)(r+1)}$ we easily get $n > r^{20}$. Also,

$$\varepsilon + \sqrt{2\theta} < 2^{-10}r^{-6}.$$

Now, Theorem 3.12 implies that G' contains an induced r-partite subgraph G_0, satisfying

$$|G_0| \geq \left(1 - 4\left(\varepsilon + \sqrt{2\theta}\right)^{1/3}\right)n \text{ and } \delta(G_0) > \left(1 - 1/r - 7\left(\varepsilon + \sqrt{2\theta}\right)^{1/3}\right)n.$$

By routine calculations we find that G differs from $T_r(n)$ in fewer than

$$\left(\theta + \left(7r^2 - 3r\right)\left(\varepsilon + \sqrt{2\theta}\right)^{1/3}\right)n^2 < \left(\varepsilon^{1/4} + c^{1/(8r+8)}\right)n^2$$

edges, and condition (b) follows, completing the proof of Theorem 3.10.   □

6   Notation and basic facts

Throughout the survey our notation generally follows [7]. Given a graph G, we write:

- $V(G)$ for the vertex set of G;
- $E(G)$ for the edge set of G and $e(G)$ for $|E(G)|$;
- $\alpha(G)$ for the independence number of G (see below);
- $\delta(G)$ and $\Delta(G)$ for the minimum and maximum degrees of G;
- $\omega(G)$ for the clique number of G (see below);
- $k_s(G)$ for the number of s-cliques of G (see below);
- $G - u$ for the graph obtained by removing the vertex $u \in V(G)$;
- $\Gamma(u)$ for the set of neighbors of a vertex u, and $d(u)$ for $|\Gamma(u)|$;
- $e(X)$ for the number of edges induced by a set $X \subset V(G)$;
- $e(X, Y)$ for the number of edges joining vertices in X to vertices in Y, where X and Y are disjoint subsets of $V(G)$;

We write $\mathcal{G}(n)$ for the set of graphs of order n and $\mathcal{G}(n, m)$ for the set of graphs of order n and size m.

Also, $[n]$ stands for the set $\{1, 2, \ldots, n\}$.

Mini glossary

clique - a subgraph that is complete. An s-clique has s vertices; $k_s(G)$ stands for the number of s-cliques of G;

clique number - the size of the largest clique of G, denoted by $\omega(G)$;

chromatic number - the minimum number of independent sets that partition $V(G)$, denoted by $\chi(G)$;

disjoint union of two graphs G and H is the union of two vertex disjoint copies of G and H. The disjoint union of G and H is denoted by $G + H$;

independent set - a set of vertices of G that induces no edges;

independence number - the size of the largest independent set of G, denoted by $\alpha(G)$;

join of two vertex disjoint graphs G and H is the union of G and H together with all edges between G and H. The join of G and H is denoted by $G \vee H$;

joint - a set of cliques of the same order sharing an edge. An r-joint of size t consists of t cliques of order r;

book of size t - a 3-joint of size t, that is to say, a collection of t triangles sharing an edge;

homomorphic graph - a graph G is said to be homomorphic to a graph H, if there exists a map $f : V(G) \rightarrow V(H)$ such that $uv \in E(G)$ implies $f(u)f(v) \in E(H)$;

graph property - a family of graphs closed under isomorphisms;

hereditary property - graph property closed under taking induced subgraphs;

monotone property - graph property closed under taking subgraphs;

H-free graph: a graph that has no subgraph isomorphic to H;

friendship graph - a collection of triangles sharing a single common vertex;

k-th power of a cycle C_n - a graph with vertices $\{1, 2, \ldots, n\}$, and (i, j) is an edge if $i - j = \pm 1, \pm 2, \cdots, \pm k \mod n$;

K_r and $\overline{K}_r$ - the complete and the edgeless graph of order r;

$K_r(s_1, s_2, ..., s_r)$ - the complete r-partite graph with class sizes $s_1, s_2, ..., s_r$. We set for short

$$K_r(p) = K_r(p, ..., p) \quad \text{and} \quad K_r(p; q) = K_r(p, ..., p, q);$$

r-uniform hypergraph - a hypergraph whose edges are subsets of r vertices;

Turán graph $T_r(n)$ - given $n \geq r \geq 2$, this is the complete r-partite graph whose class sizes differ by at most one. We let $t_r(n) = e(T_r(n))$. If t is the remainder of $n \mod r$, then

$$t_r(n) = \frac{r-1}{2r}\left(n^2 - t^2\right) + \binom{t}{2},$$

which in turn implies that

$$\frac{r-1}{2r}n^2 - \frac{r}{8} \leq t_r(n) \leq \frac{r-1}{2r}n^2;$$

Turán problem - given a family of graphs F, find the maximum number of edges in a graph of order n, having no subgraph belonging to F;

quasi-random graph - informally, an almost regular graph, in which the second largest in modulus eigenvalue is much smaller than the spectral radius;

spectral radius of a graph - in general, the spectral radius of a matrix is the largest modulus of its eigenvalues. For a graph, this is usually the spectral radius of its adjacency matrix, which is an eigenvalue itself;

Laplacian matrix - the matrix $L = D - A$, where A is the adjacency matrix and D is the diagonal matrix of the row-sums of A, that is the degrees of G;

Q-matrix, also known as **signless Laplacian** - the matrix $Q = D + A$;

Szemerédi's Regularity Lemma - an important result of analytical graph theory, which states that every graph can be approximated by graphs of bounded order. For background on this lemma we refer the reader to [7], Section IV.5;

Zarankiewicz problem - a class of problems aiming to determine the maximum number of edges in a graph with no $K_{s,t}$. There are several variations, most of which are only partially solved. See [7] for details.

Acknowledgement I am most grateful to the referee for the efficient and kind help.

References

[1] B. Andrásfai, P. Erdős and V.T. Sós, On the connection between chromatic number, maximal clique and minimum degree of a graph. *Discrete Math.* **8** (1974), 205–218.

[2] V.E. Alekseev, On the entropy values of hereditary classes of graphs, *Discrete Math. Appl.* **3** (1993), 191–199.

[3] B. Andrásfai, Über ein Extremalproblem der Graphentheorie , *Acta Math. Acad. Sci. Hungar.* **13** (1962), 443–455.

[4] N. Alon, L. Rònyai and T. Szabò, Norm-graphs: variations and applications, *J. Combin. Theory, Ser. B* **76** (1999), 280–290.

[5] L. Babai and B. Guiduli, Spectral extrema for graphs: the Zarankiewicz problem, *Electron. J. Combin.*, **16** (2009), R123.

[6] B. Bollobás, *Extremal Graph Theory*, Academic Press Inc., London-New York, 1978, xx+488 pp.

[7] B. Bollobás, *Modern Graph Theory*, Graduate Texts in Mathematics, **184**, Springer-Verlag, New York (1998), xiv+394 pp.

[8] B. Bollobás and P. Erdős, On the structure of edge graphs, *J. London Math. Soc.* **5** (1973), 317–321.

[9] B. Bollobás, P. Erdős and M. Simonovits, On the structure of edge graphs II. *J. London Math. Soc.* **12** (1976), 219–224.

[10] B. Bollobás and Y. Kohayakawa, An extension of the Erdös-Stone theorem, *Combinatorica* **14** (1994), 279–286.

[11] B. Bollobás and V. Nikiforov, Graphs and Hermitian matrices: eigenvalue interlacing, *Discrete Math.* **289** (2004), 119–127.

[12] B. Bollobás and V. Nikiforov, Books in graphs, *Eur. J. Combin.* **26** (2005), 259–270.

[13] B. Bollobás and V. Nikiforov, Cliques and the spectral radius, *J. Combin. Theory Ser. B* **97** (2007), 859–865.

[14] B. Bollobás and V. Nikiforov, Joints in graphs, *Discrete Math.* **308** (2008), 9–19.

[15] B. Bollobás and V. Nikiforov, The number of graphs with large forbidden subgraphs, *Eur. J. Combin.* **31** (2010), 1964–1968.

[16] B. Bollobás and V. Nikiforov, Large joints in graphs, to appear in *Eur. J. Combin.*

[17] B. Bollobás and A. Thomason, Projections of bodies and hereditary properties of hypergraphs, *Bull. London Math. Soc.* **27** (1995), 417–424.

[18] B. Bollobás and A. Thomason, Hereditary and monotone properties of graphs, "The mathematics of Paul Erdős, II" (R.L. Graham and J. Nešetřil, Editors), *Alg. and Combin.*, Vol. 14, Springer-Verlag, New York/Berlin (1997), 70–78.

[19] W.G. Brown, On graphs that do not contain a Thomsen graph, *Canad. Math. Bull.* **9** (1966), 281–285.

[20] S. Brandt and S. Thomassé, Dense triangle-free graphs are four-colorable: a solution to the Erdős-Simonovits problem, to appear in *J. Combin Theory Ser. B.*

[21] V. Chvátal and E. Szemerédi, Notes on the Erdős-Stone theorem, *Annals of Discrete Math.* **17** (1983), 207–214.

[22] C.C. Chen, G.P. Jin and K.M. Koh, Triangle-free graphs with large degree, *Combin. Probab. Comput.* **6** (1997), 381–396.

[23] D. Cvetković, P. Rowlinson and S. Simić, An Introduction to the Theory of Graph Spectra, *LMS Student Texts 75*, Cambridge, 2010, pp. 364+vii.

[24] D. Cvetković and S.K. Simić, Towards a spectral theory of graphs based on the signless Laplacian, I, *Publ. Inst. Math.(Beograd)*, 85(99)(2009), 19–33.

[25] D. Cvetković and S.K. Simić, Towards a spectral theory of graphs based on the signless Laplacian, II, *Linear Algebra Appl.* **432**(2010), 2257–2272.

[26] D. Cvetković and S.K. Simić, Towards a spectral theory of graphs based on the signless Laplacian, III, *Appl. Anal. Discrete Math.* 4(2010), 156–166.

[27] D. Cvetković, P. Rowlinson and S. Simić, Eigenvalue bounds for the signless Laplacian, *Publications de L'Institut Mathématique, Nouvelle série* **81(95)** (2007), 11–27.

[28] C. Edwards, A lower bound on the largest number of triangles with a common edge, *unpublished manuscript.*

[29] C. Edwards and C. Elphick, Lower bounds for the clique and the chromatic number of a graph, *Discrete Appl. Math.* **5** (1983) 51–64.

[30] P. Erdős, On a theorem of Rademacher-Turán, *Illinois J. Math.* **6** (1962), 122–127.

[31] P. Erdős, On the structure of linear graphs, *Israel J. Math.* **1** (1963), 156–160.

[32] P. Erdős, On extremal problems of graphs and generalized graphs, *Israel J. Math.* **2** (1964), 183–190.

[33] P. Erdős, Some recent results on extremal problems in graph theory (Results), in: *Theory of Graphs (Internat. Sympos., Rome, 1966)*, pp. 117–130, Gordon and Breach, New York; Dunod, Paris, 1967.

[34] P. Erdős, On some new inequalities concerning extremal properties of graphs, in: *Theory of Graphs (Proc. Colloq., Tihany, 1966)*, pp. 77–81, Academic Press, New York, 1968.

[35] P. Erdős, On the number of complete subgraphs and circuits contained in graphs, *Časopis Pěst. Mat.* **94** (1969), 290–296.

[36] P. Erdős, Z. Füredi, R.J. Gould and D.S. Gunderson, Extremal graphs for intersecting triangles, *J. Combin. Theory Ser. B* **64** (1995), 89–100.

[37] P. Erdős, P. Frankl and V. Rödl, The asymptotic number of graphs not containing a fixed subgraph and a problem for hypergraphs having no exponent, *Graphs Combin.* **2** (1986), 113–121.

[38] P. Erdős, D. J. Kleitman and B. L. Rothschild, Asymptotic enumeration of K_n-free graphs, *Colloquio Internazionale sulle Teorie Combinatorie (Rome, 1973)*, Tomo II, pp. 19–27, *Atti dei Convegni Lincei, No. 17*, Accad. Naz. Lincei, Rome, 1976.

[39] P. Erdős and T. Gallai, On maximal paths and circuits of graphs, *Acta Math. Acad. Sci. Hungar.* **10** 1959, 337–356.

[40] P. Erdős, A. Rényi, On a problem in the theory of graphs, *Publ. Math. Inst. Hungar. Acad. Sci.* **7A** (1962), 623–641.

[41] P. Erdős and M. Simonovits, On a valence problem in extremal graph theory, *Discrete Math.,* **5** (1973), 323–334.

[42] P. Erdős and A. H. Stone, On the structure of linear graphs, *Bull. Amer. Math. Soc.* **52** (1946), 1087–1091.

[43] P. Erdős and M. Simonovits, A limit theorem in graph theory, *Studia Sci. Math. Hung.* **1** (1966), 51–57.

[44] M. Fiedler and V. Nikiforov,, Spectral radius and Hamiltonicity of graphs, *Linear Algebra Appl.* **432** (2010), 2170–2173.

[45] O. Favaron, M. Mahéo and J.-F. Saclé, Some eigenvalue properties in graphs (conjectures of Graffiti. II), *Discrete Math.* **111** (1993), 197–220.

[46] Z. Füredi, An upper bound on Zarankiewicz's problem, *Comb. Probab. Comput.* **5** (1996), 29-33.

[47] Z. Füredi, New Asymptotics for Bipartite Turán Numbers, *Journal Combin. Theory Ser A* **75** (1996), 141–144.

[48] W. Goddard and J. Lyle, Dense graphs with small clique number, submitted for publication.

[49] R. Häggkvist, Odd cycles of specified length in nonbipartite graphs, *Graph theory* (Cambridge, 1981), pp. 89–99, North-Holland Math. Stud., **62**, North-Holland, Amsterdam-New York, 1982.

[50] P. Hansen and C. Lucas, Bounds and conjectures for the signless Laplacian index of graphs, *Linear Algebra Appl.* **432** (2010), 3319–3336.

[51] A.J. Hoffman, On eigenvalues and colorings of graphs, in: *Graph Theory and its Applications,* Academic Press, New York (1970), pp. 79–91.

[52] Y. Ishigami, Proof of a conjecture of Bollobás and Kohayakawa on the Erdös-Stone theorem, *J. Combin. Theory Ser. B* **85** (2002), 222–254.

[53] G.P. Jin, Triangle-free graphs with high minimal degrees, *Combin. Probab. Comput.* **2** (1993), 479–490.

[54] G. P. Jin, Triangle-free four-chromatic graphs, *Discrete Math.* **145** (1995), 151–170.

[55] N. Khadžiivanov and V. Nikiforov, The Nordhaus-Stewart-Moon-Moser inequality (in Russian), *Serdica* **4** (1978), 344–350.

[56] N. Khadžiivanov and V. Nikiforov, Solution of a problem of P. Erdős about the maximum number of triangles with a common edge in a graph (in Russian), *C. R. Acad. Bulgare Sci.* **32** (1979), 1315–1318.

[57] T. Kövari, V. Sós and P. Turán, On a problem of K. Zarankiewicz, *Colloquium Math.* **3** (1954), 50–57.

[58] L. Lovász, *Combinatorial Problems and Exercises,* North-Holland Publishing Co., Amsterdam-New York (1979), 551 pp.

[59] B. Mohar, Some applications of Laplace eigenvalues of graphs, in: *Graph Symmetry* (Montreal, PQ, 1996), NATO Adv. Sci. Inst. Ser. C Math. Phys. Sci., **497**, Kluwer Acad. Publ., Dordrecht, 1997, pp. 225–275.

[60] J. Moon and L. Moser, On a problem of Turán, *Magyar Tud. Akad. Mat. Kutató Int. Közl.* **7** (1962), 283–286.

[61] T. Motzkin and E. Straus, Maxima for graphs and a new proof of a theorem of Turán, *Canad. J. Math.,* **17** (1965), 533–540.

[62] J. Mycielski, Sur le coloriage des graphs, *Colloq. Math.* **3**, (1955), 161–162.

[63] V. Nikiforov, Graphs with many r-cliques have large complete r-partite subgraphs, *Bull. London Math. Soc.* **40** (2008), 23–25.

[64] V. Nikiforov, Graphs with many copies of a given subgraph, *Electronic J. Combin.* **15** (2008), N6.

[65] V. Nikiforov, Complete r-partite subgraphs of dense r-graphs, *Discrete Math.* **309** (2009), 4326–4331.

[66] V. Nikiforov, Stability for large forbidden subgraphs, *J. Graph Theory* **62** (2009), 362–368.

[67] V. Nikiforov, Turán's theorem inverted, *Discrete Math.* **310**, (2010), 125–131.

[68] V. Nikiforov, Chromatic number and minimum degree of K_r-free graphs, *submitted for publication, preprint available at arXiv:1001.2070v1*

[69] V. Nikiforov, Some inequalities for the largest eigenvalue of a graph, *Combin. Probab. Comput.* **11** (2002), 179–189.

[70] V. Nikiforov, Walks and the spectral radius of graphs, *Linear Algebra Appl.* **418** (2006), 257–268.

[71] V. Nikiforov, The smallest eigenvalue of K_r-free graphs, *Discrete Math.* **306** (2006), 612–616.

[72] V. Nikiforov, Bounds on graph eigenvalues II, *Linear Algebra Appl.* **427** (2007), 183–189.

[73] V. Nikiforov, Chromatic number and spectral radius, *Linear Algebra Appl.* **426** (2007), 810–814.

[74] V. Nikiforov, Eigenvalues and forbidden subgraphs I, *Linear Algebra Appl.* **422** (2007), 384–390.

[75] V. Nikiforov, More spectral bounds on the clique and independence numbers, *J. Combin. Theory Ser. B* **99** (2009), 819–826.

[76] V. Nikiforov, Spectral saturation: inverting the spectral Turán theorem, *Electronic J. Combin.* **15** (2009), R33.

[77] V. Nikiforov, The maximum spectral radius of C_4-free graphs of given order and size, *Linear Algebra Appl.* **430** (2009), 2898–2905.

[78] V. Nikiforov, A spectral Erdős-Stone-Bollobás theorem, *Combin. Probab. Comput.* **18** (2009), 455–458.

[79] V. Nikiforov, A spectral condition for odd cycles, *Linear Algebra Appl.* **428** (2008), 1492–1498.

[80] V. Nikiforov, Degree powers in graphs with forbidden even cycle, *Electronic J. Combin.* **15** (2009), R107.

[81] V. Nikiforov, The spectral radius of graphs without paths and cycles of specified length, *Linear Algebra Appl.* **432** (2010), 2243–2256.

[82] V. Nikiforov, A contribution to the Zarankiewicz problem, *Linear Algebra Appl.* **432** (2010), 1405–1411.

[83] V. Nikiforov, The Q-index and Hamiltonicity, in preparation.

[84] V. Nikiforov and C.C. Rousseau, Large generalized books are p-good, *J. Combin. Theory Ser B* **92** (2004), 85–97.

[85] E. Nosal, Eigenvalues of Graphs, Master's thesis, University of Calgary, 1970.

[86] O. Ore, Note on Hamilton circuits, *Amer. Math. Monthly* **67** (1960), 55.

[87] I. Ruzsa and E. Szemerédi, Triple systems with no six points carrying three triangles, in: Combinatorics (Keszthely, 1976), *Coll. Math. Soc. J. Bolyai 18* (1978), pp. 939–945.

[88] M. Simonovits, A method for solving extremal problems in graph theory, stability problems, in: *Theory of Graphs (Proc. Colloq., Tihany, 1966)*, pp. 279–319, Academic Press, New York, 1968.

[89] P. Turán, On an extremal problem in graph theory (in Hungarian), *Mat. és Fiz. Lapok* **48** (1941) 436–452.

[90] H. Wilf, Spectral bounds for the clique and independence numbers of graphs, *J. Combin. Theory Ser. B* **40** (1986), 113–117.

[91] M. Xu, Y. Hong, J. Shu and M. Zhai, The minimum spectral radius of graphs with a given independence number, *Linear Algebra Appl.* **431**, (2009), 937–945.

[92] B. Zhou, Signless Laplacian spectral radius and Hamiltonicity, *Linear Algebra Appl.* **432** (2010) 566–570.

Department of Mathematical Sciences
University of Memphis
Memphis, TN 38152, USA
vnikifrv@memphis.edu

The cyclic sieving phenomenon: a survey

Bruce E. Sagan

Abstract

The cyclic sieving phenomenon was defined by Reiner, Stanton, and White in a 2004 paper. Let X be a finite set, C be a finite cyclic group acting on X, and $f(q)$ be a polynomial in q with nonnegative integer coefficients. Then the triple $(X, C, f(q))$ exhibits the *cyclic sieving phenomenon* if, for all $g \in C$, we have

$$\#X^g = f(\omega)$$

where $\#$ denotes cardinality, X^g is the fixed point set of g, and ω is a root of unity chosen to have the same order as g. It might seem improbable that substituting a root of unity into a polynomial with integer coefficients would have an enumerative meaning. But many instances of the cyclic sieving phenomenon have now been found. Furthermore, the proofs that this phenomenon hold often involve interesting and sometimes deep results from representation theory. We will survey the current literature on cyclic sieving, providing the necessary background about representations, Coxeter groups, and other algebraic aspects as needed.

Acknowledgements

The author would like to thank Vic Reiner, Martin Rubey, Bruce Westbury, and an anonymous referee for helpful suggestions. And he would particularly like to thank Vic Reiner for enthusiastic encouragement. This work was partially done while the author was a Program Officer at NSF. The views expressed are not necessarily those of the NSF.

1 What is the cyclic sieving phenomenon?

Reiner, Stanton, and White introduced the cyclic sieving phenomenon in their seminal 2004 paper [58]. In order to define this concept, we need three ingredients. The first of these is a finite set, X. The second is a finite cyclic group, C, which acts on X. Given a group element $g \in C$, we denote its fixed point set by

$$X^g = \{y \in X \ : \ gy = y\}. \tag{1.1}$$

We also let $o(g)$ stand for the order of g in the group C. One group which will be especially important in what follows will be the group, Ω, of roots of unity. We let ω_d stand for a primitive dth root of unity. The reader should think of $g \in C$ and $\omega_{o(g)} \in \Omega$ as being linked because they have the same order in their respective groups. The final ingredient is a polynomial $f(q) \in \mathbb{N}[q]$, the set of polynomials in the variable q with nonnegative integer coefficients. Usually $f(q)$ will be a generating function associated with X.

Definition 1.1 The triple $(X, C, f(q))$ exhibits the *cyclic sieving phenomenon* (abbreviated CSP) if, for all $g \in C$, we have

$$\#X^g = f(\omega_{o(g)}) \tag{1.2}$$

where the hash symbol denotes cardinality.

Several comments about this definition are in order. At first blush it may seem very strange that plugging a complex number into a polynomial with nonnegative integer coefficients would yield a nonnegative integer, much less that the result would count something. However, the growing literature on the CSP shows that this phenomenon is quite wide spread. Of course, using linear algebra it is always possible to find some polynomial which will satisfy the system of equations given by (1.2). And in Section 3 we will see, via equation (3.4), that the polynomial can be taken to have nonnegative integer coefficients. But the point is that $f(q)$ should be a polynomial naturally associated with the set X. In fact, letting $g = e$ (the identity element of C) in (1.2), it follows that

$$f(1) = \#X. \tag{1.3}$$

In the case $\#C = 2$, the CSP reduces to Stembridge's "$q = -1$ phenomenon" [83, 84, 85]. Since the Reiner-Stanton-White paper, interest in cyclic sieving has been steadily increasing. But since the area is relatively new, this survey will be able to at least touch on all of the current literature on the subject. Cylic sieving illustrates a beautiful interplay between combinatorics and algebra. We will provide all the algebraic background required to understand the various instances of the CSP discussed.

The rest of the paper is organized as follows. Most proofs of cyclic sieving phenomena fall into two broad categories: those involving explicit evaluation of both sides of (1.2), and those using representation theory. In the next section, we will introduce our first example of the CSP and give a demonstration of the former type. Section 3 will provide the necessary representation theory background to present a proof of the second type for the same example. In the following section, we will develop a paradigm for representation theory proofs in general. Since much of the work that has been done on CSP involves Coxeter groups and permutation statistics, Section 5 will provide a brief introduction to them. In Section 6 we discuss the regular elements of Springer [78] which have become a useful tool in proving CSP results. Section 7 is concerned with Rhoades' startling theorem [60] connecting the CSP and Schützenberger's promotion operator [70] on rectangular standard Young tableaux. The section following that discusses work related to Rhoades' result. In Section 9 we consider generalizations of the CSP using more than one group or more than one statistic. Instances of the CSP related to Catalan numbers are discussed in Section 10. The penultimate section contains some results which do not fit nicely into one of the previous sections. And the final section consists of various remarks.

2   An example and a proof

We will now consider a simple example of the CSP and show that (1.2) holds by explicitly evaluating both sides of the equation. For $n \in \mathbb{N}$, let $[n] = \{1, 2, \ldots, n\}$. A *multiset on* $[n]$ is an unordered family, M, of elements of $[n]$ where repetition is allowed. Since order does not matter, we will always list the elements of M in weakly increasing order: $M = i_1 i_2 \ldots i_k$ with $1 \leq i_1 \leq i_2 \leq \ldots \leq i_k \leq n$. The set for our CSP will be

$$X = \left(\binom{[n]}{k} \right) \stackrel{\text{def}}{=} \{M \;:\; M \text{ is a multiset on } [n] \text{ with } k \text{ elements.}\}. \tag{2.1}$$

To illustrate, if $n = 3$ and $k = 2$ then $X = \{11, 22, 33, 12, 13, 23\}$.

For our group, we take one generated by an n-cycle

$$C_n = \langle (1, 2, \ldots, n) \rangle.$$

Then $g \in C_n$ acts on $M = i_1 i_2 \ldots i_k$ by

$$gM = g(i_1)g(i_2)\ldots g(i_k) \tag{2.2}$$

where we rearrange the right-hand side to be in increasing order. Returning to the $n = 3, k = 2$ case we have $C_3 = \{e, (1, 2, 3), (1, 3, 2)\}$. The action of $g = (1, 2, 3)$ is

$$\begin{array}{llll} (1, 2, 3)11 = 22, & (1, 2, 3)22 = 33, & (1, 2, 3)33 = 11, \\ (1, 2, 3)12 = 23, & (1, 2, 3)13 = 12, & (1, 2, 3)23 = 13. \end{array} \tag{2.3}$$

To define the polynomial we will use, consider the geometric series

$$[n]_q = 1 + q + q^2 + \cdots + q^{n-1}. \tag{2.4}$$

This is known as a *q-analogue of n* because setting $q = 1$ gives $[n]_1 = n$. Do not confuse $[n]_q$ with the set $[n]$ which has no subscript. Now, for $0 \leq k \leq n$, define the *Gaussian polynomials* or *q-binomial coefficients* by

$$\left[\begin{array}{c} n \\ k \end{array} \right]_q = \frac{[n]_q!}{[k]_q![n-k]_q!} \tag{2.5}$$

where $[n]_q! = [1]_q[2]_q \cdots [n]_q$. It is not clear from this definition that these rational functions are actually polynomials with nonnegative integer coefficients, but this is not hard to prove by induction on n. It is well known that $\#(\binom{[n]}{k}) = \binom{n+k-1}{k}$. So, in view of (1.3), a natural choice for our CSP polynomial is

$$f(q) = \left[\begin{array}{c} n + k - 1 \\ k \end{array} \right]_q.$$

We are now in a position to state our first cyclic sieving result. It is a special case of Theorem 1.1(a) in the Reiner-Stanton-White paper [58].

Theorem 2.1 *The cyclic sieving phenomenon is exhibited by the triple*

$$\left(\left(\binom{[n]}{k} \right), \; \langle (1, 2, \ldots, n) \rangle, \; \left[\begin{array}{c} n + k - 1 \\ k \end{array} \right]_q \right).$$

Before proving this theorem, let us consider the case $n = 3, k = 2$ in detail. First of all note that

$$f(q) = \begin{bmatrix} 3+2-1 \\ 2 \end{bmatrix}_q = \begin{bmatrix} 4 \\ 2 \end{bmatrix}_q = \frac{[4]_q!}{[2]_q![2]_q!} = \frac{[4]_q[3]_q}{[2]_q} = 1 + q + 2q^2 + q^3 + q^4.$$

Now we can verify that $\#X^g = f(\omega_{o(g)})$ case by case. If $g = e$ then $o(g) = 1$, so let $\omega = 1$ and compute

$$f(\omega) = f(1) = 6 = \#X = \#X^e.$$

If $g = (1, 2, 3)$ or $(1, 3, 2)$ then we can use $\omega = \exp(2\pi i/3)$ to obtain

$$f(\omega) = 1 + \omega + 2\omega^2 + \omega^3 + \omega^4 = 2 + 2\omega + 2\omega^2 = 0 = \#X^g$$

as can be seen from the table (2.3) for the action of $(1, 2, 3)$ which has no fixed points.

In order to give a proof by explicit evaluation, it will be useful to have some more notation. Another way of expressing a multiset on $[n]$ is $M = \{1^{m_1}, 2^{m_2}, \ldots, n^{m_n}\}$ where m_i is the multiplicity of i in M. Exponents equal to one are omitted as are elements of exponent zero. For example, $M = 222355 = \{2^3, 3, 5^2\}$. Define the *disjoint union* of multisets $L = \{1^{l_1}, 2^{l_2}, \ldots, n^{l_n}\}$ and $M = \{1^{m_1}, 2^{m_2}, \ldots, n^{m_n}\}$ to be

$$L \sqcup M = \{1^{l_1+m_1}, 2^{l_2+m_2}, \ldots, n^{l_n+m_n}\}.$$

Let $\mathfrak{S}_n$ denote the symmetric group of permutations of $[n]$. Note that (2.2) defines an action of any $g \in \mathfrak{S}_n$ on multisets and need not be restricted to elements of the cyclic group. We will also want to apply the disjoint union operation $\sqcup$ to cycles in $\mathfrak{S}_n$, in which case we consider each cycle as a set (which is just a multiset with all multiplicities 0 or 1). To illustrate, $(1, 4, 3) \sqcup (1, 4, 3) \sqcup (3, 4, 5) = \{1^2, 3^3, 4^3, 5\}$. To evaluate $\#X^g$ we will need the following lemma.

Lemma 2.2 *Let* $g \in \mathfrak{S}_n$ *have disjoint cycle decomposition* $g = c_1 c_2 \cdots c_t$. *Then* $gM = M$ *if and only if* M *can be written as*

$$M = c_{r_1} \sqcup c_{r_2} \sqcup \cdots \sqcup c_{r_s}$$

where the cycles in the disjoint union need not be distinct.

Proof For the reverse direction, note that if c is a cycle of g and $\bar{c}$ is the corresponding set then $g\bar{c} = \bar{c}$ since g merely permutes the elements of $\bar{c}$ amongst themselves. So g will also fix disjoint unions of such cycles as desired.

To see that these are the only fixed points, suppose that M is not such a disjoint union. Then there must be some cycle c of g and $i, j \in [n]$ such that $c(i) = j$ but i and j have different multiplicities in M. It follows that $gM \neq M$ which completes the forward direction. $\qquad\square$

As an example of this lemma, if $g = (1, 2, 4)(3, 5)$ then the multisets fixed by g of cardinality at most 5 are $\{3, 5\}$, $\{1, 2, 4\}$, $\{3^2, 5^2\}$, and $\{1, 2, 3, 4, 5\}$. We now apply the previous lemma to our case of interest when $C_n = \langle (1, 2, \ldots, n) \rangle$. In the next result, we use the standard notation $d|k$ to signify that the integer d divides evenly into the integer k.

Corollary 2.3 *If $X = \left(\binom{[n]}{k}\right)$ and $g \in C_n$ has $o(g) = d$, then*

$$\#X^g = \begin{cases} \dbinom{n/d + k/d - 1}{k/d} & \text{if } d|k, \\[2ex] 0 & \text{otherwise.} \end{cases}$$

Proof Since g is a power of $(1, 2, \ldots, n)$ and $o(g) = d$, we have that g's disjoint cycle decomposition must consist of n/d cycles each of length d. If d does not divide k, then no multiset M of cardinality k can be a disjoint union of the cycles of g. So, by Lemma 2.2, there are no fixed points and this agrees with the "otherwise" case above.

If $d|k$ then, by the lemma again, the fixed points of g are those multisets obtained by choosing k/d of the n/d cycles of g with repetition allowed. The number of ways of doing this is the binomial coefficient in the "if" case. $\square$

To evaluate $f(\omega_{o(g)})$, we need another lemma.

Lemma 2.4 *Suppose $m, n \in \mathbb{N}$ satisfy $m \equiv n \pmod{d}$, and let $\omega = \omega_d$. Then*

$$\lim_{q \to \omega} \frac{[m]_q}{[n]_q} = \begin{cases} \dfrac{m}{n} & \text{if } n \equiv 0 \pmod{d}, \\[2ex] 1 & \text{otherwise.} \end{cases}$$

Proof Let $m \equiv n \equiv r \pmod{d}$ where $0 \le r < d$. Since $1 + \omega + \omega^2 + \cdots + \omega^{d-1} = 0$, cancellation in (2.4) yields

$$[m]_\omega = 1 + \omega + \omega^2 + \cdots + \omega^{r-1} = [n]_\omega.$$

So if $r \ne 0$ then $[n]_\omega \ne 0$ and $[m]_\omega/[n]_\omega = 1$, proving the "otherwise" case.

If $r = 0$ then we can write $n = \ell d$ and $m = kd$ for certain nonnegative integers k, ℓ. It follows that

$$\frac{[m]_q}{[n]_q} = \frac{(1 + q + q^2 + \cdots + q^{d-1})(1 + q^d + q^{2d} + \cdots + q^{(k-1)d})}{(1 + q + q^2 + \cdots + q^{d-1})(1 + q^d + q^{2d} + \cdots + q^{(\ell-1)d})}.$$

Canceling the $1 + q + q^2 + \cdots q^{d-1}$ factors and plugging in ω gives

$$\lim_{q \to \omega} \frac{[m]_q}{[n]_q} = \frac{k}{\ell} = \frac{m}{n}$$

as desired. $\square$

To motivate the hypothesis of the next result, note that if $o(g) = d$ and $g \in C_n$ then $d|n$ by Lagrange's Theorem.

Corollary 2.5 *If $\omega = \omega_d$ and $d|n$, then*

$$\begin{bmatrix} n + k - 1 \\ k \end{bmatrix}_\omega = \begin{cases} \dbinom{n/d + k/d - 1}{k/d} & \text{if } d|k, \\[2ex] 0 & \text{otherwise.} \end{cases}$$

Proof In the equality above, consider the numerator and denominator of the left-hand side after canceling factorials. Since $d|n$, the product $[n]_\omega[n+1]_\omega \cdots [n+k-1]_\omega$ starts with a zero factor and has every dth factor after that also equal to zero, with all the other factors being nonzero. The product $[1]_\omega[2]_\omega \cdots [k]_\omega$ is also of period d, but starting with $d-1$ nonzero factors. It follows that the number of zero factors in the numerator is always greater than or equal to the number of zero factors in the denominator with equality if and only if $d|k$. This implies the "otherwise" case.

If $d|k$, then using the previous lemma

$$
\left[\begin{matrix} n+k-1 \\ k \end{matrix} \right]_\omega = \lim_{q \to \omega} \left(\frac{[n]_q}{[k]_q} \cdot \frac{[n+1]_q}{[1]_q} \cdot \frac{[n+2]_q}{[2]_q} \cdots \frac{[n+k-1]_q}{[k-1]_q} \right)
$$

$$
= \frac{n}{k} \cdot 1 \cdots 1 \cdot \frac{n+d}{d} \cdot 1 \cdots 1 \cdot \frac{n+2d}{2d} \cdot 1 \cdots
$$

$$
= \frac{n/d}{k/d} \cdot \frac{n/d+1}{1} \cdot \frac{n/d+2}{2} \cdots
$$

$$
= \binom{n/d+k/d-1}{k/d}
$$

as desired. $\qquad\qquad\qquad\qquad\qquad\qquad\qquad\qquad\qquad\qquad\qquad\qquad\qquad\qquad\qquad\qquad$ □

Comparing Corollary 2.3 and Corollary 2.5, we immediately have a proof of Theorem 2.1.

3 Representation theory background and another proof

Although the proof just given of Theorem 2.1 has the advantage of being elementary, it does not give much intuition about why the equality (1.2) holds. Proofs of such results using representation theory are more sophisticated but also provide more insight. We begin this section by reviewing just enough about representations to provide another demonstration of Theorem 2.1. Readers interested in more information about representation theory, especially as it relates to symmetric groups, can consult the texts of James [35], James and Kerber [34], or Sagan [65].

Given a set, X, we can create a complex vector space, $V = \mathbb{C}X$, by considering the elements of X as a basis and taking formal linear combinations. So if $X = \{s_1, s_2, \ldots, s_k\}$ then

$$
\mathbb{C}X = \{c_1\mathbf{s}_1 + c_2\mathbf{s}_2 + \cdots + c_k\mathbf{s}_k : c_i \in \mathbb{C} \text{ for all } i\}.
$$

Note that when an element of X is being considered as a vector, it is set in boldface type. If G is a group acting on X, then G also acts on $\mathbb{C}X$ by linear extension. Each element $g \in G$ corresponds to an invertible linear map $[g]$. (Although this is the same notation as for $[n]$ with $n \in \mathbb{N}$, context should make it clear which is meant.) If B is an ordered basis for V then we let $[g]_B$ denote the matrix of $[g]$ in the basis B. In particular, $[g]_X$ is the permutation matrix for g acting on X.

To illustrate these concepts, if $X = \{1, 2, 3\}$ then

$$
\mathbb{C}X = \{c_1\mathbf{1} + c_2\mathbf{2} + c_3\mathbf{3} : c_1, c_2, c_3 \in \mathbb{C}\}.
$$

The group $G = \langle (1,2,3) \rangle$ acts on X and so on $\mathbb{C}X$. For $g = (1,2,3)$ and the basis X the action is

$$(1,2,3)\mathbf{1} = \mathbf{2}, \ (1,2,3)\mathbf{2} = \mathbf{3}, \ (1,2,3)\mathbf{3} = \mathbf{1}.$$

Putting this in matrix form gives

$$[(1,2,3)]_X = \begin{bmatrix} 0 & 0 & 1 \\ 1 & 0 & 0 \\ 0 & 1 & 0 \end{bmatrix}. \tag{3.1}$$

In general, a *module for G* or *G-module* is a vector space V over $\mathbb{C}$ where G acts by invertible linear transformations. Most of our modules will be *left modules* with G acting on the left. Consider the *general linear group* $\mathrm{GL}(V)$ of invertible linear transformations of V. If V is a G-module then the map $\rho : G \to \mathrm{GL}(V)$ given by $g \mapsto [g]$ is called a *representation* of G. Equivalently, if V is a vector space, then a representation is a group homomorphism $\rho : G \to \mathrm{GL}(V)$. If G acts on a set X then the space $\mathbb{C}X$ is called the *permutation module* corresponding to X.

Given a G-module, V, the *character* of G on V is the function $\chi : G \to \mathbb{C}$ given by

$$\chi(g) = \mathrm{tr}[g]$$

where tr is the trace function. Note that χ is well defined since the trace of a linear transform is independent of the basis in which it is computed. We can now make a connection with the left-hand side of (1.2). If a group G acts on a set X, then the character of G on $\mathbb{C}X$ is given by

$$\chi(g) = \mathrm{tr}[g]_X = \#X^g \tag{3.2}$$

since $[g]_X$ is just the permutation matrix for g's action.

To see how the right side of (1.2) enters in this context, write $f(q) = \sum_{i=0}^{l} m_i q^i$ where $m_i \in \mathbb{N}$ for all i. Now suppose there is another basis, B, for $\mathbb{C}X$ with the property that every $g \in C$ is represented by a diagonal matrix of the form

$$[g]_B = \mathrm{diag}(\underbrace{1,\ldots,1}_{m_0},\underbrace{\omega,\ldots,\omega}_{m_1},\ldots,\underbrace{\omega^l,\ldots,\omega^l}_{m_l}) \tag{3.3}$$

where $\omega = \omega_{o(g)}$. (This may seem like a very strong assumption, but in the next section we will see that it must hold.) Computing the character in this basis gives

$$\chi(g) = \mathrm{tr}[g]_B = \sum_{i=0}^{l} m_i \omega^i = f(\omega). \tag{3.4}$$

Comparing (3.2) and (3.4) we immediately get the CSP. So cyclic sieving can merely amount to basis change in a C-module.

Sometimes it is better to use a C-module other than $\mathbb{C}X$ to obtain the right-hand side of (1.2). Two G-modules V, W are *G-isomorphic* or *G-equivalent*, written $V \cong W$, if there is a linear bijection $\phi : V \to W$ which preserves the action of G, i.e., for every $g \in G$ and $\mathbf{v} \in V$ we have $\phi(g\mathbf{v}) = g\phi(\mathbf{v})$. The prefix "$G$-" can be omitted if the group is understood from context. To obtain $f(\omega)$ as a character, any module isomorphic to $\mathbb{C}X$ will do. So we may pick a new module with extra structure which will be useful in the proof.

We are now ready to set up the tools we will need to reprove Theorem 2.1. Let $V^{\otimes k}$ denote the k-fold tensor product of the vector space V. If we take $V = \mathbb{C}[n]$ which has basis $B = \{\mathbf{i} : 1 \leq i \leq n\}$, then $V^{\otimes k}$ has a basis of the form

$$\{\mathbf{i}_1 \otimes \mathbf{i}_2 \otimes \cdots \otimes \mathbf{i}_k : \mathbf{i}_j \in B \text{ for } 1 \leq j \leq k\}.$$

In general, for any V, every basis B of V gives rise to a basis of $V^{\otimes k}$ consisting of k-fold tensors of elements of B.

When $V = \mathbb{C}[n]$, we consider the space of k-fold symmetric tensors, $\mathrm{Sym}_k(n)$, which is the quotient of $V^{\otimes k}$ by the subspace generated by

$$\mathbf{i}_1 \otimes \mathbf{i}_2 \otimes \cdots \otimes \mathbf{i}_k - \mathbf{i}_{g(1)} \otimes \mathbf{i}_{g(2)} \otimes \cdots \otimes \mathbf{i}_{g(k)} \qquad (3.5)$$

for all $g \in \mathfrak{S}_k$ and all tensors $\mathbf{i}_1 \otimes \mathbf{i}_2 \otimes \cdots \otimes \mathbf{i}_k$. Note that, while we are quotienting by such differences for all tensors, it would suffice (by linearity) to just consider the differences obtained using k-fold tensors from some basis. Let $\mathbf{i}_1\mathbf{i}_2\cdots\mathbf{i}_k$ denote the equivalence class of $\mathbf{i}_1 \otimes \mathbf{i}_2 \otimes \cdots \otimes \mathbf{i}_k$. These classes are indexed by the k-element multisets on $[n]$ and form a basis for $\mathrm{Sym}_k(n)$. So, for example,

$$\mathrm{Sym}_2(3) = \{c_1\mathbf{11} + c_2\mathbf{22} + c_3\mathbf{33} + c_4\mathbf{12} + c_5\mathbf{13} + c_6\mathbf{23} : c_i \in \mathbb{C} \text{ for all } i\}.$$

The cyclic group $C_n = \langle(1, 2, \ldots, n)\rangle$ acts on $\mathbb{C}[n]$ and so there is an induced action on $\mathrm{Sym}_k(n)$ given by

$$g(\mathbf{i}_1\mathbf{i}_2\cdots\mathbf{i}_k) = g(\mathbf{i}_1)g(\mathbf{i}_2)\cdots g(\mathbf{i}_k). \qquad (3.6)$$

Note that when defining $\mathrm{Sym}_k(n)$ by (3.5), one has $\mathfrak{S}_k$ acting on the subscripts to permute the places of the vectors. In contrast, the action in (3.6) has $\mathfrak{S}_n$ acting on the basis elements themselves. Comparing (3.6) with (2.2), we see that $\mathrm{Sym}_k(n) \cong \mathbb{C}(\binom{[n]}{k})$ as C_n-modules. We will work in the former module for our proof.

If $g \in C_n$ then let $[g]$ and $[g]'$ denote the associated linear transformations on $\mathbb{C}[n]$ and $\mathrm{Sym}_k(n)$, respectively. By way of illustration, when $n = 3$ and $k = 2$ we calculated the matrix $[(1, 2, 3)]_{\{1,2,3\}}$ in (3.1). On the other hand, the table (2.3) becomes

$$[(1,2,3)]'_{\{11,22,33,12,13,23\}} = \begin{bmatrix} 0 & 0 & 1 & 0 & 0 & 0 \\ 1 & 0 & 0 & 0 & 0 & 0 \\ 0 & 1 & 0 & 0 & 0 & 0 \\ 0 & 0 & 0 & 0 & 1 & 0 \\ 0 & 0 & 0 & 0 & 0 & 1 \\ 0 & 0 & 0 & 1 & 0 & 0 \end{bmatrix}.$$

Similarly, let χ and χ' be the respective characters of C_n acting on $\mathbb{C}[n]$ and $\mathrm{Sym}_k(n)$.

Now suppose we can find a basis $B = \{\mathbf{b}_1, \mathbf{b}_2, \ldots, \mathbf{b}_n\}$ for $\mathbb{C}[n]$ which diagonalizes $[g]$, say

$$[g]_B = \mathrm{diag}(x_1, x_2, \ldots, x_n).$$

Since B is a basis for $\mathbb{C}[n]$,

$$B' = \{\mathbf{b}_{i_1}\mathbf{b}_{i_2}\cdots\mathbf{b}_{i_k} : 1 \leq i_1 \leq i_2 \leq \ldots \leq i_k \leq n\}$$

is a basis for $\mathrm{Sym}_k(n)$. This is the crucial property of the space of symmetric tensors which makes us choose to work with them rather than in the original permutation module. Since each element of B is an eigenvector for $[g]$ acting on $\mathbb{C}[n]$, the same is true for B' and $[g]'$ acting on $\mathrm{Sym}_k(n)$. More precisely,

$$g(\mathbf{b}_{i_1}\mathbf{b}_{i_2}\cdots\mathbf{b}_{i_k}) = g(\mathbf{b}_{i_1})g(\mathbf{b}_{i_2})\cdots g(\mathbf{b}_{i_k}) = x_{i_1}x_{i_2}\cdots x_{i_k}\mathbf{b}_{i_1}\mathbf{b}_{i_2}\cdots\mathbf{b}_{i_k}.$$

It follows that

$$[g]'_{B'} = \mathrm{diag}(x_{i_1}x_{i_2}\cdots x_{i_k} \; : \; 1 \leq i_1 \leq i_2 \leq \ldots \leq i_k \leq n)$$

and

$$\chi'(g) = \sum_{1 \leq i_1 \leq i_2 \leq \ldots \leq i_k \leq n} x_{i_1}x_{i_2}\cdots x_{i_k}. \tag{3.7}$$

To illustrate, if $n = 3$ and $[g]_{\mathbf{a},\mathbf{b},\mathbf{c}} = \mathrm{diag}(x_1, x_2, x_3)$, then in $\mathrm{Sym}_2(3)$ we have

$$g(\mathbf{aa}) = x_1^2\mathbf{aa}, \qquad g(\mathbf{bb}) = x_2^2\mathbf{bb}, \qquad g(\mathbf{cc}) = x_3^2\mathbf{cc},$$
$$g(\mathbf{ab}) = x_1x_2\mathbf{ab}, \qquad g(\mathbf{ac}) = x_1x_3\mathbf{ac}, \qquad g(\mathbf{bc}) = x_2x_3\mathbf{bc},$$

so that

$$\chi'(g) = x_1^2 + x_2^2 + x_3^2 + x_1x_2 + x_1x_3 + x_2x_3. \tag{3.8}$$

The expression on the right-hand side of (3.7) is called a *complete homogeneous symmetric polynomial* and denoted $h_k(x_1, x_2, \ldots, x_n)$. It is called "complete homogeneous" because it is the sum of all monomials of degree k in the x_i. It is a symmetric polynomial because it is invariant under permutation of the subscripts of the variables. The theory of symmetric polynomials is intimately bound up with the representations of the symmetric and general linear groups. Equation (3.8) displays $h_2(x_1, x_2, x_3)$. To make use of (3.7), we need to related complete homogeneous symmetric functions to q-binomial coefficients. This is done by taking the *principal specialization* which sets $x_i = q^{i-1}$ for all i.

Lemma 3.1 *For $n \geq 1$ and $k \geq 0$ we have*

$$h_k(1, q, q^2, \ldots, q^{n-1}) = \begin{bmatrix} n+k-1 \\ k \end{bmatrix}_q. \tag{3.9}$$

Proof We do a double induction on n and k. For $n = 1$ we have $h_k(1) = x_1^k|_{x_1=1} = 1$ and $\begin{bmatrix} k \\ k \end{bmatrix}_q = 1$. For $k = 0$ it is also easy to see that both sides are 1.

Assume that $n \geq 2$ and $k \geq 1$. By splitting the sum for $h_k(x_1, x_2, \ldots, x_n)$ into those terms which do not contain x_n and those which do, we obtain the recursion

$$h_k(x_1, x_2, \ldots, x_n) = h_k(x_1, x_2, \ldots, x_{n-1}) + x_n h_{k-1}(x_1, x_2, \ldots, x_n).$$

Specializing yields

$$h_k(1, q, q^2, \ldots, q^{n-1}) = h_k(1, q, q^2, \ldots, q^{n-2}) + q^{n-1}h_{k-1}(1, q, q^2, \ldots, q^{n-1}).$$

Using the definition of the Gaussian polynomials in terms of q-factorials (2.5), it is easy to check that

$$\left[\begin{array}{c} n \\ k \end{array} \right]_q = \left[\begin{array}{c} n-1 \\ k \end{array} \right]_q + q^{n-k} \left[\begin{array}{c} n-1 \\ k-1 \end{array} \right]_q.$$

Substituting $n + k - 1$ for n, we see that the right-hand side of (3.9) satisfies the same recursion as the left-hand side, which completes the proof.   □

Proof (of Theorem 2.1)   Recall that $[(1, 2, \ldots, n)]$ is a linear transformation from $\mathbb{C}[n]$ to itself. The map has characteristic polynomial $x^n - 1$ with roots $1, \omega_n, \omega_n^2, \ldots, \omega_n^{n-1}$. Since these roots are distinct, there exists a diagonalizing basis B of $\mathbb{C}[n]$ with

$$[(1, 2, \ldots, n)]_B = \mathrm{diag}(1, \omega_n, \omega_n^2, \ldots, \omega_n^{n-1}).$$

Now any $g \in C_n$ is of the form $g = (1, 2, \ldots, n)^i$ for some i and so, since we have a diagonal representation,

$$[g]_B = \mathrm{diag}(1^i, \omega_n^i, \omega_n^{2i}, \ldots, \omega_n^{(n-1)i}) = \mathrm{diag}(1, \omega, \omega^2, \ldots, \omega^{n-1})$$

where $\omega = \omega_n^i$ is a primitive $o(g)$-th root of unity. The discussion leading up to equation (3.7) and the previous lemma yield

$$\chi'(g) = h_k(1, \omega, \omega^2, \ldots, \omega^{n-1}) = \left[\begin{array}{c} n+k-1 \\ k \end{array} \right]_\omega.$$

As we have already noted, $\mathrm{Sym}_k(n) \cong \mathbb{C}(\binom{[n]}{k})$, and so by (3.2)

$$\chi'(g) = \# \left(\binom{[n]}{k} \right)^g.$$

Comparing the last two equations completes the proof that the CSP holds.   □

4   A representation theory paradigm

We promised to show that the assumption of $[g]$ having a diagonalization of the form (3.3) is not a stretch. To do that, we need to develop some more representation theory which will also lead to a paradigm for proving the CSP.

A *submodule* of a G-module V is a subspace W which is left invariant under the action of G in that $g\mathbf{w} \in W$ for all $g \in G$ and $\mathbf{w} \in W$. The zero subspace and V itself are the *trivial submodules*. We say that V is *reducible* if it has a nontrivial submodule and *irreducible* otherwise. For example, the $\mathfrak{S}_3$-module $\mathbb{C}[3]$ is reducible because the 1-dimensional subspace generated by the vector $\mathbf{1} + \mathbf{2} + \mathbf{3}$ is a nontrivial submodule. It turns out that the irreducible modules are the building blocks of all other modules in our setting. The next result collects together three standard results from representation theory. They can be found along with their proofs in Proposition 1.10.1, Theorem 1.5.3, and Corollary 1.9.4 (respectively) of [65].

Theorem 4.1 *Let G be a finite group and consider G-modules which are vector spaces over $\mathbb{C}$.*

(a) *The number of pairwise inequivalent irreducible G-modules is finite and equals the number of conjugacy classes of G.*

(b) *(Maschke's Theorem) Every G-module can be written as a direct sum of irreducible G-modules.*

(c) *Two G-modules are equivalent if and only if they have the same character.* □

We note that if G is not finite or if the ground field for our G-modules is not $\mathbb{C}$ then the analogue of this theorem may not hold. Also, the forward direction of (c) is trivial (we have already been using it in the last section), while the backward direction is somewhat surprising in that one can completely characterize a G-module through the trace alone.

Let us construct the irreducible representations of a cyclic group C with $\#C = n$. The *dimension* of a G-module V is its usual vector space dimension. If $\dim V = 1$ then V must be irreducible. So what do the dimension one modules for C look like? Let g be a generator of C and let $V = \mathbb{C}\{\mathbf{v}\}$ for some vector $\mathbf{v}$. Then $g\mathbf{v} = c\mathbf{v}$ for some scalar c. Furthermore

$$\mathbf{v} = e\mathbf{v} = g^n\mathbf{v} = c^n\mathbf{v}.$$

So $c^n = 1$ and c is an nth root of unity. It is easy to verify that for each nth root of unity ω, the map $\rho(g^j) = [\omega^j]$ defines a representation of C as j varies over the integers. So we have found n irreducible G-modules of C, one for each nth root of unity: $V^{(0)}, V^{(1)}, \ldots, V^{(n-1)}$. They clearly have different characters (the trace of a 1-dimensional matrix being itself) and so are pairwise inequivalent. Finally, C is Abelian and so has $\#C = n$ conjugacy classes. Thus by (a) of Theorem 4.1, we have found all the irreducible representations.

Now given any C-module, V, part (b) of Theorem 4.1 says we have a module isomorphism

$$V \cong \bigoplus_{i=0}^{n-1} a_i V^{(i)}$$

where $a_i V^{(i)}$ denotes a direct sum of a_i copies of $V^{(i)}$. Since each of the summands is 1-dimensional, there is a basis B simultaneously diagonalizing the linear transformations $[g]$ for all $g \in C$ as in (3.3). We can also explain the multiplicities as follows. Extend the definition of $V^{(i)}$ to any $i \in \mathbb{N}$ by letting $V^{(i)} = V^{(j)}$ if $i \equiv j \pmod{n}$. Now given any polynomial $f(q) = \sum_{i \geq 0} m_i q^i$ with nonnegative integer coefficients, define a corresponding C-module

$$V_f = \bigoplus_{i \geq 0} m_i V^{(i)}. \tag{4.1}$$

Reiner, Stanton, and White identified the following representation theory paradigm for proving that the CSP holds.

Theorem 4.2 *The cyclic sieving property holds for the triple $(X, C, f(q))$ if and only if one has $\mathbb{C}X \cong V_f$ as C-modules.*

Proof Note that $\#X^g$ and $f(\omega_{o(g)})$ are the character values of $g \in C$ in the modules $\mathbb{C}X$ and V_f, respectively. So the result now follows from Theorem 4.1 (c). □

5   Coxeter groups and permutation statistics

We will now present some basic definitions and results about Coxeter groups which will be needed for later use. The interested reader can find more information in the books of Björner and Brenti [9] or Hiller [31]. A *finite Coxeter group*, W, is a finite group having a presentation with a set of generators S, and relations for each pair $s, s' \in S$ of the form

$$(ss')^{m(s,s')} = e,$$

where the $m(s, s')$ are positive integers satisfying

$$m(s, s') = m(s', s);$$
$$m(s, s') = 1 \iff s = s'.$$

One can also define infinite Coxeter groups, but we will only need the finite case and may drop "finite" as being understood in what follows. An abstract group W may have many presentations of this form, so when we talk about a Coxeter group we usually have a specific generating set S in mind which is tacitly understood. If we wish to be explicit about the generating set, then we will refer to the *Coxeter system* (W, S). Note that since $m(s, s) = 1$ we have $s^2 = 1$ and so the elements of S are involutions. It follows that one can rewrite $(ss')^{m(s,s')} = e$ by bringing half the factors to the right-hand side

$$\underbrace{s\, s'\, s\, s'\, s \, \cdots}_{m(s,s')} = \underbrace{s'\, s\, s'\, s\, s' \, \cdots}_{m(s,s')}.$$

Probably the most famous Coxeter group is the symmetric group, $\mathfrak{S}_n$. Here we take the generating set of adjacent transpositions $S = \{s_1, s_2, \ldots, s_{n-1}\}$ where $s_i = (i, i+1)$. The Coxeter relations take the form

$$s_i^2 = 1,$$
$$s_i s_j = s_j s_i \text{ for } |i - j| \geq 2,$$
$$s_i s_{i+1} s_i = s_{i+1} s_i s_{i+1} \text{ for } 1 \leq i \leq n - 1.$$

The third equation is called the *braid relation*. (Other authors also include the second equation and distinguish the two by using the terms *long* and *short* braid relations). We will often refer back to this example to illustrate Coxeter group concepts.

A Coxeter group, W, is *irreducible* if it can not be written as a nontrivial product of two other Coxeter groups. Irreducible finite Coxeter groups were classified by Coxeter [13]. A list of these groups is given in Table 1. The *rank* of a Coxeter group, $\mathrm{rk}\, W$, is the minimum cardinality of a generating set S and the subscript in each group name gives the rank. If S has this minimum cardinality then its elements are called *simple*. The middle column displays the *Coxeter graph* or *Dynkin diagram* of the group which has the set S as its vertices with an edge labeled $m(s, s')$ between vertices $s \neq s'$. By convention, if $m(s, s') = 2$ (i.e., s and s' commute) then one omits the edge, and if $m(s, s') = 3$ then the edge is displayed without a label. A Coxeter group can be realized as the symmetry group of a regular polytope if and only if its graph contains only vertices of degree one and two. So all these groups

Group	Diagram	Polytope
A_n		symmetric group $\mathfrak{S}_{n+1}$, group of the simplex
B_n		hyperoctahedral group, group of the cube/octahedron
D_n		-
E_6		-
E_7		-
E_8		-
F_4		group of the 24-cell
H_3		group of the dodecahedron/icosahedron
H_4		group of the 120-cell/600-cell
$I_2(m)$		group of the m-gon

Table 1: The irreducible finite Coxeter groups

except D_n, E_6, E_7, and E_8 have corresponding polytopes which are listed in the last column.

There is an important function on Coxeter groups which we will need to define generating functions for instances of the CSP. Given $w \in W$ we can write $w = s_1 s_2 \cdots s_k$ where the $s_i \in S$. Note that here s_i is just an element of S and not necessarily the ith generator. Such an expression is *reduced* if k is minimal among all such expressions for w and this value of k is called the *length* of w, written $\ell(w) = k$. When W is of type A_{n-1}, i.e., the symmetric group $\mathfrak{S}_n$, then there is a nice combinatorial interpretation of the length function. Write w in *one-line notation* as $w = w_1 w_2 \ldots w_n$ where $w_i = w(i)$ for $i \in [n]$. The set of *inversions* of w is

$$\operatorname{Inv} w = \{(i,j) \ : \ i < j \text{ and } w_i > w_j\}.$$

So $\operatorname{Inv} w$ records the places in w where there is a pair of out-of-order elements. The *inversion number* of w is $\operatorname{inv} w = \#\operatorname{Inv} w$. For example, if

$$w = w_1 w_2 w_3 w_4 w_5 = 31524 \tag{5.1}$$

then $\operatorname{Inv} w = \{(1,2),(1,4),(3,4),(3,5)\}$ and $\operatorname{inv} w = 4$. It turns out that for type A,

$$\ell(w) = \operatorname{inv} w. \tag{5.2}$$

We can now make a connection with the q-binomial coefficients as follows. Given a Coxeter system (W, S) and $J \subset S$, there is a corresponding *parabolic subgroup* W_J

which is the subgroup of W generated by J. It can be shown that each coset wW_J has a unique representative of minimal length. Let W^J be the set of these coset representatives and set

$$W^J(q) = \sum_{w \in W^J} q^{\ell(w)}. \tag{5.3}$$

If $W = \mathfrak{S}_n$ with $S = \{s_1, s_2, \ldots, s_{n-1}\}$ as before, then remove s_k from S to obtain $J = S \setminus \{s_k\}$ (which generates a *maximal parabolic subgroup*). So

$$(\mathfrak{S}_n)_J \cong \mathfrak{S}_k \times \mathfrak{S}_{n-k}$$

consists of all permutations permuting the sets $\{1, 2, \ldots, k\}$ and $\{k+1, k+2, \ldots, n\}$ among themselves. Thus multiplying $w \in \mathfrak{S}_n$ on the right by an element of $(\mathfrak{S}_n)_J$ merely permutes $\{w_1, w_2, \ldots, w_k\}$ and $\{w_{k+1}, w_{k+2}, \ldots, w_n\}$ among themselves. (We compose permutations from right to left.) Using (5.2), we see that the set of minimal length coset representatives is

$$(\mathfrak{S}_n)^J = \{w \in \mathfrak{S}_n \ : \ w_1 < w_2 < \ldots < w_k \text{ and } w_{k+1} < w_{k+2} < \ldots < w_n\}. \tag{5.4}$$

A straightforward double induction on n and k, much like the one used to prove Lemma 3.1, now yields the following result.

Proposition 5.1 *For $W = \mathfrak{S}_n$ and $J = S \setminus \{s_k\}$ we have*

$$W^J(q) = \begin{bmatrix} n \\ k \end{bmatrix}_q$$

for any $0 \leq k \leq n$ (where for $k = 0$ or n, $J = S$). $\square$

As has already been mentioned, the symmetry groups of regular polytopes only yield some of the Coxeter groups. However, there is a geometric way to get them all. A *reflection* in $\mathbb{R}^n$ is a linear transformation r_H which fixes a hyperplane H pointwise and sends a vector perpendicular H to its negative. A *real reflection group* is a group generated by reflections. It turns out that the finite real reflection groups exactly coincide with the finite Coxeter groups; see the papers of Coxeter [12, 13]. (A similar result holds in the infinite case if one relaxes the definition of a reflection.) Definitions for Coxeter groups are also applied to the corresponding reflection group, e.g., a *simple reflection* is one corresponding to an element of S. The text of Benson and Grove [27] gives a nice introduction to finite reflection groups. For example, to get $\mathfrak{S}_n$ one can use the reflecting hyperplanes $H_{i,j}$ with equation $x_i = x_j$. If $r_{i,j}$ is the corresponding reflection then $r_{i,j}(x_1, x_2, \ldots, x_n)$ is just the point obtained by interchanging the ith and jth coordinates and so corresponds to the transposition $(i, j) \in \mathfrak{S}_n$.

We end this section by discussing permutation statistics which are intimately connected with Coxeter groups as we have seen with the statistic inv. A *statistic* on a finite set X is a function $\mathrm{st} : X \to \mathbb{N}$. The statistic has a corresponding *weight generating function*

$$f^{\mathrm{st}}(X) = f^{\mathrm{st}}(X; q) = \sum_{y \in X} q^{\mathrm{st}\, y}.$$

w	123	132	213	231	312	321
inv w	0	1	1	2	2	3
maj w	0	2	1	2	1	3
des w	0	1	1	1	1	2
exc w	0	1	1	2	1	1

Table 2: Four statistics on $\mathfrak{S}_3$

From Table 2 we see that on $X = \mathfrak{S}_3$

$$f^{\mathrm{inv}}(\mathfrak{S}_3) = \sum_{w \in \mathfrak{S}_3} \mathrm{inv}\, w = 1 + 2q + 2q^2 + q^3 = (1+q)(1+q+q^2) = [3]_q!$$

In fact, this holds for any n (not just 3); see Proposition 5.2 below. And the reader should compare this result with Proposition 5.1 which gives the generating function for inv over another set of permutations.

There are three other statistics which will be important in what follows. The *descent set* of a permutation $w = w_1 w_2 \ldots w_n$ is

$$\mathrm{Des}\, w = \{i \ : \ w_i > w_{i+1}\}.$$

We let $\mathrm{des}\, w = \#\,\mathrm{Des}\, w$. Using the descents, one forms the *major index*

$$\mathrm{maj}\, w = \sum_{i \in \mathrm{Des}\, w} i.$$

Continuing the example in (5.1), $\mathrm{Des}\, w = \{1, 3\}$ since $w_1 > w_2$ and $w_3 > w_4$, and so $\mathrm{maj}\, w = 1 + 3 = 4$. The major index was named for Major Percy MacMahon who introduced the concept [49] (or see [50, pp. 508-549]).

We say that two statistics st and st' on X are *equidistributed* if $f^{\mathrm{st}}(X) = f^{\mathrm{st}'}(X)$. In other words, the number of elements in X with any given st value k equals the number having st' value k. Comparing the first two rows of Table 2, the reader should suspect the following result which is not hard to prove by induction on n.

Proposition 5.2 *We have*

$$\sum_{w \in \mathfrak{S}_n} q^{\mathrm{inv}\, w} = [n]_q! = \sum_{w \in \mathfrak{S}_n} q^{\mathrm{maj}\, w}.$$

So $f^{\mathrm{inv}}(\mathfrak{S}_n) = f^{\mathrm{maj}}(\mathfrak{S}_n)$. $\qquad\qquad\square$

Any statistic on $\mathfrak{S}_n$ equidistributed with inv (or maj) is said to be *Mahonian*, also in tribute to MacMahon.

The last permutation statistic we need comes from the set of *excedances* which is

$$\mathrm{Exc}\, w = \{i \ : \ w(i) > i\}.$$

One can view excedances as "descents" in the cycle decomposition of w. As usual, we let $\mathrm{exc}\, w = \#\,\mathrm{Exc}\, w$. In our running example $\mathrm{Exc}\, w = \{1, 3\}$ since $w_1 = 3 > 1$ and $w_3 = 5 > 3$. Comparing the distributions of des and exc in Table 2, the reader will see a special case of the following proposition whose proof can be found in the text of Stanley [80, Proposition 1.3.12].

Proposition 5.3 *We have* $f^{\text{des}}(\mathfrak{S}_n) = f^{\text{exc}}(\mathfrak{S}_n)$. □

Any permutation statistic equidistributed with des (or exc) is said to be *Eulerian*. The polynomial in Proposition 5.3 is called the *Eulerian polynomial*, $A_n(q)$, although some authors use this term for the polynomial $qA_n(q)$. The first few Eulerian polynomials are

$$
\begin{aligned}
A_0(q) &= 1, \\
A_1(q) &= 1, \\
A_2(q) &= 1 + q, \\
A_3(q) &= 1 + 4q + q^2, \\
A_4(q) &= 1 + 11q + 11q^2 + q^3, \\
A_5(q) &= 1 + 26q + 66q^2 + 26q^3 + q^4.
\end{aligned}
$$

The exponential generating function for these polynomials

$$
\sum_{n \geq 0} A_n(q) \frac{t^n}{n!} = \frac{1 - q}{e^{t(q-1)} - q} \tag{5.5}
$$

is attributed to Euler [39, p. 39].

6 Complex reflection groups and Springer's regular elements

Before stating Theorem 2.1, we noted that it is a special case of one part of the first theorem in the Reiner-Stanton-White paper. To state the full result, we need another pair of definitions. Let $g \in \mathfrak{S}_N$ have $o(g) = n$. Say that g acts *freely* on $[N]$ if all of g's cycles are of length n. So in this case $n|N$. For example, $g = (1,2)(3,4)(5,6)$ acts freely on $[6]$. Say that g acts *nearly freely* on $[N]$ if either it acts freely, or all of its cycles are of length n except one which is a singleton. In the latter case, $n|N-1$. So $g = (1,2)(3,4)(5,6)(7)$ acts nearly freely on $[7]$. Finally, say that the cyclic group C acts *freely* or *nearly freely* on $[N]$ if it has a generator g with the corresponding property. The next result is Theorem 1.1 in [58]. In it,

$$
\binom{[N]}{k} = \{ S \ : \ S \text{ is a } k\text{-element subset of } [N] \}. \tag{6.1}
$$

Theorem 6.1 *Suppose C is cyclic and acts nearly freely on $[N]$. The following two triples*

$$
\left(\left(\binom{[N]}{k} \right), C, \begin{bmatrix} N+k-1 \\ k \end{bmatrix}_q \right) \tag{6.2}
$$

and

$$
\left(\binom{[N]}{k}, C, \begin{bmatrix} N \\ k \end{bmatrix}_q \right) \tag{6.3}
$$

exhibit the cyclic sieving phenomenon. □

Note that Theorem 2.1 is a special case of (6.2) since $C = \langle (1, 2, \ldots, N) \rangle$ acts freely, and so also nearly freely, on $[N]$. At this point, the reader may have (at least) two questions in her mind. One might be why the multiset example was chosen to explain in detail rather than the combinatorially simpler set example. The reason for this is that the representation theory proof for the latter involves alternating tensors and so one has to worry about signs which do not occur in the symmetric tensor case. Another puzzling aspect might be why having a nearly free action is the right hypothesis on C. To clarify this, one needs to discuss complex reflection groups and Springer's regular elements.

A *complex (pseudo)-reflection* is an element of $\mathrm{GL}_N(\mathbb{C})$ $(= \mathrm{GL}(\mathbb{C}^N))$ which fixes a unique hyperplane in $\mathbb{C}^N$ and has finite order. Every real reflection can be considered as a complex reflection by extending the field. But the complex notion is more general since a real reflection must have order two. A *complex reflection group* is a group generated by complex reflections. As usual, we will only be interested in the finite case. *Irreducible* complex reflection groups are defined as in the real case, i.e., they are the ones which cannot be written as a nontrivial product of two other complex reflection groups. The irreducible complex reflection groups were classified by Shephard and Todd [75]. The book of Lehrer and Taylor [47] gives a very lucid treatment of these groups, even redoing the Shephard-Todd classification.

Call an element g in a finite complex reflection group W *regular* if it has an eigenvector which does not lie on any of the reflecting hyperplanes of W. An eigenvalue corresponding to this eigenvector is also called *regular*. In type A (i.e., in the case of the symmetric group) we have the following connection between regular elements and nearly free actions.

Proposition 6.2 *Let* $W = A_{N-1}$. *Then* $g \in W$ *is regular if and only if it acts nearly freely on* $[N]$.

Proof Assume that $o(g) = n$ and that g acts nearly freely. Suppose first that $n | N$ and

$$g = (1, 2, \ldots, n)(n+1, n+2, \ldots, 2n) \cdots .$$

Other elements of order n can be treated similarly. Now $(1, 2, \ldots, n)$ acting on $\mathbb{C}^n$ has eigenvalue $\omega^{-1} = \omega_n^{-1}$ with eigenvector $[\omega, \omega^2, \ldots, \omega^n]^t$ (t denoting transpose) all of whose entries are distinct. So

$$\mathbf{v} = [\omega, \omega^2, \ldots, \omega^n, 2\omega, 2\omega^2, \ldots, 2\omega^n, 3\omega, 3\omega^2, \ldots, 3\omega^n, \ldots]^t$$

is an eigenvector for g lying on none of the hyperplanes $x_i = x_j$. In the case that $n | N - 1$, one can insert a 0 in $\mathbf{v}$ at the coordinate of the fixed point and preserve regularity.

Now suppose g does not act nearly freely. Consider the case when g has cycles of lengths k and l where $k, l \geq 2$ and $k \neq l$. Without loss of generality, say $k < l$ and

$$g = (1, 2, \ldots, k)(k+1, k+2, \ldots, k+l) \cdots .$$

Suppose $\mathbf{v} = [a_1, a_2, \ldots, a_N]^t$ is a regular eigenvector for g. Then $[a_1, a_2, \ldots, a_k]^t$ must either be an eigenvector for $g' = (1, 2, \ldots, k)$ or be the zero vector. But if $\mathbf{v}$ lies on none of the hyperplanes then the second possibility is out because $k \geq 2$.

The eigenvectors for g' are $[\omega_k^i, \omega_k^{2i}, \ldots, \omega_k^{ki}]^t$ with corresponding eigenvalues ω_k^{-i} for $1 \le i \le k$. Since everything we have said also applies to $g'' = (k+1, k+2, \ldots, k+l)$, the eigenvalue ω of g must be a root of unity with order dividing $\gcd(k,l)$. But $\gcd(k,l) \le k < l$, so the eigenvectors of g'' with such eigenvalues will all have repeated entries, a contradiction. One can deal with the only remaining case (when g has at least two fixed points) similarly. □

In addition to the previous proposition, we will need the following general result, It is easy to prove from the definitions and so is left to the reader.

Lemma 6.3 *Let V be a G-module.*

(a) If $W \subseteq V$ is a G-submodule then the quotient space V/W is also G-module.

(b) If $H \le G$ is a subgroup, then V is also an H-module. □

Also, we generalize the notation (1.1): if V is a G-module then the *invariants* of G in V are

$$V^G = \{\mathbf{v} \in V \; : \; g\mathbf{v} = \mathbf{v} \text{ for all } g \in G\}.$$

Springer's Theorem relates two algebras. To define the first, note that a group G acts on itself by left multiplication. The corresponding permutation module $\mathbb{C}[G]$ is called the *group algebra* and it is an algebra, not just a vector space, because one can formally multiply linear combinations of group elements. The group algebra is important in part because it contains every irreducible representation of G. In particular, the following is true. See Proposition 1.10.1 in [65] for more details.

Theorem 6.4 *Let G be a finite group with irreducible modules $V^{(1)}, V^{(2)}, \ldots, V^{(k)}$ and write $\mathbb{C}[G] = \oplus_i m_i V^{(i)}$. Then, for all i,*

$$m_i = \dim V^{(i)}.$$

so every irreducible appears with multiplicity equal to its degree. Taking dimensions, we have

$$\sum_{i=1}^k \left(\dim V^{(i)}\right)^2 = \#G$$
□

The second algebra is defined for any subgroup $W \le \mathrm{GL}_N(\mathbb{C})$. Thinking of $x_1, x_2, \ldots, x_N$ as the coordinates of $\mathbb{C}^N$, W acts on the algebra of polynomials $\mathcal{S} = \mathbb{C}[x_1, x_2, \ldots, x_N]$ by linear transformations of the x_i. For example, if $W = \mathfrak{S}_N$ then W acts on $\mathcal{S}$ by permuting the variables. The *algebra of coinvariants of W* is the quotient

$$A = \mathcal{S}/\mathcal{S}_+^W \tag{6.4}$$

where $\mathcal{S}_+^W$ is the ideal generated by the invariants of W in $\mathcal{S}$ which are homogeneous of positive degree. Note that by Lemma 6.3(a), W also acts on A.

Now let g be a regular element of W of order n and let $C = \langle g \rangle$ be the cyclic group it generates. We also let $\omega = \omega_n$. Define an action of the product group $W \times C$ on the group algebra $\mathbb{C}[W]$ by having W act by multiplication on the left and C act by multiplication on the right. These actions commute because of the associative law in W, justifying the use of the direct product. We also have an action of $W \times C$ on A: We already noted in the previous paragraph how W acts on A, and we let C act by

$$g(x_i) = \omega x_i \qquad (6.5)$$

for $i \in [N]$. The following is a beautiful theorem of Springer [78] as reformulated by Kraśkiewicz and Weyman [40].

Theorem 6.5 *Let W be a finite complex reflection group with coinvariant algebra A, and let $C \leq W$ be cyclically generated by a regular element. Then A and the group algebra $\mathbb{C}[W]$ are isomorphic as $W \times C$ modules.* □

Although A and $\mathbb{C}[W]$ are isomorphic, the former has the advantage that it is graded, i.e., we can write $A = \oplus_{d \geq 0} A_d$ where A_d are the elements in A homogeneous of degree d. (This is well defined because the invariant ideal we modded out by is generated by homogeneous polynomials.) And any graded algebra over $\mathbb{C}$ has a *Hilbert series*

$$\mathrm{Hilb}(A; q) = \sum_{d \geq 0} \dim_{\mathbb{C}} A_d q^d.$$

It is this series and the previous theorem which permitted Reiner, Stanton, and White to formulate a powerful cyclic sieving result. In it and in the following corollary, the action of the cyclic group on left cosets is by left multiplication.

Theorem 6.6 *Let W be a finite complex reflection group with coinvariant algebra A, and let $C \leq W$ be cyclically generated by a regular element g. Take any $W' \leq W$ and consider the invariant algebra $A^{W'}$. Then cyclic sieving is exhibited by the triple*

$$\left(W/W', \; C, \; \mathrm{Hilb}(A^{W'}; q) \right).$$

Proof By Theorem 6.5 we have an isomorphism $\phi : A \to \mathbb{C}[W]$ of $W \times C$ modules. So by Lemma 6.3(b) they are also isomorphic as C-modules. Since the actions of C and W' commute, the invariant algebras $A^{W'}$ and $\mathbb{C}[W]^{W'}$ are also C-modules and ϕ restricts to an isomorphism between them.

By (6.5), the action of g on the dth graded piece of $A^{W'}$ is just multiplication by ω^d. But this is exactly the same as the action on the dth summand in the C-module $V_{\mathrm{Hilb}(A^{W'})}$ as defined for any generating function f by equation (4.1). So $A^{W'} \cong V_{\mathrm{Hilb}(A^{W'})}$ as C-modules.

As far as $\mathbb{C}[W]^{W'}$, consider the set of right cosets $W' \backslash W$. Note that $\sum_i c_i g_i \in \mathbb{C}[W]$ will be W'-invariant if and only if the coefficients c_i are constant on each right coset. So $\mathbb{C}[W]^{W'}$ is C-isomorphic to the permutation module $\mathbb{C}(W' \backslash W)$ with C acting on the right. But since C is Abelian, this is isomorphic to the module $\mathbb{C}(W/W')$ for the left cosets with C acting (as usual) on the left.

Putting the isomorphisms in last three paragraphs together and using Theorem 4.2 completes the proof. □

We can specialize this theorem to the case of Coxeter groups and their parabolic subgroups. One only needs the fact [31, §IV.4] that, for the length generating function defined by (5.3),

$$W^J(q) = \mathrm{Hilb}(A^{W_J}; q).$$

Corollary 6.7 *Let (W, S) be a finite Coxeter system and let $J \subseteq S$. Let $C \leq W$ be cyclically generated by a regular element g. Then the triple*

$$(W/W_J, \ C, \ W^J(q))$$

satisfies the cyclic sieving phenomenon.   □

If we specialize even further to type A, then we obtain the CSP in (6.3). To see this, note first that g being regular is equivalent to its acting nearly freely by Proposition 6.2. For $J = S \setminus \{s_k\}$, the action on left cosets $\mathfrak{S}_N/(\mathfrak{S}_N)_J$ is isomorphic to the action on $\binom{[n]}{k}$ as can be seen using the description of the minimal length representatives (5.4). Finally, Proposition 5.1 shows that the generating function is correct.

7 Promotion on rectangular standard Young tableaux

Rhoades [60] proved an amazing cyclic sieving result about rectangular Young tableaux under the action of promotion. While the theorem is combinatorially easy to state, his proof involves deep results about Kazhdan-Lusztig representations [36] and a characterization of the dual canonical basis by Skandera [77]. We will start by giving some background about Young tableaux.

A *partition of* $n \in \mathbb{N}$ is a weakly decreasing sequence of positive integers $\lambda = (\lambda_1, \lambda_2, \ldots, \lambda_l)$ such that $\sum_i \lambda_i = n$. We use the notation $\lambda \vdash n$ for this concept and call the λ_i *parts*. For example, the partitions of 4 are (4), $(3, 1)$, $(2, 2)$, $(2, 1, 1)$, and $(1, 1, 1, 1)$. We will use exponents to denote multiplicities just as with multisets. So $(1, 1, 1, 1) = (1^4)$, $(2, 1, 1) = (2, 1^2)$, and so forth. We will sometimes drop the parentheses and commas to simplify notation. Partitions play an important role in number theory, combinatorics, and representation theory. See the text of Andrews [1] for more information.

Associated with any partition $\lambda = (\lambda_1, \lambda_2, \ldots, \lambda_l)$ is its *Ferrers diagram*, also denoted λ, which consists of l left-justified rows of dots with λ_i dots in row i. We let (i, j) stand for the position of the dot in row i and column j. For example, the partition $\lambda = (5, 4, 4, 2)$ has diagram

$$\lambda = \begin{matrix} \bullet & \bullet & \bullet & \bullet & \bullet \\ \bullet & \bullet & \blacksquare & \bullet \\ \bullet & \bullet & \bullet & \bullet \\ \bullet & \bullet \end{matrix} \qquad (7.1)$$

where the $(2, 3)$ dot has been replaced by a square. Note that sometimes empty boxes are used instead of dots. Also we are using English notation, as opposed to the French version where the parts are listed bottom to top.

If $\lambda \vdash n$ is a Ferrers diagram, then a *standard Young tableau, T, of shape* λ is a bijection $T : \lambda \to [n]$ such that rows and columns increase. We let $\mathrm{SYT}(\lambda)$ denote the set of such tableaux and also

$$\mathrm{SYT}_n = \bigcup_{\lambda \vdash n} \mathrm{SYT}(\lambda).$$

Also define
$$f^\lambda = \#\,\mathrm{SYT}(\lambda).$$
To illustrate,
$$\mathrm{SYT}(3,2) = \left\{ \begin{array}{ccc} 1 & 2 & 3 \\ 4 & 5 \end{array}, \begin{array}{ccc} 1 & 2 & 4 \\ 3 & 5 \end{array}, \begin{array}{ccc} 1 & 2 & 5 \\ 3 & 4 \end{array}, \begin{array}{ccc} 1 & 3 & 4 \\ 2 & 5 \end{array}, \begin{array}{ccc} 1 & 3 & 5 \\ 2 & 4 \end{array} \right\}$$

so $f^{(3,2)} = 5$. We let $T_{i,j}$ denote the element in position (i,j) and write $\mathrm{sh}\,T$ to denote the partition which is T's shape. In Rhoades' theorem, the cyclic sieving set will be $X = \mathrm{SYT}(n^m)$, a set of standard Young tableaux of *rectangular shape*.

Partitions and Young tableaux are intimately connected with representations of the symmetric and general linear groups. Given $g \in \mathfrak{S}_n$, its *cycle type* is the partition gotten by arranging g's cycle lengths in weakly decreasing order. For example, $g = (1,5,2)(3,7)(4,8,9)(6)$ has cycle type $\lambda = (3,3,2,1)$. Since the conjugacy classes of $\mathfrak{S}_n$ consist of all elements of the same cycle type, they are naturally indexed by partitions $\lambda \vdash n$. So by Theorem 4.1 (a), the partitions λ also index the irreducible $\mathfrak{S}_n$-modules, V^λ. In fact (see Theorem 2.6.5 in [65])
$$\dim V^\lambda = f^\lambda. \tag{7.2}$$
and there are various constructions which use Young tableaux of a given shape to build the corresponding representation. Note that from Theorem 6.4 we obtain
$$\sum_{\lambda \vdash n} \left(f^\lambda \right)^2 = n! \tag{7.3}$$

If one ignores its representation theory provenance, equation (7.3) can be viewed as a purely combinatorial statement about tableaux. So one could prove it combinatorially by finding a bijection between $\mathfrak{S}_n$ and pairs (P, Q) of standard Young tableaux of the same shape λ, with λ varying over all partitions of n. The algorithm we will describe to do this is due to Schensted [68]. It was also discovered by Robinson [62] in a different form. A *partial Young tableau* will be a filling of a shape with increasing rows and columns (but not necessarily using the numbers $1, \ldots, n$). We first describe *insertion* of an element x into a partial tableau P with $x \notin P$.

1. Initialize with $i = 1$ and $p = x$.

2. If there is an element of row i of P larger than p, then remove the left-most such element and put p in that position. Now repeat this step with i replaced by $i + 1$ and p replaced with the removed element.

3. When one reaches a row where no element of that row is greater then p, then p is placed at the end of the row and insertion terminates with a new tableau, $I_x(P)$.

The removals are called *bumps* and are defined so that at each step of the algorithm the rows and columns remain increasing. For example, if
$$T = \begin{array}{cccc} 1 & 3 & 5 & 6 \\ 2 & 8 & 9 \\ 7 \end{array}$$

then inserting 4 gives

$$
\begin{array}{llll}
1 & 3 & 5 & 6 \\
2 & 8 & 9 \\
7
\end{array}
\leftarrow 4
\quad
\begin{array}{llll}
1 & 3 & 4 & 6 \\
2 & 8 & 9 \\
7
\end{array}
,
\leftarrow 5 ,
\quad
\begin{array}{llll}
1 & 3 & 4 & 6 \\
2 & 5 & 9 \\
7
\end{array}
,
\leftarrow 8
\quad
\begin{array}{llll}
1 & 3 & 4 & 6 \\
2 & 5 & 9 \\
7 & 8
\end{array}
= I_4(T).
$$

We can now describe the map $w \mapsto (P, W)$. Given $w = w_1 w_2 \ldots w_n$ in 1-line notation, we build a sequence of partial tableaux $P_0 = \emptyset, P_1, \ldots, P_n = P$ where $\emptyset$ is the empty tableau and $P_k = I_{w_k} P_{k-1}$ for all $k \geq 1$. At the same time, we construct a sequence $Q_0 = \emptyset, Q_1, \ldots, Q_n = Q$ where Q_k is obtained from Q_{k-1} by placing k in the unique new position in P_{k+1}. To illustrate, if $w = 31452$ then we obtain

$$
P_k : \quad \emptyset, \quad
\begin{array}{l} 3 \end{array}, \quad
\begin{array}{l} 1 \\ 3 \end{array}, \quad
\begin{array}{ll} 1 & 4 \\ 3 \end{array}, \quad
\begin{array}{lll} 1 & 4 & 5 \\ 3 \end{array}, \quad
\begin{array}{lll} 1 & 2 & 5 \\ 3 & 4 \end{array}
= P,
$$

$$
Q_k : \quad \emptyset, \quad
\begin{array}{l} 1 \end{array}, \quad
\begin{array}{l} 1 \\ 2 \end{array}, \quad
\begin{array}{ll} 1 & 3 \\ 2 \end{array}, \quad
\begin{array}{lll} 1 & 3 & 4 \\ 2 \end{array}, \quad
\begin{array}{lll} 1 & 3 & 4 \\ 2 & 5 \end{array}
= Q.
$$

This procedure is invertible. Given (P_k, Q_k) then we find the position (i, j) of k in Q_k. We reverse the bumping process in P_k starting with the element in (i, j). The element removed from the top row of P_k then becomes the kth entry of w. This map is called the *Robinson-Schensted correspondence* and denoted $w \stackrel{\mathrm{R-S}}{\mapsto} (P, Q)$. We have proved the following result.

Theorem 7.1 *For all $n \geq 0$, the map $w \stackrel{\mathrm{R-S}}{\mapsto} (P, Q)$ is a bijection*

$$
\mathfrak{S}_n \stackrel{\mathrm{R-S}}{\longleftrightarrow} \{(P, Q) \ : \ P, Q \in \mathrm{SYT}_n, \ \mathrm{sh}(P) = \mathrm{sh}(Q)\}. \qquad \square
$$

In order to motivate the polynomial for Rhoades' CSP, we describe a wonderful formula due to Frame, Robinson, and Thrall [23] for f^λ. The *hook* of (i, j) is the set of cells to its right in the same row or below in the same column:

$$
H_{i,j} = \{(i, j') \in \lambda \ : \ j' \geq j\} \cup \{(i', j) \in \lambda \ : \ i' \geq i\}.
$$

The corresponding *hooklength* is $h_{i,j} = \#H_{i,j}$. The hook of $(2, 2)$ in $\lambda = (5, 4, 4, 2)$ is indicated by crosses is the following diagram

$$
\begin{array}{ccccc}
\bullet & \bullet & \bullet & \bullet & \bullet \\
\bullet & \times & \times & \times \\
\bullet & \times & \bullet & \bullet \\
\bullet & \times
\end{array}
$$

and so $h_{2,2} = 5$. The next result is called the Frame-Robinson-Thrall Hooklength Formula.

Theorem 7.2 *If $\lambda \vdash n$ then*

$$
f^\lambda = \frac{n!}{\displaystyle\prod_{(i,j)\in\lambda} h_{i,j}}. \qquad \square
$$

To illustrate this theorem, the hooklengths for $\lambda = (3,2)$ are as follows

$$h_{i,j} : \begin{matrix} 4 & 3 & 1 \\ 2 & 1 \end{matrix}$$

so $f^{(3,2)} = 5!/(4 \cdot 3 \cdot 2 \cdot 1^2) = 5$ as before. The polynomial which will appear in Rhoades' cyclic sieving result is a q-analogue of the Hooklength Formula

$$f^\lambda(q) = \frac{[n]_q!}{\displaystyle\prod_{(i,j)\in\lambda} [h_{i,j}]_q} \tag{7.4}$$

where $\lambda \vdash n$.

The only thing left to define is the group action and this will be done using Schützenberger's promotion operator [70]. Define (i,j) to be a *corner* of λ if neither $(i+1,j)$ nor $(i,j+1)$ is in λ. The corners of λ displayed in (7.1) are $(1,5)$, $(3,4)$, and $(4,2)$. Given $T \in \mathrm{SYT}(\lambda)$ we define its *promotion*, ∂T, by an algorithm.

1. Replace $T_{1,1} = 1$ by a dot.

2. If the dot is in position (i,j) then exchange it with with $T_{i+1,j}$ or $T_{i,j+1}$, whichever is smaller. (If only one of the two elements exist, use it for the exchange.) Iterate this step until (i,j) becomes a corner.

3. Subtract 1 from all the elements in the array, and replace the dot in corner (i,j) by n to obtain ∂T.

The exchanges in the second step are called *slides*. Note that the slides are constructed so that at every step of the process the array has increasing rows and columns and so $\partial T \in \mathrm{SYT}(\lambda)$. By way of illustration, if

$$T = \begin{matrix} 1 & 3 & 5 \\ 2 & 4 & 6 \\ 7 & & \end{matrix} \tag{7.5}$$

then we get the sliding sequence

$$\begin{matrix} \bullet & 3 & 5 \\ 2 & 4 & 6 \\ 7 & & \end{matrix}, \quad \begin{matrix} 2 & 3 & 5 \\ \bullet & 4 & 6 \\ 7 & & \end{matrix}, \quad \begin{matrix} 2 & 3 & 5 \\ 4 & \bullet & 6 \\ 7 & & \end{matrix}, \quad \begin{matrix} 2 & 3 & 5 \\ 4 & 6 & \bullet \\ 7 & & \end{matrix}, \quad \begin{matrix} 1 & 2 & 4 \\ 3 & 5 & 7 \\ 6 & & \end{matrix} = \partial T.$$

It is easy to see that the algorithm can be reversed step-by-step, and so promotion is a bijection on $\mathrm{SYT}(\lambda)$. Thus the operator generates a group $\langle \partial \rangle$ acting on standard Young tableaux of given shape. In general, the action seems hard to describe, but things are much nicer for certain shapes. In particular, we have the following result of Haiman [30].

Theorem 7.3 *If $\lambda = (n^m)$ then $\partial^{mn}(T) = T$ for all $T \in \mathrm{SYT}(\lambda)$.* $\square$

For example, if $\lambda = (3^2)$ then ∂ has cycle decomposition

$$\partial = \begin{pmatrix} 1 & 2 & 3 \\ 4 & 5 & 6 \end{pmatrix}, \begin{matrix} 1 & 2 & 5 \\ 3 & 4 & 6 \end{matrix}, \begin{matrix} 1 & 3 & 4 \\ 2 & 5 & 6 \end{matrix} \begin{pmatrix} 1 & 2 & 4 \\ 3 & 5 & 6 \end{pmatrix}, \begin{matrix} 1 & 3 & 5 \\ 2 & 4 & 6 \end{matrix} \tag{7.6}$$

from which one sees that ∂^6 is the identity map in agreement with the previous theorem.

We can now state one of the main results in Rhoades' paper [60].

Theorem 7.4 *If* $\lambda = (n^m)$ *then the triple*

$$\Big(\ \mathrm{SYT}(\lambda), \ \langle \partial \rangle, \ f^\lambda(q) \ \Big)$$

exhibits the cyclic sieving phenomenon.   □

Rhoades also has a corresponding theorem for promotion of semistandard Young tableaux of shape λ, a generalization of standard Young tableaux where repeated entries are allowed which will be discussed in Section 9. The polynomial used is the principal specialization of the Schur function s_λ, a symmetric function which can be viewed either as encoding the character of the irreducible module V^λ or as the generating function for semistandard tableaux. Both the standard and semistandard results were originally conjectured by Dennis White.

8 Variations on a theme

We will now mention several papers which have been inspired by Rhoades' work. Stanley [81] asked if there were a more elementary proof of Theorem 7.4. A step in this direction for rectangles with 2 or 3 rows was given by Petersen, Pylyavskyy, and Rhoades [53] who reformulated the theorem in a more geometric way. We will describe how this is done in the 2-row case in detail.

A *(complete) matching* is a graph, M, with vertex set $[2n]$ and n edges no two of which share a common vertex. The matching is *noncrossing* if it does not contain a pair of edges ab and cd with

$$a < c < b < d \tag{8.1}$$

Equivalently, if the vertices are arranged in order around a circle, say counterclockwise, then no pair of edges intersect. The 5 noncrossing matchings on [6] are displayed in Figure 1 below. There is a bijection between $\mathrm{SYT}(n^2)$ and noncrossing matchings on $[2n]$ as follows.

A *ballot sequence of length* n is a sequence $B = b_1 b_2 \ldots b_n$ of positive integers such that for all prefixes $b_1 b_2 \ldots b_m$ and all $i \geq 1$, the number of i's in the prefix is at least as great as the number of $(i+1)$'s. (One thinks of counting the votes in an election where at every stage in the count, candidate i is doing at least as well as candidate $i + 1$.) If $\lambda \vdash n$ then there is a map from tableaux $T \in \mathrm{SYT}_n$ to ballot sequences of length n: let $b_m = i$ if m is in the ith row of T. The T in (7.5) has corresponding ballot sequence $B = 1212123$. It is easy to see that the row and column conditions on T force B to be a ballot sequence. And it is also a simple matter to construct the inverse of this map, so it is a bijection.

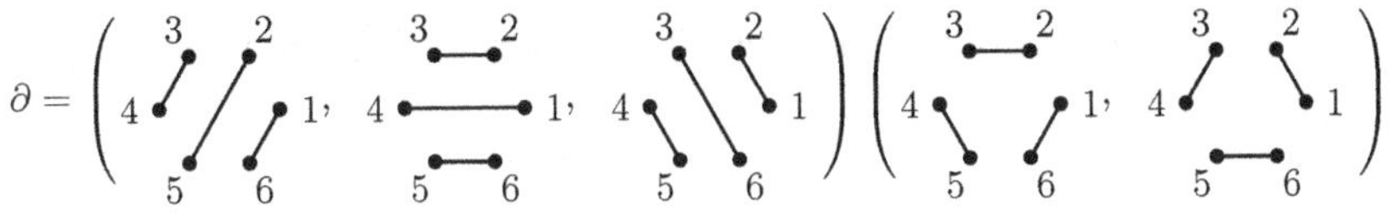

Figure 1: The action of ∂ on matchings

To make the connection with noncrossing matchings, suppose $T \in \mathrm{SYT}(n^2)$ and form a corresponding sequence of parentheses by replacing each 1 in its ballot sequence by a left parenthesis and each 2 with a right. Now match the parentheses, and thus their corresponding elements of T, in the usual way: if a left parenthesis is immediately followed by a right parenthesis they are considered matched, remove any matched pairs and recursively match what is left. The fact that the parentheses form a ballot sequence ensures that one gets a noncrossing matching. And the fact that one starts with a tableau of shape (n, n) ensures that the matching will be complete. For example,

$$T = \begin{matrix} 1 & 2 & 4 & 5 \\ 3 & 6 & 7 & 8 \end{matrix} \mapsto \begin{matrix} 1 & 2 & 3 & 4 & 5 & 6 & 7 & 8 \\ (& (&) & (& (&) &) &) \end{matrix} \mapsto M : 18, 23, 47, 56$$

where numbers are shown above the corresponding parentheses and the matching is specified by its edges. Again, an inverse is simple to construct so we have a bijection.

Applying this map to the tableaux displayed in (7.6) gives the matching description of ∂ in Figure 1. Clearly these cycles are obtained by rotating the matchings clockwise, and it is not hard to prove that this is always the case. As mentioned in [53], this interpretation of promotion was discovered, although never published, by Dennis White. Note that this viewpoint makes it clear that $\partial^{2n}(T) = T$, a special case of Theorem 7.3. Petersen, Pylyavskyy, and Rhoades use this setting and Springer's theory of regular elements to give a short proof of the following result which is equivalent to Theorem 7.4 when $m = 2$.

Theorem 8.1 Let $\mathrm{NCM}(2n)$ be the set of noncrossing matchings on $2n$ vertices and let R be rotation clockwise through an angle of π/n. Then the triple

$$\left(\mathrm{NCM}(2n), \ \langle R \rangle, \ f^{(n,n)}(q) \right)$$

exhibits the cyclic sieving phenomenon. $\square$

This trio of authors applies similar ideas to the 3-row case by replacing noncrossing matchings with A_2 webs. Webs were introduced by Kuperberg [45] to index a basis for a vector space used to describe the invariants of a tensor product of irreducible representations of a rank 2 Lie algebra.

Westbury [87] was able to generalize the Petersen-Pylyavskyy-Rhoades proof to a much wider setting. To understand his contribution, consider the coinvariant algebra, A, for a Coxeter group W as defined by (6.4). If V^λ is an irreducible module of W, then the corresponding *fake degree polynomial* is

$$F^\lambda(q) = \sum_{d \geq 0} m_d q^d \tag{8.2}$$

where m_d is the multiplicity of V^λ in the dth graded piece of A. We can extend this to any representation V of W by linearity. That is, write $V = \sum_\lambda n_\lambda V^\lambda$ in terms of irreducibles and then let

$$F^V(q) = \sum_\lambda n_\lambda F^\lambda(q).$$

Since the coinvariant algebra is W-isomorphic to the regular representation, Theorem 6.4 for $W = A_n$ and (7.2) imply that

$$F^\lambda(1) = f^\lambda \tag{8.3}$$

so that the fake degree polynomial is a q-analogue of the number of standard Young tableaux. It can be obtained from the q-Hooklength Formula (7.4) by multiplying by an appropriate power of q.

The next result, which can be found in Westbury's article [87], follows easily from Proposition 4.5 in Springer's original paper [78].

Theorem 8.2 *Let W be a finite complex reflection group and let $C \leq W$ be cyclically generated by a regular element g. Let V be a W-module with a basis B such that $gB = B$. Then the triple*

$$\left(\ B,\ C,\ F^V(q)\ \right)$$

exhibits the cyclic sieving phenomenon. □

Petersen-Pylyavskyy-Rhoades used webs of types A_1 and A_2 for the bases B, and the irreducible symmetric group modules $V^{(n,n)}$ and $V^{(n,n,n)}$ to determine the fake degree polynomials. Westbury is able to produce many interesting results by using other bases and any highest weight representation of a simple Lie algebra. Crystal bases and Lusztig's theory of based modules [48, Ch. 27–28] come into play.

A CSP related to Rhoades' was studied by Petersen and Serrano [54]. Consider the Coxeter group B_n with generating set $\{s_1, s_2, \ldots, s_n\}$ where s_1 is the special element corresponding to the endpoint of the Dynkin diagram adjacent to the edge labeled 4. Every finite Coxeter group has a *longest element* which is w_0 with $l(w_0) > l(w)$ for all $w \in W \setminus \{w_0\}$. Let $R(w_0)$ denote the set of reduced expressions for w_0 in B_n where $\ell(w_0) = n^2$. We will represent each such expression by the sequence of its subscripts. For example, in B_3 the expression $w_0 = s_1 s_3 s_2 s_3 s_1 s_2 s_3 s_1 s_2$ would become the word 132312312. Act on $R(w_0)$ by rotation, i.e., remove the first element of the sequence and move it to the end. In our example, $132312312 \mapsto 323123121$. (One has an analogous action in any Coxeter system (W, S) with S simple where, if s_i rotates from the front of w_0, then at the back it is replaced by $s_j = w_0 s_i w_0$ which is also simple. In type B_n, one has the nice property that $s_i = s_j$.) For the final ingredient, it is easy to see that the definitions of the permutation statistics from Section 5 all make sense when applied to sequences of integers.

The main theorem of Petersen and Serrano [54] can now be stated as follows. We will henceforth use C_N to denote a cyclic group of cardinality N.

Theorem 8.3 *In B_n, let C_{n^2} be the cyclic group of rotations of elements of $R(w_0)$. Then the triple*

$$\left(\ R(w_0),\ C_{n^2},\ q^{-n\binom{n}{2}} f^{\mathrm{maj}}(R(w_0); q)\right)$$

exhibits the cyclic sieving property.   □

They prove this result by using bijections due to Haiman [29, 30] to relate promotion on tableaux of shape (n^n) to rotation of words in $R(w_0)$. In fact, they also show

$$q^{-n\binom{n}{2}} f^{\text{maj}}(R(w_0; q)) = f^{(n^n)}(q)$$

where the latter is the q-analogue of the Hooklength Formula (7.4). The tableaux version of the previous theorem also appeared in [60], but the proof had gaps which Petersen and Serrano succeeded in filling.

Pon and Wang [55] have made steps towards finding an analogue of Rhoades' result for staircase tableaux. The *staircase shape* is the one corresponding to the partition $\text{sc}_n = (n, n-1, \ldots, 1)$. Note that $\text{sc}_n \vdash \binom{n+1}{2}$. The following result of Edelman and Greene [15] shows that staircase tableaux are also well behaved with respect to promotion.

Theorem 8.4 *We have* $\partial^{\binom{n+1}{2}}(T) = T^t$ *for all* $T \in \text{SYT}(\text{sc}_n)$ *where* t *denotes transpose.*   □

Haiman [30] has a theory of "generalized staircases" which considers for which partitions $\lambda \vdash N$, $\partial^N(T)$ has a nice description for all $T \in \text{SYT}(\lambda)$. It includes Theorems 7.3 and 8.4.

Note that we have $\partial^{n(n+1)}(T) = T$ for all $T \in \text{SYT}(n^{n+1})$ as well as for all $T \in \text{SYT}(\text{sc}_n)$. So it is natural to try and relate these two sets of tableaux and the action of promotion on them. In particular, Pon and Wang define an injection $\iota : \text{SYT}(\text{sc}_n) \to \text{SYT}(n^{n+1})$ commuting with ∂. To construct this map, we need another operation of Schützenberger called evacuation [69]. Given $T \in \text{SYT}(\lambda)$ where $\lambda \vdash N$, we create its *evacuation*, $\epsilon(T)$, by doing N promotions. After the ith promotion, one puts $N - i + 1$ in the ending position of the dot and this element does not move in any further promotions. The slide sequence for a promotion terminates when the position (i, j) of the dot is such that $(i+1, j)$ is either outside λ or contains a fixed element, and the same is true of $(i, j+1)$. We will illustrate this on

$$T = \begin{array}{ccc} 1 & 3 & 6 \\ 2 & 4 & \\ 5 & & \end{array} \tag{8.4}$$

where fixed elements will be typeset in boldface:

$$T = \begin{array}{ccc} 1 & 3 & 6 \\ 2 & 4 & \\ 5 & & \end{array} \overset{\partial}{\mapsto} \begin{array}{ccc} 1 & 2 & 5 \\ 3 & 6 & \\ 4 & & \end{array} \overset{\partial}{\mapsto} \begin{array}{ccc} 1 & 4 & 5 \\ 2 & 6 & \\ 3 & & \end{array} \overset{\partial}{\mapsto} \begin{array}{ccc} 1 & 3 & \mathbf{5} \\ 2 & 6 & \\ 4 & & \end{array}$$

$$\overset{\partial}{\mapsto} \begin{array}{ccc} 1 & 2 & \mathbf{5} \\ 3 & \mathbf{6} & \\ 4 & & \end{array} \overset{\partial}{\mapsto} \begin{array}{ccc} 1 & \mathbf{2} & \mathbf{5} \\ 3 & \mathbf{6} & \\ 4 & & \end{array} \overset{\partial}{\mapsto} \begin{array}{ccc} 1 & \mathbf{2} & \mathbf{5} \\ 3 & \mathbf{6} & \\ 4 & & \end{array} = \epsilon(T).$$

Now given $T \in \text{SYT}(\text{sc}_n)$ we define $\iota(T)$ as follows. Construct $\epsilon(T)$ and complement each entry, x, by replacing it with $n(n+1) + 1 - x$. Then reflect the resulting tableau in the anti-diagonal. Finally, paste T and the reflected-complemented tableau together along their staircase portions to obtain $\iota(T)$ of shape (n^{n+1}). Continuing the example (8.4), we see that the complement of $\epsilon(T)$ is

$$
\begin{array}{ccc}
12 & 11 & 8 \\
10 & 7 & \\
9 & & \\
\end{array}
$$

so that

$$
\iota(T) = \begin{array}{ccc}
1 & 3 & 6 \\
2 & 4 & 8 \\
5 & 7 & 11 \\
9 & 10 & 12 \\
\end{array} .
$$

As mentioned, Pon and Wang [55] prove the following result about their map ι.

Theorem 8.5 *We have*

$$
\partial(\iota(T)) = \iota(\partial(T)).
$$

for all $T \in \mathrm{SYT}(\mathrm{sc}_n)$. $\square$

To get a CSP for staircase tableaux, one needs an appropriate polynomial. The previous theorem permits one to obtain information about the cycle structure of the action of ∂ on staircase tableaux from what is already know about rectangular tableaux. It is hoped that this will aid in the search for the correct polynomial.

9 Multiple groups and multiple statistics

It is natural to ask whether the cyclic sieving phenomenon can be extended to other groups. Indeed, it is possible to define an analogue of the CSP for Abelian groups by considering them as products of cyclic groups. For this we will also need to use multiple statistics, one for each cyclic group. In this section we examine this idea, restricting to the case of two cyclic groups to illustrate the ideas involved.

Bicyclic sieving was first defined by Barcelo, Reiner, and Stanton [4]. Recall that Ω is the (infinite) group of roots of unity.

Definition 9.1 Let X be a finite set. Let C, C' be finite cyclic groups with $C \times C'$ acting on X, and fix embeddings $\omega : C \to \Omega$, $\omega' : C' \to \Omega$. Let $f(q,t) \in \mathbb{N}[q,t]$. The triple $(X, C \times C', f(q,t))$ exhibits the *bicyclic sieving phenomenon* or *biCSP* if, for all $(g,g') \in C \times C'$, we have

$$
\#X^{(g,g')} = f(\omega(g), \omega'(g')). \tag{9.1}
$$

Note that in the original definition of the CSP we did not insist on an embedding of C in Ω but just used any root of unity with the same order as g (although Reiner, Stanton, and White did use an embedding in the definition from their original paper). But this does not matter because if evaluating $f(q) \in \mathbb{R}[q]$ at a primitive dth root of unity gives a real number, then any dth root will give the same value. But in the definition of the biCSP the choice of embeddings can make the difference whether (9.1) holds or not. To illustrate this, we use an example from a paper of Berget, Eu, and Reiner [5].

Take $X = \{1, \omega, \omega^2\}$ where $\omega = \exp(2\pi i/3)$ and let $C = C' = X$. Define the action of any $(\alpha, \beta) \in C \times C'$ on $\gamma \in X$ by

$$(\alpha, \beta)\gamma = \alpha\beta\gamma$$

where on the right multiplication is being done in $\mathbb{C}$. Finally, consider the polynomial

$$f(q, t) = 1 + ut + u^2t^2.$$

If for the embeddings one takes identity maps, then it is easy to verify case-by-case that $(X, C \times C', f(q, t))$ exhibits the biCSP. But if one modifies the embedding of C' to be the one which takes $\omega \to \omega^{-1}$ then this is false. For example, consider (ω, ω) whose action on X is the cycle $(1, \omega^2, \omega)$. Using the first embedding pair we find $f(\omega, \omega) = 1 + \omega^2 + \omega = 0$ reflecting the fact that there are no fixed points. However, if one uses the second pair then the computation is $f(\omega, \omega^{-1}) = 1 + 1 + 1 = 3$ which does not agree with the action.

Part of the motivation for studying the biCSP was to generalize the notion of a bimahonian distribution to other Coxeter groups W. It was observed by Foata and Schützenberger [19] that certain pairs of Mahonian statistics such as $(\mathrm{maj}(w), \mathrm{inv}(w))$ and $(\mathrm{maj}(w), \mathrm{maj}(w^{-1}))$ had the same joint distribution over $\mathfrak{S}_n$. To define a corresponding bivariate distribution on W, consider a field automorphism σ lying in the Galois group $\mathrm{Gal}(\mathbb{Q}[\exp(2\pi i/m)]/\mathbb{Q})$ where m is taken large enough so that the extension $\mathbb{Q}[\exp(2\pi i/m)]$ of $\mathbb{Q}$ contains all the matrix entries of all elements in the reflection representation of W. Use the fake degree polynomials (8.2) to define the σ-bimahonian distribution on W by

$$F^\sigma(q, t) = \sum_{V^\lambda} F^{\sigma(\lambda)}(q) F^{\overline{\lambda}}(t)$$

where the sum is over all irreducible V^λ, and $V^{\sigma(\lambda)}$ and $V^{\overline{\lambda}}$ are the modules defined (via Theorem 4.1 (c)) by the characters

$$\chi^{\sigma(\lambda)}(w) = \sigma(\chi^\lambda(w)) \text{ and } \chi^{\overline{\lambda}}(w) = \overline{\chi^\lambda(w)}$$

for all $w \in W$.

To apply Springer's theory in this setting, suppose g and g' are regular elements of W with regular eigenvalues ω and ω', respectively. Consider the embeddings of $C = \langle g \rangle$ and $C' = \langle g' \rangle$ into Ω uniquely defined by mapping $g \mapsto \omega^{-1}$ and $g' \mapsto (\omega')^{-1}$. (The reason for using inverses should be clear from the proof of Proposition 6.2.) Given σ as above, pick $s \in \mathbb{N}$ so that $\sigma(\omega) = \omega^s$. In this setting, Barcelo, Reiner, and Stanton [4] prove the following result.

Theorem 9.2 *Let W be a finite complex reflection group with regular elements g, g'. With the above notation, consider the σ-twisted action of $\langle g \rangle \times \langle g' \rangle$ on W defined by*

$$(g, g')w = g^s w(g')^{-1}.$$

Then the triple

$$\left(W, \langle g \rangle \times \langle g' \rangle, F^\sigma(q, t) \right)$$

exhibits the bicyclic sieving phenomenon. □

Reiner, Stanton and White had various conjectures about biCSPS exhibited by action on nonnegative integer matrices via row and column rotation. These were communicated to Rhoades and proved by him in [61]. To talk about them, we will need some definitions. A *composition of n of length l* is a (not necessarily weakly decreasing) sequence $\mu = (\mu_1, \mu_2, \ldots, \mu_l)$ of nonnegative integers with $\sum_i \mu_i = n$. We write $\mu \models n$ and $\ell(\mu) = l$. A *semistandard Young tableau of shape λ and content μ* is a function $T : \lambda \to \mathbb{P}$ (where $\mathbb{P}$ is the positive integers) such that rows weakly increase, columns strictly increase, and μ_k is the number of k's in T. For example,

$$T = \begin{array}{cccccc} 1 & 1 & 1 & 2 & 3 & 5 \\ 2 & 3 & 3 & 3 \\ 5 & 5 \end{array}$$

is a semistandard tableau of shape $(6, 4, 2)$ and content $(3, 2, 4, 0, 3)$. We write $\operatorname{ct} T = \mu$ to denote the content and let

$$\operatorname{SSYT}(\lambda, \mu) = \{T \ : \ \operatorname{sh} T = \lambda \text{ and } \operatorname{ct} T = \mu\}.$$

The *Kostka numbers* are

$$K_{\lambda,\mu} = \# \operatorname{SSYT}(\lambda, \mu).$$

Note that if $\lambda \vdash n$ then $K_{\lambda,(1^n)} = f^\lambda$, the number of standard Young tableaux. Just like f^λ, the $K_{\lambda,\mu}$ have a nice representation theoretic interpretation. Given $\mu = (\mu_1, \mu_2, \ldots, \mu_l) \models n$ there is a corresponding *Young subgroup* of $\mathfrak{S}_n$ which is

$$\mathfrak{S}_\mu = \mathfrak{S}_{\{1,2,\ldots,\mu_1\}} \times \mathfrak{S}_{\{\mu_1+1,\mu_1+2,\ldots,\mu_1+\mu_2\}} \times \cdots \times \mathfrak{S}_{\{n-\mu_l+1,n-\mu_l+2,\ldots,n\}}$$

where for any set X we let $\mathfrak{S}_X$ be the group of permutations of X. Consider the usual action of $\mathfrak{S}_n$ on left cosets $\mathfrak{S}_n/\mathfrak{S}_\mu$ and let M^μ be the corresponding permutation module. Decomposing into $\mathfrak{S}_n$-irreducibles gives

$$M^\mu = \sum_\lambda K_{\lambda,\mu} V^\lambda,$$

so that the Kostka numbers give the multiplicities of this decomposition. The polynomials which appear in these biCSPs are a q-analogue of the $K_{\lambda,\mu}$ called the *Kostka-Foulkes polynomials*, $K_{\lambda,\mu}(q)$. We will not give a precise definition of them, but just say that they can be viewed in a couple of ways. One is as the elements of transition matrices between the Schur functions and Hall-Littlewood polynomials (another basis for the algebra of symmetric functions over the field $\mathbb{Q}(q)$). Another is as the generating function for a statistic called charge on semistandard tableaux which was introduced by Lascoux and Schützenberger [46].

Knuth [38] generalized the Robinson-Schensted correspondence in Theorem 7.1 to semistandard tableaux. Given compositions $\mu, \nu \models n$ let $\operatorname{Mat}_{\mu,\nu}$ denote the set of all matrices with nonnegative integer entries whose row sums are given by μ and whose column sums are given by ν. For example,

$$\operatorname{Mat}_{(2,1),(1,0,2)} = \left\{ \begin{bmatrix} 1 & 0 & 1 \\ 0 & 0 & 1 \end{bmatrix}, \begin{bmatrix} 0 & 0 & 2 \\ 1 & 0 & 0 \end{bmatrix} \right\}.$$

Note that $\mathrm{Mat}_{(1^n),(1^n)}$ is the set of $n \times n$ permutation matrices.

Given $M \in \mathrm{Mat}_{\mu,\nu}$ we first convert the matrix into a two-rowed array with the columns lexicographically ordered (the first row taking precedence) and column $\genfrac{}{}{0pt}{}{i}{j}$ occurring $M_{i,j}$ times. To illustrate,

$$M = \begin{bmatrix} 1 & 2 & 0 \\ 1 & 0 & 1 \end{bmatrix} \mapsto \begin{matrix} 1 & 1 & 1 & 2 & 2 \\ 1 & 2 & 2 & 1 & 3 \end{matrix}.$$

Now use the same insertion algorithm as for the Robinson-Schensted correspondence to build a semistandard tableau P from the elements of the bottom line of the array, while the elements of the top line are placed in a tableau Q to maintain equalities of shapes,

$$P_k : \quad \emptyset \;,\; \boxed{1} \;,\; \boxed{1\ 2} \;,\; \boxed{1\ 2\ 2} \;,\; \begin{matrix}\boxed{1\ 1\ 2}\\\boxed{2}\end{matrix} \;,\; \begin{matrix}\boxed{1\ 1\ 2\ 3}\\\boxed{2}\end{matrix} = P,$$

$$Q_k : \quad \emptyset \;,\; \boxed{1} \;,\; \boxed{1\ 1} \;,\; \boxed{1\ 1\ 1} \;,\; \begin{matrix}\boxed{1\ 1\ 1}\\\boxed{2}\end{matrix} \;,\; \begin{matrix}\boxed{1\ 1\ 1\ 2}\\\boxed{2}\end{matrix} = Q.$$

We denote this map by $M \overset{\mathrm{R-S-K}}{\mapsto} (P,Q)$. Reversing the steps of the algorithm is much like the standard tableau case once one realizes that during insertion equal elements enter into Q from left to right. Also, it should be clear that applying this map to permutation matrices corresponds with the original algorithm. So the full Robinson-Schensted-Knuth Theorem [38] is as follows.

Theorem 9.3 *For all $\mu, \nu \models n$, the map $M \overset{\mathrm{R-S-K}}{\mapsto} (P,Q)$ is a bijection*

$$\mathrm{Mat}_{\mu,\nu} \overset{\mathrm{R-S-K}}{\longleftrightarrow} \{(P,Q) \;:\; \mathrm{ct}\, P = \nu,\ \mathrm{ct}\, Q = \mu,\ \mathrm{sh}\, P = \mathrm{sh}\, Q\}. \qquad \square$$

To motivate Rhoades' result, note that by specializing Theorem 9.2 to type A, one can obtain the following theorem.

Theorem 9.4 *Let $C_n \times C_n$ act on $n \times n$ permutation matrices by rotation of rows in the first component and of columns in the second. Then*

$$\left(\mathrm{Mat}_{(1^n),(1^n)}, \quad C_n \times C_n, \quad \epsilon_n(q,t) \sum_{\lambda \vdash n} K_{\lambda,(1^n)}(q) K_{\lambda,(1^n)}(t) \right)$$

exhibits the bicyclic sieving phenomenon, where

$$\epsilon_n(q,t) = \begin{cases} (qt)^{n/2} & \text{if } n \text{ is even,} \\ 1 & \text{if } n \text{ is odd.} \end{cases} \qquad \square$$

It is natural to ask for a generalization of this theorem to tableaux of arbitrary content, and that is one of the conjectures demonstrated by Rhoades in [61]. The proof uses a generalization of the R-S-K correspondence due to Stanton and White [82].

Theorem 9.5 *Let $\mu, \nu \models n$ have cyclic symmetries of orders $a|\ell(\mu)$ and $b|\ell(\nu)$, respectively. Let $C_{\ell(\mu)/a} \times C_{\ell(\nu)/b}$ act on $\mathrm{Mat}_{\mu,\nu}$ by a-fold rotation of rows in the first component and b-fold rotation of columns in the second. Then*

$$\left(\mathrm{Mat}_{\mu,\nu}, \quad C_{\ell(\mu)/a} \times C_{\ell(\nu)/b}, \quad \epsilon_n(q,t) \sum_{\lambda \vdash n} K_{\lambda,\mu}(q) K_{\lambda,\nu}(t) \right)$$

exhibits the bicyclic sieving phenomenon.   □

Rather than considering two Mahonian statistics, one could take one Mahonian and one Eulerian. Given two statistics, st and st' on a set X we let

$$f^{\text{st},\text{st}'}(X;q,t) = \sum_{y \in X} q^{\text{st}\,y} t^{\text{st}'\,y}.$$

Through their study of Rees products of posets, Shareshian and Wachs [72, 73, 74] were lead to consider the pair (maj, exc). They proved, among other things, the following generalization of (5.5)

$$\sum_{n \geq 0} f^{\text{maj},\text{exc}}(\mathfrak{S}_n; q, t) \frac{x^n}{[n]_q!} = \frac{(1 - tq)\exp_q(x)}{\exp_q(qtx) - qt \exp_q(x)},$$

where

$$\exp_q(x) = \sum_{n \geq 0} \frac{x^n}{[n]_q!}.$$

(Actually, they proved a stronger result which also keeps track of the number of fixed points of w, but that will not concern us here.)

Sagan, Shareshian, and Wachs [67] used the Eulerian quasisymmetric functions as developed in the earlier papers by the last two authors, as well as a result of Désarménien [14] about evaluating principal specializations at roots of unity, to demonstrate the following cyclic sieving result. Note the interesting feature that one must take the difference maj − exc.

Theorem 9.6 *Consider the action of the symmetric group on itself by conjugation and let* $\mathfrak{S}(\lambda)$ *denote the conjugacy class of permutations of cycle type* $\lambda \vdash n$. *Then the triple*

$$\left(\ \mathfrak{S}(\lambda), \ \langle (1, 2, \ldots, n) \rangle, \ f^{\text{maj},\text{exc}}\left(\mathfrak{S}(\lambda); q, q^{-1} \right) \ \right)$$

exhibits the cyclic sieving phenomenon.   □

10   Catalan CSPs

We will now consider cyclic sieving phenomena where analogues of the Catalan numbers play a role. These will include noncrossing partitions and facets of cluster complexes. Noncrossing matchings have already come into play in Theorem 8.1.

A *set partition* of a finite set X is a collection $\pi = \{B_1, B_2, \ldots, B_k\}$ of nonempty subsets such that $\uplus_i B_i = X$ where $\uplus$ denotes disjoint union. We write $\pi \vdash X$ and the B_i are called *blocks*. A set partition $\pi \vdash [n]$ is called *noncrossing* if condition (8.1) never holds when a, b are in one block of π and c, d are in another. Equivalently, with the usual circular arrangement of $1, \ldots, n$, the convex hulls of different blocks do not intersect. Let $\text{NC}(n)$ denote the set of noncrossing partitions of $[n]$. Noncrossing partitions were introduced by Kreweras [44] and much information about them can be found in the survey article of Simion [76] and the memoir of Armstrong [2].

The noncrossing partitions are enumerated by the *Catalan numbers*

$$\# \text{NC}(n) = \text{Cat}_n \overset{\text{def}}{=} \frac{1}{n+1} \binom{2n}{n}.$$

(Often in the literature C_n is used to denote the nth Catalan number, but this would conflict with our notation for cyclic groups.) These numbers have already been behind the scenes as we also have $\#\operatorname{SYT}(n,n) = \#\operatorname{NCM}(2n) = \operatorname{Cat}_n$. There are a plethora of combinatorial objects enumerated by Cat_n, and Stanley maintains a list [79] which the reader can consult for more examples.

A q-analogue of Cat_n,

$$\operatorname{Cat}_n(q) = \frac{1}{[n+1]_q} \left[\begin{array}{c} 2n \\ n \end{array} \right]_q,$$

was introduced by Fürlinger and Hofbauer [25]. The following result follows from Theorem 7.2 in Reiner, Stanton, and White's original paper [58] where they proved a more refined version which also keeps track of the number of blocks.

Theorem 10.1 *Let* C_n *act on* $\operatorname{NC}(n)$ *by rotation. Equivalently, let* $g \in \langle(1,2,\ldots,n)\rangle$ *act on* $\pi = \{B_1, B_2, \ldots, B_k\} \in \operatorname{NC}(n)$ *by* $g\pi = \{gB_1, gB_2, \ldots, gB_n\}$ *where* gB_i *is defined by (2.2). Then the triple*

$$(\operatorname{NC}(n),\ C_n,\ \operatorname{Cat}_n(q))$$

exhibits the cyclic sieving phenomenon. □

In [8], Bessis and Reiner generalize this theorem to certain complex reflection groups. First consider a finite Coxeter group W of rank n. It can be shown that we always have a factorization

$$f^\ell(W;q) = \sum_{w\in W} q^{\ell(w)} = \prod_{i=1}^{n} [d_i]_q$$

where the positive integers $d_1 \leq d_2 \leq \ldots \leq d_n$ are called the *degrees* of W. For example, if $W = A_n \cong \mathfrak{S}_{n+1}$, then by equation (5.2) and Proposition 5.2 we have

$$f^\ell(A_n;q) = [2]_q[3]_q \cdots [n+1]_q$$

so that $d_i = i+1$ for $1 \leq i \leq n$. Table 3 lists the degrees of the irreducible finite Coxeter groups.

There is another description of the degrees which will clarify their name and is valid for complex reflection groups, not just Coxeter groups. Let W be a finite group acting irreducibly (i.e., having no invariant subspace) on $\mathbb{C}^n$ by reflections. We call n the *rank* of W and write $\operatorname{rk} W = n$. Let $X_n = \{x_1, x_2, \ldots, x_n\}$ be a set of variables. Then the invariant space $\mathbb{C}[X_n]^W$ is a free algebra. Each algebraically independent set of homogeneous generators has the same set of degrees which we will call $d_1, d_2, \ldots, d_n$. These d_i are exactly the same as in the Coxeter setting. Returning to A_n again, we have the natural action on $\mathbb{C}[X_{n+1}]$. But each of the reflecting hyperplanes $x_i = x_j$ is perpendicular to the hyperplane $x_1 + x_2 + \cdots + x_{n+1} = 0$, and so to get a space whose dimension is the rank, we need to consider the quotient $\mathbb{C}[X_{n+1}]^{A_n}/(x_1 + x_2 + \cdots + x_{n+1})$. It is well known that the algebra $\mathbb{C}[X_{n+1}]^{A_n}$ is generated freely by the complete homogeneous symmetric polynomials $h_k(X_{n+1})$ where $1 \leq k \leq n+1$. But in the quotient $h_1(X_{n+1}) = x_1 + x_2 + \cdots + x_{n+1} = 0$, so we only need to consider the generators of degrees $2, 3, \ldots, n+1$.

Group	Degrees
A_n	$2, 3, 4, \ldots, n+1$
B_n	$2, 4, 6, \ldots, 2n$
D_n	$2, 4, 6 \ldots, 2n-2, n$
E_6	$2, 5, 6, 8, 9, 12$
E_7	$2, 6, 8, 10, 12, 14, 18$
E_8	$2, 8, 12, 14, 18, 20, 24, 30$
F_4	$2, 6, 8, 12$
H_3	$2, 6, 10$
H_4	$2, 12, 20, 30$
$I_2(m)$	$2, m$

Table 3: The degrees of the irreducible finite Coxeter groups

It will also be instructive to see how one can modify the length definition in the complex case. Let R denote the set of reflections in W. So if W is a finite Coxeter group with generating set S then $R = \{wsw^{-1} \; : \; w \in W, \; s \in S\}$. In any finite complex reflection group, define the *absolute length of* $w \in W$, $\ell_R(w)$, to be the shortest length of an expression $w = r_1 r_2 \cdots r_k$ with $r_i \in R$ for all i. We have another factorization

$$f^{\ell_R}(W; q) = \sum_{w \in W} q^{\ell_R(w)} = \prod_{i=1}^{n}(1 + (d_i - 1)q)$$

where the d_i are again the degrees of W.

There is one last quantity which we will need to define the analogue of $\mathrm{NC}(n)$. It is called the *Coxeter number*, h, of a complex reflection group W. Unfortunately, there are two competing definitions of h. One is to let $h = d_n$, the largest of the degrees. The other is to set $h = (\#R + \#H)/n$ where H is the set of reflecting hyperplanes of W. But happily these two conditions coincide when W is *well generated* which means that it has a generating set of reflections of cardinality $\mathrm{rk}\,W = n$. Note that this includes the finite Coxeter groups. Under the well-generated hypothesis, W will also contain a regular element g of order h. As defined by Brady and Watt in the real case [10] and Bessis in the complex [6, 7], the *noncrossing elements* in W are

$$\mathrm{NC}(W) = \{w \in W \; : \; \ell_R(w) + \ell_R(w^{-1}g) = \ell_R(g)\}.$$

We note that $\ell_R(g) = \mathrm{rk}\,W = n$. We will let $\langle g \rangle$ act on $\mathrm{NC}(W)$ by conjugation. (One needs to check that this is well defined.)

To see how this relates to $\mathrm{NC}(n)$, map each element of $\mathrm{NC}(A_{n-1})$ to the partition whose blocks are the cycles of π considered as unordered sets. (A similar idea is behind Lemma 2.2.) Then this is a bijection with $\mathrm{NC}(n)$. To illustrate, let $n = 3$ and $g = (1, 2, 3)$. A case-by-case check using the definition yields

$$\mathrm{NC}(A_2) = \{(1)(2)(3), \; (1, 2)(3), \; (1, 3)(2), \; (1)(2, 3), \; (1, 2, 3)\}.$$

So the image of this set is all partitions of [3] which is $\mathrm{NC}(3)$ since (8.1) can not be true with only three elements.

The polynomial in the cyclic sieving result will be

$$\mathrm{Cat}(W; q) = \prod_{i=1}^{n} \frac{[h + d_i]_q}{[d_i]_q}.$$

For example

$$\mathrm{Cat}(A_{n-1}; q) = \prod_{i=1}^{n-1} \frac{[n + i + 1]_q}{[i + 1]_q} = \mathrm{Cat}_n(q).$$

It can be shown [6, 10] that

$$\mathrm{Cat}(W; 1) = \#\,\mathrm{NC}(W).$$

We can now state the Bessis-Reiner result [8].

Theorem 10.2 *Let the finite irreducible complex reflection group W be well generated. Let g be a regular element of order h and let $\langle g \rangle$ act on $\mathrm{NC}(W)$ by conjugation. Then*

$$(\ \mathrm{NC}(W),\ \langle g \rangle,\ \mathrm{Cat}(W; q)\ )$$

exhibits the cyclic sieving phenomenon. □

The Catalan numbers can be generalized to the *Fuss-Catalan numbers* which are defined by

$$\mathrm{Cat}_{n,m} = \frac{1}{mn + 1}\binom{(m + 1)n}{n}.$$

Note that $\mathrm{Cat}_{n,1} = \mathrm{Cat}_n$. The Fuss-Catalan numbers count, among other things, the *m-divisible* noncrossing partitions in $\mathrm{NC}(mn)$, i.e., those which have all their block sizes divisible by m. So, for example, $\mathrm{Cat}_{2,2} = \binom{6}{2}/5 = 3$ corresponding to the partitions $\{12, 34\}$, $\{14, 23\}$, and $\{1234\}$. (As usual, we are suppressing some set braces and commas.)

A natural q-analogue of the Fuss-Catalan numbers for any well-generated finite complex reflection group of rank n is

$$\mathrm{Cat}^{(m)}(W; q) = \prod_{i=1}^{n} \frac{[mh + d_i]_q}{[d_i]_q}.$$

Armstrong [2] has constructed a set counted by $\mathrm{Cat}^{(m)}(W; 1)$. As before, consider a regular element g of W having order h. Define

$$\mathrm{NC}^{(m)}(W) = \left\{ (w_0, w_1, \ldots, w_m) \in W^{m+1} : w_0 w_1 \cdots w_m = g,\ \sum_{i \geq 0} \ell_R(w_i) = \ell_R(g) \right\}.$$

Armstrong also defined two actions of g on $\mathrm{NC}^{(m)}(W)$ and made corresponding cyclic sieving conjectures about them. These have been proved by Krattenthaler [41] for the two infinite 2-parameter families of finite irreducible well-generated complex reflection groups, and by Krattenthaler and Müller [42, 43] for the exceptional ones. One of the actions is

$$g(w_0, w_1, w_2, \ldots, w_m) = (g w_m g^{-1}, w_0, w_1, \ldots, w_{m-1}). \tag{10.1}$$

This generates a group $C_{(m+1)h}$ acting on $\mathrm{NC}^{(m)}(W)$.

Φ	Π
A_n	$e_1 - e_2, \; e_2 - e_3, \; \ldots, \; e_{n-1} - e_n, \; e_n - e_{n+1}$
B_n	$e_1 - e_2, \; e_2 - e_3, \; \ldots, \; e_{n-1} - e_n, \; e_n$
C_n	$e_1 - e_2, \; e_2 - e_3, \; \ldots, \; e_{n-1} - e_n, \; 2e_n$
D_n	$e_1 - e_2, \; e_2 - e_3, \; \ldots, \; e_{n-1} - e_n, \; e_{n-1} + e_n$

Table 4: The simple roots in types A-D

Theorem 10.3 *Let the finite irreducible complex reflection group W be well generated. Let g be a regular element of order h and let $C_{(m+1)h}$ act on $\mathrm{NC}^{(m)}(W)$ by (10.1). Then*

$$\Big(\; \mathrm{NC}^{(m)}(W), \; C_{(m+1)h}, \; \mathrm{Cat}^{(m)}(W; q) \; \Big)$$

exhibits the cyclic sieving phenomenon.   □

We should mention that Gordon and Griffeth [26] have defined a version of the q-Fuss-Catalan polynomials for all complex reflection groups which specializes to $\mathrm{Cat}^{(m)}(W; q)$ when W is well generated. The primary ingredients of their construction are Rouquier's formulation of shift functors for the rational Cherednik algebras of W [64], and Opdam's analysis of permutations of the irreducible representations of W arising from the Knizhnik-Zamolodchikov connection [51]. Furthermore, plugging roots of unity into the Gordon-Griffeth polynomials yields nonnegative integers. But finding a corresponding CSP remains elusive.

There is another object enumerated by $\mathrm{Cat}(W; 1)$ when W is a Coxeter group, namely facets of cluster complexes. Cluster complexes were introduced by Fomin and Zelevinsky [22] motivated by their theory of cluster algebras. To define them, we need some background on root systems. Rather than take an axiomatic approach, we will rely on examples and outline the necessary facts we will need. The reader wishing details can consult the texts of Fulton and Harris [24] or Humphreys [33].

To every real reflection group W is associated a *root system*, $\Phi = \Phi_W$, which consists of a set of vectors called *roots* perpendicular to the reflecting hyperplanes. Each hyperplane has exactly two roots perpendicular to it and they are negatives of each other. We require that Φ span the space on which W acts. So, as above, when $W = A_n$ we have to restrict to the hyperplane $x_1 + x_2 + \cdots + x_{n+1} = 0$. Finally, W must act on Φ. To illustrate, if $W = A_n$ then the roots perpendicular to the hyperplane $x_i = x_j$ are taken to be $\pm(e_i - e_j)$ where e_i is the ith unit coordinate vector. Let $\Pi = \{\alpha_1, \alpha_2, \ldots, \alpha_n\}$ be a set of *simple roots* which correspond to the simple reflections $s_1, s_2, \ldots, s_n$. For groups of type A-D the standard choices for simple roots are listed in Table 4. Note that there are two root systems B_n and C_n associated with the type B group depending on whether one considers it as the set of symmetries of a hypercube or a hyperoctahedron, respectively. In fact, this group is sometimes referred to as BC_n.

One can find a hyperplane H which is not a reflecting hyperplane such that all $\alpha_i \in \Pi$ lie on the same side of H. Since the roots come in opposite pairs, half of them will lie on the same side of H as the simple roots and these are called the *positive roots*, $\Phi_{>0}$. The rest of the roots are called *negative*. The simple roots form a basis for the span $\langle \Phi \rangle$ and every positive root can be written as a linear combination of

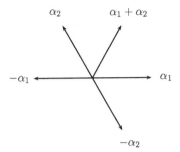

Figure 2: The roots in $\Phi_{\geq -1}$ for type A_2

the simple roots with positive coefficients. In type A one can take H to be any plane of the form $c_1 x_1 + c_2 x_2 + \cdots + c_{n+1} x_{n+1} = 0$ where $c_1 > c_2 > \ldots > c_{n+1} > 0$. In this case $\Phi_{>0} = \{e_i - e_j \ : \ i < j\}$ and we can write

$$e_i - e_j = \alpha_i + \alpha_{i+1} + \cdots + \alpha_{j-1}$$

where $\alpha_k = e_k - e_{k+1} \in \Pi$ for $1 \leq k \leq n$.

Take the union of the positive roots with the negatives of the simple roots to get

$$\Phi_{\geq -1} = \Phi_{>0} \cup (-\Pi).$$

For example, if Φ is of type A_2, then $\Phi_{\geq -1} = \{\alpha_1, \alpha_2, \alpha_1 + \alpha_2, -\alpha_1, -\alpha_2\}$ as displayed in Figure 2.

Take any partition $[n] = I_+ \uplus I_-$ such that the nodes indexed by I_+ in the Dynkin diagram of W are totally disconnected, and the same is true for I_-. Note that this means that the reflections in I_ϵ commute for $\epsilon \in \{+, -\}$. Next, define a pair of involutions $\tau_\pm : \Phi_{\geq -1} \to \Phi_{\geq -1}$ by

$$\tau_\epsilon(\alpha) = \begin{cases} \alpha & \text{if } \alpha = -\alpha_i \text{ for } i \in I_{-\epsilon}, \\ \left(\displaystyle\prod_{i \in I_\epsilon} s_i \right)(\alpha) & \text{otherwise.} \end{cases}$$

Since the roots in I_ϵ commute, the product is well defined. Returning to our example, let $I_+ = \{1\}$ and $I_- = \{2\}$, Table 5 displays the images of each root in $\Phi_{\geq -1}$ under τ_+ and τ_-.

Consider the product $\Gamma = \tau_- \tau_+$ (where maps are composed right-to-left) and the cyclic group $\langle \Gamma \rangle$ it generates acting on $\Phi_{\geq -1}$. In the running example, Γ consists of a single cycle

$$\Gamma = (\ \alpha_1, \ -\alpha_1, \ \alpha_1 + \alpha_2, \ -\alpha_2, \ \alpha_2 \ ).$$

This map induces a relation of *compatibility*, $\alpha \sim \beta$, on $\Phi_{\geq -1}$ defined by the following two conditions.

α	$\tau_+(\alpha)$	$\tau_-(\alpha)$
α_1	$-\alpha_1$	$\alpha_1 + \alpha_2$
α_2	$\alpha_1 + \alpha_2$	$-\alpha_2$
$\alpha_1 + \alpha_2$	α_2	α_1
$-\alpha_1$	α_1	$-\alpha_1$
$-\alpha_2$	$-\alpha_2$	α_2

Table 5: The images of τ_+ and τ_-

1. For $-\alpha_i \in -\Pi$ and $\beta \in \Phi_{>0}$ we have $-\alpha_i \sim \beta$ if and only if α_i does not occur in the simple root expansion of β.

2. For all $\alpha, \beta \in \Phi_{\geq -1}$ we have $\alpha \sim \beta$ if and only if $\Gamma(\alpha) \sim \Gamma(\beta)$.

In our example, $-\alpha_1 \sim \alpha_2$ by the first condition. Then repeated application of the second yields $\alpha \sim \beta$ for all $\alpha, \beta \in \Phi_{\geq -1}$ when $W = A_2$.

The *cluster complex*, $\Delta(\Phi)$, is the abstract simplicial complex (i.e., a family of sets called *faces* closed under taking subsets) consisting of all sets of pairwise compatible elements of $\Phi_{\geq 0}$. So in the A_2 case, $\Delta(\Phi)$ consists of a single facet (maximal face) which is all of $\Phi_{\geq -1}$. The faces of $\Delta(\Phi)$ can be described in terms of dissections of polygons using noncrossing diagonals. Using this interpretation, Eu and Fu [16] prove the following result.

Theorem 10.4 *Let W be a finite Coxeter group and let Φ be the corresponding root system. Let $\Delta_{\max}(\Phi)$ be the set of facets of the cluster complex $\Delta(\Phi)$. Then the triple*

$$(\Delta_{\max}(\Phi), \langle \Gamma \rangle, \mathrm{Cat}(W; q))$$

exhibits the cyclic sieving phenomenon.   □

In fact, Eu and Fu strengthened this theorem in two ways: by looking at the faces of dimension k and by considering an m-divisible generalization of the cluster complex due to Fomin and Reading [20].

11   A cyclic sieving miscellany

Here we collect some topics not previously covered. These include methods for generating new CSPs from old ones, sieving for cyclic polytopes, and extending various results to arbitrary fields.

Berget, Eu, and Reiner [5] gave various ways to construct CSPs and we will discuss one of them now. Let $f(q) = \sum_{i=0}^{l} m_i q^i \in \mathbb{N}[q]$ satisfy $f(1) = n$. Let $X_n = \{x_1, x_2, \ldots, x_n\}$ be a set of variables and let $p(X_n)$ be a polynomial symmetric in the x_i. The *plethystic substitution* of f into p is

$$p[f] = p(\underbrace{1, \ldots, 1}_{m_0}, \underbrace{q, \ldots, q}_{m_1}, \ldots, \underbrace{q^l, \ldots, q^l}_{m_l}).$$

Note that since p is symmetric, it does not matter in what order one substitutes these values. For a concrete example, let $f(q) = 1 + 2q$ and $p(X_3) = h_2(x_1, x_2, x_3)$

as in (3.8). Then $h_2[f] = h_2(1, q, q) = 1 + 2q + 3q^2$. If $f(q) = [n]_q$ then $p[f] = p(1, q, \ldots, q^{n-1})$ is the principal specialization of p. Plethysm is useful in describing the representations of wreath products of groups.

We will also need the *elementary symmetric polynomials*

$$e_k(x_1, x_2, \ldots, x_n) = \sum_{1 \leq i_1 < i_2 < \ldots < i_k \leq n} x_{i_1} x_{i_2} \cdots x_{i_k},$$

i.e., the sum of all square-free monomials of degree k in the x_i. For example, $e_2(x_1, x_2, x_3) = x_1 x_2 + x_1 x_3 + x_2 x_3$. It is well known that the algebra symmetric polynomials in the variables X_n is freely generated by $e_1(X_n), e_2(X_n), \ldots, e_n(X_n)$ and this is sometimes called the Fundamental Theorem of Symmetric Polynomials. The $e_i(X_n)$ are dual to the $h_j(X_n)$ in a way that can be made precise.

Finally, it will be convenient to extend some of our concepts slightly. We will use notations (2.1) and (6.1) replacing $[n]$ with any set. We can also consider *symmetric functions* which are formal power series in the variables $X = \{x_1, x_2, x_3, \ldots\}$ which are invariant under all permutations of variables and are of bounded degree. For example, the complete homogeneous symmetric function is

$$h_k(X) = \sum_{1 \leq i_1 \leq i_2 \leq \ldots \leq i_k} x_{i_1} x_{i_2} \cdots x_{i_k}.$$

In this case, we can still make a plethystic substitution of a polynomial $f(q)$ by letting $x_i = 0$ for $i > n = f(1)$. One of the results of Berget, Eu, and Reiner [5] is as follows.

Theorem 11.1 *If a triple $(X, C, f(q))$ exhibits the cyclic sieving phenomenon, then the triple*

$$\left(\left(\binom{X}{k} \right), \ C, \ h_k[f(q)] \right)$$

does so as well.

If, in addition, $\#C$ is odd then the triple

$$\left(\binom{X}{k}, \ C, \ e_k[f(q)] \right)$$

also exhibits the cyclic sieving phenomenon. □

Some remarks about this theorem are in order. First of all, the authors actually prove it for C a product of cyclic groups and multi-cyclic sieving, but this does not materially alter the demonstration. The proof uses symmetric tensors for the first part (much in the way they were used in Section 3) and alternating tensors in the second. So the restriction on $\#C$ is there to control the sign. Finally, it is instructive to note how the first part implies our old friend, Theorem 2.1. Indeed, it is clear that the triple $([n], \langle(1, 2, \ldots, n)\rangle, [n]_q)$ exhibits the CSP: Only the identity element of the group has fixed points and there are n of them, while for $d|n$ we have

$$[n]_{\omega_d} = \begin{cases} n & \text{if } d = 1, \\ 0 & \text{otherwise.} \end{cases}$$

Now applying Lemma 3.1 and the previous theorem completes the proof.

Eu, Fu, and Pan [17] investigated the CSP for faces of cyclic polytopes. The *moment curve in dimension d* is $\gamma : \mathbb{R} \to \mathbb{R}^d$ defined parametrically by

$$\gamma(t) = (t, t^2, \ldots, t^d).$$

Given real numbers $t_1 < t_2 < \ldots < t_n$, the corresponding *cyclic polytope* is the convex hull

$$\mathrm{CP}(n, d) = \mathrm{conv}\{\gamma(t_1), \gamma(t_2), \ldots, \gamma(t_n)\}.$$

A good reference for the theory of convex polytopes is Ziegler's book [88].

It is known that the $\gamma(t_i)$ are the vertices of $\mathrm{CP}(n, d)$. Also, its combinatorial type (i.e., the structure of its faces) does not depend on the parameters t_i. Let $f_k(n, d)$ denote the number of faces of $\mathrm{CP}(n, d)$ of dimension k which is well defined by the previous sentence. We also let $\mathrm{CP}_k(n, d)$ denote the set of such faces. Cyclic polytopes are famous, in part, because they have the maximum number of faces in all dimensions k among all polytopes with n vertices in $\mathbb{R}^d$.

In what follows, we will assume d is even. There is a formula for the face numbers (for all d). In particular, for $0 \leq k < d$ and d even

$$f_k(n, d) = \sum_{j=1}^{d/2} \frac{n}{n-j} \binom{n-j}{j} \binom{j}{k+1-j}.$$

The reader will not be surprised that we will use the q-analogue

$$f_k(n, d; q) = \sum_{j=1}^{d/2} \frac{[n]_q}{[n-j]_q} \begin{bmatrix} n-j \\ j \end{bmatrix}_q \begin{bmatrix} j \\ k+1-j \end{bmatrix}_q.$$

Let $g \in \langle (1, 2, \ldots, n) \rangle$ act on the vertices of $\mathrm{CP}(n, d)$ by sending vertex $\gamma(t_i)$ to $\gamma(t_{g(i)})$. For even d this induces an automorphism of $\mathrm{CP}(n, d)$ in that it sends faces to faces. We can now state the main result of Eu, Fu, and Pan [17].

Theorem 11.2 *Suppose d is even and $0 \leq k < d$. Then the triple*

$$(\mathrm{CP}_k(n, d), \langle (1, 2, \ldots, n) \rangle, f_k(n, d; q))$$

exhibits the cyclic sieving phenomenon.   □

For odd d the n-cycle does not necessarily induce an automorphism of $\mathrm{CP}(n, d)$ and so there can be no CSP. However, in this case there are actions of certain groups of order 2 and it would be interesting to find CSPs for them.

Reiner, Stanton, and Webb [57] considered extending Springer's theory to arbitrary fields. Broer, Reiner, Smith, and Webb [11] continued this work and also extended various invariant theory results of Chevalley, Shephard-Todd, and Mitchell such as describing the relationship between the coinvariant and group algebras. Let V be an n-dimensional vector space over a field k, and let G be a finite subgroup of $GL(V)$ generated by (pseudo)-reflections which are defined as in the complex case. Then G acts on the polynomial algebra $\mathcal{S} = k[x_1, x_2, \ldots, x_n]$. Assume that

$\mathcal{S}^G$ is a (free) polynomial algebra so that $\mathcal{S}^G = k[f_1, f_2, \ldots, f_n]$ for polynomials $f_1, f_2, \ldots, f_n$.

To define regular elements, one must work in the algebraic closure $\bar{k}$ of k. Let $\bar{V} = V \otimes_k \bar{k}$. Call an element $g \in G$ *regular* if it has an eigenvector $v \in \bar{V}$ lying on none of the reflecting hyperplanes $\bar{H} = H \otimes_k \bar{k}$ for reflections in G. It can be shown that in this case $o(g)$ is invertible in k. This implies that the $\langle g \rangle$-submodules of any G-module are *completely reducible*, meaning that they can be written as a direct sum of irreducibles. (Recall that complete reducibility is not guaranteed over arbitrary fields as it is over $\mathbb{C}$ by Maschke's Theorem, Theorem 4.1 (b).)

Now consider any subgroup $H \leq G$. The cyclic sieving set will be the cosets G/H acted upon by left multiplication of the regular element g. For the function we will take the quotient $\mathrm{Hilb}(\mathcal{S}^H; q)/\mathrm{Hilb}(\mathcal{S}^G; q)$. But one has to make sure that this is in $\mathbb{N}[q]$ and not just a rational function. Reiner, Stanton, and Webb [57] explained why this must be a polynomial with integer coefficients. But in stating their CSP they had to assume extra conditions on H so that they could prove the coefficients were nonnegative. They also asked whether it was possible to prove the CSP without these hypotheses, and this was done in by Broer, Reiner, Smith, and Webb [11] thus generalizing Theorem 6.6.

Theorem 11.3 *Let V be a finite-dimensional vector space over a field k. Let G be a finite subgroup of $\mathrm{GL}(V)$ for which $\mathcal{S}^G$ is a polynomial algebra. Let g be a regular element of G acting on G/H by left multiplication. Then for any $H \leq G$, the triple*

$$\left(G/H, \ \langle g \rangle, \ \frac{\mathrm{Hilb}(\mathcal{S}^H; q)}{\mathrm{Hilb}(\mathcal{S}^G; q)} \right)$$

exhibits the cyclic sieving phenomenon.   □

12 Remarks

12.1 Alternate definitions

In their initial paper [58], Reiner, Stanton, and White gave a second, equivalent, definition of the CSP. While this one has not come to be used as much as (1.2), we mention it here for completeness.

Given a group G acting on a set X, denote the stabilizer subgroup of $y \in X$ by

$$G_y = \{g \in G \ : \ gy = y\}.$$

If x, y are in the same orbit $\mathcal{O}$ then their stabilizers are conjugate (and if G is Abelian they are actually equal). So if $y \in \mathcal{O}$ then call $s(\mathcal{O}) = \#G_y$ the *stabilizer-order* of $\mathcal{O}$ which is well defined by the previous sentence.

Suppose $f(q) = \sum_{i \geq 0} m_i q^i \in \mathbb{N}[q]$ and define coefficients a_i for $0 \leq i < n$ by

$$f(q) \equiv a_0 + a_1 q + \cdots + a_{n-1} q^{n-1} \pmod{1 - q^n}.$$

Equivalently,

$$a_i = \sum_{j \equiv i \pmod{n}} m_j.$$

Definition 12.1 Suppose $\#X = n$, C is a cyclic group acting on X, and the a_i are as above. The triple $(X, C, f(q))$ exhibits the *cyclic sieving phenomenon* if, for $0 \le i < n$,

$$a_i = \#\{\mathcal{O} \ : \ s(\mathcal{O})|i\}. \tag{12.1}$$

Note that if (12.1) is true then a_0 counts the total number of C-orbits, while a_1 counts the number of free orbits (those of size $\#C$). In fact, one can use these equations to determine how many orbits there are of any size using Möbius inversion. Returning to our original example with $C = \langle(1,2,3)\rangle$ and 2-element multisets on [3], the orbits were

$$\mathcal{O}_1 = (11, 22, 33), \ \mathcal{O}_2 = (12, 23, 13).$$

On the other hand

$$f(q) = 1 + q + 2q^2 + q^3 + q^4 \equiv 2 + 2q + 2q^2 \pmod{1 - q^3}$$

indicating that there are 2 orbits total with both of them being free. The coefficient $a_2 = 2$ as well since a free orbit's stabilizer-order of 1 will divide any other.

The proof that these two definitions are equivalent is via the representation theory paradigm, Theorem 4.2. The main tool is Frobenius reciprocity.

In a personal communication, Reiner has pointed out that it might also be interesting to define cyclic sieving with more general polynomials. One possibility would be to allow negative integral coefficients which could be useful, for example, when considering quotients of Hilbert series. Note that this issue arose in the genesis of Theorem 11.3. Another extension could be to Laurent polynomials, that is, elements of $\mathbb{Z}[q, q^{-1}]$. Such polynomials might come up when considering a bivariate generating function $f(q, t)$ where one lets $t = q^{-1}$. We have already seen such a substitution in Theorem 9.6, although in that case it turns out that the generating function remains an ordinary polynomial.

12.2   More on Catalan CSPs

We have just begun to scratch the surface of the connection between Catalan combinatorics and cyclic sieving. We have already mentioned how polygonal dissections is behind the work of Eu and Fu on cluster complexes [16] as in Theorem 10.4. We will now describe an open problem and some ongoing work about triangulations, i.e., dissections where every face is a triangle.

Let P be a regular n-gon. Let $\mathcal{T}_n$ denote the set of triangulations T of P using nonintersecting diagonals. It is well known that

$$\#\mathcal{T}_{n+2} = \mathrm{Cat}_n.$$

We will act on triangulations by clockwise rotation. So, for example, for the pentagon these is only one cycle

Figure 3: Two triangulations, one proper (left) and one not (right)

Reiner, Stanton, and White [58] proved the following theorem in this setting. (In fact, they proved a stronger result about dissections using noncrossing diagonals where one fixes the number of diagonals.)

Theorem 12.2 *Let C_{n+2} act on T_{n+2} by rotation. Then the triple*

$$(T_{n+2}, C_{n+2}, \operatorname{Cat}_n(q))$$

exhibits the cyclic sieving phenomenon. □

Their proof was of the sort where one evaluates both sides of (1.2) directly. But as mentioned before, these proofs often lack the beautiful insights one obtains from using representation theory. It would be very interesting to find such a proof, perhaps by finding an appropriate complex reflection group along with a basis and fake degree polynomial which would permit the use of Westbury's Theorem 8.2.

One can also consider colored triangulations. Label ("color") the vertices of the polygon P clockwise $1, 2, 1, 2, \ldots$. (When n is odd, there will be an edge of P with both endpoints labeled 1.) Call a triangulation *proper* if it contains no monochromatic triangle. (This terminology is both by analogy with proper coloring of graphs and in honor of Jim Propp who first conjectured (12.2).) In Figure 3, the left-hand triangulation is proper while the one on the right is not. Let $\mathcal{P}_n$ be the set of proper triangulations of an n-gon. Sagan [66] proved that

$$
\#\mathcal{P}_{N+2} = \begin{cases} \dfrac{2^n}{2n+1}\dbinom{3n}{n} & \text{if } N = 2n \text{ where } n \in \mathbb{N}, \\[4mm] \dfrac{2^{n+1}}{2n+2}\dbinom{3n+1}{n} & \text{if } N = 2n+1 \text{ where } n \in \mathbb{N}. \end{cases}
\tag{12.2}
$$

Note that for $N = 2n$ we have $\#\mathcal{P}_{N+2} = 2^n \operatorname{Cat}_{2,n}$.

Roichman and Sagan [63] are studying CSPs for colored dissections. In the triangulation case, notice that when n is odd then there is no action of C_n on $\mathcal{P}_n$ because it is possible for the rotation of a proper triangulation to be improper as in Figure 3. So it only makes sense to consider a rotational CSP for n even. They have proved that one does indeed have such a phenomenon, although the necessary q-analogue for the factor of 2^n is somewhat surprising.

Theorem 12.3 *Let $N = 2n$ and let C_{N+2} act on $\mathcal{P}_{N+2}$ by rotation. Then the triple*

$$\left(\mathcal{P}_{N+2}, \ C_{N+2}, \ \frac{[2]_{q^2} \left([2]_q^{n-1} - [2]_q^{\lceil n/2 \rceil -1} + 2^{\lceil n/2 \rceil -1} \right)}{[2n+1]_q} \left[\begin{array}{c} 3n \\ n \end{array} \right]_q \right)$$

exhibits the cyclic sieving phenomenon. $\square$

12.3 A combinatorial proof

One could hope for purely combinatorial proofs of CSPs. Since it may not be clear exactly what this would entail, consider the following paradigm. First of all, we would need to have a combinatorial expression for $f(q)$, namely some statistic on the set X such that

$$f^{\mathrm{st}}(X; q) = f(q). \tag{12.3}$$

Suppose further that, for each $g \in C$, one has a partition of X

$$\pi = \pi_g = \{B_1, B_2, \ldots, B_k\}$$

satisfying the following criterion where $\omega = \omega_{o(g)}$:

$$f^{\mathrm{st}}(B_i; \omega) = \left\{ \begin{array}{ll} 1 & \text{if } i \leq \#X^g, \\ 0 & \text{if } i > \#X^g. \end{array} \right. \tag{12.4}$$

In other words, the initial blocks correspond to the fixed points of g and their weights evaluate to 1 when plugging in ω, while the weights of the rest of the blocks get zeroed out under this substitution. (In practice, the B_i for $i \leq \#X^g$ are singletons each with weight q^j where $o(g)|j$, while for $i > \#X^g$ the sum of the weights in the block form a geometric progression which becomes zero since $1 + \omega + \cdots + \omega^{d-1} = 0$ for any proper divisor d of $o(g)$.) In this case, one automatically has cyclic sieving because, using equations (12.3) and (12.4) as well as the fact that we have a partition

$$f(\omega) = f^{\mathrm{st}}(X; \omega) = \sum_{i \geq 0} f^{\mathrm{st}}(B_i; \omega) = \underbrace{1 + \cdots + 1}_{\#X^g} + 0 + \cdots + 0 = \#X^g.$$

Roichman and Sagan [63] have succeeded in using this method to prove Theorem 6.1. They are currently working on trying to apply it to various other cyclic sieving results.

Note added in proof

Since this article was written six new papers have appeared related to the cyclic sieving phenomenon. For completeness' sake, we briefly describe each of them here.

In [3], Armstrong, Stump, and Thomas constructed a bijection between noncrossing partitions and nonnesting partitions which sends a complementation map of Kreweras [44], **Krew**, to a function of Panyushev [52], **Pan**. Using this construction they are able to prove two CSP conjectures in the paper of Bessis and Reiner [8] about **Krew** and **Pan**. The first refines Theorem 10.2 in the case that W is a finite

Coxeter group since **Krew**^{2} coincides with the action of the regular element. The second follows from the first using the bijection.

Kluge and Rubey [37] have obtained a cyclic sieving result for rotation of Ptolemy diagrams. These diagrams were recently introduced in a paper of Holm, Jørgensen, and Rubey [32] as a model for torsion pairs in the cluster category of type A. Their result is related to the generalization of Theorem 12.2 mentioned just before its statement, except that certain diagonals are allowed to cross and one keeps track of the number of regions of various types rather than the number of diagonals. But the polynomial in both cases is a product of q-binomial coefficients, so it would be interesting to find a common generalization.

Noncrossing graphs on the vertex set $[n]$ can be defined analogously to noncrossing polygonal dissections by arranging the vertices around a circle and insisting that the resulting graph be planar. Flajolet and Noy [18] showed that the number of connected noncrossing graphs with n vertices and k edges is given by

$$\frac{1}{n-1}\binom{3n-3}{n+k}\binom{k-1}{n-2}$$

Following a personal communication of S.-P. Eu, Guo [28] has shown that one has a CSP using these graphs, rotation, and the expected q-analogue of the expression above.

Another way to generalize noncrossing dissections into triangles is to define a k-*triangulation* of a convex n-gon to be a maximal collection of diagonals such that no $k + 1$ of them mutually cross. So ordinary triangulations are the case $k = 1$. In a personal communication, Reiner has conjectured a CSP for such triangulations under rotation generalizing Theorem 12.2. In [71], Serrano and Stump reformulated this conjecture in terms of k-flagged tableaux (certain semistandard tableaux with bounds on the entries). But the conjecture remains open.

As has already been mentioned, there is no representation theory proof of Theorem 12.2. The same is true of the Eu and Fu's result, Theorem 10.4. In an attempt to partially remedy this situation, Rhoades [59] has used representation theory and cluster multicomplexes to prove related CSPs. His tools include a notion of noncrossing tableaux due to Pylyavskyy [56] and geometric realizations of finite type cluster algebras due to Fomin and Zelevinsky [21].

Westbury [86] has succeeded in generalizing Rhoades's original result [60] just as he was able to do for the special case considered by Petersen, Pylyavskyy, and Rhoades [53] for two and three rows. It turns out that the same tools (crystal bases, based modules, and regular elements) can be used.

References

[1] ANDREWS, G. E. *The theory of partitions.* Cambridge Mathematical Library. Cambridge University Press, Cambridge, 1998. Reprint of the 1976 original.

[2] ARMSTRONG, D. Generalized noncrossing partitions and combinatorics of Coxeter groups. *Mem. Amer. Math. Soc. 202*, 949 (2009), x+159.

[3] ARMSTRONG, D., STUMP, C., AND THOMAS, H. A uniform bijection between nonnesting and noncrossing partitions. Preprint arXiv:1101.1277.

[4] BARCELO, H., REINER, V., AND STANTON, D. Bimahonian distributions. *J. Lond. Math. Soc. (2) 77*, 3 (2008), 627–646.

[5] BERGET, A., EU, S.-P., AND REINER, V. Constructions for cyclic sieving phenomena. *SIAM J. Discrete Math.*. to appear, preprint arXiv:1004.0747.

[6] BESSIS, D. Finite complex reflection arrangements are $k(\pi, 1)$. Preprint arXiv:math/0610777.

[7] BESSIS, D. The dual braid monoid. *Ann. Sci. École Norm. Sup. (4) 36*, 5 (2003), 647–683.

[8] BESSIS, D., AND REINER, V. Cyclic sieving on noncrossing partitions for complex reflection groups. *Ann. Comb.*. to appear, preprint arXiv:math/0701792.

[9] BJÖRNER, A., AND BRENTI, F. *Combinatorics of Coxeter groups*, vol. 231 of *Graduate Texts in Mathematics*. Springer, New York, 2005.

[10] BRADY, T., AND WATT, C. $K(\pi, 1)$'s for Artin groups of finite type. In *Proceedings of the Conference on Geometric and Combinatorial Group Theory, Part I (Haifa, 2000)* (2002), vol. 94, pp. 225–250.

[11] BROER, A., REINER, V., SMITH, L., AND WEBB, P. Extending the coinvariant theorems of Chevalley, Shephard-Todd, Mitchell, and Sringer. Preprint arXiv:0805:3694.

[12] COXETER, H. S. M. Discrete groups generated by reflections. *Ann. of Math. (2) 35*, 3 (1934), 588–621.

[13] COXETER, H. S. M. The complete enumeration of finite groups of the form $r_i^2 = (r_i r_j)^{k_{ij}} = 1$. *J. London Math. Soc. 10* (1935), 21–25.

[14] DÉSARMÉNIEN, J. Fonctions symétriques associées à des suites classiques de nombres. *Ann. Sci. École Norm. Sup. (4) 16*, 2 (1983), 271–304.

[15] EDELMAN, P., AND GREENE, C. Balanced tableaux. *Adv. in Math. 63*, 1 (1987), 42–99.

[16] EU, S.-P., AND FU, T.-S. The cyclic sieving phenomenon for faces of generalized cluster complexes. *Adv. in Appl. Math. 40*, 3 (2008), 350–376.

[17] EU, S.-P., FU, T.-S., AND PAN, Y.-J. The cyclic sieving phenomenon for faces of cyclic polytopes. *Electron. J. Combin. 17*, 1 (2010), Research Paper 47, 17.

[18] FLAJOLET, P., AND NOY, M. Analytic combinatorics of non-crossing configurations. *Discrete Math. 204*, 1-3 (1999), 203–229.

[19] FOATA, D., AND SCHÜTZENBERGER, M.-P. Major index and inversion number of permutations. *Math. Nachr. 83* (1978), 143–159.

[20] FOMIN, S., AND READING, N. Generalized cluster complexes and Coxeter combinatorics. *Int. Math. Res. Not.*, 44 (2005), 2709–2757.

[21] FOMIN, S., AND ZELEVINSKY, A. Cluster algebras. II. Finite type classification. *Invent. Math.* *154*, 1 (2003), 63–121.

[22] FOMIN, S., AND ZELEVINSKY, A. *Y*-systems and generalized associahedra. *Ann. of Math. (2) 158*, 3 (2003), 977–1018.

[23] FRAME, J. S., ROBINSON, G. D. B., AND THRALL, R. M. The hook graphs of the symmetric groups. *Canadian J. Math. 6* (1954), 316–324.

[24] FULTON, W., AND HARRIS, J. *Representation theory*, vol. 129 of *Graduate Texts in Mathematics*. Springer-Verlag, New York, 1991. A first course, Readings in Mathematics.

[25] FÜRLINGER, J., AND HOFBAUER, J. *q*-Catalan numbers. *J. Combin. Theory Ser. A 40*, 2 (1985), 248–264.

[26] GORDON, I., AND GRIFFETH, S. Catalan numbers for complex reflection groups. *Amer. J. Math.*. to appear, preprint arXiv:0912.1578.

[27] GROVE, L. C., AND BENSON, C. T. *Finite reflection groups*, second ed., vol. 99 of *Graduate Texts in Mathematics*. Springer-Verlag, New York, 1985.

[28] GUO, A. Cyclic sieving phenomenon in non-crossing connected graphs. *Electron. J. Combin. 18*, 1 (2011), Research Paper 9, 14 pp.

[29] HAIMAN, M. D. On mixed insertion, symmetry, and shifted Young tableaux. *J. Combin. Theory Ser. A 50*, 2 (1989), 196–225.

[30] HAIMAN, M. D. Dual equivalence with applications, including a conjecture of Proctor. *Discrete Math. 99*, 1-3 (1992), 79–113.

[31] HILLER, H. *Geometry of Coxeter groups*, vol. 54 of *Research Notes in Mathematics*. Pitman (Advanced Publishing Program), Boston, Mass., 1982.

[32] HOLM, T., JØRGENSEN, P., AND RUBEY, M. Ptolemy diagrams and torsion pairs in the cluster category of Dynkin type a_n. Preprint arXiv:1010.1184.

[33] HUMPHREYS, J. E. *Introduction to Lie algebras and representation theory*, vol. 9 of *Graduate Texts in Mathematics*. Springer-Verlag, New York, 1978. Second printing, revised.

[34] JAMES, G., AND KERBER, A. *The representation theory of the symmetric group*, vol. 16 of *Encyclopedia of Mathematics and its Applications*. Addison-Wesley Publishing Co., Reading, Mass., 1981. With a foreword by P. M. Cohn, With an introduction by Gilbert de B. Robinson.

[35] JAMES, G. D. *The representation theory of the symmetric groups*, vol. 682 of *Lecture Notes in Mathematics*. Springer, Berlin, 1978.

[36] KAZHDAN, D., AND LUSZTIG, G. Representations of Coxeter groups and Hecke algebras. *Invent. Math.* *53*, 2 (1979), 165–184.

[37] KLUGE, S., AND RUBEY, M. Cyclic sieving for torsion pairs in the cluster category of Dynkin type a_n. Preprint arXiv:1101.1020.

[38] KNUTH, D. E. Permutations, matrices, and generalized Young tableaux. *Pacific J. Math.* *34* (1970), 709–727.

[39] KNUTH, D. E. *The art of computer programming. Volume 3.* Addison-Wesley Publishing Co., Reading, Mass.-London-Don Mills, Ont., 1973. Sorting and searching, Addison-Wesley Series in Computer Science and Information Processing.

[40] KRAŚKIEWICZ, W., AND WEYMAN, J. Algebra of coinvariants and the action of a Coxeter element. *Bayreuth. Math. Schr.*, 63 (2001), 265–284.

[41] KRATTENTHALER, C. Non-crossing partitions on an annulus. In preparation.

[42] KRATTENTHALER, C., AND MÜLLER, T. W. Cyclic sieving for generalized non-crossing partitions associated to complex reflection groups of exceptional type. Preprint arXiv:1001.0028.

[43] KRATTENTHALER, C., AND MÜLLER, T. W. Cyclic sieving for generalized non-crossing partitions associated to complex reflection groups of exceptional type - the details. Preprint arXiv:1001.0030.

[44] KREWERAS, G. Sur les partitions non croisées d'un cycle. *Discrete Math.* *1*, 4 (1972), 333–350.

[45] KUPERBERG, G. Spiders for rank 2 Lie algebras. *Comm. Math. Phys.* *180*, 1 (1996), 109–151.

[46] LASCOUX, A., AND SCHÜTZENBERGER, M.-P. Le monoïde plaxique. In *Noncommutative structures in algebra and geometric combinatorics (Naples, 1978)*, vol. 109 of *Quad. "Ricerca Sci."*. CNR, Rome, 1981, pp. 129–156.

[47] LEHRER, G. I., AND TAYLOR, D. E. *Unitary reflection groups*, vol. 20 of *Australian Mathematical Society Lecture Series*. Cambridge University Press, Cambridge, 2009.

[48] LUSZTIG, G. *Introduction to quantum groups*, vol. 110 of *Progress in Mathematics*. Birkhäuser Boston Inc., Boston, MA, 1993.

[49] MACMAHON, P. A. The Indices of Permutations and the Derivation Therefrom of Functions of a Single Variable Associated with the Permutations of any Assemblage of Objects. *Amer. J. Math.* *35*, 3 (1913), 281–322.

[50] MACMAHON, P. A. *Collected papers. Vol. I.* MIT Press, Cambridge, Mass., 1978. Combinatorics, Mathematicians of Our Time, Edited and with a preface by George E. Andrews, With an introduction by Gian-Carlo Rota.

[51] OPDAM, E. M.   Complex reflection groups and fake degrees.   Preprint arXiv:math/9808026.

[52] PANYUSHEV, D. I.   On orbits of antichains of positive roots.   *European J. Combin. 30*, 2 (2009), 586–594.

[53] PETERSEN, T. K., PYLYAVSKYY, P., AND RHOADES, B. Promotion and cyclic sieving via webs. *J. Algebraic Combin. 30*, 1 (2009), 19–41.

[54] PETERSEN, T. K., AND SERRANO, L. Cyclic sieving for longest reduced words in the hyperoctahedral group. *Electron. J. Combin. 17*, 1 (2010), Research Paper 67, 12 pp.

[55] PON, S., AND WANG, Q.   Promotion and evacuations on standard young tableaux of rectangle and staircase shape. *Electron. J. Combin. 18*, 1 (2011), Research Paper 18, 18 pp.

[56] PYLYAVSKYY, P. Non-crossing tableaux. *Ann. Comb. 13*, 3 (2009), 323–339.

[57] REINER, V., STANTON, D., AND WEBB, P. Springer's regular elements over arbitrary fields. *Math. Proc. Cambridge Philos. Soc. 141*, 2 (2006), 209–229.

[58] REINER, V., STANTON, D., AND WHITE, D. The cyclic sieving phenomenon. *J. Combin. Theory Ser. A 108*, 1 (2004), 17–50.

[59] RHOADES, B. Cyclic sieving and cluster multicomplexes. *Adv. in Appl. Math..* to appear, preprint arXiv:1005.2561.

[60] RHOADES, B. Cyclic sieving, promotion, and representation theory. *J. Combin. Theory Ser. A 117*, 1 (2010), 38–76.

[61] RHOADES, B. Hall-Littlewood polynomials and fixed point enumeration. *Discrete Math. 310*, 4 (2010), 869–876.

[62] ROBINSON, G. D. B. On the Representations of the Symmetric Group. *Amer. J. Math. 60*, 3 (1938), 745–760.

[63] ROICHMAN, Y., AND SAGAN, B. Combinatorial and colorful proofs of cyclic sieving phenomena. In preparation.

[64] ROUQUIER, R. *q*-Schur algebras and complex reflection groups. *Mosc. Math. J. 8*, 1 (2008), 119–158, 184.

[65] SAGAN, B. E. *The symmetric group: Representations, combinatorial algorithms, and symmetric functions*, second ed., vol. 203 of *Graduate Texts in Mathematics*. Springer-Verlag, New York, 2001.

[66] SAGAN, B. E. Proper partitions of a polygon and *k*-Catalan numbers. *Ars Combin. 88* (2008), 109–124.

[67] SAGAN, B. E., SHARESHIAN, J., AND WACHS, M. Eulerian quasisymmetric functions and cyclic sieving. *Adv. in Appl. Math..* to appear, preprint arXiv:0909.3143.

[68] SCHENSTED, C. Longest increasing and decreasing subsequences. *Canad. J. Math. 13* (1961), 179–191.

[69] SCHÜTZENBERGER, M. P. Quelques remarques sur une construction de Schensted. *Math. Scand. 12* (1963), 117–128.

[70] SCHÜTZENBERGER, M. P. Promotion des morphismes d'ensembles ordonnés. *Discrete Math. 2* (1972), 73–94.

[71] SERRANO, L., AND STUMP, C. Maximal fillings of moon polyominoes, simplicial complexes, and Schubert polynomials. Preprint arXiv:1009.4690.

[72] SHARESHIAN, J., AND WACHS, M. L. Eulerian quasisymmetric functions. Preprint arXiv:0812,0764.

[73] SHARESHIAN, J., AND WACHS, M. L. q-Eulerian polynomials: excedance number and major index. *Electron. Res. Announc. Amer. Math. Soc. 13* (2007), 33–45 (electronic).

[74] SHARESHIAN, J., AND WACHS, M. L. Poset homology of Rees products, and q-Eulerian polynomials. *Electron. J. Combin. 16*, 2, Special volume in honor of Anders Bjorner (2009), Research Paper 20, 29.

[75] SHEPHARD, G. C., AND TODD, J. A. Finite unitary reflection groups. *Canadian J. Math. 6* (1954), 274–304.

[76] SIMION, R. Combinatorial statistics on noncrossing partitions. *J. Combin. Theory Ser. A 66*, 2 (1994), 270–301.

[77] SKANDERA, M. On the dual canonical and Kazhdan-Lusztig bases and 3412-, 4231-avoiding permutations. *J. Pure Appl. Algebra 212*, 5 (2008), 1086–1104.

[78] SPRINGER, T. A. Regular elements of finite reflection groups. *Invent. Math. 25* (1974), 159–198.

[79] STANLEY, R. Catalan addendum. New problems for *Enumerative Combinatorics. Vol. 2*, available at http://math.mit.edu/ rstan/ec/catadd.pdf.

[80] STANLEY, R. P. *Enumerative Combinatorics. Vol. 1*, vol. 49 of *Cambridge Studies in Advanced Mathematics*. Cambridge University Press, Cambridge, 1997. With a foreword by Gian-Carlo Rota, Corrected reprint of the 1986 original.

[81] STANLEY, R. P. Promotion and evacuation. *Electron. J. Combin. 16*, 2, Special volume in honor of Anders Bjorner (2009), Research Paper 9, 24.

[82] STANTON, D. W., AND WHITE, D. E. A Schensted algorithm for rim hook tableaux. *J. Combin. Theory Ser. A 40*, 2 (1985), 211–247.

[83] STEMBRIDGE, J. R. On minuscule representations, plane partitions and involutions in complex Lie groups. *Duke Math. J. 73*, 2 (1994), 469–490.

[84] STEMBRIDGE, J. R. Some hidden relations involving the ten symmetry classes of plane partitions. *J. Combin. Theory Ser. A 68*, 2 (1994), 372–409.

[85] STEMBRIDGE, J. R. Canonical bases and self-evacuating tableaux. *Duke Math. J. 82*, 3 (1996), 585–606.

[86] WESTBURY, B. Invariant tensors and the cyclic sieving phenomenon. Preprint.

[87] WESTBURY, B. Invariant theory and the cyclic sieving phenomenon. Preprint arXiv:0912.1512.

[88] ZIEGLER, G. M. *Lectures on polytopes*, vol. 152 of *Graduate Texts in Mathematics*. Springer-Verlag, New York, 1995.

Department of Mathematics
Michigan State University
East Lansing, MI 48824-1027, USA
sagan@math.msu.edu

Order in building theory

Koen Thas

Abstract

A notorious open problem in Combinatorics is the so-called "Prime Power Conjecture", which states that the order of a finite projective plane is a prime power. Projective planes are important examples of Tits buildings of rank 2, and the latter class is widely considered as the central class of point-line (= rank 2) geometries. In this paper, we consider prime power and other parameter conjectures for buildings of all ranks, not restricting ourselves to the finite case. The rank 2 case will play a prominent role (due to the simple fact that they are all around the higher rank buildings). Among the topics which will pass, are Moufang sets, various aspects of finite and infinite projective planes (such as flag-transitivity), parameter conjectures for the other generalized polygons, locally finite polygons, etc. The paper also contains several new constructions and explores connections with other problems and fields. The paper ends with a discussion of recent developments in the theory of Absolute Arithmetic (over the "field with one element, $\mathbb{F}_1$"), and some mysterious interrelations with several aspects which are touched in the present text.

Acknowledgements

The author wishes to express his gratitude to Alain Connes for providing several useful suggestions and remarks.
The author is partially supported by the Fund for Scientific Research — Flanders (Belgium).

Contents

Notation

2^X: power set of X
Aut($\mathscr{S}$): automorphism group of $\mathscr{S}$, with $\mathscr{S}$ a geometry, group, etc.
$R^\times$: $R \setminus \{\mathbf{1}\}$ with R a group and $\mathbf{1}$ its identity
$\mathbb{F}_q$: finite field with q elements
$\mathbf{PG}(n, q)$: n-dimensional projective space over $\mathbb{F}_q$
$\mathbf{P}(V)$: projective space associated to the vector space V
G/H: left coset space, where H is a subgroup of G
$X(\mathbb{F})$: set of $\mathbb{F}$-rational points of the variety X
$\mathbb{F}_1$: "field with one element"
$[n]_\ell$: $1 + \ell + \cdots + \ell^{n-1}$
$\hat{\mathbb{Z}}$: profinite completion of $\mathbb{Z}$
$\mathbf{K}$: the Krasner hyperfield

1   Introduction

Initially, the purpose of the theory of buildings was primarily to understand the exceptional Lie groups from a geometrical point of view. The starting point appeared to be the observation that it is possible to associate with each complex analytic semisimple group a certain well-defined geometry, in such a way that the "basic" properties of the geometries thus obtained and their mutual relationships can be easily read from the Dynkin diagrams of the corresponding groups. The definition of these geometries was suggested by the following reconstruction method of the complex projective space $\mathbf{PG}(n, \mathbb{C})$ from the projective linear group $\mathbf{PGL}_{n+1}(\mathbb{C})$:

(i) the linear subspaces of the projective space $\mathbf{PG}(n, \mathbb{C})$ can be represented by their stabilizers in $\mathbf{PGL}_{n+1}(\mathbb{C})$ (which, by a theorem of Lie, are the maximal connected nonsemisimple subgroups of $\mathbf{PGL}_{n+1}(\mathbb{C})$);

(ii) the conjugacy classes of these subgroups represent the set of all points, the set of all lines, ..., and the set of all hyperplanes of $\mathbf{PG}(n, \mathbb{C})$ (for arbitrary complex semisimple groups, the conjugacy classes of the maximal connected nonsemisimple subgroups are called *maximal parabolic subgroups*);

(iii) two linear subspaces are incident if and only if the intersection of the corresponding subgroups contains a maximal connected solvable subgroup of $\mathbf{PGL}_{n+1}(\mathbb{C})$.

Generalizing the above example, it seemed natural to associate with an arbitrary complex semisimple group G a geometry consisting of a set (the set of maximal parabolic subgroups), partitioned into classes (the conjugacy classes), parametrized by the vertices of the Dynkin diagram M of G, and endowed with an incidence relation as follows: two maximal parabolic subgroups are incident if their intersection contains a maximal connected solvable subgroup (the "Borel subgroup"). Now we have the following two essential properties:

(D1) let M be the Dynkin diagram, Γ the associated geometry, x an object in Γ, and $v(x)$ the vertex of M corresponding to the class of x; then the *residue* of x in Γ — that is, the geometry consisting of all objects of Γ distinct from x but incident with x, with the partition and incidence relation induced by those by Γ — is the geometry associated with the diagram obtained from M by deleting $v(x)$ and all strokes containing it;

(D2) when one knows the geometries associated with the Dynkin diagrams of rank 2 (that is, those having two vertices), Assertion (i) "almost" characterizes that geometry uniquely.

As the theory of algebraic semisimple groups over an arbitrary field emerged, it then became apparent that the same geometrical theory would apply to the group $G(\mathbb{K})$ of rational points of an arbitrary isotropic algebraic semisimple group G defined over any field $\mathbb{K}$, as:

• such a group $G(\mathbb{K})$ also has "parabolic subgroups";

• the conjugacy classes of the maximal parabolic subgroups are parametrized by the vertices of the so-called "relative Dynkin diagram" of G over $\mathbb{K}$;

• for the geometry consisting of the set of all maximal parabolic subgroups partitioned into conjugacy classes and endowed with a suitable incidence relation, the relation between residues and subdiagrams which was explained in (D1) holds.

Then J. Tits noticed that all the geometries associated with a diagram consisting of a single stroke with multiplicity $n - 2$, $n \geq 3$, satisfied the following important property:

(GP$_n$) *For any two elements x and y of the geometry, there is a sequence $x = x_0, x_1, \ldots, x_m = y$ of elements so that x_i and x_{i+2} are incident and $x_i \neq x_{i+2}$ for all $i = 0, 1, \ldots, m - 2$, with $m \leq n$, and if $m < n$, then the sequence is unique.*

At that point, *geometries of type M* — which are the direct precursors of buildings — were introduced, their definition being directly inspired by the latter observation(s). Property (GP_n) is the main axiom for the *generalized polygons* or *generalized n-gons*; the concept of a generalized polygon was formally introduced in the literature by Tits in his famous paper on trialities [122]. An important class of buildings is formed by the *spherical buildings* — they include in particular all finite buildings — which have a certain rank (dimension), and precisely when the rank is 2, the concept of a thick spherical building coincides with that of a thick generalized polygon. In fact, all other rank 2 buildings are trees without vertices of valency 1.

The class of generalized 3-gons coincides with the class of axiomatic projective planes. To each such plane is associated a parameter n called the "order", which expresses the fact that any line is incident with a constant $(n+1)$ number of points, and any point is incident with a constant (also $n+1$) number of lines. In fact, such (possibly different) constants exist for any generalized polygon. When n is finite, the following conjecture needs no introduction.

Conjecture 1.1 (Prime Power Conjecture) *The order of a finite projective plane is a prime power.*

The classical examples of finite projective planes consist of those that are coordinatized over a finite field — the so-called "Desarguesian planes". They of course satisfy the statement of the conjecture. In general, finite projective planes are coordinatized by plenary ternary rings (R, T), and so it is an open question as to whether $|R|$ is a prime power. The Prime Power Conjecture (PPC) for projective planes is without a doubt the most important open problem in projective plane theory.

As projective planes are members of the class of generalized polygons, and the latter are the rank 2 members of spherical buildings, it makes sense to address a more general "Prime Power Conjecture" for general buildings. In fact, as we will describe, "PPC" will mean a whole set of conjectures related to this question (and by abuse of terminology, I will often speak about "PPC theory"). In the finite case, roughly speaking, a PPC would state something about natural parameters associated to a finite spherical building (such as the number of lines through a point), typically them being a polynomial evaluated at some prime power. But we want to consider an even larger set of conjectures: first of all, we do not want to consider only finite buildings, and second of all, we want to develop conjectures about the structural theory which underlies the arithmetic. A good example is given by the class of translation planes, which are projective planes endowed with a certain group action. In the finite case, one can associate a field called "kernel" to such a plane, and the translation group appears to be a vector space over this field. Since the number of affine points (with respect to some special line) is exactly the number of points of this vector space, we have a positive answer to the PPC for these planes. PPC theory in this case is the vector space representation. In the infinite case, one can define the kernel in the same way, ending up with a skew field instead of a field, and with a (left or right) vector space over this skew field. So there still is a valid PPC theory in the infinite case, even without mentioning the word "parameter" explicitly. If one wants to do so, observe that we now can work with a concept of dimension, and state arithmetic properties.

$$\text{translation plane} \longrightarrow \text{kernel} \longrightarrow \text{vector space} \longrightarrow \text{PPC} \qquad (1.1)$$

As soon as the dimension of an axiomatic projective space is at least 3, a famous theorem of Veblen and Young states that it can be coordinatized over a skew field (and so in the finite case over a field), in other words, we can reconstruct the space from some vector space $\mathbb{K}^n$, where $\mathbb{K}$ is a skew field. This observation is the essence of PPC theory for projective spaces. As there is lack of a good algebraic theory in the rank 2 case (many examples are known that cannot appear as projective spaces over a skew field), we will endow planes (and more general geometries) with natural group actions, and then develop PPC theory, such as in the case of translation planes. If the rank is 1, PPC theory follows from the mere definition of the projective line.

For general spherical buildings of rank at least 3, there is a Veblen-Young available which hands us an immediate PPC theory: Tits classified them in a precise way. In the rank 2 case, the same obstruction arises as before, and as we will see, there even are "free constructions" which allows one to easily construct generalized n-gons for any n. Again, extra algebraic assumptions will be needed in order to pursue our goals. And again, in the rank 1 case PPC theory follows from the appropriate definition of building.

So the low dimensional cases will be emphasized.

In all ranks (and especially rank 2), numerous characterizations exist of the (finite) classical buildings. Although all these are crucial for PPC theory (since, for one, they lead to examples which are the main motivation for prime power conjectures), we will not survey such characterizations — only will we tersely comment on that in the projective plane section. Many detailed papers are available on characterizations of finite classical polygons; so we refer to these for more.

Considering the free constructions, it could seem unlikely to the reader that interesting PPC theories can emerge for *general* generalized polygons. But in fact, this is not the case. As we will see, on the contrary an extremely rich set of PPC theories exists, with many (wide) open questions, and many interesting known results. We will show that there even is a deep PPC theory concerning thin buildings (which are often considered as the trivial examples — think of a triangle as a thin example of a projective plane); there, the underlying coordinate structure will be the "field with one element" $\mathbb{F}_1$, and a whole set of problems and conjectures will turn up in the most unexpected way.

I have omitted a very important part of the theory, namely the topological polygons. The latter are generalized n-gons for which the point set and line set are endowed with a topology, such that projection of elements at distance $n - 1$ is a continuous function. The presence of topologies allows one to study the polygons in a discretized way, and many topological polygons behave in some sense just like finite examples. A beautiful example is a result of Knarr [64] and Kramer [67], which states that for a compact connected generalized n-gon with a point space of finite topological dimension, n necessarily is in $\{3, 4, 6\}$. There is a very rich and deep

PPC theory for topological polygons (and many results are known about topological parameters), and it simply would take me too far to also handle these in the present paper. Chapter 9 of [132] is an excellent starting point. I also refer the reader to [89].

I will touch most of the objects under consideration only very briefly, although at some points, proofs are provided. The main goal is to sketch a playground which is rather small and well-understood in rank one, rich and controlled in dimension at least three, and very mysterious when the rank is two.

2   Some notions from Group Theory

We review some basic notions of Group Theory.

We usually denote a permutation group by (G, X), where G acts on X.

We denote permutation action exponentially and let elements act on the right, such that each element g of G defines a permutation $g : X \to X$ of X and the permutation defined by gh, $g, h \in G$, is given by

$$gh : X \to X : x \mapsto (x^g)^h. \tag{2.1}$$

We denote the identity element of a group often by id or $\mathbf{1}$; a group G without its identity is denoted $G^{\times}$.

Let G be a group, and let $g, h \in G$. The *conjugate* of g by h is $g^h = h^{-1}gh$. The *commutator* of g and h is equal to

$$[g, h] = g^{-1}h^{-1}gh. \tag{2.2}$$

Note that the commutator map

$$\psi : G \times G \mapsto G : (g, h) \mapsto [g, h] \tag{2.3}$$

is not symmetrical; as $[g, h]^{-1} = [h, g]$, we have that $[g, h] = [h, g]$ if and only if $[g, h]$ is an involution. The *commutator* of two subsets A and B of a group G is the subgroup $[A, B]$ generated by all elements $[a, b]$, with $a \in A$ and $b \in B$. The *commutator subgroup* of G is $[G, G] = G'$. Two subgroups A and B *centralize* each other if $[A, B] = \{\mathbf{1}\}$. The subgroup A *normalizes* B if $B^a = B$, for all $a \in A$, which is equivalent with $[A, B] \leq B$.

Inductively, we define the *nth central derivative* $L_{n+1}(G) = [G, G]_{[n]}$ of G as $[G, [G, G]_{[n-1]}]$, and the *nth normal derivative* $[G, G]_{(n)}$ as $[[G, G]_{(n-1)}, [G, G]_{(n-1)}]$. For $n = 0$, the 0th central and normal derivative are by definition equal to G itself. The series

$$L_1(G), L_2(G), \ldots \tag{2.4}$$

is called the *lower central series* of G. If, for some natural number n, $[G, G]_{(n)} = \{\mathbf{1}\}$, and $[G, G]_{(n-1)} \neq \{\mathbf{1}\}$, then we say that G is *solvable (soluble)* of length n. If $[G, G]_{[n]} = \{\mathbf{1}\}$ and $[G, G]_{[n-1]} \neq \{\mathbf{1}\}$, then we say that G is *nilpotent* of class n.

A group G is called *perfect* if $G = [G, G] = G'$.

The *center* of a group is the set of elements that commute with every other element, i.e.,

$$Z(G) = \{z \in G|[z,g] = 1, \forall g \in G\}. \tag{2.5}$$

Clearly, if a group G is nilpotent of class n, then the $(n-1)$th central derivative is a nontrivial subgroup of $Z(G)$.

For a prime number p, a *p-group* is a group of order p^n, for some natural number $n \neq 0$. A *Sylow p-subgroup* of a finite group G is a p-subgroup of some order p^n such that p^{n+1} does not divide $|G|$. Let π be a set of primes dividing $|G|$ for a finite group G. Then a *π-subgroup* is a subgroup of which the set of prime divisors is π. The following result is basic.

Theorem 2.1 ([46], Chapter 1) *A finite group is nilpotent if and only if it is the direct product of its Sylow subgroups.*

A *Hall π-subgroup* of a finite group G, where $\pi \subseteq \pi(G)$, and $\pi(G)$ is the set of primes dividing $|G|$, is a subgroup of size $\prod_{p \in \pi} p^{n_p}$, where p^{n_p} denotes the largest power of p that divides $|G|$.

Theorem 2.2 (Hall's Theorem [46], Chapter 6) *let G be a finite solvable group and π a set of primes. Then*

(a) *G possesses a Hall π-subgroup;*

(b) *G acts transitively on its Hall π-subgroups by conjugation;*

(c) *Any π-subgroup of G is contained in some Hall π-subgroup.*

Let p and q be primes. A *pq-group* is a group of order $p^a q^b$ for some natural numbers a and b. A classical result of Burnside states the following.

Theorem 2.3 ([46], Chapter 4) *A pq-group is solvable.*

Suppose (G, X) is a permutation group which satisfies the following properties:

(1) G acts transitively but not sharply transitively on X;

(2) there is no nontrivial element of G with more than one fixed point in X.

Then (G, X) is a *Frobenius group* (or G is a *Frobenius group* in its action on X). Define $N \subseteq G$ by:

$$N = \{g \in G|f(g) = 0\} \cup \{1\}, \tag{2.6}$$

where $f(g)$ is the number of fixed points of g in X. Then N is called the *Frobenius kernel* of G (or of (G, X)), and we have the following well-known result.

Theorem 2.4 (Theorem of Frobenius [46], Chapter 2) *N is a normal regular subgroup of G.*

Let R be a finite group. The *Frattini group* $\phi(R)$ of R is the intersection of all proper maximal subgroups, or is R if R has no such subgroups.

A group is *simple* if it does not contain any proper nontrivial normal subgroups. A group G is *almost simple* if

$$S \leq G \leq \operatorname{Aut}(S), \tag{2.7}$$

with S a simple group and $\operatorname{Aut}(S)$ its automorphism group.

3   Projective spaces

In this section, we want to describe a PPC theory for projective spaces. For this purpose, we need to introduce projective spaces axiomatically. Using the classic Veblen-Young theorem, PPC theory can be handled completely as soon as the rank is at least 3. If the rank is 1, PPC theory follows from the mere definition of the projective line. The obstruction will come from the rank 2 case.

3.1   Axiomatic projective spaces

Consider an axiomatic projective space of rank k, $k \geq -1$. It is defined as follows. An *axiomatic projective space* is a set $\mathscr{P}$ (the set of points), together with a set of subsets of $\mathscr{P}$ (the set of lines) — all of which have at least three elements, and a symmetric incidence relation, satisfying these axioms:

- Each two distinct points p and q are incident with exactly one line.

- AXIOM OF VEBLEN: when L contains a point of the line through p and $q \neq p$ (different from p and q), and of the line through q and $r \neq q$ (different from q and r), it also contains a point on the line through p and r.

- There is a point p and a line L that are disjoint.

The last axiom is there to prevent degenerations.

If the number of points of the space is finite, we speak of a *finite projective space*.

A *subspace* of the projective space is a subset X, such that any line containing two points of X is a subset of X. The full space and the empty space are also considered as subspaces.

The (geometric) *dimension* of the space is said to be n if that is the largest number for which there is a strictly ascending chain of subspaces of the form

$$\emptyset = X_{-1} \subset X_0 \subset \cdots \subset X_n = \mathscr{P}. \tag{3.1}$$

The following result is the classical result of Veblen-Young;

Theorem 3.1 (Veblen-Young [136]) *An axiomatic projective space of dimension at least 3 is isomorphic to a* $\mathbf{PG}(n, \mathbf{R})$ *for some natural number* $n \geq 3$ *and division ring* $\mathbf{R}$.

Here, $\mathbf{PG}(n, \mathbf{R})$ is the n-dimensional projective space over $\mathbf{R}$; it is obtained by considering the left vector space $\mathbf{R}^{n+1}$, and defining a new space

$$(\mathbf{R}^{n+1} \setminus \{0\})/ \sim \qquad (3.2)$$

by imposing the equivalent relation $\sim$, which is just left proportionality. If $\mathbf{R}$ is a finite field $\mathbb{F}_q$, $\mathbf{PG}(n, \mathbb{F}_q)$ is also denoted by $\mathbf{PG}(n, q)$.

In the finite case, the result states that axiomatic projective spaces of dimension at least three are unique.

Corollary 3.2 (PPC for projective spaces of dimension ≥ 3) *A finite axiomatic projective space of dimension at least 3 is Desarguesian, so is isomorphic to a $\mathbf{PG}(n, q)$ for some natural number $n \geq 3$ and finite field $\mathbb{F}_q$.*

For the planar case, the result is not true (in the finite nor infinite case) — projective planes have been constructed, both finite and infinite, which do not arise from vector spaces over division rings. In fact, by the idea of "free construction" (explained further in the paper for all generalized polygons), there is no hope that classification is feasible. In the finite case, the same could be said. But still, one type of general classification — and perhaps the only one — is exactly the development of PPC theory.

3.2 Affine spaces

Let $\mathbf{P}$ be an n-dimensional projective space, and Π be any hyperplane. The geometry which $\mathbf{P}$ induces on the point set of $\mathbf{P} \setminus \Pi$ is an n-dimensional *affine space*. (One can also approach affine spaces axiomatically — we will encounter such axioms for dimension 2 later on. An axiomatic affine space can then be "projectively completed" to a projective space by adjoining a hyperplane "at infinity".)

3.3 Collineations of spaces

An *automorphism* or *collineation* of a projective space is an incidence and type preserving bijection (so point set, line set, etc. are preserved) of the set of subspaces to itself.

It can be shown that any automorphism of a finite dimensional space $\mathbf{PG}(n, \mathbf{R})$, $n \geq 2$, necessarily has the following form:

$$\theta : \mathbf{x}^T \mapsto A(\mathbf{x}^\sigma)^T, \qquad (3.3)$$

where $A \in \mathbf{GL}_{n+1}(\mathbf{R})$, σ is an automorphism of $\mathbf{R}$, the homogeneous coordinate $\mathbf{x} = (x_0, x_1, \dots, x_n)$ represents a point of the space (which is determined up to a scalar), and $\mathbf{x}^\sigma = (x_0^\sigma, x_1^\sigma, \dots, x_n^\sigma)$ (recall that x_i^σ is the image of x_i under σ).

The set of automorphisms of a projective space naturally forms a group, and in case of $\mathbf{PG}(n, \mathbf{R})$, $n \geq 3$, this group is denoted by $\mathbf{P\Gamma L}_{n+1}(\mathbf{R})$. The normal subgroup of $\mathbf{P\Gamma L}_{n+1}(\mathbf{R})$ which consists of all automorphisms for which the companion automorphism σ is the identity, is the *projective general linear group*, and denoted by $\mathbf{PGL}_{n+1}(\mathbf{R})$. So

$$\mathbf{PGL}_{n+1}(\mathbf{R}) = \mathbf{GL}_{n+1}(\mathbf{R})/Z(\mathbf{GL}_{n+1}(\mathbf{R})), \qquad (3.4)$$

where $Z(\mathbf{GL}_{n+1}(\mathbf{R}))$ is the central subgroup of all scalar matrices of $\mathbf{GL}_{n+1}(\mathbf{R})$. Similarly one defines

$$\mathbf{PSL}_{n+1}(\mathbf{R}) = \mathbf{SL}_{n+1}(\mathbf{R})/Z(\mathbf{SL}_{n+1}(\mathbf{R})), \qquad (3.5)$$

where $Z(\mathbf{SL}_{n+1}(\mathbf{R}))$ is the central subgroup of all scalar matrices of $\mathbf{SL}_{n+1}(\mathbf{R})$ with unit determinant.

For notions such as "translations", "homologies", etc. we refer the reader to any standard textbook.

Now let $\mathbf{P}$ be an n-dimensional (axiomatic) projective space, Π be any hyperplane, and $\mathbf{A}$ the corresponding affine space. It can be shown that any automorphism of $\mathbf{A}$ is induced by an automorphism of $\mathbf{P}$ that stabilizes Π. (In the notation of each of the groups defined above for projective spaces, one replaces "$\mathbf{P}$" by "$\mathbf{A}$" and lowers the dimension by one. So, for example, consider $\mathbf{PG}(2,q)$, and construct $\mathbf{AG}(2,q)$ by leaving out the line L, together with its points. Then $\mathbf{AGL}_2(q) \cong \mathbf{PGL}_3(q)_L$ is the general linear automorphism group of $\mathbf{AG}(2,q)$.)

3.4 Reconstructing the space from its group

For having a better understanding of the concept of a general building, we want to say some words on how the geometry of the space is completely encapsulated in the automorphism group. (The contrast with the planar case couldn't be bigger!)

Let $\mathbf{P}$ be a finite $(n\text{-})$dimensional projective space over some division ring $\mathbf{R}$. Consider any $\mathbf{R}$-base $\mathbf{B}$. Define a simplicial complex (in the next section to be formally defined, and called "chamber") $\mathscr{C} \equiv \mathscr{C}(\mathbf{B})$, by letting it be the union of all possible subspaces of $\mathbf{P}$ generated by subsets of $\mathbf{B}$. (Let it also contain the empty set.) Define a "flag" or "apartment" in $\mathscr{C}$ as a maximal chain (so of length $n + 1$) of subspaces in $\mathscr{C}$. Let F be such a fixed flag.

Consider the special projective linear group $K := \mathbf{PSL}_{n+1}(\mathbf{R})$ of $\mathbf{P}$. Then note that K acts transitively on the pairs $(\mathscr{C}(\mathbf{B}'), F')$, where $\mathbf{B}'$ is any $\mathbf{R}$-base and F' is a flag in $\mathscr{C}(\mathbf{B}')$.

Put $B = K_{\mathscr{C}}$ and $N = K_F$; then note the following properties:

- $\langle B, N \rangle = K$;

- put $H = B \cap N \lhd N$ and $N/H = W$; then W obviously is isomorphic to the symmetric group $\mathbf{S}_{n+1}$ on $n+1$ elements. Note that a presentation of $\mathbf{S}_{n+1}$ is:

$$\langle s_i | s_i^2 = 1, (s_i s_{i+1})^3 = 1, (s_i s_j)^2 = 1, i,j \in \{1,\dots,n+1\}, j \neq i \pm 1 \rangle. \quad (3.6)$$

- $B s_i B w B \subseteq B w B \cup B s_i w B$ whenever $w \in W$ and $i \in \{1,2,\dots,n+1\}$;

- $s_i B s_i \neq B$ for all $i \in \{1,2,\dots,n+1\}$.

Now let $K \cong \mathbf{PSL}_{n+1}(\mathbf{R})$ be as above, and suppose that B and N are groups satisfying these properties. Define a geometry $\mathscr{B}(B,N)$ as follows.

- Its varieties are left cosets in K of the groups $\langle B, B^{s_i} \rangle =: P_i$, $i = 1, \ldots, n + 1$;
- two varieties gP_i and hP_j are incident if they intersect nontrivially.

Then $\mathscr{B}(B, N)$ is isomorphic to $\mathbf{PG}(n, \mathbf{R})$.

3.5 Low dimensional cases

For dimension $n = 1$, our definition of axiomatic space doesn't make much sense. Here we rather *start* from a division ring $\mathbf{R}$, and define $\mathbf{P}$, the *projective line* over $\mathbf{R}$, as being the set $(\mathbf{R}^2 \setminus \{0\})/\sim$, where $\sim$ is defined by (left) proportionality. So we can write

$$\mathbf{P} = \{(0, 1)\} \cup \{(1, \ell) | \ell \in \mathbf{R}\}. \tag{3.7}$$

Now $\mathbf{PSL}_2(\mathbf{R})$ acts naturally on $\mathbf{P}$; in fact, we have defined the projective line as a permutation group equipped with the natural doubly transitive action of $\mathbf{PSL}_2(\mathbf{R})$. Defining a geometry as we did for higher rank projective spaces, through the "(B, N)-pair structure" of $\mathbf{PSL}_2(\mathbf{R})$, one obtains the same notion of projective line.

Restricting to finite fields, we obtain the following very simple

Proposition 3.3 (PPC for projective lines) *A finite projective line has $q + 1$ points, for some prime power q.*

The 2-dimensional case is different, still. Here, other than in the 1-dimensional case, one obtains a nontrivial geometry; the axioms now boil down to just demanding that each two different points are incident with precisely one line, that, dually, any two distinct lines intersect in precisely one point, and that there exists a nonincident point-line pair ("anti-flag"). So we need not require additional algebraic structure in order to have interesting objects. Here we cannot say much about the order of the plane a priori.

We will come back to this issue in much more detail in a later section (§10).

3.6 Representation by diagram

We represent the presentation of $\mathbf{S}_{n+1}$ as above in the following way (this will be explained in more detail in the next section):

$\mathbf{A}_{n+1}$: ●————●————●——— $\cdots$ ———●————● $\qquad (n \geq 0)$

(The number of vertices is $n + 1$ — each vertex corresponding to an involution in the generating set of involutions — and we have an edge between vertices s_i and s_j if and only if $|j - i| = 1$.)

4 Buildings and BN-pairs

Recall that a (combinatorial) *simplicial complex* is a pair $(\mathscr{S}, Y)$, where Y is a set and $\mathscr{S} \subseteq 2^Y$, such that $Y \in \mathscr{S}$ and

$$U \subseteq V \in \mathscr{S} \Longrightarrow U \in \mathscr{S}. \tag{4.1}$$

We are ready to introduce buildings. We will not provide each result with a specific reference — rather, we refer the reader to [7].

4.1 Combinatorial definition

A *chamber geometry* is a geometry $\Gamma = (\mathscr{C}_1, \mathscr{C}_2, \ldots, \mathscr{C}_j, \mathbf{I})$ of rank j (so Γ has j different kinds of varieties and $\mathbf{I}$ is an incidence relation between the elements such that no two elements belonging to the same $\mathscr{C}_i$, $1 \leq i \leq j$, can be incident) so that the simplicial complex $(\mathscr{C}, X)$, where $\mathscr{C} = \cup_{i=1}^j \mathscr{C}_i$ and $S \subseteq \mathscr{C}$ is contained in X if and only if every two distinct elements of S are incident, is a chamber complex (as in, e.g., [132]). A *building* $(\mathscr{C}, X)$ is a thick chamber geometry $(\mathscr{C}_1, \mathscr{C}_2, \ldots, \mathscr{C}_j, \mathbf{I})$ of rank j, where $\mathscr{C} = \cup_{i=1}^j \mathscr{C}_i$, together with a set $\mathscr{A}$ of thin chamber subgeometries, so that:

(i) every two chambers are contained in some element of $\mathscr{A}$;

(ii) for every two elements Σ and Σ' of $\mathscr{A}$ and every two simplices F and F', contained in both Σ and Σ', there exists an isomorphism $\Sigma \mapsto \Sigma'$ which fixes all elements of both F and F'.

If all elements of $\mathscr{A}$ are finite, then the building is called *spherical*. Let us mention for the sake of completeness, that there also is a very subtle graph theoretical definition of the notion of buildings which was observed by E. E. Shult, see [93], and which also works for infinite rank. Later in this section, we will introduce buildings as a certain type of colored graph.

4.2 Coxeter groups and systems

We need to introduce the notions of "Coxeter system" and "Coxeter diagram".

4.2.1 Coxeter groups

A *Coxeter group* is a group with a presentation of type

$$\langle s_1, s_2, \ldots, s_n | (s_i s_j)^{m_{ij}} = 1 \rangle, \tag{4.2}$$

where $m_{ii} = 1$ for all i, $m_{ij} \geq 2$ for $i \neq j$, and i, j are natural numbers bounded above by the natural number n. If $m_{ij} = \infty$, no relation of the form $(s_i s_j)^{m_{ij}}$ is imposed. All generators in this presentation are involutions. The natural number n is the *rank* of the Coxeter group.

A *Coxeter system* is a pair (W, S), where W is a Coxeter group and S the set of generators defined by the presentation. Different Coxeter systems can give rise to the same Coxeter group, even if the rank is different.

Recall that a *dihedral group* of *rank* n, denoted $\mathbf{D}_n$, is the symmetry group of a regular n-gon in the real plane.

4.2.2 Coxeter matrices A square $n \times n$-matrix $(M)_{ij}$ is a *Coxeter matrix* if it is symmetric and defined over $\mathbb{Z} \cup \{\infty\}$, has only 1s on the diagonal, and if $m_{ij} \geq 2$ if $i \neq j$. Starting from a Coxeter matrix $(M)_{ij}$, one can define a Coxeter group $\langle s_1, s_2, \ldots, s_n | (s_i s_j)^{m_{ij}} = \mathbf{1} \rangle$, and conversely.

4.2.3 Coxeter diagrams Let (W, S) be a Coxeter system. Define a graph, called "Coxeter diagram", as follows. Its vertices are the elements of S. If $m_{ij} = 3$, we draw a single edge between s_i and s_j; if $m_{ij} = 4$, a double edge, and if $m_{ij} \geq 5$, we draw a single edge with label m_{ij}. If $m_{ij} = 2$, nothing is drawn. If the Coxeter diagram is connected, we call (W, S) *irreducible*. If it has a finite number of vertices, we call (W, S) *spherical*.

The irreducible, spherical Coxeter diagrams were classified by H. S. M. Coxeter [23]; the complete list is the following.

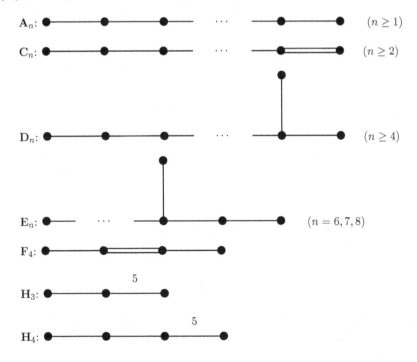

$$\mathbf{I}_2(m): \overset{m}{\bullet\!\!\!-\!\!\!-\!\!\!-\!\!\!-\!\!\!-\!\!\!\bullet} \qquad (m \geq 5)$$

4.3 BN-Pairs and buildings

A group G is said to have a *BN-pair* (B, N), where B, N are subgroups of G, if:

(BN1) $\langle B, N \rangle = G$;

(BN2) $H = B \cap N \lhd N$ and $N/H = W$ is a Coxeter group with distinct generators $s_1, s_2, \ldots, s_n$ (at some stages in this paper the value $n = \infty$ is allowed);

(BN3) $Bs_i B w B \subseteq B w B \cup B s_i w B$ whenever $w \in W$ and $i \in \{1, 2, \ldots, n\}$;

(BN4) $s_i B s_i \neq B$ for all $i \in \{1, 2, \ldots, n\}$.

The group B, respectively W, is a *Borel subgroup*, respectively the *Weyl group*, of G. The natural number n is called the *rank* of the BN-pair. If $|W| < \infty$, the BN-pair is *spherical*. It is *irreducible* if the corresponding Coxeter system is. Sometimes we call (G, B, N) also a *Tits system*.

4.3.1 Buildings as group coset geometries
To each Tits system (G, B, N) one can associate a building $\mathscr{B}_{G,B,N}$ in a natural way, through a group coset construction. For that reason we introduce the standard *parabolic subgroups* $P_i := \langle B, B^{s_i} \rangle$ for $i = 1, 2, \ldots, n$.

- VARIETIES: (or "subspaces") are elements of the left coset spaces G/P_i, $i = 1, 2, \ldots, n$.

- INCIDENCE: gP_i is incident with hP_j, $i \neq j$, if these cosets intersect nontrivially.

The building $\mathscr{B}_{G,B,N}$ is *spherical* when the BN-pair (B, N) is; note that this is in accordance with the aforementioned synthetic definition of "spherical building" (taken that there is already a BN-pair around). It is *irreducible* when (B, N) is irreducible.

4.3.2 G as an automorphism group
The group G acts as a collineation group, by multiplication on the left, on $\mathscr{B}_{G,B,N}$. The kernel K of this action is the biggest normal subgroup of G contained in B, and as such equal to

$$K = \bigcap_{g \in G} B^g. \tag{4.3}$$

The group G/K acts faithfully on $\mathscr{B}_{G,B,N}$ and the stabilizer of the flag $F = \{P_1, P_2, \ldots, P_n\}$ is B/K. If $K = \{1\}$, we say that the Tits system is *effective*.

Let Σ be an apartment of $\mathscr{B}_{G,B,N}$, and let its elementwise stabilizer be S; then NS is the global stabilizer of Σ. We can write

$$S = \bigcap_{w \in W} B^w. \tag{4.4}$$

Theorem 4.1 *Let* (G, B, N) *be a Tits system with Weyl group* W. *Then the geometry* $\mathscr{B}_{G,B,N}$ *is a Tits building. Setting*

$$K = \bigcap_{g \in G} B^g \text{ and } S = \bigcap_{w \in W} B^w, \tag{4.5}$$

we have that G/K *acts naturally and faithfully by left translation on* $\mathscr{B}_{G,B,N}$. *Also,* B *is the stabilizer of a unique flag* F *and* NS *is the stabilizer of a unique apartment containing* F, *and the triple* $(G/K, B/K, NS/K)$ *is a Tits system associated with* $\mathscr{B}_{G,B,N}$. *Moreover,* G/K *acts transitively on the sets* (A, F'), *where* A *is an apartment and* F' *is a maximal flag (chamber) in* A.

The Tits system (G, B, N) is called *saturated* precisely when $N = NS$, with S as above. Replacing N by NS, every Tits system is "equivalent" to a saturated one.

4.3.3 Bruhat decomposition Let G be a group with a spherical, saturated, effective BN-pair (B, N). Then the "Bruhat decomposition" tells us that

$$G = BWB = \coprod_{w \in W} BwB, \tag{4.6}$$

where $W = N/(B \cap N)$ is the Weyl group.

4.3.4 Classification of BN-pairs If the rank of an abstract spherical building is at least 3, Tits showed in a celebrated work [123] that it is always associated to a BN-pair in the way explained above, and this deep observation led him eventually to classify all finite BN-pairs of rank ≥ 3 (cf. [123, 11.7]).

So Tits realized a far reaching generalization of the Veblen-Young theorem for spherical buildings, which roughly could be formulated as follows.

Theorem 4.2 (Classification of spherical buildings) *An irreducible spherical building of rank at least 3 arises from a simple algebraic group (of relative rank at least 3) over an arbitrary division ring.*

4.4 Buildings as colored graphs

For our purpose, a second approach to buildings seems to be in order.

4.4.1 Colored graphs A *graph* is a pair (V, E) where V is a set (of "points" or "vertices") and E is a set of two-element subsets ("edges") of V. It is nothing else than a rank 2 geometry with thin lines. Two vertices lying on the same edge are called *neighbors*. An *edge colored graph* $\Delta = (V, E, \psi)$ is a graph (V, E) together with a surjective map

$$\psi : E \to I, \tag{4.7}$$

where I is the set of "colors"; the *rank* of the edge colored graph Δ is $|I|$. An *isomorphism* between edge colored graphs defined on the same color set is a graph isomorphism preserving colors. A *subgraph* of an edge colored graph Δ is just an ordinary subgraph endowed with the induced edge coloring. For any subset $J \subseteq I$, E_J denotes the edge set of Δ whose color is contained in J. A *J-residue* now is a connected component of the subgraph (V, E_J), and a *panel* is a J-residue for some subset J consisting of a single element of I.

4.4.2 Chamber systems We now rephrase the notion of chamber system in terms of edge colored graphs. A *chamber system* is an edge colored graph $\Delta = (V, E, \psi)$ such that each panel is a complete graph containing at least two vertices. In a chamber system, the vertices are also called *chambers*. If Δ is a chamber system, each chamber is contained in at least one edge of each color, and the color set of each J-residue is J. A chamber system is *thin* if each panel contains precisely two chambers (and then panels and edges become the same thing), and *thick* if every panel contains more than two chambers.

4.4.3 Chamber systems from certain groups Let G be a group generated by a subset $\{s_i | i \in I\}$ consisting of involutions, I being some index set, and put $V = G$. Two elements x and y of V are joined by an edge with color i whenever $x^{-1}y = r_i$. Then (V, E) becomes a thin chamber system with color set I.

4.4.4 Bipartite graphs A *bipartite graph* (V, E) is a graph such that there is a partition of V in two subsets V_1, V_2 such that every edge contains an element of each of these subsets. Let (V, E) be a bipartite graph, put $\Delta = E$, and let $I = \{1, 2\}$. Then Δ becomes an edge colored graph with color set I if two elements u and v of Δ are joined whenever $|u \cap v| = 1$, and giving this edge the color i if the single vertex in $u \cap v$ is in V_i. As such, we obtain a one-to-one correspondence between connected bipartite graphs for which every vertex has degree at least 2, and connected chamber systems of rank 2.

4.4.5 Buildings as edge colored graphs Let (W, S) be a Coxeter system, and $\Pi = (V, E)$ the corresponding Coxeter graph. Then Σ_Π denotes the thin chamber system obtained from W and S as just explained.

Now let $\Pi = (V, E)$ be a Coxeter diagram with vertex set I, and let Σ_Π be as above. A *building* of type Π is a chamber system Δ with color set I containing a collection $\mathscr{A}$ of subgraphs (called "apartments" as before), each isomorphic to Σ_Π, for which

(i) each pair of chambers is contained in some apartment;

(ii) each apartment is convex (every minimal path in Δ beginning and ending in an apartment $A \in \mathscr{A}$ is completely contained in A).

Notions such as "spherical", "rank", "irreducible", etc. are now naturally defined.

We single out the next observation for the sake of convenience.

Proposition 4.3 *The chamber system Σ_Π is a building of type Π. It is the unique thin building of this type.*

For later purposes, especially the functor

$$\mathscr{A} : \text{building of type } \Pi \to \Sigma_\Pi \tag{4.8}$$

will be essential.

4.4.6 Some more words about classification If Π is a Coxeter diagram which is spherical but not irreducible, then every building of type Π is a direct product of irreducible buildings of type Π_i, where $i \in \{1, 2, \ldots, r\}$ and $\Pi_1, \Pi_2, \ldots, \Pi_r$ are the connected components of Π.

Let Δ be a building and for each chamber C of Δ, let $E_2(C)$ be the subgraph consisting of the union of all the rank 2 residues of Δ that contain C. A famous theorem of Tits ([123, Theorem 4.1.2]) states that if Π is an irreducible spherical Coxeter diagram of rank at least 3, and if Δ and Δ' are two thick buildings of type Π, then any isomorphism

$$\alpha : E_2(C) \to E_2(C') \tag{4.9}$$

for some chamber $C \in \Delta$ and $C' \in \Delta'$, extends to an isomorphism from Δ to Δ'.

It turns out that this observation leads to the fact that once an irreducible, spherical (combinatorial) building Δ has rank at least 3, it admits an automorphism group G with a (spherical, irreducible) BN-pair (B, N) such that

$$\mathscr{B}_{G,B,N} \cong \Delta. \tag{4.10}$$

5 Buildings of rank 1

Combinatorially, a building of rank 1 is just a set without any further structure. Therefore one usually requires a set together with a certain group action if one considers rank 1 buildings. As was noted in the previous section, any building of rank at least 3 admits an automorphism group with a BN-pair of the same rank as the building. So a natural approach to rank 1 buildings seems to be to consider sets admitting a permutation group with a BN-pair of rank 1.

First suppose that (G, X) is a doubly transitive (faithful) permutation group, with $|X| > 2$. Let $x, y \in X$ be distinct points. Then with $B = G_x$ and $N = G_{\{x,y\}}$, it is easy to see that (B, N) is a BN-pair of rank 1 — the corresponding Coxeter group is just the group of order 2.

Now consider a group G with a BN-pair (B, N) of rank 1. Then putting $X := \{$left cosets of B in $G\}$, it is clear that G acts doubly transitively on X by left multiplication (as a special case of the Bruhat decomposition — see also [46, Theorem 7.2(ii)]).

So as a first try, we could consider doubly transitive groups.

5.1   The finite doubly transitive groups

Let $\mathscr{G}$ be a finite doubly transitive group. Below, if K is a group, $\mathrm{Aut}(K)$ denotes its automorphism group. Also, q denotes any prime power, unless otherwise specified. Then either I) $\mathscr{G}$ belongs to one of the following classes (the possible 2-transitive permutation representations can be found in [33]; for definitions of the groups, see [22], or also §5.2):

- Symmetric groups $\mathbf{S}_n$ $n \geq 2$:

- Alternating groups $\mathbf{A}_n$, $n \geq 4$;

- Projective special linear groups $\mathbf{PSL}_n(q) \leq \mathscr{G} \leq \mathbf{P\Gamma L}_n(q)$, $n \geq 2$;

- Symplectic groups $\mathbf{Sp}_{2m}(2)$, $m \geq 3$;

- Projective special unitary groups $\mathscr{G}$, with $\mathbf{PSU}_3(q) \leq \mathscr{G} \leq \mathbf{P\Gamma U}_3(q)$;

- Suzuki groups $\mathbf{Sz}(q) \leq \mathscr{G} \leq \mathrm{Aut}(\mathbf{Sz}(q))$ $(q = 2^{2e+1}, e \geq 1)$;

- Ree groups $\mathbf{R}(q) \leq \mathscr{G} \leq \mathrm{Aut}(\mathbf{R}(q))$ $(q = 3^{2e+1}, e \geq 0)$;

- Mathieu groups $\mathbf{M}_{11}, \mathbf{M}_{12}, \mathbf{M}_{23}$ and $\mathbf{M}_{24}$;

- the Mathieu group $\mathbf{M}_{22} \leq \mathscr{G} \leq \mathrm{Aut}(\mathbf{M}_{22})$;

- The Higman-Sims group $\mathbf{HS}$;

- The Conway group $\mathbf{CO}_3$;

or (II) $\mathscr{G}$ has a regular normal subgroup $\mathscr{N}$ which is elementary abelian of order $m = p^d$, where p is a prime. One can identify $\mathscr{G}$ with a group of affine transformations

$$x \mapsto \sigma(x) + c \tag{5.1}$$

of $\mathbb{F}_{p^d}$. One usually calls this case the "affine case".

Theorem 5.1 (PPC for finite doubly transitive groups) *The parameter of a rank 1 building associated to a doubly transitive group is related to a prime power.*■

This approach lacks several properties which one wants to incorporate in a "good" theory for rank 1 buildings. The most important one seems to be the fact that finite BN-pairs arise which are not split; as Tits showed, BN-pairs of rank at least 3 are automatically split (both in the finite and spherical case). And as we will see later, finite BN-pairs of rank 2 also are split. It turns out that, especially in the infinite case, the splitness condition yields a theory which is more interesting (less wild). In the infinite case, it seems a hopeless task to classify doubly transitive permutation groups (even for the finite case, one needs the full strength of CFSG (Classification of Finite Simple Groups) to end up with a classification); one of the reasons (also in the finite case) is the fact that we do not have the right geometric modules at hand which help to understand the groups. Also, the classification of (finite and infinite) split BN-pairs of rank 2 (= Moufang polygons) becomes easier once a good theory for (finite and infinite) split BN-pairs of rank 1 (= Moufang sets, cf. the next section) is available, cf. Fong and Seitz [39, 40] for the finite case, and Tits and Weiss [131] for the general case.

(As another minor setback, the symmetric and alternating groups occur, and those certainly do not have the geometric structure we want (here). We will see later that they appear as the natural Chevalley groups associated to projective geometries defined over the "field with one element". In some sense, they appear there as groups with maximal geometric rank, contrary to the fact that here they would have minimal geometric rank.)

5.2 Moufang sets

Nowadays, split BN-pairs of rank 1 are usually called "Moufang sets", following Tits [129]. Call a permutation group (G, X) $(|X| > 2)$ a *Moufang set* if for each $x \in X$ there is a normal subgroup U_x of G_x which acts sharply transitively on $X \setminus \{x\}$. Sometimes we also write $(X, (U_x)_{x \in X})$ for (G, X).

The groups U_x will be called *root groups*. The elements of U_x are often called *root elations*. If U_x is abelian for some $x \in X$, then U_x is abelian for all $x \in X$ and we call the Moufang set a *translation Moufang set*. (Later, we will also meet translation buildings of higher rank, cf. §10.1 and §14.)

The group S generated by the U_x is called the *little projective group* of the Moufang set. A permutation of X that normalizes the set of subgroups $\{U_y | y \in X\}$, is called an *automorphism* of the Moufang set. The set of all automorphisms of the Moufang set is a group G, called the *full projective group* of the Moufang set. Any group H, with $S \leq H \leq G$, is called a *projective group* of the Moufang set. We have the following easy lemma — see e.g. [104] for a proof.

Lemma 5.2   (i) *The little projective group S of a Moufang set $(X, (U_x)_{x \in X})$ acts doubly transitively on the set X.*

(ii) *A permutation group H (acting on X) is a projective group if and only if $U_x \trianglelefteq H_x$, for every $x \in X$.*

Finite Moufang sets were classified in the 1970's by work of Shult [92] and Hering, Kantor and Seitz [49]. It will be clear from Theorem 5.3 and the discussion following it, that a satisfying geometric theory is underlying buildings of rank 1 (although we only highlight the finite case here).

Theorem 5.3 (Classification of finite rank 1 buildings [92, 49]) *Let $\mathcal{M} = (X, (U_x)_{x \in X})$ be a finite Moufang set with little projective group S and full projective group G. Then one of the following cases occurs.*

(2T) *S acts sharply doubly transitively on X, there is a prime number p and a positive integer n such that $|X| = p^n$, and S contains a normal sharply transitive subgroup N, which is an elementary abelian p-group.*

(Ch) *S is a Chevalley group, there exists a prime number p and a positive integer n such that $|X| = p^n + 1$, and U_x is a p-group of nilpotency class at most 3. We have one of the following cases:*

— *$S \cong \mathbf{PSL}_2(p^n)$, $p^n \geq 4$, is simple and X is the projective line $\mathbf{PG}(1, p^n)$;*

— *n is a multiple of 3, $S \cong \mathbf{PSU}_3(p^{n/3})$, $p^n \geq 27$, is simple and X is the Hermitian unital $\mathcal{U}_H(p^{n/3})$;*

— *$p = 2$, $n = 2n'$ is even, n' is odd, $S \cong \mathbf{Sz}(2^{n'})$, $n' \geq 3$, is simple and X is the Suzuki-Tits ovoid $\mathcal{O}_{ST}(2^{n'})$;*

— *$p = 3$, $n = 3n'$, n' is odd, $S \cong \mathbf{R}(3^{n'})$, $n' \geq 1$, is simple for $n' \geq 3$, and X is the Ree-Tits unital $\mathcal{U}_R(3^{n'})$; if $n' = 1$, then $\mathbf{R}(3) \cong \mathbf{P\Gamma L}_2(8)$ has a simple subgroup of index 3, which coincides with the commutator subgroup of S.*

In all cases, we have that G is the full automorphism group of S. Also, the root groups are precisely the Sylow p-subgroups of S.

It should be noted that in the papers [92, 49], the term "split BN-pair of rank 1" is used instead of "Moufang set".

Concerning the sharply doubly transitive case, it follows from Frobenius' Theorem that there is a normal sharply transitive subgroup N acting on X. The group U_x, for arbitrary $x \in X$, clearly acts transitively on $N^\times$ by conjugation, hence all elements in N have the same order and N is a p-group. Since every p-group has a nontrivial center, and since U_x acts transitively on $N^\times$, we see that N must be elementary abelian. It follows that $|X|$ is a prime power. The standard example here acts on a finite field $\mathbb{F}_q$, for some prime power q, and the actions are given by

$$\mathbb{F}_q \to \mathbb{F}_q : x \mapsto ax + b, \qquad (5.2)$$

with $a \in \mathbb{F}_q \setminus \{0\}$ and $b \in \mathbb{F}_q$. The normal sharply transitive subgroup consists of the maps with $a = 1$.

A second class of examples is given by a slight modification of the previous example. If q is odd and a perfect square, then we retain the previous maps with a a square in $\mathbb{F}_q \setminus \{0\}$, and substitute the other maps with the maps

$$\mathbb{F}_q \to \mathbb{F}_q : x \mapsto ax^{\sqrt{q}} + b, \qquad (5.3)$$

with a a nonsquare in $\mathbb{F}_q$. We call it the *nonstandard example related to* $\mathbb{F}_q$.

We now turn to the examples under (Ch), the *Chevalley type groups*.

As a general remark, we would like to point out that in the following descriptions we are primarily interested in the finite case, but we choose the notation in such a way that also the general (infinite) case is covered, disregarding nevertheless noncommutative fields.

5.2.1 The case $\mathbf{PSL}_2(q)$ This is the prototype of all Moufang sets. Let $\mathbb{K}$ be any (finite) field, and denote by $\mathbf{PG}(1, \mathbb{K})$ the projective line over $\mathbb{K}$. So we may identify $\mathbf{PG}(1, \mathbb{K})$ with the set of all 1-dimensional subspaces of a given 2-dimensional vector space $\mathbb{K} \times \mathbb{K}$. The elements of $\mathbf{PG}(1, \mathbb{K})$ can be written as (a, b), with $a, b \in \mathbb{K}$, $(a, b) \neq (0, 0)$ and (a, b) identified with (ca, cb) for all $c \in \mathbb{K}^\times$. We set $O = (0, 1)$ and $\infty = (1, 0)$. We define U_O as the (multiplicative) group of matrices $(k)_O :=$ $\begin{pmatrix} 1 & k \\ 0 & 1 \end{pmatrix}$, $k \in \mathbb{K}$, U_∞ consists of the matrices $(k)_\infty := \begin{pmatrix} 1 & 0 \\ k & 1 \end{pmatrix}$, for all $k \in \mathbb{K}$, and the action of a matrix M on an element (a, b) is by right multiplication $(a, b)M$ (conceiving (a, b) as a (2×1)-matrix).

In such a way, we have a Moufang set acting on $\mathbf{PG}(1, \mathbb{K})$ [104]. Since U_∞ is abelian, we have a translation Moufang set. We denote this Moufang set by $\mathscr{M}(\mathbf{PG}(1, \mathbb{K}))$. The little projective group is equal to $\mathbf{PSL}_2(\mathbb{K})$ and is simple if $|\mathbb{K}| \geq 4$. For $|\mathbb{K}| = 2$, we have the unique Moufang set on 3 points, and for $|\mathbb{K}| = 3$, we obtain the unique Moufang set on 4 points. Both are improper Moufang sets, isomorphic to the standard examples related to $\mathbb{F}_3$ and $\mathbb{F}_4$, respectively.

5.2.2 The case $\mathbf{PSU}_3(q)$ Let $\mathbb{K}$ be a field having some quadratic Galois extension $\mathbb{F}$. This means that $\mathbb{F}$ admits some field involution σ and the elements of $\mathbb{F}$ fixed by σ are precisely those of the subfield $\mathbb{K}$. In order to treat all characteristics at the same time, we introduce the following notation.

For an arbitrary $i \in \mathbb{F} \setminus \mathbb{K}$, we can write any element $k \in \mathbb{F}$ as

$$\frac{k^\sigma i - k i^\sigma}{i - i^\sigma} + \frac{k - k^\sigma}{i - i^\sigma} i, \qquad (5.4)$$

with both $(ki^\sigma - k^\sigma i)(i^\sigma - i)^{-1}$ and $(k - k^\sigma)(i - i^\sigma)^{-1}$ in $\mathbb{K}$. So $\mathbb{F}$ can be regarded as a 2-dimensional vector space over $\mathbb{K}$ and the map $\mathbb{F} \to \mathbb{K} : x \mapsto x + x^\sigma$ is $\mathbb{K}$-linear, nontrivial, hence surjective with 1-dimensional kernel. We denote the inverse image of the element $k \in \mathbb{K}$ by $\mathbb{K}^{(k)}$.

Now we define our set X. It is the set of all projective points (x_0, x_1, x_2), up to a multiplicative nonzero factor, of the projective plane $\mathbf{PG}(2, \mathbb{F})$ satisfying the equation $x_0 x_2^\sigma + x_0^\sigma x_2 = x_1 x_1^\sigma$. We can write

$$X = \{(1, 0, 0)\} \cup \{(k, x, 1) | x \in \mathbb{F}, k \in \mathbb{K}^{(xx^\sigma)}\}, \qquad (5.5)$$

and also

$$X = \{(0,0,1)\} \cup \{(1,x,k)|x \in \mathbb{F}, k \in \mathbb{K}^{(xx^\sigma)}\}. \tag{5.6}$$

We set $O = \mathbb{F}(0,0,1)$ and $\infty = \mathbb{F}(1,0,0)$. The group of collineations $U_\infty = \{(x,k)_\infty|x \in \mathbb{F}, k \in \mathbb{K}^{(xx^\sigma)}\}$, where $(x,k)_\infty$ is the collineation of $\mathbf{PG}(2,\mathbb{F})$ defined by the linear transformation with matrix

$$\begin{pmatrix} 1 & 0 & 0 \\ x^\sigma & 1 & 0 \\ k & x & 1 \end{pmatrix}, \tag{5.7}$$

acts sharply transitively on $X \backslash \{\infty\}$ (on the right!) and fixes ∞. Likewise, the group $U_O = \{(x,k)_O|x \in \mathbb{F}, k \in \mathbb{K}^{(xx^\sigma)}\}$, where $(x,k)_O$ is the collineation of $\mathbf{PG}(2,\mathbb{F})$ defined by the linear transformation with matrix

$$\begin{pmatrix} 1 & x & k \\ 0 & 1 & x^\sigma \\ 0 & 0 & 1 \end{pmatrix}, \tag{5.8}$$

fixes O and acts sharply transitively on $X \backslash \{O\}$.

We obtain a Moufang set, which we denote by $\mathscr{M}(\mathbf{H}(2,\mathbb{F},\sigma))$, or when $|\mathbb{K}| = q$, briefly by $\mathscr{M}(\mathbf{H}(2,q^2))$, since in this case σ is uniquely determined and given by $\sigma : x \mapsto x^q$. We call it a *Hermitian Moufang set*.

The little projective group of $\mathscr{M}(\mathbf{H}(2,\mathbb{F},\sigma))$ is the unitary group $\mathbf{PSU}_3(\mathbb{F},\sigma)$, denoted just $\mathbf{PSU}_3(q)$ when $|\mathbb{K}| = q$. It is a simple group whenever $|\mathbb{K}| \geq 3$ and has order $q^3(q^3+1)(q^2-1)/(q+1,3)$. Note that $|X| = q^3+1$ in this case. For $|\mathbb{K}| = 2$, $\mathbf{PSU}_3(2)$ is isomorphic to the sharply doubly transitive nonstandard example related to $\mathbb{F}_9$, and hence is not simple (but solvable).

5.2.3 The case Sz(q) Let $\mathbb{K}$ be a field of characteristic 2, and denote by $\mathbb{K}^2$ its subfield of all squares. Suppose that $\mathbb{K}$ admits some *Tits endomorphism* θ, i.e., the endomorphism θ is such that it maps x^θ to x^2, for all $x \in \mathbb{K}$. If $|\mathbb{K}| = 2^n$, then n must be odd, say $n = 2e+1$, and θ maps x to $x^{2^{e+1}}$, for all $x \in \mathbb{K}$. Let $\mathbb{K}^\theta$ denote the image of $\mathbb{K}$ under θ. In the finite case $\mathbb{K}^\theta = \mathbb{K}$. Let L be a subspace of the vector space $\mathbb{K}$ over $\mathbb{K}^\theta$, such that $\mathbb{K}^\theta \subseteq L$ (this implies that $L \backslash \{0\}$ is closed under taking multiplicative inverse). In the finite case $L = \mathbb{K}$. We also assume that L generates $\mathbb{K}$ as a ring. We now define the *Suzuki-Tits Moufang set* $\mathscr{M}(\mathbf{Sz}(\mathbb{K},L,\theta))$.

Let X be the following set of points of $\mathbf{PG}(3,\mathbb{K})$, where the coordinates with respect to some given basis are:

$$X = \{(1,0,0,0)\} \cup \{(a^{2+\theta} + aa' + a'^\theta, 1, a', a)|a, a' \in L\}. \tag{5.9}$$

One now calculates that

$$X = \{(0,1,0,0)\} \cup \{(1, a^{2+\theta} + aa' + a'^\theta, a, a')|a, a' \in L\}. \tag{5.10}$$

We set $\infty = (1,0,0,0)$ and $O = (0,1,0,0)$. Let $(x, x')_\infty$ be the collineation of $\mathbf{PG}(3, \mathbb{K})$ determined by

$$(x_0 \ x_1 \ x_2 \ x_3) \mapsto (x_0 \ x_1 \ x_2 \ x_3) \begin{pmatrix} 1 & 0 & 0 & 0 \\ x^{2+\theta} + xx' + x'^\theta & 1 & x' & x \\ x & 0 & 1 & 0 \\ x^{1+\theta} + x' & 0 & x^\theta & 1 \end{pmatrix}, \qquad (5.11)$$

and let $(x, x')_O$ be the collineation of $\mathbf{PG}(3, \mathbb{K})$ determined by

$$(x_0 \ x_1 \ x_2 \ x_3) \mapsto (x_0 \ x_1 \ x_2 \ x_3) \begin{pmatrix} 1 & x^{2+\theta} + xx' + x'^\theta & x & x' \\ 0 & 1 & 0 & 0 \\ 0 & x^{1+\theta} + x' & 1 & x^\theta \\ 0 & x & 0 & 1 \end{pmatrix}. \qquad (5.12)$$

Define the groups

$$U_\infty = \{(x, x')_\infty \,|\, x, x' \in L\} \text{ and } U_O = \{(x, x')_O \,|\, x, x' \in L\}. \qquad (5.13)$$

Both groups U_∞ and U_O act on X, as an easy computation shows (for U_O use the second description of X above), and they act sharply transitively on $X \setminus \{(1,0,0,0)\}$ and $X \setminus \{(0,1,0,0)\}$, respectively.

We obtain a Moufang set $\mathscr{M}(\mathbf{Sz}(\mathbb{K}, L, \theta))$, called a *Suzuki-Tits Moufang set*. The group $\mathbf{Sz}(\mathbb{K}, L, \theta)$ is the *Suzuki group* generated by U_∞ and U_O. If $|\mathbb{K}| = q$, then we denote it by $\mathbf{Sz}(q)$. One has $|\mathbf{Sz}(q)| = q^2(q^2 + 1)(q - 1)$ and $|X| = q^2 + 1$. All Suzuki groups are simple groups unless $|\mathbb{K}| = 2$. The group $\mathbf{Sz}(2)$ is a sharply doubly transitive standard Moufang set related to $\mathbb{F}_5$.

In order to understand better the structure of U_∞, we can identify each point $(a^{2+\theta} + aa' + a'^\theta, 1, a', a)$ of $X \setminus \{\infty\}$ with the ordered pair (a, a'). Then the unique element of U_∞ that maps $(0,0)$ to (b, b') is given by

$$(b, b')_\infty : (a, a') \mapsto (a + b, a' + b' + ab^\theta). \qquad (5.14)$$

The root group U_∞ is hence given abstractly by the set $\{(a, a')_\infty \,|\, a, a' \in L\}$ with operation $(a, a')_\infty \oplus (b, b')_\infty = (a + b, a' + b' + ab^\theta)_\infty$.

Remark The Moufang set $\mathscr{M}(\mathbf{Sz}(\mathbb{K}, L, \theta))$ can also be defined as the set of absolute points of a polarity in a certain generalized quadrangle.

5.2.4 The case $\mathbf{R}(q)$ Let $\mathbb{K}$ be a field of characteristic 3, and denote by $\mathbb{K}^3$ its subfield of all third powers. In the finite case, $\mathbb{K}^3$ is just $\mathbb{K}$. Suppose that $\mathbb{K}$ admits some Tits endomorphism θ, i.e., the endomorphism θ is such that it maps x^θ to x^3, for all $x \in \mathbb{K}$. In the finite case, $|\mathbb{K}|$ is a power of 3 with odd exponent, say $|\mathbb{K}| = 3^{2e+1}$, and θ maps x to $x^{3^{e+1}}$. Let $\mathbb{K}^\theta$ denote the image of $\mathbb{K}$ under θ. In the finite case, again necessarily $\mathbb{K}^\theta = \mathbb{K}$.

We now define the *Ree-Tits Moufang set* $\mathscr{M}(\mathbf{R}(\mathbb{K}, \theta))$.

For $a, a', a'' \in \mathbb{K}$, we put

$$f_1(a, a', a'') = -a^{4+2\theta} - aa''^{\theta} + a^{1+\theta}a'^{\theta} + a''^2 + a'^{1+\theta} - a'a^{3+\theta} - a^2a'^2,$$
$$f_2(a, a', a'') = -a^{3+\theta} + a'^{\theta} - aa'' + a^2a',$$
$$f_3(a, a', a'') = -a^{3+2\theta} - a''^{\theta} + a^{\theta}a'^{\theta} + a'a'' + aa'^2.$$

Let X be the following set of points of $\mathbf{PG}(6, \mathbb{K})$, where the coordinates with respect to some given basis are:

$$X = \{(1, 0, 0, 0, 0, 0, 0)\} \cup \tag{5.15}$$
$$\{(f_1(a, a', a''), -a', -a, -a'', 1, f_2(a, a', a''), f_3(a, a', a''))|a, a', a'' \in \mathbb{K}\}.$$

A tedious calculation shows that

$$X = \{(0, 0, 0, 0, 1, 0, 0)\} \cup \tag{5.16}$$
$$\{(1, f_2(a, a', a''), f_3(a, a', a''), a'', f_1(a, a', a''), -a', -a)|a, a', a'' \in \mathbb{K}\}.$$

We set $\infty = (1, 0, 0, 0, 0, 0, 0)$ and $O = (0, 0, 0, 0, 1, 0, 0)$. Let $(x, x', x'')_\infty$ be the collineation of $\mathbf{PG}(6, \mathbb{K})$ determined by

$$(x_0 \ x_1 \ x_2 \ x_3 \ x_4 \ x_5 \ x_6) =: \overline{\mathbf{x}} \mapsto \tag{5.17}$$

$$\overline{\mathbf{x}} \begin{pmatrix} 1 & 0 & 0 & 0 & 0 & 0 & 0 \\ p & 1 & 0 & -x & 0 & x^2 & -x'' - xx' \\ q & x^\theta & 1 & x' - x^{1+\theta} & 0 & r & s \\ x'' & 0 & 0 & 1 & 0 & x & -x' \\ f_1(x, x', x'') & -x' & -x & -x'' & 1 & f_2(x, x', x'') & f_3(x, x', x'') \\ x' - x^{1+\theta} & 0 & 0 & 0 & 0 & 1 & -x^\theta \\ x & 0 & 0 & 0 & 0 & 0 & 1 \end{pmatrix},$$

with

$$p = x^{3+\theta} - x'^\theta - xx'' - x^2x',$$
$$q = x''^\theta + x^\theta x'^\theta + x'x'' - xx'^2 - x^{2+\theta}x' - x^{1+\theta}x'' - x^{3+2\theta},$$
$$r = x'' - xx' + x^{2+\theta},$$
$$s = x'^2 - x^{1+\theta}x' - x^\theta x'',$$

and let, with the same notation, $(x, x', x'')_O$ be the collineation of $\mathbf{PG}(6, \mathbb{K})$ determined by

$$(x_0 \ x_1 \ x_2 \ x_3 \ x_4 \ x_5 \ x_6) =: \overline{\mathbf{x}} \mapsto \tag{5.18}$$

$$\overline{\mathbf{x}} \begin{pmatrix} 1 & f_2(x, x', x'') & f_3(x, x', x'') & x'' & f_1(x, x', x'') & -x' & -x \\ 0 & 1 & -x^\theta & 0 & x' - x^{1+\theta} & 0 & 0 \\ 0 & 0 & 1 & 0 & x & 0 & 0 \\ 0 & -x & x' & 1 & -x'' & 0 & 0 \\ 0 & 0 & 0 & 0 & 1 & 0 & 0 \\ 0 & x^2 & -x'' - xx' & x & p & 1 & 0 \\ 0 & r & s & x^{1+\theta} - x' & q & x^\theta & 1 \end{pmatrix}.$$

It may be noted for the computations that $(x, x', x'')_O = (x, x', x'')_\infty^g$, with g the collineation of $\mathbf{PG}(6, \mathbb{K})$ determined by

$$(x_0, x_1, x_2, x_3, x_4, x_5, x_6) \mapsto (x_4, x_5, x_6, -x_3, x_0, x_1, x_2). \qquad (5.19)$$

Define the groups

$$U_\infty = \{(x, x', x'')_\infty | x, x', x'' \in \mathbb{K}\} \text{ and } U_O = \{(x, x', x'')_O | x, x', x'' \in \mathbb{K}\}. \qquad (5.20)$$

The groups U_∞ and U_O both act on X, as an easy computation shows (for U_O use the second description of X above), and they act sharply transitively on $X \setminus \{(1, 0, 0, 0, 0, 0, 0)\}$ and $X \setminus \{(0, 0, 0, 0, 1, 0, 0)\}$, respectively.

We obtain a Moufang set, which we denote by $\mathscr{M}(\mathbf{R}(\mathbb{K}, \theta))$ and call a *Ree-Tits Moufang set*. Its little projective group, generated by U_∞ and U_O, is the *Ree group* $\mathbf{R}(\mathbb{K}, \theta)$. If $|\mathbb{K}| = q$, then we denote the corresponding Ree group by $\mathbf{R}(q)$. In this case we have $|\mathbf{R}(q)| = q^3(q^3 + 1)(q - 1)$. All Ree groups are simple groups unless $q = 3$, in which case $\mathbf{R}(3) \cong \mathbf{P\Gamma L}_2(8)$. Hence $[\mathbf{R}(3), \mathbf{R}(3)] \cong \mathbf{PSL}_2(8)$ is simple, but not sharply doubly transitive, because it has order 504.

Following Section 7.7.7 of [132], we can now define U_∞ abstractly as follows. Define

$$(a, a', a'') := (f_1(a, a', a''), -a', -a, -a'', 1, f_2(a, a', a''), f_3(a, a', a'')). \qquad (5.21)$$

Then the unique element of U_∞ that maps $(0, 0, 0)$ to (x, x', x'') is given by

$$(x, x', x'')_\infty : (a, a', a'') \mapsto (a + x, a' + x' + ax^\theta, a'' + x'' + ax' - a'x - ax^{1+\theta}). \qquad (5.22)$$

The root group U_∞ can now be defined as the set $\{(a, a', a'')_\infty | a, a', a'' \in \mathbb{K}\}$ with operation

$$(a, a', a'')_\infty \oplus (b, b', b'')_\infty = (a + b, a' + b' + ab^\theta, a'' + b'' + ab' - a'b - ab^{1+\theta})_\infty. \qquad (5.23)$$

Remark The Moufang set $\mathscr{M}(\mathbf{R}(\mathbb{K}, \theta))$ can also be defined as the set of absolute points of a polarity in a certain generalized hexagon, see 7.7 of [132].

Corollary 5.4 (PPC for BN-pairs of rank 1) *The parameter of a rank 1 building is a prime power or a prime power minus 1.* ∎

5.3 General Moufang sets

It is certainly not the goal to go into the details of the theory of general Moufang sets. Rather, we refer the reader to the excellent and detailed lecture notes of De Medts and Segev [26], and the references therein.

6   Generalized polygons

When the rank of a finite building $\mathscr{B}$ is 2, it is, in general, not possible to associate to $\mathscr{B}$ a BN-pair in a "natural way"; this is because when the type of $\mathscr{B}$ is $\mathbf{I}_{2(3)}$ or $\mathbf{B}_2$, nonclassical examples exist that do not admit Chevalley groups as flag-transitive automorphism groups. Even more so, there exist such examples not admitting flag-transitive, and even point-transitive automorphism groups; we will encounter examples further in the paper. For the types $\mathbf{I}_{2(3)} = \mathbf{G}_2$ and $^2\mathbf{F}_4 = \mathbf{I}_{2(8)}$ (which are the only other types possible — see the result of Feit-Higman quoted below), no such examples are known. So in the rank 2 case, one wants to *endow* the building with a BN-pair structure in order to be able to classify.

Conjecture 6.1 (Tits [123]) *If a finite building Δ of irreducible type and rank 2 is such that* $\mathrm{Aut}(\Delta)$ *permutes transitively the pairs consisting of a chamber and an apartment containing it (that is, if Δ is associated with a BN-pair), then Δ is isomorphic to the building of an absolutely simple group over a finite field, or with the building of a Ree group of type $^2\mathbf{F}_4$ over a finite field.*

Using the list of finite simple groups, F. Buekenhout and H. Van Maldeghem answered Tits's question affirmatively in [9].

Theorem 6.2 ([9]) *If a thick finite generalized polygon Δ admits a BN-pair, then Δ is isomorphic to the building of an absolutely simple group over a finite field, or to the building of a Ree group of type $^2\mathbf{F}_4$ over a finite field.*

However, obviously Tits had a classification free proof in mind. The classification of finite BN-pairs of rank 2 in automorphism groups of generalized 3-gons (projective planes) is a classical result — we will come back to it later in much more detail. Only finite polygons are considered in this context; we will see later — notably in §9 — that when finiteness is not presumed, generalized n-gons admitting a group with a BN-pair exist for *any* $n \geq 4$, with the additional property that they are not classical (in the sense that they do not satisfy the *Moufang property* — which we will consider below — contrary to any other spherical building of rank more than 2).

6.1   Combinatorial definition

Combinatorially, a *generalized n-gon* $(n \geq 3)$ is a point-line geometry $\Gamma = (\mathscr{P}, \mathscr{B}, \mathbf{I})$ for which the following axioms are satisfied:

(i) Γ contains no ordinary k-gon (as a subgeometry), for $2 \leq k < n$;

(ii) any two elements $x, y \in \mathscr{P} \cup \mathscr{B}$ are contained in some ordinary n-gon (as a subgeometry) in Γ;

(iii) there exists an ordinary $(n+1)$-gon (as a subgeometry) in Γ.

A *generalized polygon* is a generalized n-gon for some n.

By (iii), generalized polygons have at least three points per line and three lines per point. Note that (i) and (ii) reflect the essence of (GP_n) of the introduction, and note also that the generalized 3-gons are precisely the projective planes. A geometry Γ which satisfies (i) and (ii) is a *weak generalized n-gon*. If (iii) is not satisfied for Γ, then Γ is called *thin*. Otherwise, it is called *thick*. Sometimes we will speak of "thick (respectively thin) generalized n-gon" instead of "thick (respectively thin) weak generalized n-gon".

The relation between buildings and generalized polygons, as observed by Tits in [129], is now as follows:

(S) *Suppose* $(\mathscr{C}, X)$, $\mathscr{C} = \mathscr{C}_1 \cup \mathscr{C}_2$, *is a spherical building of rank 2. Then* $\Gamma = (\mathscr{C}_1, \mathscr{C}_2, \mathbf{I})$ *is a generalized polygon. Conversely, suppose that* $\Gamma = (\mathscr{P}, \mathscr{B}, \mathbf{I})$ *is a generalized polygon, and let* $\mathscr{F}$ *be the set of its flags. Then* $(\mathscr{P} \cup \mathscr{B}, \emptyset \cup \{\{v\} | v \in \mathscr{P} \cup \mathscr{B}\} \cup \mathscr{F})$ *is a chamber geometry of rank 2. Declaring the thin chamber geometry corresponding to any ordinary subpolygon an apartment, we obtain a spherical building of rank 2.*

So, in view of (D1), generalized polygons truly are the essential particles of buildings, as was already emphasized by (GP_n) in the precursorial introduction of buildings.

6.2 Parameters of GPs

The following proposition is easy to prove (cf. [132, 1.5.3]).

Proposition 6.3 (Order of polygons) *For each thick generalized n-gon* Γ, $n \geq 3$, *there are (not necessarily finite) constants s and t so that each point is incident with $t + 1$ lines and each line is incident with $s + 1$ points.*

We then say that Γ has *order* or *parameters* (s, t). We also use these concepts for any rank 2 geometry.

6.3 Feit-Higman

A famous result of Feit and Higman states that the possible n-values in the notion of generalized n-gon are severely restricted when considering the finite case.

Theorem 6.4 (Feit and Higman [36]) *Let* Γ *be a finite weak generalized n-gon of order* (s, t). *Then either* $(s, t) = (1, 1)$ *(that is, Γ is an ordinary n-gon), or*

$$n \in \{3, 4, 6, 8, 12\}. \tag{6.1}$$

A finite thick generalized n-gon only exists if and only if $n \in \{3, 4, 6, 8\}$; *also, st is a perfect square if $n = 6$ and 2st is a perfect square if $n = 8$.*

Some other arithmetic information is known about the parameters, but we will encounter this in due course (e.g., in §11 and §16). By the Feit-Higman result, the (thick) finite generalized polygons are the combinatorial geometries associated to the following Coxeter graphs.

A_2:

B_2:

$$m$$

$I_2(m)$: $(m \in \{6, 8\})$

6.4 Duality

There is a *point-line duality* for GPs of order (s, t) for which in any definition or theorem the words "point" and "line" are interchanged and also the parameters. If $\Gamma = (\mathscr{P}, \mathscr{B}, \mathbf{I})$ is a GP of order (s, t), its dual

$$\Gamma^D = (\mathscr{B}, \mathscr{P}, \mathbf{I}) \tag{6.2}$$

is a GP of order (t, s).

6.5 Collinearity and concurrency

Suppose $\Gamma = (\mathscr{P}, \mathscr{B}, \mathbf{I})$ is a GP. Let x and y be (not necessarily distinct) points of the GP Γ; we write $x \sim y$ and call these points *collinear*, provided that there is some line L such that $x \mathbf{I} L \mathbf{I} y$. Dually, for $L, M \in \mathscr{B}$, we write $L \sim M$ when L and M are *concurrent*. For $X \in \mathscr{P} \cup \mathscr{B}$, put

$$X^\perp = \{Y \in \mathscr{P} \cup \mathscr{B} | X \sim Y\}. \tag{6.3}$$

By this definition we have that $X \in X^\perp$. For a set $\emptyset \neq A \subseteq \mathscr{P}$ or $\mathscr{B}$, we define $A^\perp$ as being $\cap_{X \in A} X^\perp$; also, $A^{\perp\perp} := (A^\perp)^\perp$. The latter notion only makes real sense for generalized n-gons with $n = 4$ due to (i) in the combinatorial definition mentioned above. (For instance, when $n = 3$, $A^\perp = A^{\perp\perp} = \mathscr{P}$ (or $\mathscr{B}$) for all A. If $n \geq 5$ and A contains noncollinear points then $|A^\perp| \leq 1$, etc.)

6.6 Automorphisms

An *automorphism* of a generalized polygon $\Gamma = (\mathscr{P}, \mathscr{B}, \mathbf{I})$ is a bijection of $\mathscr{P} \cup \mathscr{B}$ which preserves $\mathscr{P}, \mathscr{B}$ and incidence. The full set of automorphisms of a GP forms a group in a natural way — the *automorphism group* of Γ, denoted $\mathrm{Aut}(\Gamma)$. If B is an automorphism group of a generalized polygon $\Gamma = (\mathscr{P}, \mathscr{B}, \mathbf{I})$, and R is a subset of $\mathscr{P}$, $B_{[R]}$ is the subgroup of B fixing R pointwise (in this notation, a line is also considered to be a point set, so that for a line M, $B_{[M]}$ is the subgroup of B fixing M pointwise). Similarly, if $A \subseteq \mathscr{P} \cup \mathscr{B}$, then $B_{[A]}$ is the subgroup of A fixing A elementwise. Elements of $\mathrm{Aut}(\Gamma)_{[x]}$ for the point $x \in \mathscr{P}$ are called *whorls* (about x).

6.7 Polygons as graphs

Let $\mathscr{S} = (\mathscr{P}, \mathscr{B}, \mathbf{I})$ be a generalized n-gon. The *(point-line) incidence graph* (V, E) of $\mathscr{S}$ is defined by taking $V = \mathscr{P} \cup \mathscr{B}$, where an edge is drawn between vertices if the corresponding elements in $\mathscr{S}$ are incident; (V, E) then is a bipartite graph of diameter n and girth $2n$. Vice versa, such graphs define GPs. Note that we already met this graph theoretical viewpoint.

Let the graph corresponding to $\mathscr{S}$ be denoted by Γ. We call $(x_0, \ldots, x_k)$ a *(simple) path* if the x_i are pairwise distinct and x_i is adjacent to x_{i+1} for $i = 0, \ldots, k - 1$. The natural graph theoretic distance function on Γ is denoted by "d" or sometimes "d_n". The set of elements at distance i from some element $x \in \Gamma$ is denoted by $\Gamma_i(x)$. Elements at distance n are called *opposite*.

The following lemma is a special case of [63].

Lemma 6.5 *If Γ is a generalized n-gon and $\alpha \in \mathrm{Aut}(\Gamma)$, there exists some $x \in \Gamma$ with*

$$d(x, \alpha(x)) \geq n - 1. \tag{6.4}$$

Sometimes, we will switch from generalized polygons to their point-line incidence graphs without explicitly saying so; as such, we can use the terminology of distances, etc.

6.8 Generation

We say that a generalized n-gon $\Gamma = (\mathscr{P}, \mathscr{B}, \mathbf{I})$ is *generated* by a subset $A \subseteq \mathscr{P} \cup \mathscr{B}$ if no generalized n-gon properly contained in Γ contains A; Γ is said to be *finitely generated* if it is generated by a finite subset. Similarly, a subpolygon Γ' of Γ is *generated* by $A \subseteq \mathscr{P} \cup \mathscr{B}$ if Γ' is the smallest subpolygon of Γ containing A.

7 The classical examples and their duals

We describe the classical examples of finite thick GPs. By the Feit-Higman Theorem, we only have to consider the cases $n = 3, 4, 6, 8$. Classical polygons are defined by the fact that they are fully embedded in some finite projective space. Recall that a *full embedding* of a rank 2 geometry $\Gamma = (\mathscr{P}, \mathscr{B}, \mathbf{I})$ in a projective space $\mathbf{P}$, is an injection

$$\iota : \mathscr{P} \hookrightarrow \mathscr{P}(\mathbf{P}), \tag{7.1}$$

with $\mathscr{P}(\mathbf{P})$ the point set of $\mathbf{P}$, such that

(E1) $\langle \iota(\mathscr{P}) \rangle = \mathbf{P}$;

(E2) for any line $L \in \mathscr{B}$ (seen as a point set), L^ι is a line of $\mathbf{P}$.

From the point of view of Group Theory, no distinction can be made between a generalized polygon and its point-line dual — they have the same automorphism group.

We list the finite thick classical GPs below, together with some information about their automorphism groups. For details, the reader is referred to [132, Chapters 2-3-4].

7.1 The case $n = 3$

There is only one class of classical examples, being the Desarguesian planes $\mathbf{PG}(2,q)$ coordinatized over a finite field $\mathbb{F}_q$. The automorphism group was already introduced earlier.

7.2 The case $n = 4$

Generalized quadrangles (GQs) represent the richest class of polygons if one is considering classical examples.

7.2.1 Orthogonal quadrangles

Consider a nonsingular quadric $\mathbf{Q}$ of Witt index 2, that is, of projective index 1, in $\mathbf{PG}(3,q)$, $\mathbf{PG}(4,q)$, $\mathbf{PG}(5,q)$, respectively. So the only linear subspaces of the projective space in question lying on $\mathbf{Q}$ are points and lines. The points and lines of the quadric form a generalized quadrangle which is denoted by $\mathbf{Q}(3,q)$, $\mathbf{Q}(4,q)$, $\mathbf{Q}(5,q)$, respectively, and has order $(q,1)$, (q,q), (q,q^2), respectively. As $\mathbf{Q}(3,q)$ is a grid, its structure is trivial.

Recall that $\mathbf{Q}$ has the following canonical form:

(1) $X_0 X_1 + X_2 X_3 = 0$ if $d = 3$;

(2) $X_0^2 + X_1 X_2 + X_3 X_4 = 0$ if $d = 4$;

(3) $f(X_0, X_1) + X_2 X_3 + X_4 X_5 = 0$ if $d = 5$, where f is an irreducible binary quadratic form.

Denote the automorphism group of $\mathbf{Q}(4,q)$ by $\mathbf{P\Gamma O}_5(q)$, and put $\mathbf{PSL}_5(q) \cap \mathbf{P\Gamma O}_5(q) =: \mathbf{PSO}_5(q) =: \mathbf{O}_5(q)$. The automorphism group of $\mathbf{Q}(5,q)$ is $\mathbf{P\Gamma O}_6(q)$, and $\mathbf{PSL}_6(q) \cap \mathbf{P\Gamma O}_6(q) =: \mathbf{PSO}_6(q) =: \mathbf{O}_6^-(q)$.

7.2.2 Hermitian quadrangles

Next, let $\mathbf{H}$ be a nonsingular Hermitian variety in $\mathbf{PG}(3,q^2)$. The points and lines of $\mathbf{H}$ form a generalized quadrangle $\mathbf{H}(3,q^2)$, which has order (q^2, q). Now let $\mathbf{H}$ be a nonsingular Hermitian variety in $\mathbf{PG}(4,q^2)$. The points and lines of $\mathbf{H}$ form a generalized quadrangle $\mathbf{H}(4,q^2)$ of order (q^2, q^3).

The variety $\mathbf{H}$ has the following canonical form:

$$X_0^{q+1} + X_1^{q+1} + \cdots + X_d^{q+1} = 0. \tag{7.2}$$

Denote the automorphism group of $\mathbf{H}(3,q^2)$ by $\mathbf{P\Gamma U}_4(q)$, and put $\mathbf{PSL}_4(q) \cap \mathbf{P\Gamma U}_4(q) =: \mathbf{PSU}_4(q) =: \mathbf{U}_4(q)$. The automorphism group of $\mathbf{H}(4,q^2)$ is $\mathbf{P\Gamma U}_5(q)$, and $\mathbf{PSL}_5(q) \cap \mathbf{P\Gamma U}_5(q) =: \mathbf{PSU}_5(q) =: \mathbf{U}_5(q)$.

7.2.3 Symplectic quadrangles The points of $\mathbf{PG}(3,q)$ together with the totally isotropic lines with respect to a symplectic polarity, form a GQ $\mathbf{W}(q)$ of order (q,q).

A symplectic polarity Θ of $\mathbf{PG}(3,q)$ has the following canonical form:

$$X_0Y_3 + X_1Y_2 - X_2Y_1 - X_3Y_0. \tag{7.3}$$

The automorphism group of $\mathbf{W}(q)$ is $\mathbf{P\Gamma Sp}_4(q)$, while $\mathbf{PSL}_4(q) \cap \mathbf{P\Gamma Sp}_4(q) =:$ $\mathbf{Sp}_4(q) =: \mathbf{S}_4(q)$.

7.3 The case $n = 6$

Let $\mathbb{K}$ be a field, and consider $\mathbf{PG}(d, \mathbb{K})$, with $d \geq 2$. Choose a basis, a line L, and two distinct points x and y on L with coordinates $(x_0, x_1, \ldots, x_d)$ and $(y_0, y_1, \ldots, y_d)$. The $\binom{d+1}{2}$-tuple $(p_{ij})_{0 \leq i \leq j \leq d}$, where

$$p_{ij} = \begin{vmatrix} x_i & x_j \\ y_i & y_j \end{vmatrix} = x_i y_j - x_j y_i, \tag{7.4}$$

is up to a nonzero scalar multiple, independent of the choice of points x and y of L. So L defines a unique point $p_L = (p_{ij})_{0 \leq i \leq j \leq d}$ of $\mathbf{PG}((d+2)(d-1)/2, \mathbb{K})$. The coordinates of p_L are the *Grassmann coordinates* of L. (For $d = 3$, these points lie on a quadric called the *Klein quadric*, and then the Grassmann coordinates are also called "Plücker coordinates".)

7.3.1 Split Cayley hexagons Let $\mathbb{K} = \mathbb{F}_q$ be a finite field, and consider the parabolic quadric $\mathbf{Q}(6,q)$ in $\mathbf{PG}(6,q)$ with equation

$$X_0X_4 + X_1X_5 + X_2X_6 = X_3^2. \tag{7.5}$$

Consider the lines of $\mathbf{Q}(6,q)$ with Grassmann coordinates

$$\begin{aligned} p_{12} = p_{34} \quad p_{54} = p_{32} \quad p_{20} = p_{35} \\ p_{65} = p_{30} \quad p_{01} = p_{36} \quad p_{46} = p_{31}. \end{aligned} \tag{7.6}$$

Then the points of $\mathbf{Q}(6,q)$ together with these lines define a generalized hexagon $\mathbf{H}(q)$ of order q, called the *split Cayley hexagon*. The automorphism group of $\mathbf{H}(q)$ is induced by collineations of $\mathbf{PG}(6,q)$; moreover, $\mathrm{Aut}(\mathbf{H}(q)) \cong \mathbf{G}_2(q) \rtimes \mathrm{Aut}(\mathbb{F}_q)$, while $\mathbf{PSL}_7(q)_{\mathbf{H}(q)} \cong \mathbf{G}_2(q)$.

7.3.2 Twisted triality hexagons The absolute points and absolute lines of a triality of the hyperbolic quadric $\mathbf{Q}^+(7, q^3)$ in $\mathbf{PG}(7, q^3)$ form a generalized hexagon $\mathbf{T}(q^3, q)$ of order (q^3, q) (q any prime power), called the *twisted triality hexagon*. Its automorphism group is induced by automorphisms of $\mathbf{PG}(7, q^3)$, and $\mathrm{Aut}(\mathbf{T}(q^3, q)) \cong {}^3\mathbf{D}_4 \rtimes \mathrm{Aut}(\mathbb{F}_q)$, while $\mathbf{PSL}_8(q)_{\mathbf{T}(q^3,q)} \cong {}^3\mathbf{D}_4(q)$.

The split Cayley hexagons and the twisted triality hexagons are the only known finite hexagons up to duality.

7.4 The case $n = 8$

Only one class of classical octagons exists up to duality.

The Ree-Tits octagons   We do not aim at precisely defining the Ree-Tits octagons. The reader is referred to [132] for a detailed account. We only give a very short indirect description. Let $\mathcal{M}$ be a finite *metasymplectic space*, that is, a building of type $\mathbf{F}_4$, of which the planes are defined over the finite field $\mathbb{F}_q$, q a power of 2. This building corresponds to the following Coxeter diagram:

$\mathbf{F}_4$:

Suppose also that $\mathcal{M}$ admits a polarity (see [132, §2.5]) so that q is an odd power of 2. Then the absolute points and lines of this polarity form, together with the natural incidence, a generalized octagon of order (q, q^2) which we denote by $\mathbf{O}(q)$, and which admits a Ree group of type $\mathbf{F}_4$ as an automorphism group which acts transitively on the ordered ordinary 8-gons (that is, which admits a BN-pair structure).

Together with its point-line duals, the Ree-Tits octagons are the only finite generalized octagons known presently.

7.5 Other examples

The only examples known of generalized $(2k + 1)$-gons with $k > 1$, and of generalized $2k$-gons with $k > 4$, are defined through so-called "free constructions", which will be handled in detail in §9. For the gonalities $3, 4$ and 6 explicit constructions are known of nonclassical examples; when $n = 3$ or 4, such constructions are known in the finite case. No nonclassical generalized octagons are known without invoking at some point a free construction. We refer to [132, §3.8] for more details.

8   Split BN-pairs and the Moufang condition

In 1974, Tits published his book [123] containing the classification of all thick spherical buildings of rank at least 3. In an addendum, he introduces the *Moufang condition* for spherical buildings — and thus also for generalized polygons and quadrangles — motivated by his claim that the classification of all Moufang polygons would considerably simplify the classification of spherical buildings of higher rank. Tits started this programme himself already in the sixties, and he soon had a classification of all Moufang hexagons, although he never published this.

In the meantime, J. R. Faulkner studied certain simple groups — Chevalley groups of rank 2 — by means of their so-called "Steinberg representation". He obtained in [34] a wealth of classification results and examples of Moufang hexagons

and quadrangles under the ostensibly stronger hypotheses of these commutation relations (but they follow from the Moufang condition anyway; this was not yet known
to Faulkner, who only derived these relations if there are no involutive root elations).
In fact, Faulkner also showed how one can classify certain types of spherical buildings using his results. But since these results were not complete (characteristic 2
was missing for the quadrangles, for instance), they were not popular.

Independently, Tits worked on his programme, and he was able to classify the
Moufang octagons in [128], and to prove that no Moufang n-gons exist unless $n =
3, 4, 6, 8$ (see [125, 127]). The latter result was also proved by Richard Weiss [137],
who derived it in a more general context in a simpler way. The case $n = 3$ amounts
to projective planes and was treated long before (the terminology stems from this
case). Hence only the case of Moufang quadrangles was missing. However, Tits knew
how to derive the Steinberg relations from the Moufang property, and he considered
this as a first step in the classification. This result was published in 1994, see [130].

In the finite case, Fong and Seitz published two papers [39, 40] in which they
classify finite groups admitting a split BN-pair of rank 2. As a corollary, one also
has a classification of all finite Moufang quadrangles.

Until 1993, Tits only mentions the word "conjecture" when talking and writing
about the list of Moufang quadrangles. He emphasized the fact that no proof was
written down, and that everything was still only in his head. In the meantime,
some more results became available. Thas, Payne and Van Maldeghem proved in
[102] that every finite half Moufang GQ is a Moufang GQ, using typically finite
arguments. A similar approach was used by Van Maldeghem, Thas and Payne in
[135] to show that any finite 3-Moufang GQ is a Moufang GQ (a result that was
generalized by Van Maldeghem and Weiss to arbitrary finite polygons in [134]). In
1998, Van Maldeghem [133] proved that the 2-Moufang condition for thick GQs is
equivalent with the 3-Moufang condition. Hence, at this moment, all equivalences
were proved in the finite case (without invoking the classification of finite simple
groups).

In 1993, Tits started to lecture again about Moufang polygons, with the idea
to finish the classification, write it down (only the case of projective planes and
octagons was published) and publish it. Richard Weiss, who was in the audience,
persuaded Tits to do this jointly, and the result was that in 1997, the classification
of Moufang quadrangles was completed — with an additional new class. At first, the
conjecture of Tits was that all Moufang quadrangles were related to classical groups,
algebraic groups or mixed groups of relative rank 2, and he had an explicit list. The
new example was not only missing in the list, it did not seem to be related to any
classical, algebraic or mixed group. Until Mühlherr and Van Maldeghem [70] show
that it arises as fixed point structure in a certain mixed building of type $\mathbf{F}_4$. This
was somehow overlooked by Tits since this construction process is not captured by
the theory of algebraic groups, but it is a "mixed analogue" of it. The explicit list of
Tits turned out to be incomplete, but his conjecture that all Moufang quadrangles
arise from classical groups, algebraic groups or mixed groups of relative rank 2 was
still true; he simply overlooked that there also exist mixed groups of *relative* rank 2
(he only lists mixed groups of absolute rank 2).

In September 2002, the full classification of Moufang generalized polygons appeared in a book [131].

8.1   Split BN-pairs

Let G be a group with a BN-pair (B, N) of rank 2, and let P_1, P_2 be the two maximal parabolic subgroups containing B. For $i = 1, 2$, let $s_i \in G$ normalize N and P_i, but not P_{3-i}. Put $H = B \cap N$, as before.

A BN-pair (B, N) is called *split* if Property ($\ddagger$) below holds:

($\ddagger$)   There exists a normal nilpotent subgroup U of B such that $B = U(B \cap N)$.

In a celebrated work, P. Fong and G. M. Seitz determined all finite split BN-pairs of rank 2 (the $\mathbf{B}_2$-case being the most complicated type to handle):

Theorem 8.1 (Fong and Seitz [39, 40]) *Let G be a finite group with a split BN-pair of rank 2. Then G is an almost simple group related to one of the following classical Chevalley/Ree groups:*

(1) $\mathbf{PSL}_3(q)$;

(2) $\mathbf{PSp}_4(q) \cong \mathbf{O}_5(q)$;

(3) $\mathbf{PSU}_4(q) \cong \mathbf{O}_6^-(q)$;

(4) $\mathbf{PSU}_5(q)$;

(5) $\mathbf{G}_2(q)$;

(6) $^3\mathbf{D}_4(q)$;

(7) *a Ree group of type* $^2\mathbf{F}_4$.

Equivalently, a thick finite generalized polygon is isomorphic, up to duality, to one of the classical examples if and only if it verifies the Moufang condition (defined in the next section).

8.2   The Moufang condition

Let Γ be a generalized n-gon, $n \geq 3$.

If $G \leq \mathrm{Aut}(\Gamma)$, we denote by $G_{x_0}^{[i]}$ the subgroup of G fixing all elements of $\Gamma_i(x_0)$ and for elements $x_0, \ldots, x_k$, we set

$$G_{x_0, x_1, \ldots, x_k}^{[i]} = G_{x_0}^{[i]} \cap G_{x_1}^{[i]} \cap \ldots \cap G_{x_k}^{[i]}. \tag{8.1}$$

For every simple path $(x_0, \ldots, x_n)$ of length $n + 1$ and every i with $0 \leq i \leq n$, we have $G_{x_0, \ldots, x_n} \cap G_{x_i, x_{i+1}}^{[1]} = \{\mathbf{1}\}$; see [132, 4.4.2(v)].

For $2 \leq k \leq n$, the generalized n-gon Γ is said to be k-*Moufang* with respect to $G \leq \mathrm{Aut}(\Gamma)$ if for each simple k-path $(x_0, \ldots, x_k)$ the group $G_{x_1, \ldots, x_{k-1}}^{[1]}$ acts transitively on the set of $2n$-cycles through $(x_0, \ldots, x_k)$.

Theorem 8.2 ([132], 6.8.2) *If Γ is 4-Moufang with respect to some group G, then Γ is n-Moufang with respect to the same group G.*

If Γ is n-Moufang, we also say that Γ is a *Moufang polygon*, and that Γ satisfies the *Moufang condition*. If G_{x_0,x_1} acts transitively on the set of $2n$-cycles through (x_0, x_1) for all paths (x_0, x_1) (sometimes referred to as the *1-Moufang condition*), then G acts transitively on the set of ordered $2n$-cycles of Γ. This is equivalent to G having a spherical BN-pair of rank 2, which we will see is in general too weak to allow a classification.

We call a generalized n-gon Γ *almost 2-Moufang* with respect to G if for every finite set $A \subseteq \Gamma_1(x_1)$, and any path (x_0, x_1, x_2) the group G_A acts transitively on the $2n$-cycles containing (x_0, x_1, x_2). Similarly, one can define *almost 3-Moufang* for paths (x_0, x_1, x_2, x_3) and finite subsets $A \subseteq \Gamma_1(x_1) \cup \Gamma_1(x_2)$. It was shown in [97] that the 2-Moufang condition implies the Moufang condition for generalized n-gons with $n \leq 6$.

It is not our intention to survey what is known about global Moufang conditions; we refer to [103, 118] for this purpose. We just mention one generalization of the Fong-Seitz theorem in the finite quadrangular case.

Theorem 8.3 ([119]) *A thick finite generalized quadrangle is isomorphic, up to duality, to one of the classical examples if and only if for each point there exists an automorphism group fixing it linewise and acting transitively on the set of its opposite points.*

Later, we will axiomatize generalized quadrangles which resemble this property locally, that is, having a point for which there exists an automorphism group fixing it linewise and acting (sharply) transitively on the set of its opposite points. It appears that almost all finite GQs have such a point (or line), and that a beautiful PPC theory already emerges from this general local condition.

8.3   The power of primes

Inspired by techniques developed initially only for BN-pairs of type $\mathbf{B}_2$, the author proved the following result for general finite BN-pairs of rank 2.

Theorem 8.4 ([112, 115]) *Let G be a finite group with an irreducible effective BN-pair of rank 2, and suppose that the parameters of the associated building are powers of 2. Then the BN-pair is split, hence G is an almost simple group related to one of the classical Chevalley/Ree groups mentioned in Theorem 8.1. Equivalently, a thick finite generalized n-gon for which the associated parameters are powers of 2 is isomorphic, up to duality, to one of the classical examples if and only if $\mathrm{Aut}(\mathscr{S})$ acts transitively on its ordered ordinary n-gons.*

In the odd case similar results are available under extra hypotheses.

The theorem shows how PPC theory can have strong implications for global assumptions. (While usually global assumptions lead to extra information about PPC theory.)

9 Free construction

In this section we will show that classification of generalized polygons is not a feasible problem, due to the existence of so-called "free constructions". The idea is very simple: start with a configuration A of points and lines which does not violate the axioms of generalized n-gon (for given n), and iterate a chain of geometries

$$A_0 = A \subseteq A_1 \subseteq A_2 \subseteq \ldots \tag{9.1}$$

of which the first one is A, and such that the union $\cup_i A_i$ completes A to a generalized n-gon. For all natural $n \geq 3$, it will turn out that in such a way generalized n-gons can be "freely constructed" over any such set A.

Then, we will describe an iterating process taken from Model Theory, and recently described by Tent, by which one can easily construct, for each natural $n \geq 3$, generalized n-gons admitting a BN-pair which are *not* Moufang (so for which the BN-pair is *not* split). Such examples were first found by Tits in [126].

These constructions underline (again) the stark contrast between the finite and the infinite case (since all polygons in the aforementioned constructions are, of course, infinite). The global assumption of BN-pair is not enough to classify generalized n-gons — there is not even a Feit-Higman/Tits-Weiss theorem possible which restricts the values of n. (So when searching for good local properties to develop a reasonable PPC theory, one has to be very careful so as to not end up in a more general local version of the situation given by, say, BN-pairs.)

9.1 Free constructions

For our purpose, it is convenient to pass to the graph theoretic version of generalized polygons (considering the point-line incidence graph); a generalized n-gon is then a bipartite graph with valencies at least 3, diameter n and girth $2n$. (Without the assumption on the valencies, such a graph is a weak n-gon.)

We here construct generalized n-gons for all $n \geq 3$ which are almost 2-Moufang, but not Moufang (and in fact not even almost 3-Moufang). Recall that Moufang generalized n-gons exist only for $n = 3, 4, 6, 8$.

The following well-known construction shows the existence of many generalized n-gons for any $n \geq 3$.

9.1.1 Free n-completion Let Γ_0 be a connected bipartite graph not containing any k-cycles for $k < 2n$. Then we obtain the *free n-completion* of Γ_0 in stages, in the following way:

- at stage $i \geq 1$ we obtain Γ_i from Γ_{i-1} by adding a new path of length $n - 1$ for each pair of elements x, y at distance $n + 1$ in Γ_{i-1}.

Then

$$\Gamma = \bigcup \Gamma_i \qquad (9.2)$$

is called the *free n-completion* of Γ_0 and we say that Γ is *freely generated* over Γ_0. If Γ_0 contains at least two pairs x_1, y_1 and x_2, y_2 of elements with $d(x_1, y_1) = d(x_2, y_2) = n + 1$ in Γ_0 and $d(x_1, x_2)$ is prime to n, then Γ is a generalized n-gon — see [132, 1.3.13].

The following is obvious.

Theorem 9.1 *If $\Gamma_0 \cong \Delta_0$ then also their free n-completions are isomorphic.*

The converse need not hold: the free n-completions of Γ_0 and Γ_i are obviously the same for any $i \in \mathbb{N}$. However, there is a necessary criterion for the free completions of connected bipartite graphs to be isomorphic, which can be stated in terms of the *rank function δ_n*.

For any finite graph $\Gamma = (V, E)$ with vertex set V and edge set E we define

$$\delta_n(\Gamma) = (n - 1)|V| - (n - 2)|E|. \qquad (9.3)$$

We say that the finite graph Γ_0 is *n-strong* in some graph Γ (and we write $\Gamma_0 \leq_n \Gamma$) if $\Gamma_0 \subseteq \Gamma$, and if for each finite graph $A \subseteq \Gamma$ for which $\Gamma_0 \subseteq A$ we have $\delta_n(\Gamma_0) \leq \delta_n(A)$. If $A \leq_n B$ and $C \subseteq B$, then $A \cap C \leq_n C$.

Proposition 9.2 (Tent [98]) *The following are equivalent for a generalized n-gon Γ generated by a finite connected set Γ_0:*

(i) Γ *is the free n-completion of Γ_0;*

(ii) $\Gamma \leq_n \Gamma$.

Corollary 9.3 ([98]) *The class of generalized n-gons freely generated over a finite connected subgraph is countably infinite.* ∎

9.2 Amalgamation

In [98], Tent uses Fraïssé's Amalgamation technique for a first order language to obtain new generalized polygons. Fraïssé's Theorem states the following (see [55, 7.1.2] for a proof, and more details on the terminology):

Theorem 9.4 *Suppose* **L** *is a first-order language and $\mathscr{C}$ is a class of finitely generated* **L**-*structures which is closed under finitely generated substructures, satisfying the following additional properties:*

- JOINT EMBEDDING PROPERTY. *For $\mathscr{A}, \mathscr{B} \in \mathscr{C}$, there is some $\mathscr{E} \in \mathscr{C}$ such that both $\mathscr{A}$ and $\mathscr{B}$ are embeddable in $\mathscr{E}$;*

- AMALGAMATION PROPERTY. *For $\mathscr{A}, \mathscr{B}, \mathscr{E} \in \mathscr{C}$, and embeddings $e : \mathscr{A} \hookrightarrow \mathscr{B}$, $f : \mathscr{A} \hookrightarrow \mathscr{E}$, there is some $\mathscr{D} \in \mathscr{C}$ and embeddings $g : \mathscr{B} \hookrightarrow \mathscr{D}$, $h : \mathscr{E} \hookrightarrow \mathscr{D}$ such that the following diagram commutes*

$$
\begin{array}{ccc}
\mathscr{A} & \xrightarrow{e} & \mathscr{B} \\
\downarrow f & & \downarrow g \\
\mathscr{E} & \xrightarrow{h} & \mathscr{D}
\end{array}
\tag{9.4}
$$

*Then there is a countable **L**-structure $\mathscr{M}$, unique up to isomorphism, satisfying:*

(i) *every finitely generated substructure of $\mathscr{M}$ is isomorphic to an element of $\mathscr{C}$;*

(ii) *every element of $\mathscr{C}$ embeds into $\mathscr{M}$;*

(iii) *if $\mathscr{A} \in \mathscr{C}$, then $\mathrm{Aut}(\mathscr{M})$ acts transitively on the set of substructures of $\mathscr{M}$ isomorphic to $\mathscr{A}$.*

The model $\mathscr{M}$ is also called the *Fraïssé limit* of the class $\mathscr{C}$. Note that Fraïssé limits are existentially closed (that is, if there is some existential sentence which is true in some extension of the structure, then it is already true in that structure).

We now fix a first order language

$$
\mathbf{L} = \{f_k | k \in \mathbb{N}\}
\tag{9.5}
$$

containing binary functions f_k. Any (partial) generalized n-gon becomes an **L**-structure if we interpret these functions as follows:

$$
f_k(x, y) = x_k
\tag{9.6}
$$

if $(x = x_0, \ldots, x_k, \ldots, y)$ is the unique shortest path from x to y. If there is no unique such path, then we let

$$
f_k(x, y) = x.
\tag{9.7}
$$

Notice that edges of the graph can be defined in this language: in a generalized n-gon Γ the pair (x, y) is an edge if and only if

$$
f_1(x, y) = y.
\tag{9.8}
$$

Thus, the axioms of a generalized n-gon are expressible in this language. Clearly, the language **L** has the following property: If Γ is a generalized n-gon and $A \subseteq \Gamma$, then the **L**-substructure $\langle A \rangle$ of Γ generated by A is the same as the (possibly weak) sub n-gon generated by A in Γ.

Let $\mathscr{C}_n$ be the class of all finitely generated **L**-substructures of all free n-completions of finite connected bipartite graphs not containing any k-cycles for $k < 2n$.

Then $\mathscr{C}_n$ is countable by Corollary 9.3 and closed under finitely generated **L**-substructures.

Remark $\mathscr{C}_n$ contains in particular the following structures:

(i) the empty structure (making the Joint Embedding Property a special case of the Amalgamation Property);

(ii) all paths of length at most n, and more generally any "hat-rack", i.e., any path $(x_0, \ldots, x_k)$, $k \leq n$, together with finite subsets of $\Gamma_1(x_i)$, $i = 1, \ldots, k - 1$;

(iii) a $2n$-cycle, and more generally all finite weak n-gons containing at most 2 thick elements (at distance n);

(iv) any finite generalized n-gon;

(v) arbitrarily large finite discrete sets if n is even and a discrete set of order two if n is odd.

In order to show that one may apply Fraïssé's Theorem to our class $\mathscr{C}_n$, it suffices to show that the Amalgamation Property holds for this class (as the empty structure is included in $\mathscr{C}_n$, which makes the Joint Embedding Property a special case of the Amalgamation Property).

Lemma 9.5 (Tent [98]) $\mathscr{C}_n$ *has the Amalgamation Property.*

9.3 BN-Pairs

We are now ready to state the result by Tent about BN-pairs.

Theorem 9.6 (Tent [98]) *For all $n \geq 3$ there is a countable generalized n-gon Γ_n whose automorphism group acts transitively on all finitely generated $\mathbf{L}_n$-substructures of given isomorphism type.*

Corollary 9.7 ([98]) *In particular, $G = \mathrm{Aut}(\Gamma_n)$ has the following transitivity properties.*

(i) *G acts transitively on ordered $2n$-cycles, so G has a BN-pair.*

(ii) *For any $x \in \Gamma_n$, G_x acts highly transitively on $\Gamma_1(x)$, i.e., G_x acts k-transitively for any $k \in \mathbb{N}_0$ on $\Gamma_1(x)$.*

(iii) *G acts transitively on finite weak n-gons of the same cardinality. In particular, Γ_n is almost 2-Moufang.*

(iv) *Let $\gamma = (x_0, x_1, \ldots, x_{2n} = x_0)$ be a $2n$-cycle. Then the pointwise stabilizer of γ acts highly transitively on $\Gamma_1(x_1) \setminus \{x_0, x_2\}$.*

(v) *For any finite set A of vertices of Γ, the elements $\mathrm{Fix}(G_A)$ fixed by G_A form exactly the substructure of Γ_n generated by A.*

(vi) *Γ_n is not almost-3-Moufang.*

In particular, Γ is not Moufang.

10 The prime power conjecture for projective planes

The class of generalized 3-gons coincides with that of projective planes. The prime power conjecture for projective planes states:

Conjecture 10.1 (PPC for planes) *The order of a finite projective plane is a prime power.*

The following seminal theorem is available.

Theorem 10.2 (Bruck and Ryser [8]) *If n is the order of a finite projective plane and $n \equiv 1 \mod 4$ or $n \equiv 2 \mod 4$, then n is the sum of two integer squares.*

Besides the result of Bruck-Ryser, no other restrictions on the parameters are known for *general* axiomatic planes. Usually, one requires further algebraic (group theoretic) conditions.

10.1 Translation planes

A *translation plane* Π is a projective plane which has some line L for which there is an automorphism group T which fixes L pointwise and which acts sharply transitively on the set of points not incident with L. The group T is the *translation group* (corresponding to L), L is a translation line. (As we will see below, we can indeed speak of "the" translation group, since it is unique.) An affine plane for which there is a sharply transitive group on the points that fixes all parallel classes is also called *translation plane*. In that case, the line at infinity is a translation line of the projective completion.

Most of what we recall can be found in the monograph of Knarr [65].

10.2 Finite order

Suppose that the order of the plane is finite. Recall that any translation group is abelian (we will prove this below for general planes). Let z be any point of the finite (projective) translation plane Π, which has order n, where z is supposed to be not incident with the translation line L. Denote the $n+1$ lines on z by $L_0, L_1, \ldots, L_n$, let T denote the translation group, and for any $i \in \{0, 1, \ldots, n\}$, put $T_{L_i} =: L_i$. Let $\mathbf{R}$ be the ring of endomorphisms α of T for which $T_j^\alpha \le T_j$ for every $j \in \{0, 1, \ldots, n\}$. (Addition and scalar multiplication are just the expected ones.) We call $\mathbf{R}$ the *kernel* of the translation plane. A well-known fact is the following:

Proposition 10.3 $\mathbf{R}$ *is a (commutative) field, and T is an $\mathbf{R}$-vector space.*

Corollary 10.4 T *is an elementary abelian p-group for some prime p, so that n is a prime power.* ∎

The fact that T is a vector space allows us to interpret $\mathscr{T} = \{T_i | i = 0, 1, \ldots, n\}$ in the corresponding projective space $\mathbf{PG}(2k - 1, q) = \eta$; here $|\mathbf{R}| = q = p^h$, and $n^2 = q^{2k}$. This interpretation yields the André-Bruck-Bose representation for finite translation planes; $\mathscr{T}$ becomes a $(k-1)$-spread of η, and if one embeds η as a hyperplane in a $\mathbf{PG}(2k, q)$, say ξ, then Π is isomorphic to the projective plane for which the points are the points of $\xi \setminus \eta$, and the lines are the k-subspaces of ξ which meet η in an element of $\mathscr{T}$. Now T becomes the translation group of ξ with axis η, in its usual action, and so it is unique.

10.3   Infinite order

In the infinite case, we can do pretty much the same as in the finite case. Let Π be a (projective) translation plane with translation line U and translation group H. Call a collineation of a plane *axial* if it fixes some line pointwise, and call it *central* if it fixes a point linewise. Now note the following.

Proposition 10.5 *A collineation of a projective plane is axial if and only if it is central.*

(The proof is easy; let α be an axial collineation of some plane ξ, and let M be its *axis* (the line fixed pointwise by α — it is unique if and only if α is not trivial). If u is a point which is not fixed by α, uu^{α} is a fixed line, so any point either is fixed, or incident with a fixed line. If α is not central, then it now easily follows that the entire plane must be elementwise fixed, contradiction.)

(A-C)-Obstruction

As we will later see, similar properties do not hold for generalized quadrangles, not even in the finite case. This obstruction is the essential reason why several problems concerning parameters or classification of quadrangles still are open, or become very hard to handle, while they can be handled smoothly for planes. We call it "(A-C)-obstruction" throughout these notes.

Let H be as above. It easily follows by the observation that if $V \neq U$ is a line, H_V fixes $U \cap V$ linewise, that is, H_V is a group of central collineations with *center* $U \cap V$, and H_V acts sharply transitively on $V \setminus \{U \cap V\}$. Denote the group of central collineations in H with center cIU by $H(c)$. Then obviously we have the following two facts:

- if u, v are distinct points on U, $[H(u), H(v)] = \{1\}$;

- with u and v as above, we have that $\langle H(u), H(v) \rangle = H(u)H(v) = H$. (Written additively, we have that $H = H(u) \oplus H(v)$.)

Combining both properties, it follows that H is abelian.

Now define $\mathbf{R}$ as above; then it can be easily shown that $\mathbf{R}$ is a skew field.

Theorem 10.6 R *is a skew field, and T is a left* **R**-*vector space.*   ∎

In the same way as in the finite case, we have an André-Bruck-Bose construction of Π through the use of spreads, and vice versa.

10.4   Generalized translation planes

Motivated by the satisfactory treatment of PPC theory for translation planes, it makes one wonder whether generalizations come at little cost. For instance, let Π be a finite projective plane, and suppose that E is an automorphism group of Π which fixes some line L, and acts sharply transitively on the points not incident with L. Is it known that the order n of Π is a prime power? The answer is *no*. The strongest result known is probably the following.

Theorem 10.7 (Blokhuis, Jungnickel and Schmidt [5]) *Let G be an abelian automorphism group of order n^2 of a projective plane of order n. Then n is a prime power.*

Under these assumptions either the plane is a translation plane (up to duality), or it is a "plane of type (b)" (up to duality); there are three point orbits, being a singleton $\{x\}$, the n points different from x of some line L incident with x, and the n^2 points off L. More generally, let a permutation group (H, X) be *quasiregular* if H acts sharply transitively on each orbit (modulo the kernel). So in particular, any abelian automorphism group has this property. The following theorem distinguishes eight classes of "large" quasiregular permutation groups acting on finite projective planes.

Theorem 10.8 (Dembowski and Piper [25]) *Let G be a collineation group acting quasiregularly on the points and lines of a projective plane of order n, and assume*

$$|G| > \frac{n^2 + n + 1}{2}. \tag{10.1}$$

Let t denote the number of point orbits and F denote the incidence structure consisting of the fixed points and fixed lines. Then one of the following holds.

(a) $|G| = n^2 + n + 1$, $t = 1$, $F = \emptyset$. *Here G is transitive.*

(b) $|G| = n^2$, $t = 3$, F *is a flag, that is, an incident point-line pair.*

(c) $|G| = n^2$, $t = n + 2$, F *is either a line and all its points or, dually, a point together with all its lines.*

(d) $|G| = n^2 - 1$, $t = 3$, F *is an antiflag — a nonincident point-line pair.*

(e) $|G| = n^2 - \sqrt{n}$, $t = 2$, $F = \emptyset$. *In this case one of the point orbits is precisely the set of points of a Baer subplane (a subplane of order $\sqrt{n}$).*

(f) $|G| = n^2 - n$, $t = 5$, F *consists of two points, the line joining them and another line through one of the two points.*

(g) $|G| = n^2 - 2n + 1$, $t = 7$, F consists of the vertices and sides of a triangle.

(h) $|G| = (n^2 - \sqrt{n} + 1)^2$, $t = 2\sqrt{n} + 1$, $F = \emptyset$. In this case there are $t - 1$ disjoint subplanes of order $\sqrt{n} - 1$ whose point sets constitute $t - 1$ orbits, each of length $n - \sqrt{n} + 1$.

So the planes of Theorem 10.7 fall under (b) and (c) of Theorem 10.8. The planes of (c) are (dual) translation planes.

The paper [44] extensively considers the projective planes described in the latter theorem, with an emphasis on the status of the PPC for the respective classes. We refer to that paper, and also to its companion paper [43], for the details.

10.5 Local versus global dichotomy

A theme that will occur over and over in the present paper is the dichotomy between local and global theory. Till now, we considered groups which act transitively on the points of some affine plane, so that we have a "global action" from the viewpoint of the affine plane. Still, from the viewpoint of its projective completion, we have a "local action" — the group fixes some line. Let us thus review some basic global actions for projective planes. Before doing this, the reader is already noticed for the fact that, at least conjecturally, it will become clear that global theory probably always is only leading to Desarguesian planes (when restricting to finite structures). Any existing conjecture on finite planes with a point-transitive automorphism group states that it is eventually Desarguesian. Still, on the other hand, it makes perfectly clear how difficult the theory really is — many of these conjectures are wide open, even with respect to the mere fact that a simple corollary of such a conjecture is that the PPC would follow.

10.6 Ostrom-Wagner

One of the first results on global theory was the now famous Ostrom-Wagner theorem.

Theorem 10.9 (Ostrom and Wagner [71]) *Let Π be a finite projective plane of order n admitting a doubly transitive automorphism group G on the point set. Then n is a prime power, G contains $\mathbf{PSL}_3(n)$ and Π is Desarguesian.*

The strong global assumption indeed leads to the fact that the order is a prime power, but not at little cost: the plane is coordinatized over a commutative field. So we need to relax the action as much as possible in order to handle more classes of planes. (However, as already mentioned, conjecturally "global" leads to "classical".) A first try is the assumption of "flag-transitivity", which is a natural property lying in between point-transitivity and point-2-transitivity.

Koen Thas

10.7 Flag-transitive and point-primitive planes

A projective plane is called *flag-transitive* if it admits an automorphism group which acts transitively on its flags; such a groups is a *flag-transitive group*.

We start with some "elementary" results (prior to the classification of finite simple groups).

Theorem 10.10 (Higman and McLaughlin [52]) *If q is a prime-power different from 2 and 8, then any flag-transitive collineation group of $\mathbf{PG}(2, q)$ contains all elations of $\mathbf{PG}(2, q)$.*

Theorem 10.11 (Roth [88]) *Suppose either $n^2 + n + 1$ or $n + 1$ is a prime. Then a flag-transitive group G is either doubly-transitive (on points and lines) or it contains a sharply flag-transitive subgroup.*

In the doubly-transitive case, the plane is Desarguesian by the Ostrom-Wagner result. If G is nonsoluble, then it is doubly-transitive; this is a consequence (both for $n^2 + n + 1$ and $n + 1$ a prime) of a result of Burnside which states that a nonsoluble transitive permutation group of prime degree is doubly-transitive. If G is soluble, then a result of Galois is applied, saying that a soluble transitive permutation group of prime degree is soluble if and only it is either regular (= sharply transitive), or a Frobenius group [33].

A group G is called a *geometric ABA-group* [52] if G contains the groups A and B for which $G = ABA$, if $AB \cap BA = A \cup B$, and if $A \not\leq B$ and $B \not\leq A$.

Remark Any group admitting a BN-pair is a geometric ABA-group, due to its Bruhat decomposition. More on this matter can be found in [52], especially §8 of *loc. cit.*

A (Steiner) 2-*design* is a point-line incidence structure, consisting of two nonempty disjoint sets called *points* and *lines*, together with an incidence relation **I**, such that

(1) each point is incident with a constant number (≥ 2) of lines and

(2) each line is incident with a constant number (≥ 2) of points, and

(3) through every two distinct points there is precisely one line.

A *flag-transitive representation* of a group H on a design Γ is a homomorphism of G onto an automorphism group of Γ acting flag-transitively. Higman and McLaughlin proved in [52] that a group G admits a flag-transitive representation on a 2-design if and only if G is a geometric ABA-group (through a very simple coset geometry representation). Such a group always acts *primitively* (i.e. there are no nontrivial blocks of imprimitivity) on the points of the design, which translates to the fact that in an ABA-group A is maximal. Using these observations, the following was proved.

Theorem 10.12 (Higman and McLaughlin [52]) *If* Π *is a finite flag-transitive projective plane of order* n *with* n *odd and* n *not a fourth power, and if* n *is a square of a natural number* m *for which* $m \equiv -1 \mod 4$, *or* n *is not a square and* $n^2 + n + 1$ *is not a prime, then* Π *is Desarguesian.*

McLaughlin noted in Dembowski [24] that, using the ideas of [52], in fact the following is also true.

Theorem 10.13 (Higman and McLaughlin [52, 24]) *If* Π *is a finite flag-transitive projective plane of order* n *with* n *odd and* n *not a fourth power, or* n *is not a square and* $n^2 + n + 1$ *is not a prime, then* Π *is Desarguesian.*

Theorem 10.14 (Ott [73]) *Let* (Π, G) *be a finite flag-transitive projective plane of order* n *(where the notation is obvious). Then the following are equivalent.*

(1) *The size of* G *is odd.*

(2) G *acts sharply flag-transitively.*

(3) G *acts on the points of* Π *as a Frobenius group.*

If one of these conditions hold, then n *is even and* $n^2 + n + 1$ *is a prime number.*

Ott's results are very geometrical in nature, and some strong results have very elementary proofs. We recall some more theorems of Ott in this direction.

Theorem 10.15 (Ott [73]) *Let* (Π, G) *be a finite flag-transitive projective plane of order* n. *Then* n *is the power of a prime or* G *acts sharp-transitively on the flags.*

Theorem 10.16 (Ott [72]) *Let* (Π, G) *be a finite flag-transitive projective plane of order* n. *If* n *is odd, then* n *is the power of a prime.*

So, by Theorems 10.14, 10.15 and 10.16, we have that for a finite flag-transitive projective plane Π of order n admitting a flag-transitive group G, either n is the power of a prime, or n is even, G acts flag-regularly and $n^2 + n + 1$ is a prime number.

A *cyclic projective plane* is a projective plane admitting a cyclic transitive collineation group.

Theorem 10.17 (Fink [37]) *Let* Π *be a finite projective plane of order* n, *which admits a group* G *acting transitively on the flags and which has odd order. Then either* $n \in \{2, 8\}$ *or else* Π *is a nonDesarguesian cyclic plane determined by a difference set* $\mathscr{D}$ *in the cyclic group* $(\mathbb{F}_p, +)$, *where* $p = n^2 + n + 1$ *is prime and* n *even. The set* $\mathscr{D}$ *may be taken to be the set of* n-*th powers in the multiplicative group of* $\mathbb{F}_p$.

Difference sets are defined later in this section.

Note. If Π is the Desarguesian projective plane of order 8, then the Singer group G together with the "multipliers" generate a flag-regular automorphism group, where $\mathscr{D}$ is the set of 8-th powers in the multiplicative group of $\mathbb{F}_{73}$. More details appear below.

Fink also obtained the following result.

Theorem 10.18 (Fink [38]) *Let Π be a finite flag-transitive projective plane of order n where n is not a fourth power. If Π admits a collineation group G such that G is flag-transitive but not regular on flags, then Π is Desarguesian and G contains the little projective group.*

The following theorem is perhaps the strongest result in the study of finite flag-transitive projective planes; the proof provides a heavily group theoretical analysis to obtain a list of the odd degree primitive permutation representations of all non-sporadic nearly simple groups. The classification of finite simple groups is not used to obtain the latter result, but for applications, such as Theorem 10.19 stated below, a tedious case-by-case analysis of the sporadic simple groups is needed.

Theorem 10.19 (Kantor [60]) *Let Π be a finite point-primitive projective plane of order n (that is, suppose there is a group G acting primitively on the points of Π). If n is odd, then Π is Desarguesian. If Π is not Desarguesian, then n is even, $n^2 + n + 1$ is a prime, G is a Frobenius group and $|G|$ divides $(n+1)(n^2+n+1)$ or $n(n^2+n+1)$.*

As a flag-transitive group of a finite projective plane is point-primitive, Theorem 10.19 applies. In that case, if Π is not Desarguesian, $|G| = (n+1)(n^2+n+1)$, and hence the action is regular. The proof of Theorem 10.19 is independent of Theorems 10.14, 10.15 and 10.16; as such, those theorems are completely covered by it. By e.g. Theorem 10.19, G contains a (necessarily unique) normal subgroup of prime order $n^2 + n + 1$ acting sharply transitively on the points of Π.

Theorem 10.20 (Feit [35]) *Let Π be a finite flag-transitive projective plane of order n, and suppose Π is not Desarguesian. Then $n \equiv 0 \mod 8$, n is not the power of a prime, $n^2 + n + 1$ is a prime, and if d is a divisor of n, then $d^{n+1} \equiv 1 \mod n^2 + n + 1$. Furthermore, $q > 14,400,008$.*

To prove that if d is a divisor of n, then $d^{n+1} \equiv 1 \mod n^2 + n + 1$, Feit makes the well-known observation that, under the assumptions of Theorem 10.20,

(*) *the unique group of order $n+1$ of the multiplicative group of $\mathbb{F}_p$, $p = n^2+n+1$, can be taken as a difference set in $(\mathbb{F}_p, +)$,*

see, e.g., Theorem 10.17, and then applies the multiplier theorem of Hall, Jr. [48].

In [35], Feit also made the following number theoretical conjecture:

Conjecture 10.21 (W. Feit) *Let n be an even natural number so that $n^2 + n + 1$ is a prime, and for which $2^{n+1} \equiv 1 \mod n^2 + n + 1$. Suppose also that $n + 1 \equiv 0 \mod 3$. Then n is a power of 2.*

Remark In [35], W. Feit claims that if Π and n are as in Theorem 10.20, then n is not a power of 2, and in [74], U. Ott claims that any flag-transitive finite projective plane has prime power order. Together with the above theorem, these two results would imply the nonexistence of nonDesarguesian flag-transitive finite projective planes. Unfortunately, both proofs appear to contain mistakes: Feit uses a lemma of B. Gordon, W. H. Mills and L. R Welch [45] (in the proof of [35, Theorem A]) which is proved only under much more restrictive hypotheses in [45], and there is a mistake in [74] in deriving [74, Formula (18)] from [74, Formula (17)], as is pointed out in [138].

In the paper of Feit cited above, it is proved that under the assumptions of Theorem 10.20 every divisor d of n must satisfy $d^{n+1} \equiv 1 \pmod{n^2 + n + 1}$, and also that n must be larger than $14,400,008$. An elementary proof of the first assertion is given in a recent paper by the author [107], which also contains a survey of the most important results on finite flag-transitive projective planes since 1961 and some related problems.

10.8   Fermat surfaces and Fermat curves

A general construction of potential examples of finite projective planes, known in the literature as the method of *difference sets*, is as follows. Suppose we have a finite (not necessarily abelian) group F containing a subset $\mathbf{D}$ for which the map

$$\begin{aligned}
\mathbf{D} \times \mathbf{D} \setminus \{\text{diagonal}\} &\to F^{\times} \\
(x, y) &\mapsto xy^{-1}
\end{aligned} \tag{10.2}$$

is bijective, so that $|F| = n^2 + n + 1$, where $|\mathbf{D}| = n + 1$. Then we obtain a finite projective plane $\Pi = \Pi(F, \mathbf{D})$ of order n by taking both the set of points and the set of lines of Π to be the elements of F, with the incidence relation that a point x and a line y are incident if and only if yx^{-1} belongs to $\mathbf{D}$. We will be concerned with the special case of this described by the following proposition, which is essentially a restatement of a result of J. Fink [37].

We call a prime number or prime power *special* if it has the form $q = n^2 + n + 1$ and every element of the finite field $\mathbb{F}_q$ is a difference of two nonzero nth powers. We call a finite projective plane *flag-regular* if it has a group of automorphisms that acts regularly (simply transitively) on the flags.

Proposition 10.22 (Fink [37]) *If $q = n^2+n+1$ is a special prime or prime power with $n > 1$, then $\Pi(\mathbb{F}_q, (\mathbb{F}_q^{\times})^n)$ is a flag-regular finite projective plane. Conversely, if Π is a flag-regular finite projective plane of order n, and if the number $p = n^2+n+1$ is prime, then p is special and $\Pi \cong \Pi(\mathbb{F}_p, (\mathbb{F}_p^{\times})^n)$.*

We also have the following stronger result:

Theorem 10.23 (Thas and Zagier [120]) *Let n be the order of a flag-transitive finite projective plane Π. Then at least one of the following holds:*

(a) n is a prime power and $\Pi \cong \mathbf{PG}(2, \mathbb{F}_n)$;

(b) $p = n^2 + n + 1$ is a special prime and $\Pi \cong \Pi(\mathbb{F}_p, (\mathbb{F}_p^\times)^n)$.

Notice that the two alternatives occurring in the theorem are not necessarily exclusive: it is possible that the number n is both a prime power and is associated to a special prime $p = n^2 + n + 1$, and in this case the projective plane Π of this order, while still unique, has both forms $\mathbf{PG}(2, \mathbb{F}_n)$ and $\Pi \cong \Pi(\mathbb{F}, (\mathbb{F}_p^\times)^n)$. This can happen for only two values of n, namely $n = 2$ $(p = 7)$ and $n = 8$ $(p = 73)$. Let us look in detail at the exceptional case $n = 2$ to see how the isomorphism between the two differently-defined projective plane structures works.

So consider the case $n = 2$. We define an automorphism A of $\mathbf{PG}(2, \mathbb{F}_2)$ of order 7 by

$$A : (x : y : z) \mapsto (y : z : x + y). \tag{10.3}$$

Then every point of $\mathbf{PG}(2, \mathbb{F}_2)$ has the form $p_i = A^i(p_0)$ for a unique $i \in \mathbb{Z}/7\mathbb{Z}$, where p_0 is the point $(1 : 0 : 0)$:

i	0	1	2	3	4	5	6
p_i	$(1:0:0)$	$(0:0:1)$	$(0:1:0)$	$(1:0:1)$	$(0:1:1)$	$(1:1:1)$	$(1:1:0)$

$$\tag{10.4}$$

and every line in $\mathbf{PG}(2, \mathbb{F}_2)$ has the form $L_j = A^j(L_0)$ for a unique $j \in \mathbb{Z}/7\mathbb{Z}$, where L_0 is the line $\{x = 0\}$:

j	0	1	2	3	4	5	6
L_j	$x = 0$	$x = z$	$x + y + z = 0$	$y = z$	$x = y$	$z = 0$	$y = 0$

$$\tag{10.5}$$

Then $L_j = \{p_{j+1}, p_{j+2}, p_{j+4}\}$ for every j, so the correspondence

$$(p_i, L_j) \mapsto (i, j) \tag{10.6}$$

defines an isomorphism between the Desarguesian projective plane {points in $\mathbf{PG}(2, \mathbb{F}_2)$, lines in $\mathbf{PG}(2, \mathbb{F}_2)$, usual incidence} and the special projective plane $\{i \in \mathbb{F}_7, j \in \mathbb{F}_7, i - j \in \mathbf{D}\}$, where $\mathbf{D} = (\mathbb{F}_7^\times)^2 = \langle 2 \rangle = \{1, 2, 4\}$. The automorphism

$$B : (x : y : z) \mapsto (x : y + z : y) \tag{10.7}$$

of $\mathbf{PG}(2, \mathbb{F}_2)$ fixes p_0 and L_0 and sends p_i to p_{2i} and L_j to L_{2j}, and the group of automorphisms generated by A and B, with the relations $A^7 = B^3 = 1$, $BAB^{-1} = A^2$, acts regularly on the flags of $\mathbf{PG}(2, \mathbb{F}_2)$.

Let $n > 8$. We have seen that a nonDesarguesian finite flag-transitive projective plane of order n exists if and only if $p = n^2 + n + 1$ is a special prime, i.e., if and only if p is prime and every element of the finite field $\mathbb{F}_p$ is the difference of two elements of $\mathbf{D} = (\mathbb{F}_p^\times)^n$. In this section we give a number of elementary number-theoretical statements about n and p which are equivalent to this property. All are taken from [120]. These involve the *Fermat surface*

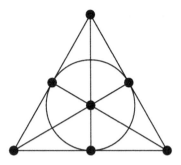

Figure 1: The unique projective plane of order 2 ("Fano plane").

$$\mathbf{S} \; : \; X_0^n + X_1^n \; = \; X_2^n + X_3^n \,, \tag{10.8}$$

the *Fermat curves*

$$\mathscr{F}_\eta \; : \; X_0^n - X_1^n \; = \; \eta X_2^n \qquad (\eta \in \mathbb{F}_p^\times), \tag{10.9}$$

and the *Gaussian periods*

$$\omega \; = \; \sum_{a \in \mathbf{D}} \zeta^a \; = \; \frac{1}{n} \sum_{x \in \mathbb{F}_p^\times} \zeta^{x^n} \,, \qquad \Omega = \sum_{x \in \mathbb{F}} \zeta^{x^n} = 1 + n\omega \,, \tag{10.10}$$

where $\zeta = \zeta_p$ denotes a primitive pth root of unity. All of these are classical objects, much studied in Number Theory. In particular, the Gaussian periods, which are defined for any prime number p and divisor n of $p-1$, generate the unique subfield of degree n of the cyclotomic field $\mathbb{Q}(\zeta)$ and were introduced for essentially this purpose by Gauss.

It appears that the prime $p = n^2 + n + 1$ is special if and only if the Fermat surface $\mathbf{S}$ has no nontrivial $\mathbb{F}_p$-rational points (by "trivial points" of $\mathbf{S}$ over $\mathbb{F}_p$ we mean points $(x_0, x_1, x_2, x_3) \in \mathbf{S}(\mathbb{F}_p)$ with either $x_0 x_1 x_2 x_3 = 0$ or $\{x_0^n, x_1^n\} = \{x_2^n, x_3^n\}$); if and only if the Fermat curves $\mathscr{F}_\eta$ all have the same number of $\mathbb{F}_p$-rational points; and if and only if the absolute value of the Gaussian period ω is the square-root of a rational integer.

We denote by $X(\mathbb{F})$ the set of $\mathbb{F}$-rational points of any variety X defined over a finite field $\mathbb{F}$ and by $|X(\mathbb{F})|$ its cardinality.

Theorem 10.24 ([120]) *Suppose that $p = n^2 + n + 1$ is prime. Then the following are equivalent:*

(a) *p is special;*

(b) *the map $\phi : \mathbf{D} \times \mathbf{D} \smallsetminus (\text{diagonal}) \to \mathbb{F}_p^\times$ sending (x, y) to $x - y$ is bijective;*

(c) *the surface $\mathbf{S}$ has no nontrivial points over $\mathbb{F}_p$;*

(d) *$|\mathscr{F}_\eta(\mathbb{F}_p)| > 3n$ for every $\eta \in \mathbb{F}_p^\times$;*

(e) $|\mathscr{F}_\eta(\mathbb{F}_p)| < 2n^2 + n$ for every $\eta \in \mathbb{F}_p^\times$;

(f) $|\mathscr{F}_\eta(\mathbb{F}_p)| = n^2 + n$ or $n^2 + 2n$ for every $\eta \in \mathbb{F}_p^\times$;

(g) $|\mathbf{S}(\mathbb{F}_p)| < 2n^4 + 5n^3 + 4n$;

(h) $|\mathbf{S}(\mathbb{F}_p)| = 2n^4 + n^3 + 4n^2 + 4n$;

(i) $|\omega|^2 \in \mathbb{Q}$;

(j) $|\omega| = \sqrt{n}$;

(k) $\mathrm{tr}_{\mathbb{Q}(\zeta)/\mathbb{Q}}(|\Omega|^4) = n^8 + n^7 - 2n^4 - 4n^3 - 5n^2 - 3n$.

More results of this kind can be found in [120].

The above considerations were generalized in the following way in [120]. Let p be an arbitrary prime number. Then any subgroup of $\mathbb{F}_p^\times$ has the form

$$\mathbf{D}_n = \{x^n \mid x \in \mathbb{F}_p^\times\} = \{x \in \mathbb{F}_p^\times \mid x^k = 1\} \tag{10.11}$$

for some divisor n of $p-1$ and $k = (p-1)/n$. We define the Gaussian period ω_n as before by

$$\omega_n = \sum_{x \in \mathbf{D}_n} \zeta^x = \frac{1}{n} \sum_{a \in \mathbb{F}_p^\times} \zeta^{a^n} = \mathrm{tr}_{\mathbb{Q}(\zeta)/\mathbb{K}_n}(\zeta), \tag{10.12}$$

where ζ is a primitive pth root of unity and $\mathbb{K}_n$ is the unique subfield of $\mathbb{Q}(\zeta)$ of degree n over $\mathbb{Q}$. We further define $t_n(\eta)$ for $\eta \in \mathbb{F}_p$ as the number of representations of η as the difference of two elements of $\mathbf{D}_n$, and call the pair (p, n) *special* if this number is independent of η for $\eta \neq 0$. Since

$$\sum_{\eta \neq 0} t_n(\eta) = |\mathbf{D}_n|^2 - |\mathbf{D}_n| = k^2 - k, \tag{10.13}$$

this common value must then be equal to $(k-1)/n$, which must therefore be an integer. In particular, except in the trivial case when $k = 1$ an $n = p-1$, one always has $k \geq n+1$ and $p \geq n^2 + n + 1$, so that the case considered before is extremal.

Theorem 10.25 ([120]) *Let $p = nk + 1$ be prime. Then the following are equivalent:*

(i) *the pair (p, n) is special;*

(ii) $t_n(\eta) = (k-1)/n$ *for all $\eta \neq 0$;*

(iii) *the surface $\mathbf{S}$ has precisely $n^3 k + n^2(k-1)^2 + 4nk$ $\mathbb{F}_p$-rational points;*

(iv) $|\omega_n|^2$ *is a rational number;*

(v) $|\omega_n|^2 = k - (k-1)/n$.

Suppose (p, n) is special, $k = (p - 1)/n$. Define a point-line incidence structure Γ as follows:

- the POINTS of Γ are the elements of $\mathbb{F}_p$;

- the LINES or *blocks* of Γ also consist of the elements of $\mathbb{F}_p$; and

- a point $\alpha \in \mathbb{F}_p$ is INCIDENT with a block $\beta \in \mathbb{F}_p$ if and only if $\alpha - \beta \in \mathbf{D}_n$.

Thus there are p points and p blocks, each point is incident with k blocks and each block is incident with k points, any two distinct points are contained in exactly $(k-1)/n$ distinct blocks, and any two distinct blocks intersect in exactly $(k-1)/n$ distinct points. Hence Γ is a 2-$(p, k, (k-1)/n)$ *symmetric block design*. It is clear that for $a \in \mathbf{D}_n$ and $b \in \mathbb{F}_p$ the map

$$x \mapsto ax + b \tag{10.14}$$

from $\mathbb{F}_p$ to itself defines an automorphism of Γ in a natural way and that the group of these automorphisms acts regularly on the flags ($=$ incident point-block pairs) of Γ. The only known examples of such designs other than finite projective spaces (of dimension at least 3) follow from [14, 68, 75]. These constructions are essentially covered by Theorem 10.26 (and Theorem 10.27) of [120] stated below, where among other results the existence results of [14, 68, 75] are re-proved in an alternative fashion.

Theorem 10.26 ([120]) *Let p be a prime and $n | (p - 1)$. Then (p, n) is special in each of the following five cases:*

(i) $n = 1$, p arbitrary, $|\omega_n|^2 = 1$,

(ii) $n = 2$, $p \equiv 3 \pmod 4$, $|\omega_n|^2 = (p + 1)/4$,

(iii) $n = 4$, $p = 4b^2 + 1$ with b odd, $|\omega_n|^2 = (3p + 1)/16$,

(iv) $n = 8$, $p = 64b^2 + 9 = 8d^2 + 1$ with b and d integral, $|\omega_n|^2 = (7p + 1)/64$,

(v) $n = p - 1$, p arbitrary, $|\omega_n|^2 = 1$,

the corresponding values of ω_n being given by

(i) $\omega_1 = -1$,

(ii) $\omega_2 = \dfrac{-1 + i\sqrt{p}}{2}$,

(iii) $\omega_4 = \dfrac{\sqrt{p} - 1}{4} \pm i \sqrt{\dfrac{p + \sqrt{p}}{8}}$,

(iv) $\omega_8 = \dfrac{\sqrt{p} - 1}{8} + \sqrt{\dfrac{p + 3\sqrt{p}}{32}} + i \sqrt{\dfrac{\sqrt{p} - 1}{16}} \sqrt{\sqrt{p} - \sqrt{\dfrac{p + 3\sqrt{p}}{2}}}$.

(v) $\omega_{p-1} = \zeta$.

Cases (i) and (v) of this theorem are trivial and do not lead to interesting designs, but the families (ii), (iii) and (iv) yield three infinite or potentially interesting classes of flag-regular symmetric designs. Note that the family (iii) is quite sparse: the only primes up to 40000 belonging to this class are 5, 37, 101, 197, 677, 2917, 4357, 5477, 8101, 8837, 12101, 15877, 16901, 17957, 21317, 22501, and 28901. Family (iv), corresponding to the prime solutions of a Pell's equation, is even thinner, though conjecturally still infinite: the first prime of this form is our old acquaintance $p = 73$, with $b = 1$ and $d = 3$; the next two are 104411704393 ($b = 40391$, $d = 114243$) and 160459573394847767113 ($b = 1583407981$, $d = 4478554083$), with 12 and 21 digits, respectively, and the next four have 103, 119, 425, and 615 decimal digits, respectively.

Theorem 10.27 ([120]) *Assume that $p = kn+1$ is prime and that (p, n) is special. Then*

(i) *If $n > 1$, then n is even and k is odd.*

(ii) *If $n \leq 8$, then (p, n) belongs to one of the families of Theorem 10.26.*

10.9 Singer groups

Let Γ be any rank r geometry. (The rank is not important here, so one can put $r = 2$ for starters.) A *Singer group* G on the elements of type $i \leq r - 1$ is an automorphism group of Γ which acts sharply transitively on the elements of type i. We require that there is no kernel of the action, that is, no element of $G^\times$ fixes all i-elements. Here, we only consider Singer groups with respect to points (without loss of generality, by duality). We already encountered such groups in various guises: affine translation planes and cyclic projective planes are examples. (In the latter case, one usually speaks of "Singer cycles".) In fact, affine planes admitting a Singer group are precisely the planes of type (b).

For finite projective planes admitting Singer groups, many results are known; see, e.g., [43, 44]. The Main Conjecture is the following:

Conjecture 10.28 *A finite projective plane admitting a Singer group is Desarguesian, and whence satisfies the PPC.*

However, to obtain a solution of this conjecture, one would *first* want to obtain that the order of the plane be a prime power (think also of the discussion about sharply flag-transitive planes in relation to the parameters). Although transitivity is a much weaker condition than doubly transitivity, cf. the Ostrom-Wagner result, the regularity of the action adds a strong extra condition. Still, even for Singer cycles (and hence also for abelian Singer groups), the conjecture remains unsolved. So the next problem might be a lot harder to handle.

Conjecture 10.29 *A finite projective plane admitting a transitive group is Desarguesian, and whence satisfies the PPC.*

Note that when the group is abelian, the action must be sharply transitive.

When the planes are infinite, the theory is entirely different; they need not be coordinatized over a skew field when admitting a Singer group. We will mention examples in the final section of this paper (on Absolute Arithmetic), where the classification of projective planes admitting sharply transitive automorphism groups will occur as a crucial element in a problem about so-called *hyperfield extensions* related to the "field with one element".

10.10 Further references

This section only gives a first taste on the status of the theory for planes. Much more has been done. An excellent survey consists of the two papers by Ghinelli and Jungnickel [43, 44].

11 Generalized quadrangles

We start this section with a combinatorial definition of generalized quadrangles with an order.

11.1 Combinatorial definition

A *generalized quadrangle* (GQ) *of order* (s,t), $s,t \in \mathbb{N}_0$, is a point-line incidence geometry $\mathscr{S} = (\mathscr{P}, \mathscr{B}, \mathbf{I})$ satisfying the following axioms:

(i) each point is incident with $t + 1$ lines and two distinct points are incident with at most one line;

(ii) each line is incident with $s + 1$ points;

(iii) if p is a point and L is a line not incident with p, then there is a unique point-line pair (q, M) such that $p\mathbf{I}M\mathbf{I}q\mathbf{I}L$.

In this definition, s and t are allowed to be infinite cardinals.

A thin GQ of order $(s, 1)$ is also called a *grid*, while a thin GQ of order $(1, t)$ is a *dual grid*. A GQ of order $(1, 1)$ is both a grid and a dual grid — it is an ordinary quadrangle. If $s = t$, then $\mathscr{S}$ is also said to be *of order s*.

Suppose $(p, L) \notin \mathbf{I}$. Then by $\text{proj}_L p$, we denote the unique point on L collinear with p. Dually, $\text{proj}_p L$ is the unique line incident with p concurrent with L.

11.2 Regularity and antiregularity

Let x and y be distinct points. Then $|\{x, y\}^\perp| = s + 1$ or $t + 1$, according as $x \sim y$ or $x \not\sim y$, respectively. When $x \not\sim y$, we have that $|\{x, y\}^{\perp\perp}| = s + 1$ or $|\{x, y\}^{\perp\perp}| \le t + 1$ according as $x \sim y$ or $x \not\sim y$, respectively. If $x \sim y$, $x \ne y$, or if $x \not\sim y$ and $|\{x, y\}^{\perp\perp}| = t + 1$, we say that the pair $\{x, y\}$ is *regular*. The point x is *regular* provided $\{x, y\}$ is regular for every $y \in \mathscr{P} \setminus \{x\}$, $z \not\sim x$. Regularity for lines is defined dually.

We will need the following easy lemma, the proof of which we leave to the reader.

Lemma 11.1 *If $\mathscr{S}$ is a thick GQ of order (s, t) and $\mathscr{S}$ has a regular pair of lines, then $s \leq t$.*   ∎

The pair of points $\{x, y\}$, $x \not\sim y$, is *antiregular* if $|\{x, y\}^{\perp} \cap z^{\perp}| \leq 2$ for all $z \in \mathscr{P} \setminus \{x, y\}$. The point x is *antiregular* if $\{x, y\}$ is antiregular for each $y \in \mathscr{P} \setminus x^{\perp}$. The same terminology is used in the dual setting.

11.3 Subquadrangles

A *subquadrangle*, or also *subGQ*, $\mathscr{S}' = (\mathscr{P}', \mathscr{B}', \mathbf{I}')$ of a GQ $\mathscr{S} = (\mathscr{P}, \mathscr{B}, \mathbf{I})$ is a GQ for which $\mathscr{P}' \subseteq \mathscr{P}$, $\mathscr{B}' \subseteq \mathscr{B}$, and where $\mathbf{I}'$ is the restriction of $\mathbf{I}$ to $(\mathscr{P}' \times \mathscr{B}') \cup (\mathscr{B}' \times \mathscr{P}')$.

As soon as finite GQs have certain subquadrangles, one can obtain extra information about parameters. The next theorem sums up some of this information, which is required later when considering construction processes of infinite ovoids.

Theorem 11.2 ([83], 2.2.2) *Let $\mathscr{S}'$ be a proper subquadrangle of the GQ $\mathscr{S}$, where $\mathscr{S}$ has order (s, t) and $\mathscr{S}'$ has order (s, t') (so $t > t'$). Then the following hold.*

(1) $t \geq s$; *if $s = t$, then $t' = 1$.*

(2) *If $s > 1$, then $t' \leq s$; if $t' = s \geq 2$, then $t = s^2$.*

(3) *If $s = 1$, then $1 \leq t' < t$ is the only restriction on t'.*

(4) *If $s > 1$ and $t' > 1$, then $\sqrt{s} \leq t' \leq s$ and $s^{3/2} \leq t \leq s^2$.*

(5) *If $t = s^{3/2} > 1$ and $t' > 1$, then $t' = \sqrt{s}$.*

(6) *Let $\mathscr{S}'$ have a proper subquadrangle $\mathscr{S}''$ of order (s, t''), $s > 1$. Then $t'' = 1$, $t' = s$ and $t = s^2$.*

For the sake of convenience, we provide the interested reader with some basic recognition theorems of subquadrangles which are especially handy when considering fixed points structures of automorphisms. All these theorems have natural analogues for infinite quadrangles, which are just as handy. The proofs are left to the reader as an exercise.

Theorem 11.3 ([83], 2.3.1) *Let $\mathscr{S}' = (\mathscr{P}', \mathscr{B}', \mathbf{I}')$ be a substructure of the GQ $\mathscr{S}$ of order (s, t) so that the following two conditions are satisfied:*

(i) *if $x, y \in \mathscr{P}'$ are distinct points of $\mathscr{S}'$ and L is a line of $\mathscr{S}$ such that $x \mathbf{I} L \mathbf{I} y$, then $L \in \mathscr{B}'$;*

(ii) *each element of $\mathscr{B}'$ is incident with $s + 1$ elements of $\mathscr{P}'$.*

Then there are four possibilities:

(1) $\mathscr{S}'$ is a dual grid, so $s = 1$;

(2) the elements of $\mathscr{B}'$ are lines which are incident with a distinguished point of $\mathscr{P}$, and $\mathscr{P}'$ consists of those points of $\mathscr{P}$ which are incident with these lines;

(3) $\mathscr{B}' = \emptyset$ and $\mathscr{P}'$ is a set of pairwise noncollinear points of $\mathscr{P}$;

(4) $\mathscr{S}'$ is a subquadrangle of order (s, t').

The following result is now easy to prove.

Theorem 11.4 ([83], 2.4.1) *Let θ be an automorphism of the GQ $\mathscr{S} = (\mathscr{P}, \mathscr{B}, \mathbf{I})$ of order (s,t). The substructure $\mathscr{S}_\theta = (\mathscr{P}_\theta, \mathscr{B}_\theta, \mathbf{I}_\theta)$ of $\mathscr{S}$ which consists of the fixed elements of θ must be given by (at least) one of the following:*

(i) $\mathscr{B}_\theta = \emptyset$ and $\mathscr{P}_\theta$ is a set of pairwise noncollinear points;

(i)' $\mathscr{P}_\theta = \emptyset$ and $\mathscr{B}_\theta$ is a set of pairwise nonconcurrent lines;

(ii) $\mathscr{P}_\theta$ contains a point x so that $y \sim x$ for each $y \in \mathscr{P}_\theta$, and each line of $\mathscr{B}_\theta$ is incident with x;

(ii)' $\mathscr{B}_\theta$ contains a line L so that $M \sim L$ for each $M \in \mathscr{B}_\theta$, and each point of $\mathscr{P}_\theta$ is incident with L;

(iii) $\mathscr{S}_\theta$ is a grid;

(iii)' $\mathscr{S}_\theta$ is a dual grid;

(iv) $\mathscr{S}_\theta$ is a subGQ of $\mathscr{S}$ of order (s', t'), $s', t' \geq 2$.

Finally, we recall a result on fixed elements structures of whorls.

Theorem 11.5 ([83], 8.1.1) *Let θ be a nontrivial whorl about p of the thick GQ $\mathscr{S} = (\mathscr{P}, \mathscr{B}, \mathbf{I})$ of order (s,t). Then one of the following must hold for the fixed element structure $\mathscr{S}_\theta = (\mathscr{P}_\theta, \mathscr{B}_\theta, \mathbf{I}_\theta)$.*

(1) $y^\theta \neq y$ for each $y \in \mathscr{P} \setminus p^\perp$.

(2) There is a point y, $y \not\sim p$, for which $y^\theta = y$. Put $V = \{p, y\}^\perp$ and $U = V^\perp$. Then $V \cup \{p, y\} \subseteq \mathscr{P}_\theta \subseteq V \cup U$, and $L \in \mathscr{B}_\theta$ if and only if L joins a point of V with a point of $U \cap \mathscr{P}_\theta$.

(3) $\mathscr{S}_\theta$ is a subGQ of order (s', t), where $2 \leq s' \leq s/t \leq t$, and hence $t < s$.

12 Parameters of generalized quadrangles

In this section we mention several basic results on parameters of finite GQs.

12.1 Standard restrictions on parameters

Let $\mathscr{S} = (\mathscr{P}, \mathscr{B}, \mathbf{I})$ be a (finite) GQ of order (s,t). Then $\mathscr{S}$ has $v = |\mathscr{P}| = (1+s)(1+st)$ points and $b = |\mathscr{B}| = (1+t)(1+st)$ lines; see [83, 1.2.1]. Also, we have that

$$st(1+s)(1+t) \equiv 0 \mod s+t, \tag{12.1}$$

and, for $s \neq 1 \neq t$, we have $t \leq s^2$ and, dually, $s \leq t^2$ (inequalities of Higman [51]); see [83, 1.2.2, 1.2.3].

12.2 Orders of the known finite GQs

The orders (s,t) of the known finite GQs are

$$\begin{cases} (s,1), & s \in \mathbb{N}_0, \\ (1,t), & t \in \mathbb{N}_0, \\ (q,q), & q \text{ any prime power}, \\ (q,q^2), & q \text{ any prime power}, \\ (q^2,q), & q \text{ any prime power}, \\ (q^2,q^3), & q \text{ any prime power}, \\ (q^3,q^2), & q \text{ any prime power}, \\ (q-1,q+1), & q \text{ any prime power}, \\ (q+1,q-1), & q \text{ any prime power}. \end{cases} \tag{12.2}$$

The prime power conjecture for finite GQs states the converse.

Conjecture 12.1 (PPC for finite quadrangles) *For a thick GQ of order (s,t) with $s \leq t$, we have that $t \in \{s, s+2, \sqrt{s^3}, s^2\}$, with t a prime power in the first and last two cases, and $t-1$ a prime power when $t = s+2$.*

Most of the known results on Conjecture 12.1 are contained in the present notes. For st infinite, we will see that the situation is very different.

12.3 Generalized quadrangles with small parameters

The proofs of all the results in this section are contained in Chapter 6 of Payne and Thas [83]. We will need the next theorem in the discussion below.

Theorem 12.2 ([83], 3.2.1, 3.2.2 and 3.2.3) *The following isomorphisms hold.*

(i) $\mathbf{Q}(4,q) \cong \mathbf{W}(q)^D$;

(ii) $\mathbf{Q}(4,q) \cong \mathbf{W}(q)$ *if and only if q is even;*

(iii) $\mathbf{Q}(5,q) \cong \mathbf{H}(3,q^2)^D$.

Let $\mathscr{S} = (\mathscr{P}, \mathscr{B}, \mathbf{I})$ be a finite GQ of order $(s,t), 1 < s \leq t$.

12.3.1 $s = 2$ By Section 12.1, $s + t$ divides $st(s + 1)(t + 1)$ and $t \leq s^2$. Hence $t \in \{2, 4\}$. Up to isomorphism there is only one GQ of order 2 and only one GQ of order (2,4). It follows that the GQs $\mathbf{W}(2)$ and $\mathbf{Q}(4, 2)$ are self-dual and mutually isomorphic. It is easy to show that the GQ of order 2 is unique.

The uniqueness of the GQ of order (2,4) was proved independently at least five times, by Seidel [90], Shult [91], Thas [99], Freudenthal [41] and Dixmier and Zara [32].

12.3.2 $s = 3$ Again by 12.1 we have $t \in \{3, 5, 6, 9\}$. Any GQ of order (3,5) must be isomorphic to the GQ $\mathbf{T}_2^*(\mathscr{O})$ [83] arising from the unique hyperoval in $\mathbf{PG}(2, 4)$, any GQ of order (3, 9) must be isomorphic to $\mathbf{Q}(5, 3)$, and a GQ of order 3 is isomorphic to either $\mathbf{W}(3)$ or to its dual $\mathbf{Q}(4, 3)$. Finally, there is no GQ of order (3,6).

The uniqueness of the GQ of order (3,5) was proved by Dixmier and Zara [32], the uniqueness of the GQ of order (3,9) was proved independently by Dixmier and Zara [32] and Cameron (see Payne and Thas [82]), the determination of all GQs of order 3 is due independently to Payne [77] and to Dixmier and Zara [32]. Dixmier and Zara [32] proved that there is no GQ of order (3,6).

12.3.3 $s = 4$ Using 12.1 it is easy to check that $t \in \{4, 6, 8, 11, 12, 16\}$. Nothing is known about $t = 11$ or $t = 12$. In the other cases unique examples are known, but the uniqueness question is settled only in the case $t = 4$. The proof of this uniqueness that appears in Payne and Thas [83] is that of Payne [78, 79], with a gap filled in by Tits.

12.4 From quadrangles to planes

Often, it is possible to relate different generalized polygons to each other through certain combinatorial properties. In such cases, the respective PPCs are induced.

In this section, we give two standard examples: the first one relates nets (which are generalizations of affine planes) to generalized quadrangles via the regularity property; the second one via antiregularity. Other such connections can be found in [132, §1.9] (for instance, between hexagons and quadrangles).

12.4.1 Nets and GQs A *net of order* k and *degree* r is a point-line incidence geometry $\mathscr{N} = (\mathscr{P}, \mathscr{B}, \mathbf{I})$ satisfying the following axioms:

(i) each point is incident with r lines ($r \geq 2$, $r \in \mathbb{N}$) and two distinct points are incident with at most one line;

(ii) each line is incident with k points ($k \geq 2$, $r \in \mathbb{N}$);

(iii) if p is a point and L is a line not incident with p, then there is a unique line M incident with p and not concurrent with L.

Theorem 12.3 ([83], 1.3.1) *Let x be a regular point of a thick GQ $\mathscr{S} = (\mathscr{P}, \mathscr{B}, \mathbf{I})$ of order (s,t). Then the following rank 2 incidence structure is the dual of a net of order s and degree $t + 1$.*

- *The* POINT SET *is $x^{\perp} \setminus \{x\}$;*

- *the* LINE SET *is the set of spans $\{u, v\}^{\perp\perp}$, where u and v are noncollinear points of $x^{\perp} \setminus \{x\}$, and*

- INCIDENCE *is the natural one.*

If in particular $s = t$, there arises a dual affine plane of order s.

In the case $s = t$, the incidence structure π_x with point set $x^{\perp}$, with line set the set of spans $\{u, v\}^{\perp\perp}$, where u and v are different points in $x^{\perp}$, and with the natural incidence, is a projective plane of order s.

The construction naturally generalizes to the infinite case. The same can be said about the construction in the next section.

12.4.2   Antiregularity and planes   Let Y and X be different concurrent lines in a thick GQ Γ of order s, and suppose X is antiregular. Put $X \cap Y = z$, and define a rank 2 incidence structure $\Pi(X, Y) = (\mathscr{P}, \mathscr{B}, \mathbf{I})$ as follows.

- POINTS are the elements of $X^{\perp}$ not through z.

- LINES are of two types:

  - sets $\{Z, X\}^{\perp} \setminus \{Y\}$, with $Z \in Y^{\perp}$, $Z \not\mathbf{I} z$;
  - the points of $X \setminus \{z\}$.

- INCIDENCE is the expected one.

Then $\Pi(X, Y)$ is an affine plane of order s. Generalizations of this observation can be found in [114].

13   Elation quadrangles

In this section we consider a large class of finite GQs, called *elation generalized quadrangles* (EGQs). EGQs are the natural analogues for generalized quadrangles as translation planes are for projective planes. One difference is that elation groups need not be abelian, as is the case for planes. This obstructs the definition of a kernel (or representation in projective space). In the next section, we will therefore consider a subclass of EGQs, namely those with an abelian elation group, called "translation generalized quadrangles".

Each known finite generalized quadrangle is, up to duality, an EGQ, or can be constructed from one by "Payne derivation" [113][1], which goes as follows. Let

---

[1]Except for the Hermitian quadrangles $\mathbf{H}(4, q^2)$, the same can be said about STGQs, which we will encounter later on.

$\mathscr{S} = (\mathscr{P}, \mathscr{B}, \mathbf{I})$ be a thick GQ of finite order s, with regular point x. Define a rank 2 geometry

$$\mathscr{P}(\mathscr{S}, x) = \partial_x(\mathscr{S}). \tag{13.1}$$

- POINTS are the points of $\mathscr{P} \setminus x^\perp$.

- LINES are the elements of $\mathscr{B}$ not incident with x, and sets $\{x, z\}^{\perp\perp} \setminus \{x\}$, $z \not\sim x$.

- INCIDENCE is "containment".

Then $\mathscr{S}(\mathscr{P}, x)$ is a GQ of order $(s-1, s+1)$, which is called the *Payne derivative* of $\mathscr{S}$ with respect to x. (If $\mathscr{S}$ is an infinite GQ, then asking that $\mathscr{N}_x$ is a projective plane is necessary and sufficient for $\mathscr{P}(\mathscr{S}, x)$ to be a GQ.) We say that

$$\mathscr{S} = \int \mathscr{S}(\mathscr{P}, x) dx \tag{13.2}$$

is the *Payne integral* of $\mathscr{S}(\mathscr{P}, x)$ with respect to x. A GQ can have nonisomorphic Payne integrals [28].

A specific (but general by the remarks above) case of the prime power conjecture for generalized quadrangles is the following version posed in the 1980's:

Conjecture 13.1 (W. M. Kantor) *The parameters (s, t) of an EGQ are powers of the same prime.*

Frohardt solved it affirmatively in [42] when $s \leq t$. In this section, we will overview his elegant combinatorial proof.

13.1 Elation generalized quadrangles

Let $\mathscr{S} = (\mathscr{P}, \mathscr{B}, \mathbf{I})$ be a GQ. If there is an automorphism group H of $\mathscr{S}$ which fixes some point $x \in \mathscr{P}$ linewise and acts sharply transitively on $\mathscr{P} \setminus x^\perp$, we call x an *elation point*, and H *elation group*. (Note that several — even nonisomorphic — elation groups can be associated to the same elation point [113].) If a GQ has an elation point, it is called an *elation generalized quadrangle* or, shortly, "EGQ". We frequently will use the notation $(\mathscr{S}^x, H)$ to indicate that x is an elation point with associated elation group H. Sometimes we also write $\mathscr{S}^x$ if we don't need to specify the elation group.

13.2 Kantor families

Suppose $(\mathscr{S}^x, H) = (\mathscr{P}, \mathscr{B}, \mathbf{I})$ is a finite thick EGQ of order (s, t), and let z be a point of $\mathscr{P} \setminus x^\perp$. Let $L_0, L_1, \ldots, L_t$ be the lines incident with x, and define r_i and M_i by $L_i \mathbf{I} r_i \mathbf{I} M_i \mathbf{I} z$, $0 \leq i \leq t$. Define, for $i = 0, 1, \ldots, t$, $H_i = H_{L_i}$ and $H_i^* = H_{r_i}$, and set $\mathscr{J} = \{H_i | 0 \leq i \leq t\}$.

Then we have the following properties:

- $|H| = |\mathscr{P} \setminus x^{\perp}| = s^2 t$;

- $\mathscr{J}$ is a set of $t + 1$ subgroups of H, each of order s;

- for each $i = 0, 1, \ldots, t$, H_i^* is a subgroup of H of order st containing H_i as a subgroup.

Moreover, the following two conditions are satisfied:

(K1) $H_i H_j \cap H_k = \{\mathbf{1}\}$ for distinct i, j and k;

(K2) $H_i^* \cap H_j = \{\mathbf{1}\}$ for distinct i and j.

Conversely, let H be a group of order $s^2 t$ and $\mathscr{J}$ (respectively $\mathscr{J}^*$) be a set of $t + 1$ subgroups H_i (respectively H_i^*) of H of order s (respectively of order st), and suppose (K1) and (K2) are satisfied. We call H_i^* the *tangent space* at H_i, and $(\mathscr{J}, \mathscr{J}^*)$ is said to be a *Kantor family* or *4-gonal family* of type (s, t) in H. Sometimes we will also say that $\mathscr{J}$ is a (Kantor, 4-gonal) family of type (s, t) *in* H.

Notation. If $(\mathscr{J}, \mathscr{J}^*)$ is a Kantor family in H, and $A \in \mathscr{J}$, then A^* denotes the tangent space at A.

Let $(\mathscr{J}, \mathscr{J}^*)$ be a Kantor family of type (s, t) in the group H of order $s^2 t$, taken that $s \neq 1 \neq t$. Define a rank 2 incidence structure $\mathscr{S}(H, \mathscr{J})$ as follows.

- POINTS of $\mathscr{S}(H, \mathscr{J})$ are of three kinds:

  (i) elements of H;

  (ii) left cosets $g H_i^*$, $g \in H$, $i \in \{0, 1, \ldots, t\}$;

  (iii) a symbol (∞).

- LINES are of two kinds:

  (a) left cosets $g H_i$, $g \in H$, $i \in \{0, 1, \ldots, t\}$;

  (b) symbols $[H_i]$, $i \in \{0, 1, \ldots, t\}$.

- INCIDENCE.   A point g of type (i) is incident with each line $g H_i$, $0 \leq i \leq t$. A point $g H_i^*$ of type (ii) is incident with $[H_i]$ and with each line $h H_i$ contained in $g H_i^*$. The point (∞) is incident with each line $[H_i]$ of type (b). There are no further incidences.

It is easy to prove that $\mathscr{S}(H, \mathscr{J})$ is a GQ of order (s, t), and H acts by left multiplication as an elation group for the point (∞). The next theorem is now clear.

Theorem 13.2 ([58]) *If we start with an EGQ $(\mathscr{S}^x, H)$ to obtain $\mathscr{J}$ as above, then we have that $\mathscr{S}^x \cong \mathscr{S}(H, \mathscr{J})$. So a group of order $s^2 t$ admitting a 4-gonal family is an elation group for a suitable elation generalized quadrangle.*

13.3   Parameters of elation quadrangles

In this section, we use the following notation: if p is a prime divisor of the natural number m, then m_p is the largest power of p dividing m, and $m_{p'}$ is defined as m/m_p. Also, $\pi(m)$ is the set of primes dividing m. Similarly, define m_π and $m_{\pi'}$ for any set of primes π. If G is a finite group, $\pi(G)$ is defined as $\pi(|G|)$. If R is a finite group, we denote its set of Sylow p-subgroups by $\mathrm{Syl}_p(R)$ (this set could be the empty set).

Let $(\mathscr{S}^x, G)$ be a thick EGQ of order (s,t). Starting from a point $z \not\sim x$, construct the 4-gonal family $(\mathscr{J}, \mathscr{J}^*)$ as above, and put $\mathscr{J} = \{G_0, G_1, \ldots, G_t\}$, while $\mathscr{J}^* = \{G_0^*, G_1^*, \ldots, G_t^*\}$.

Lemma 13.3   (i) *If A and B are distinct elements of $\mathscr{J}$, and $g \in G$, then $A^* \cap B^g = \{\mathbf{1}\}$.*

(ii) *Let $\mathbb{S} \leq \cap_{A \in \mathscr{J}} A^*$, and suppose $\mathbb{S} \trianglelefteq G$. For any subgroup $K \leq G$, define $\overline{K} = K\mathbb{S}/\mathbb{S}$, and if $g \in G$, put $\overline{g} = g\mathbb{S}$. If A and B are distinct elements of $\mathscr{J}$, and $g \in G$, then $\overline{A^*} \cap \overline{B}^{\overline{g}} = \{\mathbf{1}\}$. In particular, with $\{\mathbf{1}\} = \mathbb{S}$, we obtain (i).*

Proof.   (i).   As $|G| = s^2 t$, $|A^*| = st$, $|B| = s$ and $A^* \cap B = \{\mathbf{1}\}$, we have that $G = A^*B$. Let $a \in A^*$ and suppose $a \in B^g$. Write $g^{-1} = hb$ with $h \in A^*$ and $b \in B$. Put $c = a^h$, so that $c \in A^*$ and $c^b \in B$. The latter expression implies that $c \in B$, so $c = 1$. It follows that $a = 1$.

(ii).   Suppose $\overline{A^*} \cap \overline{B}^{\overline{g}} \neq \{\mathbf{1}\}$ for some $A, B \neq A \in \mathscr{J}$ and $g \in G$. Then there are $a \in A^*$ and $b \in B$ for which $A^g \mathbb{S} = b\mathbb{S}$. As $\mathbb{S}$ is normal in G, it is a group of symmetries about x. It follows that $b \in A^{*g}$. By (i) this is only possible when $b = 1$. ∎

The following lemma has a surprisingly easy proof.

Lemma 13.4 *Let p be a prime, and assume that $t_p > 1$. Then*

$$t_{p'} < s_p. \tag{13.3}$$

Proof.   Take $A \in \mathscr{J}$, and let $P \in \mathrm{Syl}_p(G)$ contain a Sylow p-subgroup A_p^* of A^*. For each $B \in \mathscr{J} \setminus \{A\}$, let $B_p \in \mathrm{Syl}_p(B)$. Then B is G-conjugate to a subgroup Q_B of P. By Lemma 13.3, we have that the groups A_p^*, Q_B and B are mutually disjoint. So

$$|A_p^*| + \sum_{B \in \mathscr{J} \setminus \{A\}} |Q_B^\times| \leq |P|. \tag{13.4}$$

Since $|A_p^*| = s_p t_p$, $|Q_B^\times| = s_p - 1$ for all $B \in \mathscr{J} \setminus \{A\}$, and since $|P| = s_p^2 t_p$, this implies that

$$s_p t_p + t(s_p - 1) \leq s_p^2 t_p. \tag{13.5}$$

Hence $t_p t_{p'}(s_p - 1) \leq s_p t_p(s_p - 1)$. As $t_p > 0$ and $s_p > 1$, we have $t_{p'} < s_p$. ∎

Lemma 13.5 *One of the following occurs:*

(i) *s is a prime power;*

(ii) *$t < s$, $(s,t) = \gcd(s,t) \neq 1$, s has exactly two prime divisors and every element of $\mathscr{J}$ is solvable.*

Proof. Let $k = |\pi(s)|$. By the previous lemma, taking products over all $p \in \pi(s)$ yields

$$t^k = \prod(t_p, t_{p'}) < \prod s_p t_p = s \prod t_p \leq st. \tag{13.6}$$

So $t^{k-1} < s$, and Higman's inequality now leads us to $k \leq 2$.

Suppose now that $k = 2$, and that $\pi(s) = \{p, q\}$. Then any element of $\mathscr{J}$ is a pq-group, so solvable by Burnside's $p^a q^b$-theorem.

Finally, since $t^2 < t_{\pi(s)} s \leq t_{\pi(s)} t^2$, it follows that $(s,t) \neq 1$. ∎

The following intermediate result is interesting.

Lemma 13.6 *Assume that G has a normal Hall π-subgroup H. Then either $s_\pi = 1$ or $\pi(t) \subseteq \pi$. In particular, if G is nilpotent, then G is a p-group.*

Proof. For each $A \in \mathscr{J}$, define $A_H = A \cap H$ and $A_H^* = A^* \cap H$. Set

$$\mathscr{J}_H = \{A_H | A \in \mathscr{J}\} \quad \text{and} \quad \mathscr{J}_H^* = \{A_H^* | A^* \in \mathscr{J}^*\}. \tag{13.7}$$

Then either $s_\pi = 1$, or $(\mathscr{J}_H, \mathscr{J}_H^*)$ is a Kantor family of type (s_π, t_π) in H. Since in the latter case $\mathscr{S}(H, \mathscr{J}_H)$ is a subGQ of order (s_π, t_π), by Theorem 11.2 we have

$$|\mathscr{J}| - 1 = |\mathscr{J}_H| - 1 \leq t_\pi, \tag{13.8}$$

so that $t = t_\pi$ and $\pi(t) \subseteq \pi$.

Suppose now that p divides s and that G has a normal Sylow p-subgroup. Then $\pi(t) = \{p\}$. ∎

We are ready to obtain Frohardt's proposition on Kantor's conjecture. The proof differs a bit from the original one, and is taken from [113].

Theorem 13.7 *If either $|\pi(s)| = 1$ or G is solvable, then $\pi(G) \subseteq \pi(s)$. In particular, if $\pi(s) = \{p\}$, then G is a p-group.*

Proof. For every $A \in \mathscr{J}$ choose a Hall (or Sylow) π-subgroup S_A of G such that $A \leq S_A$ and S_A contains a Hall π-subgroup of A^*. Since G has at most $|G|_{\pi'} = t_{\pi'}$ distinct Hall π-subgroups and $|\mathscr{J}| = t_\pi t_{\pi'} + 1$, the pigeonhole principle shows that there is a Hall π-subgroup S of G with $S = S_A$ for at least $t_\pi + 1$ members of $\mathscr{J}$. Fix such and S, and let

$$\mathscr{J}_S = \{A \in \mathscr{J} \,|\, S_A = A\}. \tag{13.9}$$

If we set $A^+ = A^* \cap S$ for all $A \in \mathscr{J}_S$ and

$$\mathscr{J}_S^* = \{A^+ \,|\, A \in \mathscr{J}_S\}, \tag{13.10}$$

then $(\mathscr{J}_S, \mathscr{J}_S^*)$ is a Kantor family of type (s, t'), where $t' \geq t_\pi$. So $t' = t_\pi$. Let $\mathscr{S}'$ be the corresponding subGQ of order (s, t_π). Of course, $\mathscr{S}'^x$ is an EGQ with elation group S. Consider a $C \in \mathscr{J}$ which is not contained in $\mathscr{J}_S$; then we use Lemma 13.4 to obtain

$$|SC| \leq |G| = s^2 t = s^2 t_\pi t_{\pi'} < s^2 t_\pi s_\pi \leq |S| \times |C|. \tag{13.11}$$

Whence $|S \cap C| \geq 2$ for any such C. Since $\mathscr{S}'$ is a subGQ, this implies readily that $C \in \mathscr{J}_S$ — in other words, $\mathscr{S}' = \mathscr{S}$ and $t_\pi = t$. ∎

Putting Lemma 13.5 and Theorem 13.7 together, we obtain

Theorem 13.8 (PPC for EGQs) *If $\mathscr{S}$ is a thick EGQ of order (s, t) with $s \leq t$, then st is a prime power.* ∎

13.4 Infinite EGQs

It is easy to generalize the concept of Kantor family to the infinite case. The correspondence between EGQs and Kantor families remains.

Let E be a multiplicatively written group. Let $\mathscr{F}$ be a set of nontrivial subgroups of E and let $*$ be a map from $\mathscr{F}$ to the set of subgroups of E mapping each $A \in \mathscr{F}$ to a subgroup A^* containing A properly. Let $\mathscr{F}^* = \{A^* \,|\, A \in \mathscr{F}\}$. The triple $(E, \mathscr{F}, \mathscr{F}^*)$ is a *Kantor family* if the following axioms hold.

- $A^*B = E$ for all $A, B \in \mathscr{F}$ with $A \neq B$.

- $A^* \cap B = \{\mathbf{1}\}$ for all $A, B \in \mathscr{F}$ with $A \neq B$.

- $AB \cap C = \{\mathbf{1}\}$ for all $A, B, C \in \mathscr{F}$ with $A \neq C \neq B \neq A$.

- $E = A^* \bigcup (\cup_{B \in \mathscr{F}} AB)$ for each $A \in \mathscr{F}$.

It is easy to see that one can construct an EGQ $\mathscr{S}(E, \mathscr{F})$ from a Kantor family, for which E acts as an elation group; everything works as in the finite case.

As already mentioned, we want to associate a (skew) field to a quadrangle just as we did for projective planes. This is what we will do in the next section for a subclass of the class of EGQs.

14   Translation quadrangles

If the elation group of an EGQ is abelian, we speak of a *translation generalized quadrangle* (TGQ). The elation group is then usually called "translation group", the elation point "translation point". A first major difference with EGQs is that the elation group of a finite TGQ is unique, see [84, 104]. (For finite EGQs, we already remarked that this is not necessarily the case.) For infinite TGQs, the same is true.

Theorem 14.1 *A thick TGQ* $(\mathscr{S}^x, T)$ *has a unique translation group.*

Proof. Since T is abelian, it is straightforward to see that any line incident with x is an axis of symmetry (if $(\mathscr{F}, \mathscr{F}^*)$ is the associated Kantor family in T, $A \trianglelefteq T$ for any $A \in \mathscr{F}$). So for each line U on x, all symmetries with axis U are contained in T, and T is generated by the symmetries with axis a line through x. ∎

Such as was the case for translation planes, the fact that the translation group of a TGQ is abelian allows us to define a kernel in much the same way as for planes. The interesting thing here is that it immediately leads to a solution of a PPC for finite TGQs, but, moreover, in even characteristic it is *precise*, in the sense that we know what t/s is. Also, it allows one to represent TGQs in projective spaces in an André-Bruck-Bose type setting. We will give sketches of some proofs for the sake of convenience, as we did in the previous section.

When the number of points of the TGQ is infinite, we will see that although to some extent the theory is analogous to that of (infinite) translation planes, some obstructions do arise.

14.1   The kernel

Each finite TGQ $\mathscr{S}$ of order (s,t) with translation point (∞) has a *kernel* $\mathbb{K}$, which is a field with multiplicative group isomorphic to the group of all collineations of $\mathscr{S}$ fixing the point (∞) and any given point not collinear with (∞) linewise. We will introduce the kernel in detail in this section.

Let $(\mathscr{S}^x, G)$ be a TGQ with translation group G and with $H_i, H_i^*, \mathscr{J}$, etc., as before. The *kernel* $\mathbb{K}$ of $\mathscr{S}^x$ is the set of all endomorphisms α of G for which $H_i^\alpha \leq H_i$, $0 \leq i \leq t$. With the usual addition and multiplication of endomorphisms $\mathbb{K}$ is a ring.

Theorem 14.2 ([83], 8.5.1) *The ring* $\mathbb{K}$ *is a field, so that* $H_i^\alpha = H_i$, $(H_i^*)^\alpha = H_i^*$ *for all* $i = 0, 1, \ldots, t$ *and all* $\alpha \in \mathbb{K}^0 = \mathbb{K} \backslash \{0\}$.

Proof. The only GQs with $s = 2$ and $t > 1$ are $\mathbf{W}(2)$ and $\mathbf{Q}(5,2)$, in which cases we can check the theorem (in these cases $\mathbb{K} = \{0,1\}$). So from now on we may assume that $s > 2$.

If each $\alpha \in \mathbb{K}^0$ is an automorphism of G, then clearly $\mathbb{K}$ is a field. So suppose some $\alpha \in \mathbb{K}^0$ is not an automorphism. Then

$$\langle H_0, H_1, \ldots, H_t \rangle = G \geq G^\alpha = \langle H_0^\alpha, H_1^\alpha, \ldots, H_t^\alpha \rangle, \tag{14.1}$$

with $G \neq G^{\alpha}$, implying $H_i^{\alpha} \neq H_i$ for some i. Let $g^{\alpha} = 1$, $G \in H_i^{\times}$. If i, j, k are mutually distinct and $g' \in H_j$ with $\{g'\} \neq H_j \cap H_k^* g^{-1}$, then we have $gg' = hh'$ with $h \in H_k$, $h' \in H_l$, for a uniquely defined l, with $l \neq k, j$. Hence $h^{\alpha} h'^{\alpha} = g'^{\alpha}$, implying that $h^{\alpha} = h'^{\alpha} = g'^{\alpha} = 1$ (by (K1)). Since g' was any one of $s - 1$ elements of H_j, $|\ker(\alpha) \cap H_j| \geq s - 1 > s/2$, implying $H_j \leq \ker(\alpha)$. This implies $H_j \leq \ker(\alpha)$ for each j, with $j \neq i$, so that $G_i \leq \ker(\alpha)$, where

$$G_i = \langle H_j | j \in \{0, 1, \ldots, t\} \setminus \{i\} \rangle. \tag{14.2}$$

Each $\sigma \in H_i$ can be written as

$$\sigma_1 \sigma_2 \sigma_3, \tag{14.3}$$

with $\sigma_1, \sigma_2, \sigma_3$ elements of respectively $H_j, H_{j'}, H_{j''}$ for some $j, j', j'' \in \{0, 1, \ldots, t\} \setminus \{i\}$ (exercise), so we have $G = G_i$. This says $\alpha = 0$, a contradiction. Hence we have shown that $\mathbb{K}$ is a field and $H_i^* = H_i$ for $i = 0, 1, \ldots, t$ and $\alpha \in \mathbb{K}^0$. Since H_i^* is the set theoretic union of H_i together with all those cosets of H_i disjoint from $\bigcup \{H_i | 0 \leq i \leq t\}$, we also have $(H_i^*)^{\alpha} = H_i^*$. So for each line U on x, all symmetries with axis U are contained in T, and T is generated by the symmetries with axis a line through x. ∎

For each subfield $\mathbb{F}$ of $\mathbb{K}$ there is a vector space $(G, \mathbb{F})$ whose vectors are the elements of G, and whose scalars are the elements of $\mathbb{F}$. Vector addition is the group operation in G, and scalar multiplication is defined by $ga = g^{\alpha}, g \in G, \alpha \in \mathbb{F}$. It is easy to verify that $(G, \mathbb{F})$ is indeed a vector space. As H_i is a subspace of $(G, \mathbb{F})$, we have $|H_i| \geq |\mathbb{F}|$. It follows that $s \geq |\mathbb{K}|$.

Theorem 14.3 ([83], 8.5.2 — PPC for finite TGQs) *The group G is elementary abelian, so s and t must be powers of the same prime.*

Proof. Let $|\mathbb{F}| = q$, so q is a prime power. As H_i and H_i^* may be viewed as subspaces of the vector space $(G, \mathbb{F})$, we have $s = |H_i| = q^n$ and $st = |H_i^*| = q^{n+m}$, hence $t = q^m$, $n, m \in \mathbb{N}$. ∎

We also mention the following more precise result for the even case.

Theorem 14.4 ([104], 3.10.1(ii) — PPC for finite TGQs, II) *For a thick TGQ of order (s, t) with st even, we have $s \in \{t, t^2\}$.*

Geometric interpretation

Let $\mathscr{S}^{(\infty)}$ be a finite TGQ with translation group G. We keep using the notation of above. Since every element of $\mathbb{K}^0$ is an automorphism of G fixing the 4-gonal family $\mathscr{J}$ elementwise, it is straightforward to see that such an element induces a whorl of the TGQ about (∞). Moreover, each such element must fix the identity of G, so it induces an element stabilizing some fixed point y not collinear with the translation point. Let κ be any whorl about (∞), and σ any symmetry about a line through (∞). Then clearly σ^{κ} is also a symmetry about that line. As G is generated by the symmetries about the lines incident with (∞), it follows that G is a normal subgroup of the group of all whorls about (∞). Now consider any whorl φ about (∞) and y. Then $G^\varphi = G$, and so φ induces an automorphism of G that fixes $\mathcal{J}$ elementwise. Hence $\varphi \in \mathbb{K}$.

We have proved the following theorem.

Theorem 14.5 *The multiplicative group* $\mathbb{K}^0$ *induces the group of all whorls about* (∞) *and* y.   ∎

14.2 $\mathbf{T}(n, m, q)$s and translation quadrangles

In this section, we introduce the notion of $\mathbf{T}(n, m, q)$.

Suppose $H = \mathbf{PG}(2n + m - 1, q)$ is the finite projective $(2n + m - 1)$-space over $\mathbb{F}_q$. Now define a set $\mathcal{O} = \mathcal{O}(n, m, q)$ of subspaces as follows: $\mathcal{O}$ is a set of $q^m + 1$ $(n - 1)$-dimensional subspaces of H, denoted $\mathbf{PG}^{(i)}(n - 1, q)$, and often also by π_i, so that

(i) every three generate a $\mathbf{PG}(3n - 1, q)$;

(ii) for every $i = 0, 1, \ldots, q^m$, there is a subspace $\mathbf{PG}^{(i)}(n + m - 1, q)$, also denoted by τ_i, of H of dimension $n + m - 1$, which contains $\mathbf{PG}^{(i)}(n - 1, q)$ and which is disjoint from any $\mathbf{PG}^{(j)}(n - 1, q)$ if $j \neq i$.

If $\mathcal{O}$ satisfies these conditions for $n = m$, then $\mathcal{O}$ is called a *pseudo-oval* or a *generalized oval* or an $[n - 1]$-*oval* of $\mathbf{PG}(3n - 1, q)$. A $[0]$-oval of $\mathbf{PG}(2, q)$ is just an oval of $\mathbf{PG}(2, q)$. For $n \neq m$, $\mathcal{O}(n, m, q)$ is called a *pseudo-ovoid* or a *generalized ovoid* or an $[n - 1]$-*ovoid* or an *egg* of $\mathbf{PG}(2n + m - 1, q)$. A $[0]$-ovoid of $\mathbf{PG}(3, q)$ is just an ovoid of $\mathbf{PG}(3, q)$.

The space $\mathbf{PG}^{(i)}(n + m - 1, q)$ is the *tangent space* of $\mathcal{O}(n, m, q)$ at $\mathbf{PG}^{(i)}(n - 1, q)$; it is uniquely determined by $\mathcal{O}(n, m, q)$ and $\mathbf{PG}^{(i)}(n - 1, q)$. Sometimes we will call an $\mathcal{O}(n, n, q)$ also an "egg" or a "generalized ovoid" for the sake of convenience.

From any egg $\mathcal{O}(n, m, q)$ arises a GQ $\mathbf{T}(n, m, q) = \mathbf{T}(\mathcal{O})$ which is a TGQ of order (q^n, q^m) for some base-point (∞). This goes as follows. Let H be embedded in a $\mathbf{PG}(2n + m, q) = H'$.

- The POINTS are of three types.

    (i) The points of $H' \setminus H$.

    (ii) The subspaces $\mathbf{PG}(n + m, q)$ of H' which intersect H in a $\mathbf{PG}^{(i)}(n + m - 1, q)$.

    (iii) A symbol (∞).

- The LINES are of two types.

    (a) The subspaces $\mathbf{PG}(n, q)$ of $\mathbf{PG}(2n + m, q)$ which intersect H in an element of the egg.

(b) The elements of the egg $\mathscr{O}(n, m, q)$.

- INCIDENCE is defined as follows. The point (∞) is incident with all the lines of type (b) and with no other lines. A point of type (ii) is incident with the unique line of type (b) contained in it and with all the lines of type (a) contained in it. Finally, a point of type (i) is incident with the lines of type (a) containing it.

Conversely, any TGQ can be seen in this way, that is, as a $\mathbf{T}(n, m, q)$ associated to an $\mathscr{O}(n, m, q)$ in $\mathbf{PG}(2n + m - 1, q)$.

Theorem 14.6 ([83], 8.7.1) *The geometry* $\mathbf{T}(n, m, q)$ *is a TGQ of order* (q^n, q^m) *with translation point* (∞) *and for which* $\mathbb{F}_q$ *is a subfield of the kernel. Moreover, the translations of* $\mathbf{T}(n, m, q)$ *induce translations of the affine space* $\mathbf{AG}(2n + m, q) = \mathbf{PG}(2n+m, q) \backslash \mathbf{PG}(2n+m-1, q)$. *Conversely, every TGQ for which* $\mathbb{F}_q$ *is a subfield of the kernel is isomorphic to a* $\mathbf{T}(n, m, q)$.

Proof. It is routine to show that $\mathbf{T}(n, m, q)$ is a GQ of order (q^n, q^m). A translation of $\mathbf{AG}(2n+m, q)$ defines in a natural way an elation about (∞) of $\mathbf{T}(n, m, q)$. It follows that $\mathbf{T}(n, m, q)$ is an EGQ with abelian elation group G, where G is isomorphic to the translation group of $\mathbf{AG}(2n + m, q)$, and hence $\mathbf{T}(n, m, q)$ is a TGQ with translation group G. It also follows that $\mathbb{F}_q$ is a subfield of the kernel of $\mathbf{T}(n, m, q)$: with the group of all homologies of $\mathbf{PG}(2n + m, q)$ having a center y not in $\mathbf{PG}(2n + m - 1, q)$ and axis $\mathbf{PG}(2n + m - 1, q)$ corresponds in a natural way the multiplicative group of a subfield of the kernel (recall Theorem 14.5).

Conversely, consider a TGQ $\mathscr{S}^x$ with translation group G for which $\mathbb{F}_q = \mathbb{F}$ is a subfield of the kernel. If $s = q^n$ and $t = q^m$, then $[(G, \mathbb{F}) : \mathbb{F}] = 2n + m$. Hence with $\mathscr{S}^x$ "corresponds" an affine space $\mathbf{AG}(2n + m, q)$. The cosets $H_i g$ of a fixed H_i are the elements of a parallel class of n-dimensional subspaces of $\mathbf{AG}(2n+m, q)$, and the cosets $H_i^* g$ of a fixed H_i^* are the elements of a parallel class of $(n + m)$-dimensional subspaces of $\mathbf{AG}(2n + m, q)$. The interpretation in $\mathbf{PG}(2n + m, q)$ together with (K1) and (K2) prove the last part of the theorem. ∎

Corollary 14.7 *For any* $\mathscr{O}(n, m, q)$ *we have* $n \leq m \leq 2n$.

Proof. The GQ $\mathbf{T}(n, m, q)$ is a TGQ, and so any line incident with the translation point is regular. By Lemma 11.1, $t \geq s$, hence $m \geq n$. By the inequality of Higman we have that $t \leq s^2$, and so $m \leq 2n$. ∎

14.3 Infinite TGQs

If one is considering infinite TGQs, things get harder. We still can define the kernel as in the finite case, but it is a long-standing open problem as to whether it is a division ring (although some special cases are known). Once one knows that the kernel of some TGQ *is* a division ring, the TGQ can be represented in projective space in the same way as in the finite case, through generalizations of ovoids. So a first step in PPC theory for infinite TGQs should read as follows:

Conjecture 14.8 *The kernel of a TGQ is a division ring.*

TGQs satisfying this conjecture are called "linear" — we have seen that in the finite case any TGQ is linear. Let Γ^x be a linear TGQ with kernel $\mathbb{K}$, and let $(\mathscr{F}, \mathscr{F}^*)$ be the associated Kantor family in the abelian translation group E; we see E as a vector space over $\mathbb{K}$. Then each element of $\mathscr{F}$, respectively $\mathscr{F}^*$, can be seen as a linear subspace of E, and each element of $\mathscr{F}$, respectively $\mathscr{F}^*$, has the same (possibly infinite) dimension over $\mathbb{K}$. We call these dimensions the "$\mathbb{K}$-dimensions" or "$\mathbb{K}$-parameters" (and use the same terminology relative to subfields of the kernel). Without any other (major) restriction, there is no hope of making a more subtle Higman-type PPC (on the dimensions of s and t over the division ring, taken that the TGQ is linear) than this. There are many reasons. For one, Niels Rosehr obtained the next theorem:

Theorem 14.9 ([87]) *Let $\mathbb{K}$ be any infinite division ring, and let $n, m \in \mathbb{N}_0$ be arbitrary, but such that $n \leq m$. Then there exists a linear TGQ Γ^x of order (s, t) with kernel containing $\mathbb{K}$, such that the $\mathbb{K}$-dimensions of s and t are n and m, respectively.*

Rosehr obtains this theorem by constructing an egg in $\mathbf{PG}(2n + m - 1, \mathbb{K})$ by transfinite recursion. A corollary of Theorem 14.9 is:

Corollary 14.10 *For every $n, m \in \mathbb{N}_0$ with $s \leq t$ and every infinite skew field $\mathbb{K}$, there is a linear TGQ with $\mathbb{K}$-parameters (n, m) and kernel $\mathbb{K}$.*

For other related transfinite constructions of ovoid-like objects, we refer to [4]. We note that in many constructions of this type (and also for instance in [87]), the following lemma (and variations) is used.

Lemma 14.11 *Let S be any subset of a projective space $\mathbf{PG}(n, \mathbb{K})$ over an infinite skew field $\mathbb{K}$, with n finite. If $|S| < |\mathbb{K}| + 1^2$, then for any point u outside S, there is a hyperplane containing u and disjoint from S.*

The proof is easy, and uses induction on n, combined with passing to some appropriate quotient space. The fact that n is finite is not really essential, but the inequality

$$|S| < |\mathbb{K}| + 1 \tag{14.4}$$

is.

Remark Note that similar transfinite constructions for ovoids cannot work for infinite dimensional projective spaces $\mathbf{P}$ over finite fields (in the hope to construct "locally finite" TGQs — see a later section). Any finite dimensional linear subspace ζ would intersect the ovoid $\mathcal{O}$ in an ovoid of ζ, or would be contained in a tangent space, and then one easily obtains a contradiction by considering three distinct finite dimensional spaces (of dimension at least 2)

$$\mathbf{P}_1 \hookrightarrow \mathbf{P}_2 \hookrightarrow \mathbf{P}, \tag{14.5}$$

---

[2]Here $|\cdot|$ denotes the cardinality.

all meeting $\mathscr{O}$, respectively, in an ovoid $\mathbf{P}_i \cap \mathscr{O}$, and hence giving rise to a tower of three finite thick full TGQs. This contradicts Theorem 11.2 and the fact that any line incident with the translation point is regular.

In fact, we will later show that it is not difficult to construct linear TGQs of order $(\aleph, \aleph')$ with $\aleph$ and $\aleph'$ different cardinals. In fact, we will mention a construction of infinite classical (orthogonal) quadrangles $\mathbf{Q}(\mathbb{Q}^{|\mathbb{N}|}, \iota)$ with the latter property (for some quadratic form ι), and these even have the property that *every* point is a translation point.

14.4 PPC for general TGQs

Recently, the author of the present text showed that Conjecture 14.8 is indeed true, up to possibly some exceptional examples (that hopefully are killed soon), hence obtaining a rather complete PPC theory for general TGQs. We refer to the upcoming paper [116] for the details.

15 Skew translation quadrangles

Recall for the sake of convenience that a point u of a GQ $\mathscr{S}$ is a *center of symmetry* if there exists a group $\mathbb{S}$ of automorphisms of $\mathscr{S}$ which fixes $u^\perp$ elementwise, such that for any two noncollinear points $v, w \in u^\perp$, $\mathbb{S}$ acts transitively on $\{v, w\}^\perp \setminus \{u\}$. This action necessarily is sharply transitive, and in the finite case, when the parameters are (s, t), this requirement is equivalent to demanding that $|\mathbb{S}| = t$. The elements of $\mathbb{S}$ are *symmetries* about u. (Note that any dual root on a center of symmetry is Moufang.)

An EGQ $(\mathscr{S}^x, G)$ is called a *skew translation generalized quadrangle* (STGQ) provided the point x is a center of symmetry, the symmetries about which are contained in G. As we will see, for this type of EGQ Kantor's conjecture is true. Call an STGQ $(\mathscr{S}^x, G)$ *central* if the symmetries about the elation point are contained in the center of G. No noncentral STGQs are known. The conjecture that there are no such STGQs is the "Centrality conjecture", and is showed in [114] to be the most important problem when classifying finite STGQs.

We first introduce a more general concept.

15.1 F-Factors

Let $\mathscr{S}$ be a thick EGQ, and let $(\mathscr{F}, \mathscr{F}^*)$ be the associated Kantor family. Let $\mathbf{F} = \mathscr{F} \cup \mathscr{F}^*$. A nontrivial subgroup X of the elation group G is an $\mathbf{F}$-*factor* of G if

$$(U \cap X)(V \cap X) = X \quad \text{for all } U, V \in \mathbf{F} \text{ satisfying } UV = G. \tag{15.1}$$

Define $\mathscr{F}_X = \{U \cap X | U \in \mathscr{F}\}$ and $\mathscr{F}_X^* = \{U^* \cap X | U^* \in \mathscr{F}^*\}$. We say that X is "of type (σ, τ)" if $|X| = \sigma^2 \tau$, $|A \cap X| = \sigma$ and $|A^* \cap X| = \sigma\tau$ for all $A \in \mathscr{F}$ (in [47] it is shown that such integers σ, τ always exist).

Theorem 15.1 ([47]) *Let X be an* **F**-*factor of type (σ, τ) in G. Then necessarily one of the following cases occurs:*

(a) $\sigma = 1$, $|X| = \tau \leq t$ *and X is a subgroup of $\cap_{A \in \mathscr{F}} A^*$;*

(b) $\sigma > 1$, $\tau = t$ *and $(\mathscr{F}_X, \mathscr{F}_X^*)$ is a Kantor family in X of type (σ, τ).*

If we are in Case (b) of Theorem 15.1, we call X a *thick* **F**-factor. An **F**-factor X in G is *normal* if X is a normal subgroup of G. In [47] Hachenberger obtained the following partial classification of normal **F**-factors in Kantor families. Since Case (a) of Theorem 15.1 is not of particular interest for now, one may suppose essentially without loss of generality that $\sigma = t$.

Theorem 15.2 ([47]) *Let G be a group of order $s^2 t$ admitting a Kantor family $(\mathscr{F}, \mathscr{F}^*)$ of type (s, t), with $s, t > 1$, and having a normal* **F**-*factor X of type (σ, τ) with $\tau = t$. Then one of the following cases occurs:*

(a) G *is a group of prime power order;*

(b) $\sigma > 1$, $|G|$ *has exactly two prime divisors, and X is a Sylow subgroup of G for one of these primes.*

Theorem 15.2 led Hachenberger to prove a well-known conjecture of S. E. Payne, which amounted to showing that G is a p-group if X is of type $(1, t)$ — see [47] — in other words, the parameters of any thick STGQ are powers of a prime. In [47] Hachenberger conjectured that Case (b) of Theorem 15.2 cannot occur. In [110], we completed his classification by proving that this conjecture is indeed true.

In the next section, we focus on the independent proof of X. Chen of Payne's conjecture.

15.2 Parameters of STGQs

We consider finite EGQs $(\mathscr{S}^x, G)$ with Kantor family $(\mathscr{F}, \mathscr{F}^*)$ and parameters (s, t).

Lemma 15.3 *Suppose $\mathbb{S} = \cap_{A \in \mathscr{F}} A^*$, and that $|\mathbb{S}| = r$. If $s_p > 1$, then $k_{p'} r \leq s_p$ where $k = t/r$.*

Proof. We know that $|G| = s^2 t$ and $|\overline{G}| = s^2 k$. If P is a Sylow p-group of $\overline{G}$ then its size is $s_p^2 k_p$. If B is an element of $\mathscr{F}$, clearly $B \cap \mathbb{S} = \{1\}$, so

$$\overline{B} = B\mathbb{S}/\mathbb{S} \cong B/(B \cap \mathbb{S}) \cong B. \tag{15.2}$$

So if R is a Sylow p-subgroup of $\overline{B}$, then its size is s_p. Take $A \in \mathscr{F}$, and now let P be a Sylow p-subgroup of $\overline{G}$ that contains a Sylow p-subgroup P_{A^*} of $\overline{A^*}$. For each $B \in \mathscr{F} \setminus \{A\}$ let P_B be a subgroup of P that is conjugate to a Sylow p-subgroup of $\overline{B}$. By Lemma 13.3(ii) the members of $\{P_{A^*}\} \cup \{P_B | B \in \mathscr{F} \setminus \{A\}\}$ have pairwise trivial intersection. Then

$$|P_{A^*}| + \sum_{B \in \mathscr{F} \setminus \{A\}} |P_B^\times| \leq |P|, \tag{15.3}$$

so that $|P_{A^*}| = s_p k_p, |P_B| = s_p$, and $|P| = s_p^2 t_p$ lead us to

$$s_p k_p + t(s_p - 1) \leq s_p^2 k_p. \tag{15.4}$$

Whence $r k_p k_{p'}(s_p - 1) \leq s_p k_p(s_p - 1)$. If $s_p > 1$, then $r k_{p'} \leq s_p$. ∎

Theorem 15.4 *Let G be a group of order $s^2 t$ admitting a Kantor family $(\mathscr{F}, \mathscr{F}^*)$ of type (s,t). If $\mathbb{S} = \cap_{A \in \mathscr{F}} A^*$ is a normal subgroup of G and $|\mathbb{S}| \geq \sqrt{s}$, then G is a p-group.*

Proof. Suppose that G is not a p-group; then we already know that s has at least two prime divisors. Let p and q be such primes. Then from Lemma 15.3 we have $r k_{p'} \leq s_p$ and $r k_{q'} \leq s_q$. One of the equalities holds only if $r = s_p$ and $k_{p'} = 1$, or $r = s_q$ and $k_{q'} = 1$. So one of the equalities is strict. It follows that

$$r^2 k_{p'} k_{q'} < s_p s_q \leq s. \tag{15.5}$$

Note that $r^2 k_{p'} k_{q'} = tr k_{p,q'}$, so that $tr < s$. But then $r \geq \sqrt{s}$ would contradict Higman's inequality. ∎

Corollary 15.5 (PPC for finite STGQs) *If $(\mathscr{S}^x, G)$ is an STGQ, then G is a p-group.* ∎

15.3 Singer groups for quadrangles

We met (point) Singer groups of projective and affine planes as automorphism groups acting sharply transitively on the points. In §10.9, we introduced the concept of Singer groups for general incidence geometries. There is a deep theory available for GQs, especially developed in [27, 28, 30, 31, 29]. Erroneously, the theory seems to be closer to classification than that of projective planes, certainly in the case of PPC type problems. In [27], the authors showed that when a finite GQ admits an abelian Singer group, it must be isomorphic to some $\partial_x(\mathscr{S})$, where $\mathscr{S}$ is a TGQ of order s, s a power of 2, and x the (necessarily regular) translation point. The Singer group is uniquely determined as the translation group of $\mathscr{S}$. The next theorem immediately follows:

Theorem 15.6 *A finite cyclic thick generalized quadrangle does not exist.*

Proof. Since a cyclic group is abelian, the quadrangle should be isomorphic to some $\partial_x(\mathscr{S})$, with $\mathscr{S}^x$ a TGQ. Since the Singer group should be induced by the translation group of $\mathscr{S}^x$, it should be elementary abelian, implying that it is a group of prime order, contradiction (the order of the translation group should be a cube). ∎

So when the Singer group is abelian, we know that the order is $(s-1, s+1)$ for s some power of 2. (Note that even when one assumes a cyclic Singer group acting on a finite projective plane, it is not known yet whether the order of the plane is a prime power.) In general, such a result is not known for finite GQs, but as especially the recent preprint [29] shows, the classical Payne derived GQ $\partial_x(\mathbf{W}(q))$, where x is any point of $\mathbf{W}(q)$, even contains "many" nonisomorphic (nonabelian) Singer groups. Also, in [29], many classes of nonclassical Payne derived GQs are displayed with Singer groups.

Remark With regards to PPC theory, one has to be careful, since in [30] an example is given of a GQ of order $(5, 3)$ admitting a point-Singer group (which is not a p-group — its order is 96). It could be an exceptional example.

The reader notices that when a thick GQ $\mathscr{S}$ of order s with regular point x admits an automorphism group K which fixes x and acts sharply transitively on the points opposite x, then K induces a (point) Singer group on $\partial_x(\mathscr{S})$. (The converse is not necessarily true, as [29] shows.) Since $\mathscr{S}$ is the GQ-analogue (with respect to $\partial_x(\mathscr{S})$) of the projective completion of an affine plane (noting that projective completion in the GQ sense need not be unique), this situation of such a Singer group occurring for $\partial_x(\mathscr{S})$ is the GQ analogue of the planes of type (b) which we encountered earlier. The role of the translation planes in the classification of planes of type (b) is now played by elation quadrangles. We have the following correspondence:

$$
\begin{array}{ccc}
\text{affine plane} & \xrightarrow{\text{projective completion}} & \text{projective plane} \\
\text{(with Singer group)} & & \text{(type (b) plane)} \\
\\
\downarrow & & \downarrow \qquad (15.6) \\
\\
\text{Payne derivation } \partial_x(\mathscr{S}) & \xrightarrow{\text{Payne integration}} & \mathscr{S} \\
\text{(with induced Singer group)} & & \text{(type (b) quadrangle)}
\end{array}
$$

The reader is referred to the survey [30] and the preprint [29] for the details.

15.4   Appendix: Translation and elation polygons

One possible definition for "elation generalized n-gon" could be an n-gon Γ containing a point x and an automorphism group H that fixes each line incident with x and acts sharply transitively on $\Gamma_n(x)$. For $n \geq 5$, almost no interesting results are known, and no nonclassical polygons have been constructed, see §4.9 of [132]. One

could then proceed to define "translation polygons" by demanding that the group
be abelian. It is easy to see that such structures only can exist for $n \in \{3, 4\}$. (The
fact that the group is abelian forces the polygon to have ordinary quadrangles, see
[132, 4.9.7].)

16   Buildings with mixed parameters

The phenomenon that certain natural parameters (such as the number of points
in some subspace of prescribed type and the number of subspaces of that same type
through a point) associated to buildings occur as different cardinals seems to be
isolated if one of the parameters is assumed to be finite. Still, it is not so rare once
these parameters are both taken to be infinite. In this section, we give some remarks
and examples.

By no means we aim to be complete — rather, the examples should serve as a
starting point for the interested reader.

16.1   Locally finite polygons

Consider a thick generalized n-gon Γ with $s + 1$ points on each line and $t + 1$
lines through each point, where st is allowed to be infinite.

Theorem 16.1 ([132], 1.5.3) *If n is odd, then $s = t$. (Meaning that they are equal
if finite, and otherwise have the same cardinality.)*

Proof.   Let Γ' be an ordinary $(n + 1)$-gon in Γ. Let x be a point of Γ', and let L be
the line in Γ' opposite x. Then by projection, the lines incident with x (in Γ) are in
bijective correspondence with the points of L.   ∎

If n is even, though, we have seen that there are finite examples where $s \neq t$,
a most striking example being $n = 8$ in which case the theorem of W. Feit and G.
Higman [36] implies that if st is finite, $2st$ is a perfect square and so s is never equal
to t. If both s and t are finite, they are bounded by each other; to be more specific,
$s \leq t^2 \leq s^4$ for $n = 4$ and $n = 8$ and $s \leq t^3 \leq s^9$ for $n = 6$ (see [132, 1.7.2]). (For
other even values of n, Γ cannot exist by the Feit-Higman result.)

An old and notorious question, first posed by Jacques Tits in the 1960's, now
asks about the existence of locally finite generalized polygons. In other words, do
there exist, up to duality, (thick) generalized polygons with a finite number of points
incident with a line, and an infinite number of lines through a point? (It can be
found as Problem 5 in the "Ten Most Famous Open Problems" chapter of Van
Maldeghem's book [132], see also §10 of [100], etc.) Note that in Van Maldeghem's
book [132], such generalized polygons are called *semi-finite*.

There is only a very short list of results on Tits's question. All of them comprise
the case $n = 4$. P. J. Cameron [10] showed in 1981 that if $n = 4$ and $s = 2$, then t is
finite. In [6] A. E. Brouwer shows the same thing for $n = 4$ and $s = 3$ and the proof
is purely combinatorial (unlike an unnpublished but earlier proof of Kantor, see

[132]). More recently, G. Cherlin used Model Theory (in [13]) to handle the generalized 4-gons with five points on a line. For other values of n and s, nothing is known.

We will provide some more details about Cherlin's approach in the next section.
The surprising thing is that although we start from a purely combinatorial problem,
automorphism groups naturally come into play.

16.2 Indiscernibles

Let Γ be a generalized polygon. An ordered set $\mathscr{L}$ of lines is *indiscernible* if
for any two increasing sequences $M_1, M_2, \ldots, M_n$ and $M_1', M_2', \ldots, M_n'$ (of the same
length n) of lines of $\mathscr{L}$, there is an automorphism of Γ mapping M_i onto M_i' for
each i. It is indiscernible *over* D, if D is a finite set of points and lines fixed by the
automorphisms just described.

By combining the Compactness Theorem and Ramsey's Theorem [54] (in a theory which has a model in which a given definable set is infinite), one can prove the
following.

Theorem 16.2 (Cherlin [13]) *Suppose there is an infinite locally finite generalized n-gon with finite lines. Then there is an infinite locally finite generalized n-gon
Γ containing an indiscernible sequence $\mathscr{L}$ of parallel lines, of any specified order
type. The sequence may be taken to be indiscernible over the set D of all points
incident with one fixed line L of Γ.*

So as soon as locally finite polygons exist, there must exist examples with interesting automorphism groups (in [13] this is only stated for quadrangles, but the
observation is independent of the gonality). In [117], the author uses this observation as a starting point of an isomorphism theory for locally finite polygons.

Let $n = 4$, and Γ be as in Theorem 16.2. Choose fixed labels $\{1, 2, \ldots, k\}$ for the
points incident with L. By projection, each point on a line of $\mathscr{L}$ has a well-defined
label. Let M, M' be distinct lines of $\mathscr{L}$. Then $\sigma(M, M') \in \mathbf{S}_k$ is defined (on the
labels) by projecting each point of M to M'.

Proposition 16.3 ([13])   (i) $\sigma(M, M') = \sigma^{-1}(M', M)$.

(ii) σ *is independent of the choice of $M < M'$ (this follows from the fact that
for any (M, M') and (M'', M''') with $M < M'$ and $M'' < M'''$, there is an
automorphism in A mapping (M, M') onto (M'', M''')).*

(iii) $\sigma(M, M')$ *cannot have a fixed letter (here indiscernibility is used).*

We proceed with an observation from Cherlin [13]. Let $q < r$, and consider lines

$$M_q < M_r < \widehat{M_r} < \widehat{M_q}. \tag{16.1}$$

Fix a label i, and suppose M_q^* is the line which connects the point on M_q with
label i with a (unique) point of $\widehat{M_q}$, say with label j. Define M_r^* similarly. Then
$\sigma(M_q^*, M_r^*)$ involves the transposition (ij) [13].

Cherlin uses this observation to obtain short proofs (by way of contradiction) of
the next

Theorem 16.4 ([13]) *Let* Γ *be a thick GQ of order* $(k-1,t)$ *with* $k \in \{4,5\}$. *Then* $t < \infty$.

As soon as $k \geq 6$, things get messy.

If A and B are different subsets of $\mathscr{L}$, then it can be shown (see [117]) that

$$\langle \{L\} \cup A \rangle \neq \langle \{L\} \cup B \rangle. \tag{16.2}$$

Let $\mathscr{U} \subset \mathscr{L}$ have cardinality at least 4, and let U, V, W, X be distinct lines in $\mathscr{U}$. By considering the chain of full subGQs

$$\langle L, U \rangle \subset \langle L, U, V \rangle \subset \langle L, U, V, W \rangle \subset \langle \{L\} \cup \mathscr{U} \rangle, \tag{16.3}$$

it follows by Theorem 11.2 that the latter must be locally finite, while being generated by a finite number of points if we choose $\mathscr{U}$ to be finite. For future reference, we call such GQs (GPs) "Burnside quadrangles" ("Burnside polygons").

Remark The Burnside problem, posed by William Burnside in 1902 and one of the oldest and most influential questions in Group Theory, asks whether a finitely generated group in which every element has finite order must necessarily be a finite group. Such groups indeed exist, and are usually called *Burnside groups*. Since the quadrangles we encountered are geometric analogues of such groups, the term Burnside polygon seems in place. For more on Burnside's problem, we refer to [139]. Note that since Burnside groups do exist, a straightforward adaptation of Theorem 16.2 holds true for this class of groups.

16.3   Quadrangles of orthogonal type

Let I be an infinite uncountable set, and consider the vector space $\mathbb{Q}^{|I|}$ consisting of all $|I|$-tuples $(q_i)_{i \in I}$, $q_i \in \mathbb{Q}$, with only a finite number of nonzero entries. We assume w.l.o.g. that I contains the symbols 0 and 1. Define the following quadratic form:

$$\iota : X_0^2 + X_1^2 - \sum_{i \in I \setminus \{0,1\}} X_i^2. \tag{16.4}$$

Then ι has Witt index 2, and the corresponding classical orthogonal quadrangle $\mathbf{Q}(|I|, \mathbb{Q}, \iota)$ (cf. [132, Chapter 2]) is a Moufang generalized quadrangle with $|\mathbb{Q}|$ points per line and $|I|$ lines on a point. It is fully embedded in the projective space $\mathbf{P}(\mathbb{Q}^{|I|})$ (cf. [132, Chapter 2]).

Remark   (i) Note that orthogonal quadrangles are (linear) TGQs for any point (cf. [132, 3.4.8] and [132, 4.9.8]). So the above construction can be easily translated in terms of eggs.

(ii) Clearly, this example is a prototype of a class of quadrangles with a wealth of members having parameters with similar arithmetic properties.

It should not be too hard to inductively construct eggs which produce TGQs with similar arithmetic properties as the orthogonal quadrangles described above (note that Lemma 14.11 can not be used in its present form).

The example presented in this section was communicated to me by Hendrik Van Maldeghem, who is gratefully acknowledged.

16.4 Projective spaces $\mathbf{PG}(|\mathbb{N}|, q)$

Consider the following chain of projective spaces over $\mathbb{F}_q$ (where each of the inclusions is a natural full embedding):

$$\mathbf{PG}(-1, q) \subset \mathbf{PG}(0, q) \subset \mathbf{PG}(1, q) \subset \ldots \subset \mathbf{PG}(i, q) \subset \ldots \qquad (16.5)$$

with $i \in \mathbb{N}$. The union of these spaces defines a projective space $\mathbf{PG}(|\mathbb{N}|, q)$ which is of countably infinite dimension over $\mathbb{F}_q$. It has a finite number of subspaces of given finite dimension r in each finite ℓ-dimensional subspace (over $\mathbb{F}_q$), $\ell \geq r$, but a countably infinite number of subspaces of a given dimension $\ell \geq r + 1$ (and infinite co-dimension) through any r-space. (So, for instance, the point-line geometry of $\mathbf{PG}(|\mathbb{N}|, q)$ is a locally finite geometry with parameters $(q, |\mathbb{N}|)$.))

Again, many similar projective spaces can be constructed yielding parameters of different cardinal type. For instance, more generally, one can consider an $\mathbb{F}_q$-vector space

$$\mathbb{F}_q^\omega = \bigoplus_{i \in \omega} \mathbb{F}_q \qquad (16.6)$$

for a specified cardinal number ω, and construct $\mathbf{P}(\mathbb{F}_q^\omega)$, etc.; $\mathbb{F}_q$ can also be replaced by any skew field in this construction.

Theorem 16.5 *For any two cardinal numbers $\aleph$ and $\aleph'$, there exists a projective space for which the point-line geometry has parameters $(\aleph, \aleph')$.*

Proof. The only thing we have to show is that there are fields of any given cardinality. First note that for an integral domain $\mathbf{R}$, we have the natural injections

$$\mathbf{R} \hookrightarrow \mathbb{Q}(\mathbf{R}) \hookrightarrow \mathbf{R} \times \mathbf{R}, \qquad (16.7)$$

where $\mathbb{Q}(\mathbf{R})$ is the field of fractions of $\mathbf{R}$. So

$$|\mathbf{R}| = |\mathbb{Q}(\mathbf{R})|. \qquad (16.8)$$

Now let ω be any infinite cardinal. Consider any commutative ring $\mathbf{C}$, and a set $\{X_i\}_i$ of indeterminates of size ω. Then

$$|\mathbf{C}[\{X_i\}_i]| = \max(\omega, |\mathbf{C}|). \qquad (16.9)$$

The theorem now follows, by putting, e.g., $\mathbf{C} := \mathbb{Q}$. ∎

Note that it also follows that for any cardinal ω, there exist buildings of rank 1 with ω points (by considering, e.g., the natural action of $\mathbf{PSL}_2(\mathbb{Q}(\mathbb{Q}[\{X_i\}_{i\in\omega}]))$).

For polar spaces, similar considerations can be made.

16.5 Other buildings of higher rank

As was highlighted throughout this paper, GPs are the corner stones of spherical buildings of higher rank, and the most essential problems of the type considered in this section can be reduced to parameter problems in the rank 2 case. We leave details to the reader.

17 Thin buildings and the field $\mathbb{F}_1$

Till now, we have only considered and encountered parameter problems for thick buildings. Still, as we will see, there is also a rich theory available for thin buildings. Rather than thinking about the number of lines through a point or similar arithmetic properties, more essential here is the structural theory below these questions (think again of developing the theory of TGQs just to háve information about the parameters). In this section, we want to consider (certain) thin buildings as limits of buildings defined over, say, a finite field, where the number of field elements tends to 1, instead of defining them as "just" buildings with, e.g., thin lines.

In a paper which was published in 1957 [121], Tits made a seminal and provocative remark which alluded to the fact that through a certain analogy between the groups $\mathbf{GL}_n(q)$ (or $\mathbf{PGL}_n(q)$, q any prime power) and the symmetric groups $\mathbf{S}_n$, one should interpret $\mathbf{S}_n$ as a Chevalley group "over the field of characteristic one":

$$\lim_{q\to 1}\mathbf{PGL}_n(q) = \mathbf{S}_n. \tag{17.1}$$

Only much later serious considerations were made about Tits's point of view, and nowadays a deep theory is being developed on the philosophy over $\mathbb{F}_1$.

In fact, underlying this idea is the fact that thin (spherical) buildings are well-defined objects, and with a natural definition of automorphism group, the latter would become Weyl groups in thick buildings of the same type defined over "real fields" if one considers the appropriate building. So although for instance a thin building of type $\mathbf{A}_n$ is present in any thick building $\mathbf{PG}(n,q)$ over any finite field $\mathbb{F}_q$, we cannot define it as an incidence geometry over any field, since the cardinality of the latter should be one. Still, the automorphism group of the underlying geometry (which is just 2^X if X is the point set of the thin geometry) would precisely be the Weyl group of the associated thick building — namely the symmetric group on $n+1$ letters.

The geometry of algebraic curves underlying the structure of global fields of positive characteristic lies at the base of the solution of several deep and fundamental questions in Number Theory. As we will see below, several formulas of combinatorial nature (such as the number of subspaces of a finite projective space), still keep a meaningful value if evaluated at $q = 1$. Such results seem to suggest the existence of a mathematical object which is a nontrivial limit of finite fields $\mathbb{F}_q$ with $q \to 1$. The goal would be to define an analogue, for number fields, of the geometry underlying the arithmetic theory of function fields, cf. [19]. In [95], C. Soulé associated a zeta function to any sufficiently regular counting-type function $N(q)$ by considering the limit

$$\zeta_N(s) := \lim_{q \to 1} Z(q, q^{-s})(q-1)^{N(1)}, \quad s \in \mathbb{R}. \tag{17.2}$$

Here $Z(q, q^{-s})$ is the evaluation at $T = q^{-s}$ of the Hasse-Weil zeta function

$$Z(q, T) = \exp(\sum_{r \geq 1} N(q^r)\frac{T^r}{r}). \tag{17.3}$$

For the consistency of (17.2), one requires that $N(q)$ is defined over all real numbers $q \geq 1$ and not just only for prime powers. For many examples of algebraic varieties (such as projective spaces), it is known that $N(q)$ extends unambiguously to the real positive numbers, and often the associated zeta function is easy to compute [19]. As mentioned in [19], another basic example which is easy to handle is provided by a Chevalley group diagram. In [21] it was shown that they are varieties over $\mathbb{F}_1$ in the sense of Soulé [95]. We will consider this class in more detail below.

Another motivation for the introduction of $\mathbb{F}_1$-geometry stems from the search for a proof of the Riemann hypothesis. In the early 90s, Deninger gave criteria for a category of motives that would provide a geometric framework for translating Weil's proof of the Riemann hypothesis for global fields of positive characteristic to number fields. (One wants to see $\mathrm{Spec}(\mathbb{Z})$ as a curve over $\mathbb{F}_1$, so as to be able to define

$$\mathrm{Spec}(\mathbb{Z}) \times_{\mathrm{Spec}(\mathbb{F}_1)} \mathrm{Spec}(\mathbb{Z}); \tag{17.4}$$

Weil's proof of the Riemann hypothesis for a curve over a finite field makes essential use of such a product $\mathscr{C} \times_{\mathbb{F}_q} \mathscr{C}$ of two copies of a curve.) In particular, the Riemann zeta function $\zeta(s)$ should have a cohomological interpretation, where an H^0, an H^1 and an H^2-term are involved. Manin proposed in [69] to interpret the H^0-term as the zeta function of the "absolute point" $\mathrm{Spec}(\mathbb{F}_1)$ and the H^2-term as the zeta function of the "absolute Tate motive" or the "affine line over $\mathbb{F}_1$".

In this final section, we mention some aspects of (Algebraic and Incidence) geometry over $\mathbb{F}_1$, and we especially look for certain aspects of prime power/parameter conjectures. Needless to say, we will be sketchy on some parts, and only very few aspects of this emerging theory are mentioned; the reader is referred to the papers cited below (and the references therein) for details and more.

17.1 Bad approach

There is a well-known recipe which relates to any generalized n-gon a thin generalized $2n$-gon in a canonical way. It can be described for general rank 2 geometries, as follows. Let $\Gamma = (\mathscr{P}, \mathscr{B}, \mathbf{I})$ be a rank 2 geometry. Let $\mathscr{F}$ be its set of flags:

$$\mathscr{F} := \{\{x, L\} | (x, L) \in \mathbf{I}\}. \tag{17.5}$$

The *double* of Γ is the geometry $2\Gamma := (\mathscr{P} \cup \mathscr{B}, \mathscr{F}, \in)$. The following proposition is obvious.

Proposition 17.1   (i) *For any rank* 2 *geometry* Γ, 2Γ *is a thin rank* 2 *geometry.*

(ii) *We have a natural injection* $\mathrm{Aut}(\Gamma) \hookrightarrow \mathrm{Aut}(2\Gamma)$.   ∎

Although we obtain, in a functorial way, a thin generalized $2n$-gon from a thick generalized n-gon Γ, it is not the natural way to associate to Γ a polygon "over $\mathbb{F}_1$". The reason follows directly from the desired properties such an $\mathbb{F}_1$-polygon should have, cf. the next section. (It is not a universal object in the sense explained below, since it is a function of the parameters of the initial object, and as such it is not a well-defined limit object. Also, the injection of (ii) should rather be a projection for obvious reasons.) For similar reasons, degenerate polygons (such as the plane without order displayed in the figure) also are bad candidates.

As we will see, the natural way to do it will be through the functor

$$\mathscr{A} : \mathbb{B} \to \mathbb{A}, \tag{17.6}$$

from the category of (spherical) buildings to the category of apartments of such buildings.

In the next section, we take a closer look at projective spaces.

17.2   The projective space $\mathbf{PG}(n, \mathbb{F}_1)$, and $\mathbf{PGL}_{n+1}(\mathbb{F}_1)$

So for instance, projective geometry in this context would be something like

$$\lim_{q \to 1} \mathbf{PG}(n, q) =: \mathbf{PG}(n, 1). \tag{17.7}$$

The object $\mathbf{PG}(n, 1)$ should have several properties:

— all lines should have precisely 2 different points;

— it should be a "universal object", in the sense that it should be a subgeometry of any thick projective space (defined over *any* field, if at all defined over one) of dimension at least n;

— it should, of course, still carry the axiomatic structure of a projective space;

— it should give a geometric meaning to (certain) arithmetic formulas which express (certain) combinatorial properties of the finite thick geometries, evaluated at $q = 1$.

Let $n, q \in \mathbb{N}$, and define $[n]_q = 1 + q + \cdots + q^{n-1}$. (For q a prime power, $[n]_q = |\mathbf{PG}(n, q)|$.) Put $[0]_q! = 1$, and define

$$[n]_q! := [1]_q [2]_q \ldots [n]_q. \tag{17.8}$$

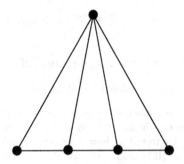

Figure 2: A degenerate projective plane.

Let **R** be a ring, and let x, y, q be "variables" for which $yx = qxy$. Then there are polynomials $\begin{bmatrix} n \\ k \end{bmatrix}_q$ in q with integer coefficients, such that

$$(x + y)^n = \sum_{k=0}^{n} \begin{bmatrix} n \\ k \end{bmatrix}_q x^k y^{n-k}. \tag{17.9}$$

Then

$$\begin{bmatrix} n \\ k \end{bmatrix}_q = \frac{[n]_q!}{[k]_q![n-k]_q!}, \tag{17.10}$$

and if q is a prime power, this is the number of $(k-1)$-dimensional subspaces of **PG**$(n-1, q)$.

It is clear that a projective plane defined over $\mathbb{F}_1$ should be an ordinary triangle (since that is the only thin plane containing an anti-flag); we should only be admitting thin lines, and then we are already done (in this case!). So it is nothing else than a chamber in the building of any thick projective plane. Adopting this point of view, it is easily seen that, more generally, projective n-spaces over $\mathbb{F}_1$ are just sets X of cardinality $n + 1$ endowed with the geometry of 2^X: any subset (of cardinality $0 \leq r + 1 \leq n + 1$) is a subspace (of dimension r). It is important to note that these spaces still satisfy the Veblen-Young axioms, and that they are the only such incidence geometries with thin lines.

Proposition 17.2 *Let* $n \in \mathbb{N} \cup \{-1\}$. *The projective space* **PG**$(n, \mathbb{F}_1)$ *is the complete graph on* $n + 1$ *vertices endowed with the induced geometry of subsets, and* **PGL**$_{n+1}(\mathbb{F}_1) \cong \mathbf{S}_{n+1}$. ∎

Note that over $\mathbb{F}_1$,

$$\mathbf{P\Gamma L}_{n+1}(\mathbb{F}_1) \cong \mathbf{PGL}_{n+1}(\mathbb{F}_1) \cong \mathbf{PSL}_{n+1}(\mathbb{F}_1). \tag{17.11}$$

The number of k-dimensional linear subspaces of **PG**$(n, \mathbb{F}_1)$, with $k \leq n \in \mathbb{N}$, equals

$$\begin{bmatrix} n+1 \\ k+1 \end{bmatrix}_1 = \begin{bmatrix} n+1 \\ k+1 \end{bmatrix}. \tag{17.12}$$

In this setting, affine spaces (or vector spaces) over $\mathbb{F}_1$, say of dimension $n \in \mathbb{N}$, now are naturally defined as follows:

Proposition 17.3 *Let $n \in \mathbb{N}$. The affine space $\mathbf{AG}(n, \mathbb{F}_1) \equiv \mathbb{F}_1^n$ is a single distinguished point $\mathbf{0}$, together with n edges consisting only of that point, endowed with the induced geometry of subsets of edges. We have that $\mathbf{AGL}_{n+1}(\mathbb{F}_1) \cong \mathbf{S}_n$.* ∎

So affine n-spaces over $\mathbb{F}_1$ can be regarded as data

$$(\mathbf{0}, (\mathbf{S}_n, X)), \tag{17.13}$$

where $\mathbf{0}$ is a distinguished point, and $(\mathbf{S}_n, X)$ represents the natural action of the symmetric group $\mathbf{S}_n$ on a set X of n letters; X corresponds to the set of "directions" determined by the space. (In some texts, affine spaces over $\mathbb{F}_1$ are identified with complete graphs, but in that case, we do not see any structural difference between affine and projective spaces; in fact, an affine space cannot be constructed anymore in the usual sense by deleting a hyperplane from a projective space — only the point $\mathbf{0}$ remains.)

Projective completion boils down to adding a second point on each edge, and then naturally extending the subgeometry structure. In this point of view, translation groups of $\mathbf{AG}(n, \mathbb{F}_1)$ are trivial.

17.3 Algebraic varieties over $\mathbb{F}_1$

We first describe Soulé's approach to algebraic varieties.

17.3.1 Soulé's approach Let $\mathscr{R}$ be a certain category of rings and $\mathscr{E}$ the category of sets. An *affine variety* over $\mathbb{F}_1$ is defined to be a pair $X = (\underline{X}, A_X)$, where $\underline{X}$ is a subfunctor of the functor

$$\mathscr{R} \to \mathscr{E} \tag{17.14}$$

represented by an affine variety $X_{\mathbb{Z}}$ over $\mathbb{Z}$ and A_X is a $\mathbb{C}$-algebra with an evaluation map $A_X \to \mathbb{C}$ for each $\mathbb{C}$-valued point of X. For any object R of $\mathscr{R}$, $X(R)$ is finite and the inclusion $X \hookrightarrow X_{\mathbb{Z}}$ admits a universal property for morphisms from X to functors represented by varieties over $\mathbb{Z}$.

Let us denote by $\mathscr{A}$ the category of affine varieties over $\mathbb{F}_1$ and by $\mathrm{Spec}(R)$ the affine variety over $\mathbb{F}_1$ represented by an object R of $\mathscr{R}$. A general *variety* over $\mathbb{F}_1$ is defined to be a pair $X = (\underline{X}, A_X)$, where $\underline{X}$ is a contravariant functor

$$\mathscr{A} \to \mathscr{E} \tag{17.15}$$

that is a subfunctor of the one represented by a variety $X_{\mathbb{Z}}$ over $\mathbb{Z}$ and A_X is a $\mathbb{C}$-algebra with a map $A_X \to A_B$ for each element of $\underline{X}(B)$. For any object R of $\mathscr{R}$, $X(\mathrm{Spec}(R))$ is finite and the injection $X \hookrightarrow X_{\mathbb{Z}}$ has the same universal property as in the affine case. The variety $X_{\mathbb{Z}}$ corresponds to the base change of X over $\mathbb{Z}$.

17.3.2 Varieties according to Connes-Consani In the above definition of variety, a problem arises when one is already considering finite projective spaces; it seems that in this approach the definition of $\mathbf{PG}(d, 1)$ of above is not compatible with that given by the previous subsection. In [95], the cardinality of $\mathbf{PG}(d, \mathbb{F}_{1^n})$ is shown to be $N(2n + 1)$, where $N(x) = [d + 1]_x$. But for $n = 1$, we then obtain the obstruction

$$|\mathbf{PG}(d, 1)| = N(3) = \frac{3^{d+1} - 1}{2}. \tag{17.16}$$

In Connes and Consani [19], a slightly more refined definition of variety is introduced so as to make both approaches compatible again. Their definition of algebraic variety is described by the following data:

(a) a covariant functor from the category of finite abelian groups to the category of graded sets

$$\underline{X} = \coprod_{k \geq 0} \underline{X}^{(k)} : \mathscr{F}_{ab} \to \mathscr{S}; \tag{17.17}$$

(b) an affine variety $X_{\mathbb{C}}$ over $\mathbb{C}$;

(c) a natural transformation e_X connecting $\underline{X}$ to the functor

$$\mathscr{F}_{ab} \to \mathscr{S} : D \to \mathrm{Hom}(\mathrm{Spec}(\mathbb{C}[D]), X_{\mathbb{C}}). \tag{17.18}$$

A triple $(\underline{X}, X_{\mathbb{C}}, e_X)$ is a *gadget over* $\mathbb{F}_1$. Such a gadget is called *graded* if

$$\underline{X} = \coprod_{k \geq 0} \underline{X}^{(k)} : \mathscr{F}_{ab} \to \mathscr{S} \tag{17.19}$$

takes values in the category of $\mathbb{Z}_{\geq 0}$-graded sets. It is *finite* when the set $\underline{X}(D)$ is finite for all $D \in \mathscr{F}_{ab}$.

An *affine variety* in the sense of Connes-Consani is a finite, graded gadget satisfying some extra technical condition, and varieties over $\mathbb{F}_{1^n}$ can then be naturally defined. The authors show in *loc. cit.* that this viewpoint is in accordance with the synthetic approach of above.

17.4 Algebraic groups over $\mathbb{F}_1$ and $\mathbb{F}_1$-buildings

Note again that since $\mathbb{F}_1$ expresses the idea of an Absolute Arithmetic, it is clear that the buildings of a certain prescribed type $\mathbf{T}$ over $\mathbb{F}_1$ should be present in any thick building of the same type.

Motivated by the properties which a building over $\mathbb{F}_1$ of type $\mathbf{T}$ should have, we are ready to define such geometries (and their groups) in general. Let $\mathscr{B} = (\mathscr{C}_1, \mathscr{C}_2, \ldots, \mathscr{C}_j, \mathbf{I})$ be a thick building of rank j and type $\mathbf{T}$ (given by one of the Coxeter diagrams below), and let $\mathscr{A}$ be its set of thin chamber subgeometries. Suppose (B, N) is a saturated effective BN-pair associated to $\mathscr{B}$; its Weyl group W is a Coxeter group defined by one of the Coxeter graphs below.

Proposition 17.4 *A building of rank j and type $\mathbf{T}$ defined over $\mathbb{F}_1$ is isomorphic to any element of $\mathscr{A}$. Its automorphism group is isomorphic to the Coxeter group W.* ∎

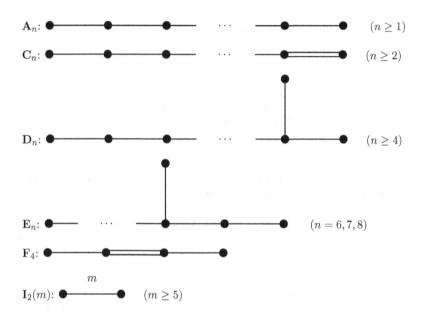

$\mathbf{A}_n$: ⬤————⬤————⬤—— $\cdots$ ——⬤————⬤ $(n \geq 1)$

$\mathbf{C}_n$: ⬤————⬤————⬤—— $\cdots$ ——⬤⬤ $(n \geq 2)$

$\mathbf{D}_n$: ⬤————⬤————⬤—— $\cdots$ ——⬤————⬤ $(n \geq 4)$

$\mathbf{E}_n$: ⬤—— $\cdots$ ————⬤————⬤————⬤ $(n = 6, 7, 8)$

$\mathbf{F}_4$: ⬤————⬤⬤————⬤

$\mathbf{I}_2(m)$: ⬤————m————⬤ $(m \geq 5)$

17.4.1   Generalized polygons   In particular, generalized n-gons over $\mathbb{F}_1$ are ordinary n-gons, and their automorphism groups are dihedral groups $\mathbf{D}_n$. It follows that the corner stones of the (spherical) buildings of rank at least 3 over $\mathbb{F}_1$ are the ordinary n-gons with $n = 3, 4, 6, 8$ (since these gonalities are the only ones which do occur in the corresponding thick buildings). Still, it is important to note that in the rank 2 examples, *all* positive integer values for n occur (except $n = 0, 1, 2$).

17.4.2   Quadrics   We give one final explicit example — it concerns quadrics.

Let $n \in \mathbb{N}_0$. A *quadric* of projective dimension $2n$ or $2n + 1$ over $\mathbb{F}_1$ is a set $\mathbf{Q}$ of $2(n + 1)$ points arranged in pairs $x_0, y_0, x_1, y_1, \ldots, x_n, y_n$, and its subspaces are the subsets not containing any couple (x_i, y_i). The Witt index of the so defined quadrics is n. The quadrics in dimension $2n$ have the further property that the

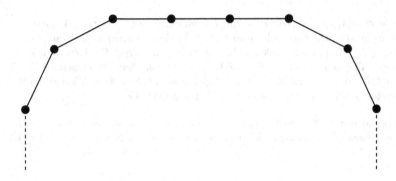

Figure 3: Generalized ∞-gon over $\mathbb{F}_1$.

maximal singular subspaces (n-spaces consisting of $n + 1$ points) are partitioned in two types, namely those containing an even number of points of $\{a_0, a_1, \ldots, a_n\}$ and those containing an odd number. Automorphisms are permutations of the set $\mathbf{Q}$ (which preserve the given pairing in the $2n$-dimensional case).

17.4.3   Trees as $\mathbb{F}_1$-geometries   If we allow the value $\infty = n$ in the definition of generalized n-gon, we obtain a point-line geometry Γ without closed paths, such that any two points or lines are contained in a path without end points. So Γ becomes a tree (allowing more than 2 points per line) without end points. Its apartments are paths without end points, and the Weyl group is an infinite dihedral group (generated by the reflections about two different adjacent vertices of such an apartment). So in this setting, a generalized ∞-gon over $\mathbb{F}_1$ is a tree of valency 2 without end points. (Note that in the nondiscrete setting — that is, generalized $\aleph$-gons for any cardinal number $\aleph > |\mathbb{N}|$ — similar apartments arise with more points.)

Consider, for instance, $G = \mathbf{SL}_2(\mathbb{F}_q((t^{-1})))$. Then G has a BN-pair (B, N), where

$$B = \{ \begin{pmatrix} a & b \\ c & d \end{pmatrix} \in \mathbf{SL}_2(\mathbb{F}_q[[t^{-1}]]) | c \equiv 0 \mod t^{-1} \}, \qquad (17.20)$$

and N is the subgroup of G consisting of elements with only 0 on the diagonal or only 0 on the antidiagonal. Its Weyl group is an infinite dihedral group generated by

$$s_1 = \begin{pmatrix} 0 & -1 \\ 1 & 0 \end{pmatrix} \quad \text{and} \quad s_2 = \begin{pmatrix} 0 & -t \\ 1/t & 0 \end{pmatrix}. \qquad (17.21)$$

The corresponding building (defined in the same way as before) is a generalized ∞-gon with $q + 1$ points per line and $q + 1$ lines per point. Its apartments are exactly the trees we introduced earlier in this section.

17.5   Zeta functions over $\mathbb{F}_1$

In [69], Manin proposes to develop Algebraic Geometry over $\mathbb{F}_1$, and predicts that zeta functions of varieties over $\mathbb{F}_1$ should have a simple form. For the projective space $\mathbf{PG}(d,1)$, $d \in \mathbb{N}_0$, it should look something like

$$\zeta_{\mathbf{PG}(d,1)}(s) = \frac{1}{\prod_{i=0}^{d}(s-i)}. \tag{17.22}$$

Recall that for an irreducible, smooth and projective algebraic curve X over a prime field $\mathbb{F}_p$, the counting function is of the form

$$\#X(\mathbb{F}_q) = N(q) = q - \sum_{\alpha} \alpha^r + 1, \quad q = p^r \tag{17.23}$$

where the α's are the complex roots of the characteristic polynomial of the Frobenius endomorphism acting on the étale cohomology $H^1(X \otimes \overline{\mathbb{F}_p}, \mathbb{Q}_\ell)$ of the curve ($\ell \neq p$).

In [95], C. Soulé, inspired by Manin's paper [69], associated a zeta function to any sufficiently regular counting-type function $N(q)$ by considering the limit

$$\zeta_N(s) := \lim_{q \to 1} Z(q, q^{-s})(q-1)^{N(1)}, \quad s \in \mathbb{R}. \tag{17.24}$$

Here $Z(q, q^{-s})$ is the evaluation at $T = q^{-s}$ of the Hasse-Weil zeta function

$$Z(q, T) = \exp(\sum_{r \geq 1} N(q^r)\frac{T^r}{r}). \tag{17.25}$$

One computes that if $N(X) = a_0 + a_1 X + \cdots + a_n X^n$, then

$$\zeta_{X|\mathbb{F}_1}(s) = \frac{1}{\prod_{i=0}^{n}(s-i)^{a_i}}, \tag{17.26}$$

which is in accordance with the aforementioned example for projective $\mathbb{F}_1$-spaces.

Remark   In [18, 19], the authors show that there is a unique "counting distribution" $N(q)$ whose associated zeta function (à la Soulé) is the complete Riemann zeta function. This distribution has all the desired properties, like being positive for $q > 1$ and having the value

$$N(1) = -\infty \tag{17.27}$$

as required by the Euler characteristic $\chi(\mathscr{C})$ of the curve $\mathscr{C}$. It is given by the following equation

$$N(q) = q - \frac{d}{dq}\left(\sum_{\rho \in Z} \text{order}(\rho)\frac{q^{\rho+1}}{\rho+1}\right) + 1, \tag{17.28}$$

where Z is the set of nontrivial zeros of the Riemann zeta function and the derivative is taken in the sense of distributions. The above formula also provides a strong indication that the hypothetical curve $\mathscr{C} = \overline{\text{Spec}(\mathbb{Z})}$ is related to the interpretation of the explicit formulae as a trace formula using the noncommutative geometric framework of the adèle class space (cf. below).

17.6 The hyperring of adèle classes

In a recent paper [20], the authors discovered unexpected connections between hyperrings and (axiomatic) projective geometry, foreseen with certain group actions. Denoting the profinite completion of $\mathbb{Z}$ by $\widehat{\mathbb{Z}}$ (and noting that it is isomorphic to the product $\prod_p \mathbb{Z}_p$ of all p-adic integer rings), the *integral adèle ring* is defined as

$$\mathbb{A}_{\mathbb{Z}} = \mathbb{R} \times \widehat{\mathbb{Z}}, \tag{17.29}$$

endowed with a suitable topology. Let $\mathbb{K}$ be a global field (that is, a finite extension of $\mathbb{Q}$, or the function field of an algebraic curve over $\mathbb{F}_q$ — the latter is a finite field extension of the field of rational functions $\mathbb{F}_q(X)$). The *adèle ring* of $\mathbb{K}$ is given by the expression

$$\mathbb{A}_{\mathbb{K}} = \prod_{\nu}' \mathbb{K}_{\nu}, \tag{17.30}$$

which is the restricted product of local fields $\mathbb{K}_{\nu}$, labeled by the places of $\mathbb{K}$.

If $\mathbb{K}$ is a number field, the adèle ring of $\mathbb{K}$ can be defined to be the tensor product

$$\mathbb{A}_{\mathbb{K}} = \mathbb{K} \otimes_{\mathbb{Z}} \mathbb{A}_{\mathbb{Z}}. \tag{17.31}$$

The adèle class space is important especially for two reasons:

- it gives a spectral realization of zeros of L-functions;

- it gives a trace formula interpretation of the explicit formulas.

We need a few more definitions.

17.6.1 Hyperrings and hyperfields Let H be a set, and "+" be a "hyperoperation" on H, namely a map

$$+ : H \times H \to (2^H)^{\times}, \tag{17.32}$$

where $(2^H)^{\times} = 2^H \setminus \{\emptyset\}$. For $U, V \subseteq H$, denote $\{\cup(u+v)|u \in U, v \in V\}$ by $U + V$. Then $(H, +)$ is an *abelian hypergroup* provided the following properties are satisfied:

- $x + y = y + x$ for all $x, y \in H$;

- $(x + y) + z = x + (y + z)$ for all $x, y, z \in H$;

- there is an element $0 \in H$ such that $x + 0 = 0 + x$ for all $x \in H$;

- for all $x \in H$ there is a unique $y \in H$ $(=: -x)$ such that $0 \in x + y$;

- $x \in y + z \implies z \in x - y$.

Proposition 17.5 ([19]) *Let $(G, \cdot)$ be an abelian group, and let $K \le \operatorname{Aut}(G)$. Then the following operation defines a hypergroup structure on $H = \{g^K | g \in G\}$:*

$$g_1^K \cdot g_2^K := (g_1^K \cdot g_2^K)/K. \tag{17.33}$$

A *hyperring* $(\mathbf{R}, +, \cdot)$ is a nonempty set $\mathbf{R}$ endowed with a hyperaddition $+$ and the usual multiplication $\cdot$ such that:

- $(\mathbf{R}, +)$ is an abelian hypergroup with neutral element 0;

- $(\mathbf{R}, \cdot)$ is a monoid with multiplicative identity 1;

- for all $u, v, w \in \mathbf{R}$ we have that $u \cdot (v+w) = u \cdot v + u \cdot w$ and $(v+w) \cdot u = v \cdot u + w \cdot u$;

- for all $u \in \mathbf{R}$ we have that $u \cdot 0 = 0 = 0 \cdot u$;

- $0 \neq 1$.

A hyperring $(\mathbf{R}, +, \cdot)$ is a *hyperfield* if $(\mathbf{R} \setminus \{0\}, \cdot)$ is a group.

17.6.2   The Krasner hyperfield   The *Krasner hyperfield* $\mathbf{K}$ is the hyperfield $(\{0, 1\}, +, \cdot)$ with additive neutral element 0, usual multiplication with identity $\mathbf{1}$, and satisfying the hyperrule

$$1 + 1 = \{0, 1\}. \tag{17.34}$$

In the category of hyperrings, $\mathbf{K}$ can be seen as the natural extension of the commutative pointed monoid $\mathbb{F}_1$, that is, $(\mathbf{K}, \cdot) = \mathbb{F}_1$. As remarked in [21], the Krasner hyperfield encodes the arithmetic of zero and nonzero numbers, just as $\mathbb{F}_2$ does for even and odd numbers. From this viewpoint, it is of no surprise that projective geometry will come into play.

17.7   Hyperfield extensions of the Krasner hyperfield $\mathbf{K}$

Let $\mathbf{R}$ be a commutative ring, and let G be a subgroup of its multiplicative group. The following operations define a hyperring on the set $\mathbf{R}/G$ of G-orbits in $\mathbf{R}$ under multiplication.

- HYPERADDITION. $x + y := (xG + yG)/G$ for $x, y \in \mathbf{R}/G$.

- MULTIPLICATION. $xG \cdot yG := xyG$ for $x, y \in \mathbf{R}/G$.

Proposition 17.6 ([20]) *Let $\mathbb{K}$ be a field with at least three elements. Then the hyperring $\mathbb{K}/\mathbb{K}^\times$ is isomorphic to the Krasner hyperfield. If, in general, $\mathbf{R}$ is a commutative ring and $G \subset \mathbb{K}^\times$ is a proper subgroup of the group of units of $\mathbf{R}$, then the hyperring $\mathbf{R}/G$ defined as above contains $\mathbf{K}$ as a subhyperfield if and only if $\{0\} \cup G$ is a subfield of $\mathbf{R}$.*

Remark   Consider a global field $\mathbb{K}$. Its adèle class space $\mathbb{H}_{\mathbb{K}} = \mathbb{A}_{\mathbb{K}}/\mathbb{K}^\times$ is the quotient of a commutative ring $\mathbb{A}_{\mathbb{K}}$ by $G = \mathbb{K}^\times$, and $\{0\} \cup G = \mathbb{K}$, so it is a hyperring extension of $\mathbf{K}$.

A **K**-*vector space* is a hypergroup E provided with a compatible action of **K**. As $0 \in \mathbf{K}$ acts by retraction (to $\{0\} \subset E$) and $1 \in \mathbf{K}$ acts as the identity on E, the **K**-vector space structure is completely determined by the hypergroup structure. It follows that a hypergroup E is a **K**-vector space if and only if

$$x + x = \{0, x\} \quad \text{for} \quad x \neq 0. \tag{17.35}$$

Let E be a **K**-vector space, and define $\mathscr{P} := E \setminus \{0\}$. For $x, y \neq x \in \mathscr{P}$, define the *line* $L(x, y)$ as

$$x + y \cup \{x, y\}. \tag{17.36}$$

It can be easily shown — see [86] — that $(\mathscr{P}, \{L(x,y) | x, y \neq x \in \mathbf{P}\})$ is a projective space. Conversely, if $(\mathscr{P}, \mathscr{L})$ is the point-line geometry of a projective space with at least 4 points per line, then a hyperaddition on $E := \mathscr{P} \cup \{0\}$ can be defined as follows:

$$x + y = xy \setminus \{x, y\} \quad \text{for} \quad x \neq y, \quad \text{and} \quad x + x = \{0, x\}. \tag{17.37}$$

Now let $\mathbb{H}$ be a hyperfield extension of **K**, and let $(\mathscr{P}, \mathscr{L})$ be the point-line geometry of the associated projective space; then A. Connes and C. Consani [20] show that $\mathbb{H}^{\times}$ induces a so-called "two-sided incidence group" (and conversely, starting from such a group G, there is a unique hyperfield extension $\mathbb{H}$ of **K** such that $\mathbb{H} = G \cup \{0\}$). By the Veblen-Young result, this connection is reflected by the next theorem.

Proposition 17.7 ([20]) *Let $\mathbb{H} \supset \mathbf{K}$ be a finite commutative hyperfield extension of* **K**. *Then one of the following cases occurs:*

(i) $\mathbb{H} = \mathbf{K}[G]$ *for a finite abelian group G.*

(ii) *There exists a finite field extension $\mathbb{F}_q \subseteq \mathbb{F}_{q^m}$ such that $\mathbb{H} = \mathbb{F}_{q^m}/\mathbb{F}_q^{\times}$.*

(iii) *There exists a finite nonDesarguesian projective plane admitting a sharply point-transitive automorphism group G, and G is the abelian incidence group associated to $\mathbb{H}$.*

Sketch of Proof. Let $\mathbb{H} \supset \mathbf{K}$ be a finite commutative hyperfield extension of **K**; we know that to this extension there is associated a thick projective n-space. If its dimension is at least 3, the space is Desarguesian by the Veblen-Young result, and we are in (ii) since the extension is commutative (see [20] for details). If the dimension is 2, we are in (ii) or (iii). If the dimension is 1 we are in (i). ∎

In case (i), there is only one line (otherwise we have to be in the other cases), so for all $x, y, x', y' \in \mathbb{H} \setminus \{0\}$ with $x \neq y$ and $x' \neq y'$, we must have

$$L(x, y) = (x + y) \cup \{x, y\} = (x' + y') \cup \{x', y'\} = L(x' + y') = \mathbb{H} \setminus \{0\}. \tag{17.38}$$

In other words, hyperaddition is completely determined:

$$\begin{cases} x + 0 = x & \text{for } x \in \mathbb{H} \\ x + x = \{0, x\} & \text{for } x \in \mathbb{H}^\times \\ x + y = \mathbb{H} \setminus \{0, x, y\} & \text{for } x \neq y \in \mathbb{H}^\times \end{cases} \qquad (17.39)$$

Remark Note that there exist infinite hyperfield extensions $\mathbb{H} \supset \mathbf{K}$ for which $\mathbb{H}^\times \cong \mathbb{Z}$ and not coming from Desarguesian projective spaces in the above sense, see M. Hall [48].

References

[1] A. A. ALBERT. Finite division algebras and finite planes, in: *1960 Proc. Sympos. Appl. Math.* **10**, pp. 53–70, American Mathematical Society, Providence, R.I.

[2] J. BAMBERG, T. PENTTILA AND C. SCHNEIDER. Elation generalized quadrangles for which the number of lines on a point is the successor of a prime, *J. Aust. Math. Soc.* **85** (2008), 289–303.

[3] T. BETH, D. JUNGNICKEL AND H. LENZ. *Design Theory, Vol. I, Second Edition*, Encyclopedia of Mathematics and its Applications **69**, Cambridge University Press, Cambridge, 1999.

[4] A. BEUTELSPACHER AND P. J. CAMERON. Transfinite methods in geometry, in: *A tribute to J. A. Thas (Gent, 1994), Bull. Belg. Math. Soc. — Simon Stevin* **1** (1994), 337–347.

[5] A. BLOKHUIS, D. JUNGNICKEL AND B. SCHMIDT. Proof of the prime power conjecture for projective planes of order n with abelian collineation groups of order n^2, *Proc. Amer. Math. Soc.* **130** (2002), 1473–1476.

[6] A. E. BROUWER. A nondegenerate generalized quadrangle with lines of size four is finite, in: J. W. P. Hirschfeld, D. R. Hughes, J. A. Thas (Eds.), *Advances in Finite Geometries and Designs*, Proceedings of Isle of Thorns 1990, Oxford University Press, Oxford, 1991, pp. 47–49.

[7] P. ABRAMENKO AND K. S. BROWN. *Buildings. Theory and applications*, Graduate Texts in Mathematics **248**, Springer, New York, 2008.

[8] R. H. BRUCK AND H. J. RYSER. The nonexistence of certain finite projective planes, *Canadian J. Math.* **1** (1949), 88–93.

[9] F. BUEKENHOUT AND H. VAN MALDEGHEM. Finite distance-transitive generalized polygons, *Geom. Dedicata* **52** (1994), 41–51.

[10] P. J. CAMERON. Orbits of permutation groups on unordered sets II, *J. London Math. Soc.* **23** (1981), 249–264.

[11] X. CHEN. On the groups that generate skew translation generalized quadrangles, *Unpublished Manuscript*, 1990.

[12] X. CHEN AND D. FROHARDT. Normality in a Kantor family, *J. Combin. Theory Ser. A* **64** (1993), 130–136.

[13] G. CHERLIN. Locally finite generalized quadrangles with at most five points per line, *Discrete Math.* **291** (2005), 73–79.

[14] S. A. CHOWLA. A property of biquadratic residues, *Proc. Nat. Acad. Sci. India. Sect. A.* **14** (1944), 45–46.

[15] H. COHN. Projective geometry over $\mathbb{F}_1$ and the Gaussian binomial coefficients, *Amer. Math. Monthly* **111** (2004), 487–495.

[16] A. CONNES, C. CONSANI AND M. MARCOLLI. The Weil proof and the geometry of the adèles class space, in: *Algebra, Arithmetic, and Geometry: in honor of Yu. I. Manin, Vol. I*, Progr. Math. **269** (2009), Birkhäuser Boston, Inc., Boston, MA, pp. 339–405.

[17] A. CONNES, C. CONSANI AND M. MARCOLLI. Fun with $\mathbb{F}_1$, *J. Number Theory* **129** (2009), 1532–1561.

[18] A. CONNES AND C. CONSANI. Schemes over $\mathbb{F}_1$ and zeta functions, *Compositio Math.* **146** (2010), 1383–1415.

[19] A. CONNES AND C. CONSANI. From monoids to hyperstructures: in search of an absolute arithmetic, in: *Casimir Force, Casimir Operators and the Riemann Hypothesis*, de Gruyter (2010), pp. 147–198.

[20] A. CONNES AND C. CONSANI. The hyperring of adèle classes, *J. Number Theory* **131** (2011), 159–194.

[21] A. CONNES AND C. CONSANI. On the notion of geometry over $\mathbb{F}_1$, *J. Algebraic Geometry*, To appear.

[22] J. H. CONWAY, R. T. CURTIS, S. P. NORTON, R. A. PARKER AND R. A. WILSON. *Atlas of Finite Groups. Maximal subgroups and ordinary characters for simple groups*, With computational assistance from J. G. Thackray, Oxford University Press, Eynsham, 1985.

[23] H. S. M. COXETER. The complete enumeration of finite groups of the form $R_i^2 = (R_iR_j)^{k_{ij}} = 1$, *J. London Math. Soc.* **10** (1935), 21–25.

[24] P. DEMBOWSKI. *Finite Geometries*, Springer-Verlag, New York, 1968.

[25] P. DEMBOWSKI AND F. PIPER. Quasiregular collineation groups of finite projective planes, *Math. Z.* **99** (1967), 53–75.

[26] T. DE MEDTS AND Y. SEGEV. A course on Moufang sets, *Innov. Incidence Geom.* **9** (2009), 79-122.

[27] S. DE WINTER AND K. THAS. Generalized quadrangles with an abelian Singer group, *Des. Codes Cryptogr.* **39** (2006), 81–87.

[28] S. DE WINTER AND K. THAS. The automorphism group of Payne derived generalized quadrangles, *Adv. Math.* **214** (2007), 146–156.

[29] S. DE WINTER AND K. THAS. A criterion concerning Singer groups of generalized quadrangles, Preprint.

[30] S. DE WINTER, E. E. SHULT AND K. THAS. *Singer Quadrangles*, Oberwolfach Preprint **OWP 2009-07**, MFO Oberwolfach, 2009.

[31] S. DE WINTER, E. E. SHULT AND K. THAS. The Singer quadrangles amongst the known finite quadrangles, Preprint.

[32] S. DIXMIER & F. ZARA. Etude d'un quadrangle généralisé autour de deux de ses points non liés, Preprint, 1976.

[33] J. DIXON AND B. MORTIMER. *Permutation Groups*, Springer-Verlag, New York, 1996.

[34] J. R. FAULKNER. Groups with Steinberg relations and coordinatization of polygonal geometries, *Mem. Am. Math. Soc.* **10** (1977).

[35] W. FEIT. Finite projective planes and a question about primes, *Proc. Amer. Math. Soc.* **108** (1990) 561–564.

[36] W. FEIT AND G. HIGMAN. The nonexistence of certain generalized polygons, *J. Algebra* **1** (1964), 114–131.

[37] J. B. FINK. A note on sharply flag-transitive projective planes, in: Norman L. Johnson, Michael J. Kallaher and Calvin T. Long., eds, *Finite Geometries, Proceedings of a Conference in honor of T. G. Ostrom*, Washington State University, Pullman, Wash., April 7–11, 1981, Lecture Notes in Pure and Applied Mathematics **82** (New York, 1983), pp. 161–164.

[38] J. B. FINK. Flag-transitive projective planes, *Geom. Dedicata* **17** (1985) 219–226.

[39] P. FONG AND G. M. SEITZ. Groups with a (B,N)-pair of rank 2, I, *Invent. Math.* **21** (1973), 1–57.

[40] P. FONG AND G. M. SEITZ. Groups with a (B,N)-pair of rank 2, II, *Invent. Math.* **24** (1974), 191–239.

[41] H. FREUDENTHAL. Une étude de quelques quadrangles généralisés, *Ann. Mat. Pura Appl.* **102** (1975), 109–133.

[42] D. FROHARDT. Groups which produce generalized quadrangles, *J. Combin. Theory Ser. A* **48** (1988), 139–145.

[43] D. GHINELLI AND D. JUNGNICKEL. Some geometric aspects of finite abelian groups, *Rend. Mat. Appl.* **26** (2006), 29–68.

326 Koen Thas

[44] D. GHINELLI AND D. JUNGNICKEL. Finite projective planes with a large abelian group, in: *Surveys in Combinatorics, 2003 (Bangor)*, pp. 175–237, London Math. Soc. Lecture Note Ser. **307**, Cambridge Univ. Press, Cambridge, 2003.

[45] B. GORDON, W. H. MILLS AND L. R. WELCH. Some new difference sets, *Canad. J. Math.* **14** (1962), 614–625.

[46] D. GORENSTEIN. *Finite Groups. Second Edition*, University Series in Mathematics, Plenum Publishing Corp., New York, 1982.

[47] D. HACHENBERGER. Groups admitting a Kantor family and a factorized normal subgroup, *Des. Codes Cryptogr.* **8** (1996), 135–143.

[48] M. HALL, JR. Cyclic projective planes, *Duke Math. J.* **14** (1947), 1079–1090.

[49] C. HERING, W. M. KANTOR AND G. M. SEITZ. Finite groups with a split BN-pair of rank 1, *J. Algebra* **20** (1972), 435–475.

[50] D. G. HIGMAN. Flag-transitive collineation groups of finite projective spaces, *Illinois J. Math.* **6** (1962), 434–446.

[51] D. G. HIGMAN. Partial geometries, generalized quadrangles and strongly regular graphs, in: *Atti Convegno di Geometriae Combinatorica e Sue Applicazioni* (Univ. Perugia, Perugia, 1970) Ist. Mat., Univ. Perugia, Perugia (1971), pp. 263–293.

[52] D. G. HIGMAN AND J. E. MCLAUGHLIN. Geometric ABA-groups, *Illinois J. Math* **5** (1961), 382–397.

[53] Y. HIRAMINE. Automorphisms of p-groups of semifield type, *Osaka J. Math.* **20** (1983), 735–746.

[54] W. HODGES. *Model Theory*, Encyclopedia of Mathematics and its Applications **42**, Cambridge University Press, Cambridge, 1993.

[55] W. HODGES. *A Shorter Model Theory*, Cambridge University Press, Cambridge, 1997.

[56] D. HUGHES AND F. C. PIPER. *Projective Planes*, Graduate Texts in Mathematics **6**, Springer-Verlag, New York—Berlin, 1973.

[57] W. M. KANTOR. On 2-transitive collineation groups of finite projective spaces, *Pacific J. Math.* **48** (1973), 119–131.

[58] W. M. KANTOR. Generalized quadrangles associated with $G_2(q)$, *J. Combin. Theory Ser. A* **29** (1980), 212–219.

[59] W. M. KANTOR. Some generalized quadrangles with parameters q^2, q, *Math. Z.* **192** (1986), 45–50.

[60] W. M. KANTOR. Primitive groups of odd degree and an application to finite projective planes, *J. Algebra* **106** (1987), 15–45.

[61] W. M. KANTOR. Automorphism groups of some generalized quadrangles, in: *Advances in Finite Geometries and Designs*, Edited by J. W. P. Hirschfeld et al., Proceedings of the Third Isle of Thorns Conference 1990, Oxford University Press, New York (1991), pp. 251–256.

[62] W. M. KANTOR. Finite semifields, in: *Finite Geometries, Groups, and Computation*, pp. 103–114, Walter de Gruyter GmbH & Co. KG, Berlin, 2006.

[63] B. KLEINER AND B. LEEB. Rigidity of quasi-isometries for symmetric spaces and Euclidean buildings, *Inst. Hautes Études Sci. Publ. Math.* **86** (1997), 115–197.

[64] N. KNARR. The nonexistence of certain topological polygons, *Forum Math.* **2** (1990), 603–612.

[65] N. KNARR. *Translation Planes. Foundations and construction principles*, Lecture Notes in Mathematics **1611**, Springer-Verlag, Berlin, 1995.

[66] N. KNARR. *Private Communication*, October 2004 and June 2005.

[67] L. KRAMER. *Compact Polygons*, Ph.D. Thesis, Universität Tübingen, Tübingen, Germany, 1994.

[68] E. LEHMER. On residue difference sets, *Canadian J. Math.* **5** (1953), 425–432.

[69] YU. MANIN. Lectures on zeta functions and motives (according to Deninger and Kurokawa), Columbia University Number Theory Seminar (New York, 1992), *Astérisque* **228** (1995), 121–163.

[70] B. MÜHLHERR AND H. VAN MALDEGHEM. Exceptional Moufang quadrangles of type ßF₄, *Canad. J. Math.* **51** (1999), 347–371.

[71] T. G. OSTROM AND A. WAGNER. On projective and affine planes with transitive collineation groups, *Math. Z* **71** (1959), 186–199.

[72] U. OTT. Fahnentransitive Ebenen ungerader Ordnung, *Geom. Dedicata* **8** (1979), 219–252.

[73] U. OTT. Fahnentransitive Ebenen gerader Ordnung, *Arch. Math. (Basel)* **28** (1977), 661–668.

[74] U. OTT. Flag-transitive projective planes and power residue difference sets, *J. Algebra* **276** (2004), 663–673.

[75] R. E. A. C. PALEY. On orthogonal matrices, *J. Math. Phys.* **12** (1933), 311–320.

[76] S. E. PAYNE. Skew-translation generalized quadrangles, *Congress. Numer.* **XIV**, Proc. 6th S. E. Conf. Comb., Graph Th. and Comp. (1975), 485–504.

[77] S. E. PAYNE. All generalized quadrangles of order 3 are known, *J. Combin. Theory Ser. A* **18** (1975), 203–206.

[78] S. E. PAYNE. Generalized quadrangles of order 4, I, *J. Combin. Theory Ser. A* **22** (1977), 267–279.

[79] S. E. PAYNE. Generalized quadrangles of order 4, II, *J. Combin. Theory Ser. A* **22** (1977), 280–288.

[80] S. E. PAYNE. Generalized quadrangles as group coset geometries, *Proceedings of the Eleventh Southeastern Conference on Combinatorics, Graph Theory and Computing (Florida Atlantic Univ., Boca Raton, Fla., 1980), Vol II, Congr. Numer.* **29** (1980), 717–734.

[81] S. E. PAYNE. An essay on skew translation generalized quadrangles, *Geom. Dedicata* **32** (1989), 93–118.

[82] S. E. PAYNE AND J. A. THAS. Generalized quadrangles with symmetry, Part II, *Simon Stevin* **49** (1975/76), 81–103.

[83] S. E. PAYNE AND J. A. THAS. *Finite Generalized Quadrangles*, Research Notes in Mathematics **110**, Pitman Advanced Publishing Program, Boston/London/Melbourne, 1984.

[84] S. E. PAYNE AND J. A. THAS. *Finite Generalized Quadrangles*. Second edition, EMS Series of Lectures in Mathematics, European Mathematical Society, 2009.

[85] D. PERIN. On collineation groups of finite projective spaces, *Math. Z.* **126** (1972), 135–142.

[86] W. PRENOWITZ. Projective geometries as multigroups, *Amer. J. Math.* **65** (1943), 235–256.

[87] N. ROSEHR. Parameters of translation generalized quadrangles, *Bull. Belg. Math. Soc. — Simon Stevin* **12** (2005), 329–340.

[88] R. ROTH. Flag-transitive planes of even order, *Proc. Amer. Math. Soc.* **15** (1964), 585-490.

[89] H. SALZMANN, D. BETTEN, T. GRUNDHÖFER, H. HÄHL, R. LÖWEN, AND M. STROPPEL. *Compact Projective Planes. With an introduction to octonion geometry*, de Gruyter Expositions in Mathematics **21**, Walter de Gruyter & Co., Berlin, 1995.

[90] J. J. SEIDEL. Strongly regular graphs with $(-1, 1, 0)$ adjacency matrix having eigenvalue 3, *Linear Algebra Appl.* **1** (1968), 281–298.

[91] E. E. SHULT. Characterizations of certain classes of graphs, *J. Combin. Theory Ser. B.* **13** (1972), 142–167.

[92] E. E. SHULT, On a class of doubly transitive groups, *Illinois J. Math.* **16** (1972), 434 – 455.

[93] E. SHULT. Aspects of buildings, *Groups and Geometries (Siena, 1996), Trends Math.*, Birkhäuser, Basel (1998), 177–188.

[94] J. SINGER. A theorem in finite projective geometry and some applications to number theory, *Trans. Amer. Math. Soc.* **43** (1938), 377–385.

[95] C. SOULÉ. Les variéatés sur le corps à un élément, *Mosc. Math. J.* **4** (2004), 217–244, 312.

[96] K. TENT. Very homogeneous generalized n-gons of finite Morley rank, *J. London Math. Soc.* **62** (2000), 1–15.

[97] K. TENT. A weak Moufang condition suffices, *European J. Combin.* **26** (2005), 1207–1215.

[98] K. TENT. Free polygons, twin trees, and CAT(1)-spaces, *Pure Appl. Math. Quart.* **7** (2011), 1031–1052.

[99] J. A. THAS. On 4-gonal configurations with parameters $r = q^2 + 1$ and $k = q + 1$, *Geom. Dedicata* **3** (1974), 365–375.

[100] J. A. THAS. *Generalized Polygons, Handbook of Incidence Geometry*, 383–431, North-Holland, Amsterdam, 1995.

[101] J. A. THAS AND H. VAN MALDEGHEM. Generalized quadrangles and the axiom of Veblen, *Geometry, Combinatorial Designs and Related Structures*, Edited by J. W. P. Hirschfeld, Math. Soc. Lecture Note Ser. **245**, Cambridge University Press, London (1997), 241–253.

[102] J. A. THAS, S. E. PAYNE AND H. VAN MALDEGHEM. Half Moufang implies Moufang for finite generalized quadrangles, *Invent. Math.* **105** (1991), 153–156.

[103] J. A. THAS, K. THAS AND H. VAN MALDEGHEM. *Moufang Quadrangles: Characterizations, Classification, Generalizations*, Capita Selecta in Geometry — A series of 27 lectures, i+87pp., Ghent University, Ghent, Spring 2003.

[104] J. A. THAS, K. THAS AND H. VAN MALDEGHEM. *Translation Generalized Quadrangles*, Series in Pure Mathematics **26**, World Scientific, Singapore, 2006.

[105] K. THAS. Automorphisms and characterizations of finite generalized quadrangles, *Generalized Polygons*, Proceedings of the Academy Contact Forum "Generalized Polygons" 20 October, Palace of the Academies, Brussels, Belgium, Edited by F. De Clerck et al., Universa Press (2001), 111–172.

[106] K. THAS. A theorem concerning nets arising from generalized quadrangles with a regular point, *Des. Codes Cryptogr.* **25** (2002), 247–253.

[107] K. THAS. Finite flag-transitive projective planes: a survey and some remarks, *Discrete Math.* **266** (2003), 417–429.

[108] K. THAS. *Symmetry in Finite Generalized Quadrangles*, Frontiers in Mathematics **1**, Birkhäuser, 2004.

[109] K. THAS. Solution of a question of Knarr, *Proc. Amer. Math. Soc.* **136** (2008), 1409–1418.

[110] K. THAS. A Kantor family admitting a normal **F**-factor constitutes a p-group, *Adv. Geom.* **8** (2008), 155–159.

[111] K. THAS. A generalized quadrangle of order (s, t) with center of transitivity is an elation quadrangle if $s \leq t$, *Des. Codes Cryptogr.* **47** (2008), 221–224.

[112] K. THAS. Finite BN-pairs of rank 2 in characteristic 2, I, *Innov. Incidence Geom.* **9** (2009), 189–202.

[113] K. THAS. *Lectures on Elation Quadrangles*, Monograph, 135 pp., Submitted.

[114] K. THAS. Central aspects of skew translation quadrangles, Preprint.

[115] K. THAS. Finite BN-pairs of rank 2 in characteristic 2, II, Preprint.

[116] K. THAS. Linearity of translation generalized quadrangles, In preparation.

[117] K. THAS. Locally finite polygons, In preparation.

[118] K. THAS AND H. VAN MALDEGHEM. Moufang-like conditions for generalized quadrangles and classification of all quasi-transitive generalized quadrangles, *Discrete Math.* **294** (2005), 203–217.

[119] K. THAS AND H. VAN MALDEGHEM. Geometrical characterizations of Chevalley groups of type **B**$_2$, *Trans. Amer. Math. Soc.* **360** (2008), 2327–2357.

[120] K. THAS AND D. ZAGIER. Finite projective planes, Fermat curves, and Gaussian periods, *J. Eur. Math. Soc. (JEMS)* **10** (2008), 173–190.

[121] J. TITS. Sur les analogues algébriques des groupes semi-simples complexes, *Centre Belge Rech. Math., Colloque d'Algèbre supérieure, Bruxelles du 19 au 22 déc. 1956*, pp. 261-289, 1957.

[122] J. TITS. Sur la trialité et certains groupes qui s'en déduisent, *Inst. Hautes Etudes Sci. Publ. Math.* **2** (1959), 13–60.

[123] J. TITS. *Buildings of Spherical Type and Finite BN-Pairs*, Lecture Notes in Mathematics **386**, Springer-Verlag, Berlin—New York, 1974.

[124] J. TITS. Classification of buildings of spherical type and Moufang polygons: a survey, *Coll. Intern. Teorie Combin. Acc. Naz. Lincei, Roma 1973, Atti dei convegni Lincei* **17** (1976), 229-246.

[125] J. TITS. Non-existence de certains polygones généralisés, I, *Invent. Math.* **36** (1976), 275–284.

[126] J. TITS. Endliche Spiegelungsgruppen, die als Weylgruppen auftreten, *Invent. Math.* **43** (1977), 283–295.

[127] J. TITS. Non-existence de certains polygones généralisés, II, *Invent. Math.* **51** (1979), 267–269.

[128] J. TITS. Moufang octagons and the Ree groups of type $^2F_4$, *Amer. J. Math.* **105** (1983), 539–594.

[129] J. TITS. Twin buildings and groups of Kac-Moody type, *Proceedings of a Conference on Groups, Combinatorics and Geometry (Durham 1990)*, Edited by M. Liebeck and J. Saxl, London Math. Soc. Lecture Note Ser. **165**, Cambridge University Press, Cambridge (1992), 249–286.

[130] J. TITS. Moufang polygons, I. Root data, *Bull. Belg. Math. Soc. Simon Stevin* **1** (1994), 455–468.

[131] J. TITS AND R. M. WEISS. *Moufang Polygons*, Springer Monographs in Mathematics, Springer-Verlag, Berlin, 2002.

[132] H. VAN MALDEGHEM. *Generalized Polygons*, Monographs in Mathematics **93**, Birkhäuser-Verlag, Basel, 1998.

[133] H. VAN MALDEGHEM. Some consequences of a result of Brouwer, *Ars Combin.* **48** (1998), 185–190.

[134] H. VAN MALDEGHEM AND R. WEISS. On finite Moufang polygons, *Israel J. Math.* **79** (1992), 321–330.

[135] H. VAN MALDEGHEM, J. A. THAS AND S. E. PAYNE. Desarguesian finite generalized quadrangles are classical or dual classical, *Designs, Codes, Cryptogr.* **1** (1992), 299–305.

[136] O. VEBLEN AND J. W. YOUNG. A set of assumptions for projective geometry, *Amer. J. Math.* **30** (1908), 347–380.

[137] R. WEISS. The nonexistence of certain Moufang polygons, *Invent. Math.* **51** (1979), 261–266.

[138] P. YUAN AND H. YAHUI. A note on power residue difference sets, *J. Algebra* **291** (2005), 269–273.

[139] E. I. ZELMANOV. On the restricted Burnside problem, in: *Proceedings of the International Congress of Mathematicians, Vol. I, II* (Kyoto, 1990), pp. 395–402, Math. Soc. Japan, Tokyo, 1991.

Department of Mathematics, Ghent University, Krijgslaan 281, S25, Ghent, Belgium
kthas@cage.UGent.be

Graphs, colours, weights and hereditary properties

Andrew Thomason

Abstract

Graphs whose edges are coloured with two colours are naturally related to induced subgraphs of ordinary graphs. This leads to an extremal theory for coloured graphs, in which the edges are given weights. We describe the theory, explaining its connections with the study of hereditary graph properties and noting recent progress.

1 Coloured graphs

Every graph H corresponds in a natural way to a complete graph on the same set $V(H)$ of vertices, whose edges are coloured red and blue: an edge of the complete graph is coloured red if it is present in H and is coloured blue if it is absent from H. In the normal way of things there is nothing to be gained by thinking about the 2-coloured complete graph instead of the original graph, but there are circumstances in which the 2-coloured representation has merit. This is particularly true if H is an induced subgraph of some other graph, because the presence of the blue edges in the coloured complete graph emphasises that the absence of an edge from H is just as important as the presence of an edge.

Various questions about induced subgraphs have been studied quite intensively over the last couple of decades, and it turns out that some of these can be reduced to extremal questions about coloured graphs. This is what motivates the work that is surveyed in this article. The relationship with induced subgraphs is the reason why we use only two colours on the edges. It will be clear once we give some of the definitions that we could just as well consider more colours, and we shall point out later some applications where more than two colours are appropriate. But some of the more significant results hold only in the case of two colours, and so by and large we shall stick with just two.

Having said that, it will be helpful in the discussion to introduce a third colour, green. This colour is used in a different way to the two main colours, though: roughly speaking, it is used on edges where we are not bothered whether the colour is red or blue. Once again, this usage is motivated by the applications. In order to make what we mean precise, it is convenient to extend our perspective to multigraphs.

We begin, then, by defining 2-coloured multigraphs and their extremal functions. We then discuss 'types', which describe the extremal structures. At that point, in §4, we describe the applications which have motivated the earlier definitions. In §5 and §6 we highlight some theoretical and some practical aspects of the general theory, whereas in §7 we give more detail about how the theory applies in a few specific examples. We finish with two conjectures and some remarks on proofs.

Much of the extremal theory of coloured graphs was worked out by Marchant and the author in [39]. However, that paper is necessarily dense and detailed, and our aim here is to give a readable introduction to the area.

2   2-coloured multigraphs

Definition 2.1 A *2-coloured multigraph* H is a multigraph whose edge set is the disjoint union of two simple graphs H_r and H_b on the same vertex set $V(H)$. The underlying simple graph of H is denoted by H_u. The *order* of H, denoted $|H|$, is the number of vertices, namely $|V(H)|$.

Thus two vertices are adjacent in H_u if and only if they are joined in at least one of H_r and H_b. We think of the edges of H_r as being coloured *red* and the edges of H_b as being coloured *blue*.

The coloured graphs described at the start of this article were 2-coloured multigraphs in which H_u is a complete graph and the edge sets of H_r and H_b are disjoint. Often, but not always, it will be the case for the coloured multigraphs we consider that the underlying graph is complete. When we refer to a *complete* coloured multigraph this is what we mean, that the underlying graph is complete.

However, it will not always be the case that H_r and H_b are disjoint, and indeed it is important that they are sometimes allowed to overlap. One way to think of such a 2-coloured multigraph H is that it is a colouring of the underlying simple graph H_u with three colours red, blue and *green*: red edges of H_u are those in H_r but not H_b, blue edges are those in H_b but not H_r, and green edges are those in both H_r and H_b.

Definition 2.2 A 2-coloured multigraph G *contains* H, written $H \subset G$, if there is an injection $f : V(H) \to V(G)$ such that $f(x)f(y) \in E(G_r)$ whenever $xy \in E(H_r)$ and $f(x)f(y) \in E(G_b)$ whenever $xy \in E(H_b)$.

This definition is the natural definition of a subgraph in the world of 2-coloured multigraphs. But let us see what it means if we think of G and H as red-blue-green coloured simple graphs. From this point of view, G contains H if H is a subgraph of G in such a way that the colouring is respected, except that green edges of G act as a kind of "wildcard", because a green edge of G can represent an edge of H of any colour, red, blue or green (or no edge at all, of course). In particular, if G is a complete graph with every edge green then G contains every H with $|H| \le |G|$.

2.1   Extremal multigraphs in former times

The natural extremal question for 2-coloured multigraphs is, given a 2-coloured multigraph H and an integer n, what is the largest size of a 2-coloured multigraph G with $|G| = n$ and $H \not\subset G$?

But this statement makes no sense unless we decide what is meant by the size of G. It is how we do this that lies at the heart of all that is to follow. Before giving our definition, we comment briefly on some alternatives that have appeared in the literature.

The most straightforward way to define the size of G is as the total number of edges, namely $e(G_r) + e(G_b)$. If the colours on the edges are ignored, this equates to the definition of Brown, Erdős and Simonovits, who in a series of papers [19, 20, 21] (surveyed by Brown and Simonovits in [22]) studied the maximum number of edges in a multigraph G that contains no fixed multigraph H and whose edge multiplicity is bounded by some number k. Clearly a bound on the edge multiplicity is needed, since otherwise G could have any number of edges between just two vertices without containing H; of course, in the situation of 2-coloured multigraphs, the edge multiplicity is at most two by definition.

But the work in [19, 20, 21] takes no account of colours, and for our purposes the colours are essential. It is sometimes still enough just to count the total number of edges; for example, Füredi and Simonovits [33], in the course of their solution of the extremal problem for the Fano plane, made use of the maximum total number of edges in a 4-coloured multigraph containing no properly 3-coloured K_4. An alternative approach to recognizing the colours was taken by Diwan and Mubayi [23], who obtained interesting results when defining the size to be the *minimum* of $e(G_r)$ and $e(G_b)$.

A well-known work in the extremal theory of multigraphs is that of Füredi and Kundgen [32], who allowed integral weights to be placed on the edges of a multigraph, the size being defined as the total weight. Extending work of Bondy and Tuza [17], they determined the maximum size of a multigraph if the total weight of edges within any set of k vertices is bounded. But once again this work takes no account of colours.

2.2 Weights

Our approach here, motivated as ever by the applications, is to use just two edge weights, one for each colour, and to take the size of a 2-coloured multigraph to be the total weight.

Definition 2.3 Let $0 \le p \le 1$ and let $p + q = 1$. The *p-weight* of a 2-coloured multigraph G is

$$w_p(G) = pe(G_r) + (1 - p)e(G_b) = pe(G_r) + qe(G_b)$$

where $e(G_r)$ and $e(G_b)$ are the numbers of edges in G_r and G_b.

It is important to see how this definition relates to the idea of a 2-coloured multigraph as a simple graph coloured red, blue and green. In this case, red edges have weight p, blue edges have weight q, and green edges, which represent a red edge parallel to a blue edge, have weight 1.

2.3 The two extremal functions

We can now define the natural extremal function for 2-coloured multigraphs. In general, rather than forbid just one graph H, we can forbid any member of a class $\mathcal{H}$ of graphs. (From now on we shall often refer in this way to a 2-coloured multigraph as a graph when there is no danger of confusion. If we really are referring to a graph in the standard sense, we might use the term "ordinary" graph for emphasis.)

Definition 2.4 Given a class $\mathcal{H}$ of 2-coloured multigraphs, we define

$$\mathrm{ex}_p(\mathcal{H}, n) = \max \{ w_p(G) : |G| = n, H \not\subseteq G \text{ for all } H \in \mathcal{H} \}.$$

If $\mathcal{H} = \{H\}$ we may write $\mathrm{ex}_p(H, n)$ instead of $\mathrm{ex}_p(\{H\}, n)$.

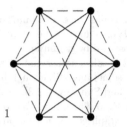

Figure 1: The 2-coloured multigraph C_6^*; dashed edges are red and solid edges blue.

This extremal function is relevant to some of the applications we have in mind but it turns out that, for the more interesting and important applications, we need instead to take the maximum taken over all *complete* graphs G.

Definition 2.5 Given a class $\mathcal{H}$ of 2-coloured multigraphs, we define

$$\mathrm{kex}_p(\mathcal{H}, n) = \max\{\, w_p(G) : |G| = n,\ G \text{ complete},\ H \not\subset G \text{ for all } H \in \mathcal{H} \,\}.$$

The letter 'k' in 'kex' is to emphasise that the maximum is taken over complete graphs. Clearly $\mathrm{kex}_p(\mathcal{H}, n) \le \mathrm{ex}_p(\mathcal{H}, n)$, and the inequality can be strict. One example is $H = C_7$, described in §7.6, but here is another that is simpler to analyse.

Example 2.6 Let C_6^* be the 2-coloured multigraph of order 6 in which the red graph is a 6-cycle with a diagonal and the blue graph is the complement of the red, as shown in Figure 1.

Evidently C_6^* is not contained in a complete 5-partite graph coloured green, and so $\mathrm{ex}_p(C_6^*, n) \ge \frac{4}{5}\binom{n}{2} + O(n)$ holds for all p. But this example fails to give a lower bound for $\mathrm{kex}_p(C_6^*, n)$. One colouring of the complete graph of order n that does give a lower bound for $\mathrm{ex}_p(C_6^*, n)$ is that obtained by splitting the vertices into two classes of (close to) $n/2$ vertices each, colouring the edges within the classes red and those between the classes green. Since the blue subgraph of C_6^* is not bipartite (it contains a triangle), this colouring gives the lower bound $\mathrm{kex}_p(C_6^*, n) \ge (1 - q/2)\binom{n}{2} + O(n)$. Another colouring is to split the vertices into three classes, the edges in the first being blue and in the other two being red, with edges between the two red classes being blue and the edges between the blue and red classes being green. It is not too difficult to check that this 2-coloured multigraph does not contain C_6^*. Its weight depends on the relative sizes of the classes, and again it is easily checked that the weight is greatest when the classes are in proportions (close to) $q : p : p$. Thus we obtain the lower bound $\mathrm{kex}_p(C_6^*, n) \ge (1 - p/(1 + 2p))\binom{n}{2} + O(n)$.

The first lower bound for $\mathrm{kex}_p(C_6^*, n)$ is better when $p > 1/2$ and the second is better when $p < 1/2$. It can be shown (see §7.4.2) that these bounds give the correct value of $\mathrm{kex}_p(C_6^*, n)$ (at least, to within the term $O(n)$). Since both bounds are smaller than $\frac{4}{5}\binom{n}{2} + O(n)$ for $1/3 < p < 3/5$, we have $\mathrm{kex}_p(C_6^*, n) < \mathrm{ex}_p(C_6^*, n)$ for p in this range (for large n).

As suggested by this example, we will find life simpler if we take limiting weights as $n \to \infty$, in the following way.

Definition 2.7 Given a class $\mathcal{H}$ of 2-coloured multigraphs, we define

$$\mu_p(\mathcal{H}) = \lim_{n \to \infty} \mathrm{ex}_p(\mathcal{H}, n) \binom{n}{2}^{-1} \quad \text{and} \quad \kappa_p(\mathcal{H}) = \lim_{n \to \infty} \mathrm{kex}_p(\mathcal{H}, n) \binom{n}{2}^{-1}.$$

The limits are easily shown to exist. From now on, our interest will be in finding $\kappa_p(\mathcal{H})$ and $\mu_p(\mathcal{H})$ for classes $\mathcal{H}$ of 2-coloured multigraphs.

2.4 General properties of the extremal functions

Before proceeding, we make some general remarks about these extremal functions for 2-coloured multigraphs.

First, note that they contain ordinary graph extremal functions. If $\mathcal{G}$ is a collection of ordinary graphs then the classical extremal function for $\mathcal{G}$ is $\mathrm{ex}(\mathcal{G}, n) = \max\{e(F) : |F| = n, G \not\subseteq F \text{ for all } G \in \mathcal{G}\}$. Let $\mathcal{H}$ be the graphs in $\mathcal{G}$ with all their edges coloured red, so $\mathcal{H}$ is a class of 2-coloured multigraphs. Then evidently $\mathrm{ex}_1(\mathcal{H}, n) = \mathrm{kex}_1(\mathcal{H}, n) = \mathrm{ex}(\mathcal{G}, n)$, since extremal graphs for $\mathcal{G}$ correspond, when coloured red, to extremal graphs for $\mathcal{H}$ (and we can add blue edges without creating forbidden subgraphs or changing the weight). Now the result of Erdős and Stone [31] and the work of Erdős and Simonovits [25, 26, 28, 52] show that, while the limit $\lim_{n \to \infty} \mathrm{ex}(\mathcal{G}, n) \binom{n}{2}^{-1}$ can be given explicitly, the exact value of $\mathrm{ex}(\mathcal{G}, n)$ is often hard to come by. For general classes $\mathcal{H}$ the calculation of $\mathrm{kex}_p(\mathcal{H}, n)$ and $\mathrm{ex}_p(\mathcal{H}, n)$ can get intricate even for very small graphs, as shown in §7.1. These observations strengthen our resolve to concentrate on the limits $\kappa_p(\mathcal{H})$ and $\mu_p(\mathcal{H})$.

In fact, if $\mathcal{H}$ contains both a monochromatic red graph and a monochromatic blue graph, then Ramsey's theorem implies that $\mathrm{kex}_p(\mathcal{H}, n)$ doesn't exist when n is large. From now on we exclude this pathological case.

Suppose a class $\mathcal{H}$ contains only monochromatic graphs of the same colour, say red. Then $\kappa_0(\mathcal{H}) = \mu_0(\mathcal{H}) = 1$, because an all-blue complete graph is extremal, and $\kappa_1(\mathcal{H}) \leq \mu_1(\mathcal{H}) < 1$ by the previous observations. The classes of most interest, though, do not contain monochromatic graphs, and for such classes it is evident that $\kappa_0(\mathcal{H}) = \mu_0(\mathcal{H}) = \kappa_1(\mathcal{H}) = \mu_1(\mathcal{H}) = 1$. From the definition we have $\kappa_p(\mathcal{H}) \leq \mu_p(\mathcal{H}) \leq 1$ for all p.

How do the functions κ_p and μ_p vary with p? The following result was noted by Alon and Stav [5], by Balogh and Martin [10] and in [39].

Theorem 2.8 *For every class $\mathcal{H}$, both $\kappa_p(\mathcal{H})$ and $\mu_p(\mathcal{H})$ are continuous and convex as functions of p for $p \in [0, 1]$.*

3 Types

The colourings given in Example 2.6 as lower bounds for $\mathrm{kex}_p(C_6^*)$ have a simple structure: the graph is partitioned into a small number of classes of equivalent vertices. Colourings of this kind are important and we discuss them now.

Definition 3.1 A *type* is a complete graph, whose vertices are coloured red and blue, and whose edges are coloured red, blue and green. A *weak type* is the same, except that the colour white may be used both on vertices and on edges.

The idea of a type is that it is a shorthand for describing a partition and colouring of a large complete graph. A weak type similarly describes a partition of a graph that need not be complete.

Definition 3.2 Let τ be a type. The *basic property* $\mathcal{Q}(\tau)$ consists of all complete 2-coloured graphs G whose vertex set has a partition $\{V_v : v \in V(\tau)\}$ indexed by the vertices of τ, such that the edges within V_v are all the same colour as v and, if uv is red or blue, the edges between V_u and V_v are all the same colour as uv: if uv is green then the edges between V_u and V_v may be coloured in any way (and not necessarily the same colour).

If τ is a weak type then $\mathcal{Q}(\tau)$ is defined similarly, except that if v or uv is white then there are no edges within V_v or between V_u and V_v. Thus the graphs in $\mathcal{Q}(\tau)$ are complete except for where edges are forbidden by the white colour.

The 2-coloured multigraphs in Example 2.6, giving lower bounds on $\kappa_p(C_6^*)$, can now be described as lying in the classes $\mathcal{Q}(\rho)$ and $\mathcal{Q}(\sigma)$, where ρ is the type consisting of two red vertices joined by a green edge, and σ has one blue and two red vertices, with two green edges and one blue edge (that between the red vertices).

Definition 3.3 The 2-coloured graph H is said to be τ-*colourable* if it is contained in some member of $\mathcal{Q}(\tau)$.

If H is not τ-colourable then graphs in $\mathcal{Q}(\tau)$ can be used to give lower bounds on $\mathrm{kex}_p(H, n)$ (or on $\mathrm{ex}_p(H, n)$ if τ is a weak type). The obvious question then is: which graphs in $\mathcal{Q}(\tau)$ have greatest p-weight? Clearly, whenever uv is a green edge of τ we should make all the edges between V_u and V_v green, so the question remains, if $|G| = n$, how should the relative sizes of the partition classes $\{V_v : v \in V(\tau)\}$ be chosen so as to maximize the p-weight?

Suppose that $|V_v| = y_v n$, where $0 \le y_v \le 1$ and $\sum_{v \in V(\tau)} y_v = 1$. Then

$$
\begin{aligned}
w_p(G) \;=\; & p \sum_{v \text{ is red}} \binom{y_v n}{2} + q \sum_{v \text{ is blue}} \binom{y_v n}{2} \\
& + p \sum_{uv \text{ is red}} y_u y_v n^2 + q \sum_{uv \text{ is blue}} y_u y_v n^2
\end{aligned}
$$

Definition 3.4 The *weight* $w_p(v)$ of a vertex $v \in V(\tau)$ is defined to be p if v is red and q if v is blue; likewise the weight $w_p(uv)$ of an edge uv is defined to be 0, p, q or 1 according as uv is white, red, blue or green. Let $W_p(\tau)$ be the symmetric $|\tau| \times |\tau|$ matrix whose uu entry is $w_p(u)$ and whose uv entry is $w_p(uv)$. The *p-value* of τ is

$$
\lambda_p(\tau) = \max_{z \in \Delta} z^{\mathrm{t}} W_p(\tau) z = \max_{z \in \Delta} \sum_{u \in V(\tau)} z_u^2 w_p(u) + 2 \sum_{uv \in E(\tau)} z_u z_v w_p(uv),
$$

where $\Delta = \Delta(|\tau|)$ is the simplex $\{z \in [0,1]^{|\tau|} : z_1 + \ldots + z_{|\tau|} = 1\}$.

It can be seen that, in the previous calculation, $w_p(G) = y^{\mathrm{t}} W_p(\tau) y \binom{n}{2} + O(n)$, where $y = (y_v)_{v \in V(\tau)}$. Hence the maximum p-weight of a graph of order n in $\mathcal{Q}(\tau)$ is $\lambda_p(\tau)\binom{n}{2} + O(n)$. Therefore, if no graph in the class $\mathcal{H}$ is τ-colourable, we have $\kappa_p(\mathcal{H}) \ge \lambda_p(\tau)$ if τ is a type, or $\mu_p(\mathcal{H}) \ge \lambda_p(\tau)$ if τ is a weak type.

3.1 Types and extremal functions

The above observations are straightforward enough, but do they have anything to do with the actual value of the extremal functions $\kappa_p(\mathcal{H})$ and $\mu_p(\mathcal{H})$ in general? They do. The next theorem is a natural one in extremal graph theory, and it states that the extremal functions $\kappa_p(\mathcal{H})$ and $\mu_p(\mathcal{H})$ can be approximated arbitrarily closely by the p-weights of graphs in $\mathcal{Q}(\tau)$, for suitably chosen types (or weak types) τ.

Theorem 3.5 *Let $\mathcal{H}$ be a class of 2-coloured multigraphs. Then*

$$\kappa_p(\mathcal{H}) \;=\; \sup\{\lambda_p(\tau) : \tau \text{ is a type, no graph in } \mathcal{H} \text{ is } \tau\text{-colourable}\} \text{ and}$$
$$\mu_p(\mathcal{H}) \;=\; \sup\{\lambda_p(\tau) : \tau \text{ is a weak type, no graph in } \mathcal{H} \text{ is } \tau\text{-colourable}\}.$$

The theorem shows that the extremal functions can be estimated by considering 2-coloured graphs G of a very simple kind, namely those in $\mathcal{Q}(\tau)$ for some type or weak type τ. We remind the reader, though, that the exact extremal graphs need not be of this precise form: see 7.1 for some examples.

Theorem 3.5 can be found in Balogh and Martin [10, Theorem 11]; an equivalent form was given by Bollobás and the author [16, Theorem 1.1], and it is implicit in Alon and Stav [6, Lemma 3.4]. These authors all used Szemerédi's regularity lemma. But in [39] a direct proof was given, in terms of "extensions", an idea analogous to that of the "augmentations" of Brown, Erdős and Simonovits [19, 20, 21]. We describe the notion briefly.

Let $(G^n)_{n=1}^{\infty}$ be a sequence of 2-coloured complete multigraphs with $|G^n| \geq n$. We say that (G^n) *contains* the type τ if every member of $\mathcal{Q}(\tau)$ is contained in some G^n. We also define the degree of a vertex $v \in V(G^n)$ to be p times the number of red edges incident with v plus q times the number of blue edges, and we write $\delta_p(G^n)$ for the minimum of the vertex degrees.

Lemma 3.6 *Let $\delta = \liminf_{n\to\infty}(\delta_p(G^n)/|G^n|)$. Let τ be a type contained in (G^n). Let $x = (x_1,\ldots,x_{|\tau|}) \in \Delta$. Then (G^n) contains a type σ, having a vertex v such that $\sigma - v = \tau$ and*

$$\sum_{u\in V(\tau)} x_u w_p(uv) \geq \delta.$$

The type σ, consisting of τ plus an extra vertex, is called an *extension* of τ. To prove the first claim in Theorem 3.5, take a sequence (G^n) of extremal graphs for $\kappa_p(\mathcal{H})$. It can be arranged that $\delta = \kappa_p(\mathcal{H})$. If we have a type τ contained in (G^n) but with $\lambda_p(\tau) < \delta$ (there is always such a type with $|\tau| = 1$), we find an extension σ with the property stated in Lemma 3.6. It is then easily shown that $\lambda_p(\sigma) > \lambda_p(\tau)$. Repeated applications lead to a proof of the claim. Lemma 3.6 holds too if types are replaced by weak types, and so the second claim of Theorem 3.5 follows in the same way.

As mentioned earlier, Theorem 3.5 is of a somewhat generic nature in extremal graph theory. But the following theorem, though superficially similar, is quite different in kind. It states that, for κ, the supremum is attained. It is the fundamental theorem in regard to 2-coloured multigraphs.

Theorem 3.7 ([39, Theorem 3.23]) *Let $\mathcal{H}$ be a class of 2-coloured multigraphs. Then*

$$\kappa_p(\mathcal{H}) = \max\{\lambda_p(\tau) : \ \tau \ \text{is a type, no graph in } \mathcal{H} \ \text{is } \tau\text{-colourable}\}.$$

An analogous theorem in the extremal theory of multigraphs was proved for multigraphs of maximum multiplicity 2 by Brown, Erdős and Simonovits [21, Theorem 5], and the proof of Theorem 3.7, which uses extensions, is based on that proof. It is not known whether the multigraph theorem holds for multiplicity greater than 2. Likewise it is not known whether the supremum for μ in Theorem 3.5 is attained.

Of course, what we want to be able to do is to calculate $\kappa_p(\mathcal{H})$ for a given class $\mathcal{H}$. In the light of Theorem 3.7, this comes down to finding a suitable type τ that yields the exact value. Unfortunately the proof of Theorem 3.7 seems to give no clue as to how large such a τ must be, and there is no known algorithm for finding $\kappa_p(\mathcal{H})$ in general.

Nevertheless it is possible to calculate $\kappa_p(\mathcal{H})$ in some cases. Sometimes this can be done directly, but more often Theorem 3.7 is the start of the argument. Later we shall discuss how this is done, and give some examples, including one (in §7.8.2) where Lemma 3.6 is used. But first we describe some of the applications that motivated the study of 2-coloured multigraphs.

4 Applications

Here we describe two applications for the function $\kappa_p(\mathcal{H})$, namely the probability of hereditary graph properties, and the edit distance of hereditary graph properties. We also give an application of $\mu_p(\mathcal{H})$ to Ramsey-type games.

The applications to be discussed were investigated before a theory of 2-coloured multigraphs was developed. However, that earlier work is not superseded by the present material, but is rather complemented by it. Previous authors in effect showed that the questions they were interested in can be converted into questions about what we are calling here 2-coloured multigraphs. Hence by studying these multigraphs we can feed back insight into the original applications.

4.1 Hereditary graph properties

A graph property $\mathcal{P}$ is a class of graphs closed under isomorphism; it is hereditary if it is closed under the taking of induced subgraphs. It is customary to say that graphs in the class $\mathcal{P}$ "have the property $\mathcal{P}$".

In the last two decades there has been a significant amount of study of the *size* of hereditary graph properties, by which is meant the number of graphs on n vertices having the property. More precisely, let $\mathcal{P}^n$ be the set of labelled graphs on vertex set $\{1, 2, \ldots, n\}$ that are in $\mathcal{P}$; then the size is $|\mathcal{P}^n|$.

Bollobás [12] wrote an excellent and wide-ranging survey of hereditary properties of graphs and other structures for the last but one British Combinatorial Conference. We refer the reader to that survey for fuller information and more detail, but we recap just a little in order to make clear the application of 2-coloured multigraphs to hereditary graph properties.

Given a class $\mathcal{F}$ of ordinary graphs, the class Forb($\mathcal{F}$), consisting of graphs having no induced subgraph isomorphic to a member of $\mathcal{F}$, is a hereditary property. Conversely every hereditary property $\mathcal{P}$ can be written as $\mathcal{P} = \text{Forb}(\mathcal{F})$ for some $\mathcal{F}$ (for example, take $\mathcal{F}$ to be the graphs not in $\mathcal{P}$).

There is a simple connection between hereditary graph properties and 2-coloured multigraphs.

Definition 4.1 Given a graph F, let $cc(F)$ be the complete 2-coloured multigraph with vertex set $V(F)$, in which the edges of F are coloured red and the edges of the complement of F are coloured blue (so there are no green edges). Moreover, given a graph property $\mathcal{P}$, let $\mathcal{H}(\mathcal{P}) = \{cc(F) : F \notin \mathcal{P}\}$.

Thus the graph F corresponds to the 2-coloured complete graph $cc(F)$ in the way described at the outset of this article. It is useful to have a notation for the graphs corresponding to members of $\mathcal{Q}(\tau)$ (defined in Definition 3.2).

Definition 4.2 Let τ be a type. Then $\mathcal{P}(\tau)$ is the class of (ordinary, simple, uncoloured) graphs G for which $V(G)$ has a partition $\{V_v : v \in V(\tau)\}$, such that $G[V_u]$ is complete or empty according as u is red or blue, and the bipartite graph with vertex classes V_u and V_v is complete or empty according as uv is red or blue (if uv is green then there is no condition on the bipartite graph).

So the set of 2-coloured graphs in $\mathcal{Q}(\tau)$ that are simple, i.e., have no green edges, is precisely $\{cc(F) : F \in \mathcal{P}(\tau)\}$.

Remark Notice that $\mathcal{P}(\tau)$ is defined for types and not for weak types. The notion of a type naturally captures the notion of presence or absence of edges in ordinary graphs. Hence, in the light of Theorem 3.5, it is κ_p rather than μ_p that is more relevant to the discussion of hereditary properties.

4.1.1 The number of graphs with a given property (The material in this subsection, apart from the 2-coloured aspect, is covered more fully in [12, Section 5].) Prömel and Steger [43] were the first to study the size of certain hereditary properties. Let F be a single graph and let $\mathcal{P} = \text{Forb}(\{F\})$. Let t be the maximum integer for which the following statement is true: there is some integer s, $0 \leq s \leq t$, such that the vertices of F cannot be partitioned into t sets, s of which span complete subgraphs and the other $t - s$ of which span independent sets. It follows that there are at least $2^{(1-1/t+o(1))\binom{n}{2}}$ graphs in $\mathcal{P}^n$ because, if we partition $\{1, \ldots, n\}$ into t parts as near equal as possible, filling in s of them with edges and leaving the remaining $t - s$ empty, then any way of adding edges between the classes will give rise to a graph in $\mathcal{P}$. Prömel and Steger [43] showed that this bound is tight, namely, $|\mathcal{P}^n| = 2^{(1-1/t+o(1))\binom{n}{2}}$.

We can state this nicely in terms of the simplest kind of types.

Definition 4.3 The type $\tau(a,b)$ has a red vertices and b blue vertices, all edges being green.

The set of $2^{(1-1/t+o(1))\binom{n}{2}}$ graphs just described is therefore $\mathcal{P}(\tau(s, t - s))^n$. Moreover, t can be defined as the maximum order of a type $\tau(a, b)$ such that $\mathcal{P}(\tau(a, b)) \subset \mathcal{P}$. Equivalently, it is the maximum order of a type $\tau(a, b)$ such that $\mathrm{cc}(F)$ is not $\tau(a, b)$-colourable. (In [12, Section 5], $\mathcal{P}(\tau(a, b))$ is denoted $\mathcal{C}(a + b, a)$.)

Alekseev [2] and Bollobás and the author [15] extended this result to general hereditary properties: for any hereditary property $\mathcal{P}$, we have $|\mathcal{P}^n| = 2^{(1-1/t+o(1))\binom{n}{2}}$, where

$$\begin{aligned} t &= \max\{a + b : \mathcal{P}(\tau(a, b)) \subset \mathcal{P}\} \\ &= \max\{a + b : H \text{ is not } \tau(a, b)\text{-colourable for any } H \in \mathcal{H}(\mathcal{P})\}. \end{aligned}$$

4.1.2 The probability of a hereditary property Let $G(n, p)$ be a random graph on n vertices in which edges are chosen independently at random with probability p. The previous statement that $|\mathcal{P}^n| = 2^{(1-1/t+o(1))\binom{n}{2}}$ is equivalent to the statement $\Pr[G(n, 1/2) \in \mathcal{P}] = 2^{(-1/t+o(1))\binom{n}{2}}$. It is natural to ask what happens if $p \neq 1/2$. This was investigated by Bollobás and the author [16], who defined the function $c_{n,p}(\mathcal{P})$ by $\Pr[G(n, p) \in \mathcal{P}] = 2^{-c_{n,p}(\mathcal{P})\binom{n}{2}}$. The limit $c_p(\mathcal{P}) = \lim_{n \to \infty} c_{n,p}(\mathcal{P})$ exists (see Alekseev [1]) and so

$$\Pr[G(n, p) \in \mathcal{P}] = 2^{(-c_p(\mathcal{P})+o(1))\binom{n}{2}}.$$

The original motivation in [16] for studying $c_p(\mathcal{P})$ was to find the $\mathcal{P}$-chromatic number of a random graph. The $\mathcal{P}$-*chromatic number* $\chi_\mathcal{P}(G)$ of a graph G is the smallest number of subsets in a partition of $V(G)$ such that the subgraph induced by each subset lies in $\mathcal{P}$. Thus if $\mathcal{P}$ is the property of having no edges then $\chi_\mathcal{P}(G)$ is the ordinary chromatic number. Scheinerman [49] conjectured that $\chi_\mathcal{P}(G(n, p))$ is concentrated, and this was proved in [16], where it was shown that $\chi_\mathcal{P}(G(n, p)) = (c_p(\mathcal{P}) + o(1))n/2 \log_2 n$. The case $p = 1/2$ was settled earlier in [14].

How can we determine the value of $c_p(\mathcal{P})$? From the previous remarks we know that if $p = 1/2$ then $c_p(\mathcal{P}) = 1/t$, where $t = \max\{a + b : \mathcal{P}(\tau(a, b)) \subset \mathcal{P}\}$. However, as shown in [14], the types $\tau(a, b)$ with all green edges do not suffice to capture $c_p(\mathcal{P})$ if $p \neq 1/2$, and this was the reason for introducing the more general types with red and blue edges also. It was then shown in [16] that

$$\begin{aligned} c_p(\mathcal{P}) &= \inf\{H_p(\tau) : \mathcal{P}(\tau) \subset \mathcal{P}\} \\ \text{where} \quad H_p(\tau) &= \min_{x \in \Delta} \left[-\sum_{v \in V(\tau)} x_v^2 \log_2 w_p(v) - 2 \sum_{uv \in E(\tau)} x_u x_v \log_2 w_p(uv) \right], \end{aligned}$$

and an example was put forward to show that the infimum cannot be replaced by a minimum. (The notation $H_p(\tau)$ was used for the function because of its relation to entropy; it should not be confused with the notation H_r and H_b used elsewhere in this article for the red and blue subgraphs of a 2-coloured multigraph H.)

The definition of $H_p(\tau)$ appears close to that of $\lambda_p(\tau)$ in Definition 3.4, and indeed a little manipulation (see Marchant and the author [40]) gives

$$H_p(\tau) = -(\log_2 p + \log_2 q)\left[1 - \lambda_{p'}(\tau)\right] \quad \text{where} \quad p' = \frac{\log q}{\log p + \log q}.$$

The equation for $c_p(\mathcal{P})$ can therefore be rewritten

$$c_p(\mathcal{P}) = -(\log_2 p + \log_2 q)\left[1 - \sup\{\lambda_{p'}(\tau) : \mathcal{P}(\tau) \subset \mathcal{P}\}\right].$$

Recall now from §4.1.1 that, in the language of 2-coloured multigraphs, we have $\{\tau : \mathcal{P}(\tau) \subset \mathcal{P}\} = \{\tau : \text{no graph in } \mathcal{H}(\mathcal{P}) \text{ is } \tau(a, b)\text{-colourable}\}$. Theorem 3.5 shows that $\sup\{\lambda_{p'}(\tau) : \mathcal{P}(\tau) \subset \mathcal{P}\} = \kappa_{p'}(\mathcal{H}(\mathcal{P}))$. But Theorem 3.7 shows that this supremum is attained: that is, it is a maximum. Here is a summary in the language of probability.

Theorem 4.4 *Let $\mathcal{P}$ be a hereditary property and let $0 \leq p \leq 1$. Then*

$$c_p(\mathcal{P}) = -(\log_2 p + \log_2 q)\left[1 - \kappa_{p'}(\mathcal{H}(\mathcal{P}))\right] \quad \text{where} \quad p' = \frac{\log q}{\log p + \log q}.$$

Moreover, $\kappa_{p'}(\mathcal{H}(\mathcal{P})) = \max\{\lambda_{p'}(\tau) : \mathcal{P}(\tau) \subset \mathcal{P}\}$.

The import of Theorem 3.7 is that the equation $c_p(\mathcal{P}) = \inf\{H_p(\tau) : \mathcal{P}(\tau) \subset \mathcal{P}\}$, written above, could in fact have been written $c_p(\mathcal{P}) = \min\{H_p(\tau) : \mathcal{P}(\tau) \subset \mathcal{P}\}$. But, as mentioned before, an example was given in [16] to show that the infimum is not always a minimum. How is this contradiction to be resolved? Well, the example in [16] was when $\mathcal{P}$ is the property of being a union of complete graphs and $p < 1/2$. It was proved in [16] that if $\mathcal{P}(\tau) \subset \mathcal{P}$ and if τ has at least two vertices then $c_p(\mathcal{P}) < H_p(\tau)$, which is correct. But it does not imply $c_p(\mathcal{P}) \neq \min\{H_p(\tau) : \mathcal{P}(\tau) \subset \mathcal{P}\}$. As pointed out by Uri Stav (personal communication), it was overlooked that there is a τ with a single vertex in the set $\{\tau : \mathcal{P}(\tau) \subset \mathcal{P}\}$, namely the type $\tau(0, 1)$ with a single blue vertex. Then $\mathcal{P}(\tau(0, 1))$ consists of graphs with no edges at all, an apparently very small sub-property of $\mathcal{P}$. However it is readily checked that $c_p(\mathcal{P}(\tau(0, 1))) = c_p(\mathcal{P}) = -\log_2 q$.

The relationship between hereditary properties and 2-coloured complete graphs makes it possible to compute actual probabilities that were unknown before: for example,

if $p \geq 1/5$ then $\Pr[G(n, p) \text{ contains no induced } K_{3,3}] = p^{\left\lceil\frac{\log q}{\log p + 2\log q}\right\rceil\binom{n}{2} + o(n^2)}$.

Because of Theorem 4.4 this calculation comes down to computing $\kappa_p(\mathcal{H})$ for some $\mathcal{H}$, and ways to do this are described later in this article.

4.2 Edit distance

The notion of edit distance referred to here is that between graphs. The distance from a graph G to a property $\mathcal{P}$ is

$$\text{Dist}(G, \mathcal{P}) = \min\{|E(J)\triangle E(G)| : J \in \mathcal{P}, V(J) = V(G)\}$$

and the *edit distance* from the class of all n-vertex graphs to $\mathcal{P}$ is

$$\text{Dist}(n, \mathcal{P}) = \max\{\text{Dist}(G, \mathcal{P}) : |G| = n\}.$$

As usual, we normalize and define

$$\text{ed}(\mathcal{P}) = \lim_{n\to\infty} \text{Dist}(n, \mathcal{P})\binom{n}{2}^{-1}.$$

It is not obvious, *a priori*, that this limit exists, but it does, as we shall see.

The study of edit distance in this form was initiated in recent times by Axenovich, Kézdy and Martin [8] because of applications in biology, and by Alon and Stav [5] because of applications in computer science, especially property testing. (We remark that, in computer science, the term "edit distance" is sometimes used for the Levenshtein distance between two sequences, whereas the distance here is closer to the Hamming distance, insofar as the Levenshtein distance allows insertions and deletions [35].)

It is difficult to say much about properties in general and so we restrict our attention to hereditary properties. The fundamental fact about the edit distance of a hereditary property $\mathcal{P}$ is that the furthest graph from $\mathcal{P}$ is a *random graph*. To be more precise, there is some probability $p^* = p^*(\mathcal{P})$ for which $\mathrm{Dist}(n, \mathcal{P}) = \mathrm{Dist}(G(n, p^*), \mathcal{P}) + o(n^2)$ holds almost surely. This striking result was proved by Alon and Stav [5], building on ideas of Alon and Shapira [4]. But the proof gives no hint as to the value of p^*.

It is also shown in [5] that the limit of the expectation

$$e_p(\mathcal{P}) = \lim_{n \to \infty} \mathbb{E} \, \mathrm{Dist}(G(n, p), \mathcal{P}) \binom{n}{2}^{-1}$$

exists. Hence $\mathrm{ed}(\mathcal{P})$, defined above, exists and $\mathrm{ed}(\mathcal{P}) = e_{p^*}(\mathcal{P})$. Now the definition of edit distance means that $e_p(\mathcal{P}) \leq \mathrm{ed}(\mathcal{P})$, and it follows that $e_{p^*}(\mathcal{P}) = \max_p e_p(\mathcal{P})$. In other words, p^* is the value of p that maximizes $e_p(\mathcal{P})$, and the maximum value is the quantity we are seeking, namely $\mathrm{ed}(\mathcal{P})$.

Where do 2-coloured multigraphs fit into all this? Consider $\mathcal{H}(\mathcal{P})$ as defined in Definition 4.1. Let G be a complete 2-coloured multigraph of order n with $H \not\subset G$ for all $H \in \mathcal{H}(\mathcal{P})$. Given a random graph $G(n, p)$ on the same vertex set as G, edit it to a graph J by deleting those edges in $G(n, p)$ which are blue (not green) in G, and by adding those edges not already in $G(n, p)$ which are red (not green) in G. Thus $\mathrm{cc}(J) \subset G$ (it is here that the completeness of G is crucial, and that green edges of G contain both red and blue edges), and so $\mathrm{cc}(J) \notin \mathcal{H}(\mathcal{P})$. It follows that $J \in \mathcal{P}$, and thus $\mathrm{Dist}(n, \mathcal{P}) \leq |E(J) \triangle E(G(n, p))|$. Now the expected value of $|E(J) \triangle E(G(n, p))|$ is p times the number of blue (not green) edges plus q times the number of red (not green) edges, which equals $\binom{n}{2} - w_p(G)$. This holds for any G; by choosing G with the maximum possible weight $\mathrm{kex}_p(\mathcal{H}(\mathcal{P}), n)$ we find, in the limit, that $e_p(\mathcal{P}) \leq 1 - \kappa_p(\mathcal{H}(\mathcal{P}))$.

The inequality is in fact tight. This is implicit in [5] and explicit in Balogh and Martin [10, Theorem 11] (though their terminology differs from that here).

Theorem 4.5 *Let $\mathcal{P}$ be a hereditary property. Then $e_p(\mathcal{P}) = 1 - \kappa_p(\mathcal{H}(\mathcal{P}))$ holds for all p. In particular, $\mathrm{ed}(\mathcal{P}) = \max_p e_p(\mathcal{P}) = 1 - \min_p \kappa_p(\mathcal{H}(\mathcal{P}))$.*

Proof For those readers familiar with Szemerédi's Regularity Lemma we outline the reason why the theorem holds, though such readers will also spot that there are significant technical issues to be overcome before the outline becomes a proof. Because $\mathrm{Dist}(G(n, p), \mathcal{P})$ is concentrated near its mean, as shown in [5], there is with high probability some $J \in \mathcal{P}$ with $|E(J) \triangle E(G(n, p))| = e_p(\mathcal{P}) \binom{n}{2} + o(n^2)$. Given a partition of J into k parts, the density of $G(n, p)$ between each pair of sets in this partition will be close to p. Now construct the 2-coloured multigraph G whose vertices are the k parts and whose edges are blue/red/green according as the pairs have density in J close to zero/close to one/neither of these. Then with high probability $|E(J) \triangle E(G(n,p))| \geq ((\binom{k}{2}) - w_p(G))(\frac{n}{k})^2 + o(n^2)$. But G cannot contain a member of $\mathcal{H}(\mathcal{P})$, that is, $\mathrm{cc}(F)$ where $F \notin \mathcal{P}$, for otherwise, by the properties of the Regularity Lemma, we could find a copy of F induced in J. Thus $w_p(G) \leq \mathrm{kex}_p(\mathcal{H}(\mathcal{P}), k) = \kappa_p(\mathcal{H}(\mathcal{P}))\binom{k}{2} + o(k^2)$. Consequently

$$e_p(\mathcal{P})\binom{n}{2} + o(n^2) \geq (1 - \kappa_p(\mathcal{H}(\mathcal{P})))\binom{k}{2}\left(\frac{n}{k}\right)^2 + o(n^2)$$

which is what we need.   □

Both Axenovich, Kézdy and Martin [8] and Alon and Stav [6] have evaluated ed($\mathcal{P}$) for several hereditary properties. In general, to evaluate ed($\mathcal{P}$) we must find the minimum value of $\kappa_p(\mathcal{H}(\mathcal{P}))$ as a function of p. Fortunately, this does not mean we need to evaluate $\kappa_p(\mathcal{H}(\mathcal{P}))$ for all p. Theorem 2.8 shows that a local minimum is the global minimum, so it suffices to guess p^* and to evaluate $\kappa_p(\mathcal{H}(\mathcal{P}))$ for p near to p^*. Balogh and Martin [10] made use of this to show that ed($\mathcal{P}$) = $3 - 2\sqrt{2}$ if $\mathcal{P} = \mathrm{Forb}(\{K_{3,3}\})$, the property of containing no induced $K_{3,3}$ (an outstanding unresolved question in [6]).

4.3   Ramsey games

Richer's *colour-label* game [44] developed originally out of the Canonical Ramsey Theorem of Erdős and Rado [27], and it gave rise to the first study of the extremal properties of 2-coloured multigraphs. The connection between graph games and Ramsey theory in general is a beautiful one, well set out by Beck [11]. Richer's game in its final form moved away from the standard pattern of Ramsey games, but the title of this subsection is retained for historical reasons.

The game is played by two players on the edges of a complete graph K_n. Before the game, some 2-coloured complete graph H is fixed, which has no green edges (this is equivalent to specifying an ordinary graph F with $H = \mathrm{cc}(F)$). The players play alternately, the first player generating a red-blue colouring of the edges of K_n and the second player marking some of the edges. When it is the first player's turn he chooses some pre-agreed number of edges and colours them. The second player, in turn, selects any edge (coloured or not) and marks it. The first player wins if, at the end of the game (when all of K_n is coloured), there is a copy of H in K_n, all of whose edges are marked. Whether the second player can win depends, of course, on the pre-agreed number of edges that the first player colours at each turn.

A simplified, two-move, version of this game has the first player selecting outright a red/blue colouring of K_n and the second player then marking some prescribed number of edges. The first player wins if the second player cannot avoid marking a copy of H. The quantity of interest here is $\ell(H)$, the minimum value of ℓ such that if the second player must mark $\ell\binom{n}{2}$ edges then the first player is guaranteed a win (for large n).

More precisely, let $\ell_p(H, n)$ be the minimum number of edges the second player must be made to mark in order that the first player can find a winning colouring with $\lceil p\binom{n}{2}\rceil$ red edges. Let $\ell_p(H) = \lim_{n\to\infty} \ell_p(H,n)\binom{n}{2}^{-1}$ (the existence of of which follows from the proof of the next theorem) and then define $\ell(H) = \min_p \ell_p(H)$.

The next theorem is reminiscent of Theorem 4.5, but it is the incomplete extremal function $\mu_p(H)$, rather than $\kappa_p(H)$, that comes in to play.

Theorem 4.6 ([39, Theorem 4.1]) *In the simplified game,* $\ell_p(H) = \mu_p(H)$, *and so* $\ell(H) = \min_p \mu_p(H)$.

Proof Again, we just outline the proof in order to give the flavour of the connection with 2-coloured multigraphs. A little more detail is given [39]. Once again we make use of Szemerédi's lemma. We assume the first player must colour $p\binom{n}{2}$ edges red and that the second player must mark $\ell\binom{n}{2}$ edges.

If $\ell < \mu_p(H)$, player two picks a (not necessarily complete) 2-coloured graph G with $|G| = n$, $H \not\subseteq G$ and $w_p(G) \geq \ell\binom{n}{2}$. After player one has 2-coloured the K_n, player two chooses a bijection $f : V(G) \to V(K_n)$ maximizing the number of edges in $E(G_r) \cup E(G_b)$ that map to an edge of the same colour. Since K_n has $p\binom{n}{2}$ red edges, in a random bijection an edge in $E(G_r)$ has probability p of mapping to a red edge, so the number of agreeing edges in f is at least the expected number $w_p(G)$. For every edge uv that agrees, player two marks $f(u)f(v)$. Since the marked edges have the same colour as a subgraph of G, they do not contain a copy of H, so player two wins. Hence $\ell_p(H,n) \geq \mu_p(H)\binom{n}{2}$.

If $\ell > \mu_p(H) + 6\epsilon$, then player one colours K_n randomly with red probability p. Player two then marks $\ell\binom{n}{2}$ edges; call the red and blue marked subgraphs M_r and M_b. Partition $V(K_n)$ into $k > 2/\epsilon$ classes so that the partition is regular both in M_r and M_b. Let J be the 2-coloured multigraph whose vertices are the k parts and whose edges are red/blue/green according as the pair is regular and of density not close to zero in M_r/M_b/both (so J might not be complete). The number of marked edges is at most $(n/k)^2 w_p(J) + 4\epsilon\binom{n}{2}$, the second (error) term accounting for sparse or irregular pairs and edges inside classes. Thus $w_p(J) \geq (\mu_p(H) + \epsilon)\binom{k}{2} \geq \mathrm{ex}(H,k)$ if k is large, so J contains a copy of H. A corresponding copy of H can now be recovered in $M_r \cup M_b$, giving player one the win. Hence $\ell_p(H,n) \leq (\mu_p(H) + 6\epsilon)\binom{n}{2}$ for large n. $\square$

Random colourings are always good for player one, even for the full colour-label game, as the proof shows, though sometimes other colourings are just as good.

The colour-label game is actually a special case of the game originally studied by Richer, involving an extra parameter $0 \leq t \leq \binom{|H|}{2}$ specified in advance, in which player one wins if player two marks at least t edges of some copy of H. This is equivalent to saying that player two must avoid marking the edges of a copy of a graph in $\mathcal{H}_t$, which is the set of spanning 2-coloured subgraphs of H with exactly t edges (so the graphs in $\mathcal{H}_t$ are not complete unless $t = \binom{|H|}{2}$). We can then define $\ell_p(H,t)$ and $\ell(H,t)$ analogously to $\ell_p(H)$ and $\ell(H)$.

Theorem 4.7 *For the above game,* $\ell_p(H,t) = \mu_p(\mathcal{H}_t)$ *and* $\ell(H,t) = \min_p \mu_p(\mathcal{H}_t)$.

Proof The previous proof carries over for the new game more or less verbatim until the very end, at which point, rather than J containing a copy of H, it contains a copy of a spanning subgraph H' of H having t edges. The pairs corresponding to edges of H' are regular in M_r or M_b, and the pairs corresponding to edges of $H \setminus H'$ are regular in the red and blue edges of K_n, because the colouring is random. Thus we can recover a copy of H in K_n in which t edges, the edges common to H', are marked. □

5   Finiteness and stability

Here we address a couple of natural questions about finiteness and stability.

5.1   Finiteness

Given a class $\mathcal{H}$ of 2-coloured multigraphs, must there be a finite sub-class $\mathcal{H}_0$ such that $\kappa_p(\mathcal{H}) = \kappa_p(\mathcal{H}_0)$? The corresponding question in classical extremal graph theory has a positive answer, as does the question for multigraphs of multiplicity 2, as proved by Brown, Erdős and Simonovits [21]. But Rödl and Sidorenko [46] showed that the answer is negative for multigraphs of multiplicity at least four. So how do 2-coloured multigraphs behave?

The answer depends on the value of p. For some values of p, the answer is negative.

Theorem 5.1 ([39, Theorem 3.28]) *Let $l \geq 5$ be an integer and let $p = 1/l$. Then there exists a family $\mathcal{H}$ of 2-coloured multigraphs for which there is no finite sub-family $\mathcal{H}_0 \subset \mathcal{H}$ either with $\kappa_p(\mathcal{H}_0) = \kappa_p(\mathcal{H})$ or with $\mu_p(\mathcal{H}_0) = \mu_p(\mathcal{H})$.*

The proof of this theorem is based on the beautiful counterexamples given by Rödl and Sidorenko [46], and the fact that the proof works only for some $p \leq 1/5$ is due to the relationship between the edge multiplicities in the counterexamples and the weights of the edges in the 2-coloured multigraphs.

In fact, when $p = 1/2$, there is always a suitable finite $\mathcal{H}_0 \subset \mathcal{H}$. This is because the only types that are relevant to the case $p = 1/2$ are types $\tau(a,b)$ of Definition 4.3, with just green edges (this claim, implicit in §4.1.1, is proved in §7.3). Theorem 3.7 then becomes

$$\kappa_{1/2}(\mathcal{H}) = \max\left\{\lambda_{1/2}(\tau(a,b)) : \text{no graph in } \mathcal{H} \text{ is } \tau(a,b)\text{-colourable}\right\}.$$

It is easily shown (see Example 6.3) that $\lambda_{1/2}(\tau(a,b)) = 1 - 1/2(a+b)$. If $t = \max\{a+b : \text{no graph in } \mathcal{H} \text{ is } \tau(a,b)\text{-colourable}\}$ then, for each j with $0 \leq j \leq t+1$, we can choose $H_j \in \mathcal{H}$ which is $\tau(t+1-j,j)$-colourable. Put $\mathcal{H}_0 = \{H_0,\ldots,H_{t+1}\}$. Then $\mathcal{H}_0 \subset \mathcal{H}$ and $\kappa_{1/2}(\mathcal{H}_0) = \kappa_{1/2}(\mathcal{H})$.

5.2   Stability

Stability is the name given to the phenomenon where all graphs that are extremal or close to extremal must have similar structure. The archetypal occurrence of stability is in classical extremal graph theory (see Simonovits [52]).

By Theorem 3.7, for any class of 2-coloured multigraphs $\mathcal{H}$ there is a type τ such that no graph in $\mathcal{H}$ is τ-colourable and $\kappa_p(\mathcal{H}) = \lambda_p(\tau)$. Let $(G_n)_{n=1}^{\infty}$ be an extremal sequence for $\mathcal{H}$; that is, G_n is a 2-coloured multigraph with $|G_n| = n$, such that G_n contains no member of $\mathcal{H}$ and $w_p(G_n) = \kappa_p(\mathcal{H})\binom{n}{2} + o(n^2)$. By Theorem 3.7 we can find such a sequence with $G_n \in \mathcal{Q}(\tau)$; a stability result would assert that for any extremal sequence, G_n is close to a member of $\mathcal{Q}(\tau)$.

No stability result can hold if there are two minimal types τ_1 and τ_2 that both satisfy Theorem 3.7, since both $\mathcal{Q}(\tau_1)$ and $\mathcal{Q}(\tau_2)$ contain extremal sequences. (It is not immediate that an extremal sequence in $\mathcal{Q}(\tau_1)$ is not close to a sequence in $\mathcal{Q}(\tau_2)$ but it is true — c.f. the proof of [39, Theorem 3.23]).

Alon and Stav [7] have proved certain stability results in the case where $p = 1/2$ and there is a unique $\tau(a, b)$ satisfying Theorem 3.7; we don't give more detail because the results are phrased in terms of edit distance and another metric.

More generally, we might settle for a version of stability in which any extremal sequence must be close to one of a finite number of structures (rather than some unique structure). However, all hope that even such a weakened stability might hold in general, either for κ_p or for μ_p, is dashed by the next theorem. The term 'p-core' in the theorem, meaning minimal, is made precise in Definition 6.1.

Theorem 5.2 *Let $t \geq 4$ be an integer and let $p = 1/t$. Let H be the complete 2-coloured multigraph of order $t + 1$ consisting of a red star $K_{1,t}$ with all other edges being blue. Then there are infinitely many p-core types τ such that H is not τ-colourable and $\lambda_p(\tau) = \kappa_p(H) = \mu_p(H)$.*

We shall see in §7.5.2 why this theorem is true. The example is modified from a counterexample given by Sidorenko [51] to a stability conjecture for multigraphs with edge multiplicity at least three.

6   Working with types

How can we work out the value of $\kappa_p(\mathcal{H})$ for some given class $\mathcal{H}$ of 2-coloured multigraphs? We know from Theorem 3.7 that there is some type τ, no graph in $\mathcal{H}$ being τ-colourable, with $\kappa_p(\mathcal{H}) = \lambda_p(\tau)$. Now it might be, as seen in Theorem 5.2, that there are infinitely many genuinely different such types. But it is necessary only to find one type in order to evaluate $\kappa_p(\mathcal{H})$. The results in this section are aimed at simplifying the search for such a type.

First of all, it is clear that if there is some sub-type $\sigma \subset \tau$, obtained by removing vertices of τ, with $\lambda_p(\sigma) = \lambda_p(\tau)$ then we need not worry about τ and can use σ instead.

Definition 6.1 A type or weak type τ is called p-core if $\lambda_p(\sigma) < \lambda_p(\tau)$ for every proper $\sigma \subset \tau$.

It is clear that Theorem 3.7 is true for a p-core type τ.

Remark We say p-core rather than just 'core' because a type can be p-core for some p but not all. For example, the types $\tau(a, b)$ are core for all p, whereas the type having two red vertices joined by a blue edge is p-core only for $p < 1/2$.

Remark The notion of H_p-core was used in [16] but the definition differs slightly for reasons related to the discussion in §4.1.2; a type is H_p-core in [16] if and only if it is p'-core here, where $p' = \log q / (\log p + \log q)$.

The following lemma about the optimal weighting of a p-core type has been observed by several authors.

Lemma 6.2 *Let τ be a p-core type or weak type with p-value $\lambda = \lambda_p(\tau)$. Let $x \in \Delta$ satisfy $x^{\mathrm{t}} W x = \lambda$, where $W = W_p(\tau)$. Then $W x = \lambda e$, where e is the all one vector. In other words, $x_u w_p(u) + \sum_{uv \in E(\tau)} x_v w_p(uv) = \lambda$ holds for all $u \in V(\tau)$.*

Example 6.3 Let $\tau = \tau(a, b)$. Suppose τ is p-core. An optimal weighting x satisfies $x_u w_p(u) + (1 - x_u) = \lambda$. Thus $x_u(1 - w_p(u)) = 1 - \lambda$, so $x_u = (1 - \lambda)/q$ if u is red and $x_u = (1 - \lambda)/p$ if u is blue. Since $\sum x_u = 1$ we find $\lambda = 1 - pq/(ap + bq)$. If τ is not p-core then $\lambda \geq 1 - pq/(ap + bq)$ by using this weighting x. But then τ contains some p-core sub-type $\tau(a', b')$, so $\lambda = 1 - pq/(a'p + b'q) < 1 - pq/(ap + bq)$, a contradiction. Hence $\tau(a, b)$ is p-core for all p and $\lambda_p(\tau) = 1 - pq/(ap + bq)$.

Sidorenko has given an elegant characterization of p-core types and weak types.

Theorem 6.4 (Sidorenko [51]) *A type or weak type τ is p-core if and only if its matrix $W = W_p(\tau)$ satisfies*

- W *is non-singular and all components of $W^{-1}e$ are positive, and*

- $y^{\mathrm{t}} W y < 0$ *for every non-zero vector y with $e^{\mathrm{t}} y = 0$.*

Theorem 6.4 is strong but can be delicate to apply. A simpler observation that can be helpful is that, if τ consists of two disjoint types ρ and σ joined by green edges, then τ is p-core if and only if both ρ and σ are p-core. For example, this implies immediately that $\tau(a, b)$ is p-core.

But the theorem about p-core types of greatest practical use is the next one. Its proof is via extensions.

Theorem 6.5 ([39, Theorem 3.23]) *Let τ be a p-core type. Then all edges of τ are green, apart from*

- *if $p < 1/2$, when some edges joining two red vertices might be blue, or*

- *if $p > 1/2$, when some edges joining two blue vertices might be red.*

It follows immediately, for example, that the only $1/2$-core types are $\tau(a, b)$. The proof of Theorem 6.5 is based on that of Brown, Erdős and Simonovits [21, Theorem 5]. It is possible to prove something for weak types but the result is less strong, though it is still true that if, say, $p < 1/2$ then blue vertices are joined only by green edges.

7 Examples

To what extent is it possible to calculate $\kappa_p(\mathcal{H})$ and $\mu_p(\mathcal{H})$ in cases of interest? We look now at some specific examples and see how the general theory can be applied to reduce the work. Most of the examples except §7.3 (the simple case when $p = 1/2$) involve just one forbidden graph, but in §7.7 an example is given in which $p \neq 1/2$ and $\mathcal{H}$ contains two graphs. The example in §7.8.2 involves the use of extensions.

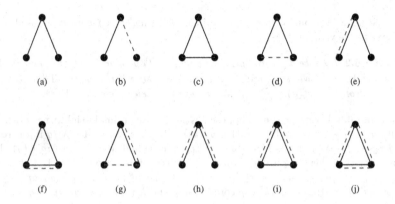

Figure 2: The 2-coloured multigraphs with three vertices and no isolates.

Remark In many cases, the 2-coloured multigraphs H in the examples are complete and have no green edges. In other words, $H = \mathrm{cc}(F)$ where $H_r = F$ and $H_b = \overline{F}$; for instance we consider $H = \mathrm{cc}(C_n)$ and $H = \mathrm{cc}(K_{t,t})$. In such cases we shall generally just write $H = C_n$ for brevity, so for instance $\kappa_p(K_{t,t})$ should, strictly speaking, be $\kappa_p(\mathrm{cc}(K_{t,t}))$.

7.1 Graphs with three vertices

There are ten 2-coloured multigraphs H of order 3 without isolated vertices (to within a swap of red and blue), and these are shown in Figure 2. Richer [44] evaluated $\mathrm{ex}(H,n)$ for them.

H	$\mathrm{ex}(H,n)$	$\mu_p(H) = \kappa_p(H)$
(a)	$p\binom{n}{2} + q\lfloor\frac{n}{2}\rfloor$	p
(b)	$\max\{p\binom{n}{2}, q\binom{n}{2}\}$	$\max\{p,q\}$
(c)	$p\binom{n}{2} + q\lfloor\frac{n^2}{4}\rfloor$	$p + q/2$
(d)	$\max\{q\binom{n}{2}, \lfloor\frac{n^2}{4}\rfloor, p\binom{n}{2} + q\lfloor\frac{n}{2}\rfloor\}$	$\max\{p,q\}$
(e)	$\max\{q\binom{n}{2}, p\binom{n}{2} + q\lfloor\frac{n}{2}\rfloor\}$	$\max\{p,q\}$
(f)	$\max\{q\binom{n}{2}, p\binom{n}{2} + q\lfloor\frac{n^2}{4}\rfloor\}$	$\max\{p + q/2, q\}$
(g)	$\max\{p\binom{n}{2} + q\lfloor\frac{n}{2}\rfloor, q\binom{n}{2} + p\lfloor\frac{n}{2}\rfloor\}$	$\max\{p,q\}$
(h)	$\max\{p\binom{n}{2} + q\lfloor\frac{n}{2}\rfloor, q\binom{n}{2} + p\lfloor\frac{n}{2}\rfloor\}$	$\max\{p,q\}$
(i)	$\max\{p\binom{n}{2} + q\lfloor\frac{n^2}{4}\rfloor, q\binom{n}{2} + p\lfloor\frac{n}{2}\rfloor\}$	$\max\{p + q/2, q\}$
(j)	$\max\{p\binom{n}{2} + q\lfloor\frac{n^2}{4}\rfloor, q\binom{n}{2} + p\lfloor\frac{n^2}{4}\rfloor\}$	$\max\{p + q/2, q + p/2\}$

The form of the extremal graphs G can readily be guessed from the graph H and from the form of the function $\mathrm{ex}(H,n)$. For example, the extremal graph for (a) is a red complete graph with $\lfloor n/2\rfloor$ blue edges added; we might write $G_r = K_n$ and

$G_b = \lfloor n/2 \rfloor K_2$ (these blue edges become green in the red/blue/green viewpoint). The two possible extremal graphs for (i) are $G_r = K_n$, $G_b = K_{\lfloor n/2 \rfloor, \lceil n/2 \rceil}$ and $G_r = \lfloor n/2 \rfloor K_2$ and $G_b = K_n$. The values of $\mu_p(H)$ are immediate by taking the limit $\lim_{n \to \infty} \mathrm{ex}(H, n) \binom{n}{2}^{-1}$.

The most interesting example is perhaps (d), where there are three extremal graphs: for most values of p, one of the two graphs $G_r = \overline{K}_n$, $G_b = K_n$ or $G_r = K_n$, $G_b = \lfloor n/2 \rfloor K_2$ is optimal, but for the very narrow range $(1 - 1/(2\lceil n/2 \rceil - 1))/2 < p < 1/2$ the graph $G_r = G_b = K_{\lfloor n/2 \rfloor, \lceil n/2 \rceil}$ is better. This third extremal graph has no effect on $\mu_p(H)$, however.

Because the extremal graphs in every case except (d) are complete, it follows that $\mathrm{kex}(H, n) = \mathrm{ex}(H, n)$ in these cases. In case (d), the extremal graphs are complete except when $(1 - 1/(2\lceil n/2 \rceil - 1))/2 \leq p \leq 1/2$. Thus $\kappa_p(H) = \mu_p(H)$ for $p \neq 1/2$, and by Theorem 2.8 the same is true when $p = 1/2$. In fact, for case (d), if G is complete and $H \not\subseteq G$ then it is easy to see that G_b must be a union of complete graphs, and none of these except K_2 components can contain a red edge. From this it quickly follows that $G_r = \overline{K}_n$, $G_b = K_n$ and $G_r = K_n$, $G_b = \lfloor n/2 \rfloor K_2$ are the only two extremal graphs, so $\mathrm{kex}(H, n) = \max\{q\binom{n}{2}, p\binom{n}{2} + q\lfloor \frac{n}{2} \rfloor\}$.

7.1.1 Finding $\kappa(H)$ and $\mu(H)$ directly The examples just given show that, even for very small H, finding the exact values of $\mathrm{ex}(H, n)$ and $\mathrm{kex}(H, n)$ can be quite intricate.

However, it is much easier to find the limiting values $\kappa_p(H)$ and $\mu_p(H)$. Let us look again at H in case (d). By Theorem 3.7 there is some p-core type τ such that $\kappa_p(H) = \lambda_p(\tau)$ and H is not τ-colourable. Now τ cannot have a green edge else H is τ-colourable (by placing one edge of H in one class and the other vertex in another class). Since τ has only red or blue vertices and edges we have at once from the definition that $\lambda_p(\tau) \leq \max\{p, q\}$. Moreover Theorem 6.5 shows that the only p-core types τ with no green edges that satisfy $\lambda_p(\tau) = \max\{p, q\}$ are the single vertex types $\tau(1, 0)$ and $\tau(0, 1)$.

Finding $\mu_p(H)$ is trickier in general because Theorems 3.7 and 6.5 don't apply, but for this particular H it is not difficult. We do know by Theorem 3.5 that $\mu_p(H)$ can be approximated arbitrarily closely with weak types τ such that H is not τ-colourable. If τ has a green edge then its end-vertices must be white, else H would be τ-colourable. Hence τ can have some white vertices, joined by green, and all other edges and vertices must be red or blue. It is not hard to see that such a weak type must have $\lambda_p(\tau) = \max\{p, q, 1/2\} = \max\{p, q\}$. Moreover the green graph cannot contain a triangle. The only such weak type of maximum weight with more than one vertex has two white vertices joined by a green edge, corresponding to the extremal graph $G_r = G_b = K_{\lfloor n/2 \rfloor, \lceil n/2 \rceil}$.

7.2 Graphs with four vertices

There are 140 2-coloured multigraphs of order 4 without isolates, so we do not attempt to find their extremal functions. Richer [44] picks out the example of the complete H with $H_r = K_4 - e$ and $H_b = 2K_2$, for which he shows $\kappa_p(H) = \mu_p(H) = \max\{p + q/2, 1 - pq\}$. The extremal types are $\tau(2, 0)$ and $\tau(1, 1)$.

If we restrict ourselves just to complete H with no green edges, in other words,

$H = \mathrm{cc}(F)$ for some F, then there are only six graphs to consider (to within a swap of red and blue). Alon and Stav [6] and Balogh and Martin [10] found $\mathrm{ed}(\mathrm{Forb}(F)) = 1 - \min_p \kappa_p(H)$ for all these six graphs. The value of $\kappa_p(H)$ for all such H and for all p was given in [39, Example 5.9]. Richer [44], in studying the function $\ell(H, t)$ described in §4.3, found $\min_p \mu_p(H)$ for each graph and also $\min_p \mu_p(\mathcal{H})$ for various collections $\mathcal{H}$ of them.

We remark that in all cases the relevant types are $\tau(a, b)$ for small a and b.

7.3   The case $p = 1/2$

By Theorem 6.5, a $1/2$-core type τ has only green edges, and so it is of the form $\tau = \tau(a, b)$. Example 6.3 shows $\lambda_{1/2}(\tau(a, b)) = 1 - 1/2(a + b) = 1 - 1/2|\tau|$. Hence, for any class $\mathcal{H}$, we have $\kappa_{1/2}(\mathcal{H}) = 1 - 1/2t$ where $t = \max\{a + b :$ no graph in $\mathcal{H}$ is $\tau(a, b)$-colourable$\}$.

If $\mathcal{P}$ is a hereditary property then we can apply the above to $\mathcal{H} = \mathcal{H}(\mathcal{P})$, as in §4.1. Theorem 4.4 then gives $c_{1/2}(\mathcal{P}) = 1 - 1/t$, which is the result of §4.1.1.

7.4   Types other than $\tau(a, b)$: two specimen cases

As mentioned in §4.1.2, the types $\tau(a, b)$ are not always enough to determine the value of $c_p(\mathcal{P})$, and in the light of Theorem 4.4 this means that Theorem 3.7 is not always satisfied by a type with only green edges. Here are two specific examples.

7.4.1   A graph of order 9   The first example of a graph H for which $\min_p \kappa_p(H)$ is given by a type other than $\tau(a, b)$ was exhibited by Balogh and Martin [10] in their study of edit distance. Their graph is $H = \mathrm{cc}(F)$, where F is the graph with vertex set $\{0, 1, 2, 3, 4, 5, 6, 7, 8\}$, in which ij is an edge if $i - j \in \{\pm 1, \pm 2\}$ (mod 9) or $i \equiv j \equiv 0$ (mod 3). Martin [41] has subsequently proved that $\kappa_p(H) = \max\{1 - p/3, 1 - p/(1 + 4p), 1 - q/2\}$. The types involved are $\tau(0, 3)$, $\tau(2, 0)$, and a third type with 5 blue vertices, two independent red edges and the remaining edges green. This third type has the greatest p-value of the three when $1/2 < p < (1 + \sqrt{17})/8$. It follows that $\min_p \kappa_p(H)$ occurs at $p^* = (1 + \sqrt{17})/8$ and $\mathrm{ed}(\mathrm{Forb}(F)) = 1 - \kappa_{p^*}(H) = (7 - \sqrt{17})/16$.

7.4.2   A graph of order 6   A simpler example of a graph H for which the types $\tau(a, b)$ are insufficient is $H = C_6^*$ described in Example 2.6. As an illustration of some of the techniques that can be used, we prove $\kappa_p(C_6^*) = \max\{1 - pq, 1 - p/(1 + 2p)\}$. We skip routine checks; more detail is in [39, Example 5.10].

To find a lower bound on $\kappa_p(C_6^*)$ we need types τ for which C_6^* is not τ-colourable. Let σ be the type with one blue and two red vertices, the red vertices being joined by a blue edge and the other edges being green. It is not difficult to check that σ is p-core for $0 < p < 1/2$ and that $\lambda_p(\sigma) = 1 - p/(1 + 2p)$. As described in Example 2.6, C_6^* is neither σ-colourable nor $\tau(2, 0)$-colourable. Thus $\kappa_p(C_6^*) \geq \max\{1 - pq, 1 - p/(1 + 2p)\}$.

To establish an upper bound on $\kappa_p(C_6^*)$ we begin by checking that if $\tau \in \{\tau(0, 2), \tau(2, 1), \rho_1, \rho_2, \rho_3\}$ then C_6^* is τ-colourable. Here, ρ_1 is the type with three red vertices joined by two green and one blue edges, ρ_2 has one blue vertex joined to three red vertices by green edges, other edges being blue, and ρ_3 has a red vertex joined to two blue vertices by green edges, the other edge being red.

Now let τ be a p-core type satisfying Theorem 3.7 for C_6^*. If $p < 1/2$ then, by Theorem 6.5, all edges of τ are green except perhaps for blue edges joining red vertices. Since $\tau(0,2) \not\subset \tau$, τ has at most one blue vertex. If τ has a blue vertex then it has at most two red vertices, else $\rho_2 \subset \tau$. The red vertices cannot be joined by green because $\tau(2,1) \not\subset \tau$. Thus $\tau \subset \sigma$. If, on the other hand, τ has no blue vertices, then each vertex can have at most one incident green edge, else $\rho_1 \subset \tau$.

We can now use the following simple but useful lemma.

Lemma 7.1 *Let τ be a p-core type whose vertices are all red, each joined to at most d others by green edges. Then $\lambda_p(\tau) \le \max\{q, 1 - q/(d+1)\}$.*

So if $p < 1/2$ we have $\lambda_p(\tau) \le \max\{q, 1 - q/2\} < 1 - p/(1 - 2p)) = \lambda_p(\sigma)$.

Now suppose $p > 1/2$. In this case, Theorem 6.5 shows that red vertices of τ meet only green edges. Since $\rho_1 \not\subset \tau$ and $\tau(2,1) \not\subset \tau$, this means if τ has two red vertices then $\tau = \tau(2,0)$. If τ has only blue vertices then all the edges are red, since $\tau(0,2) \not\subset \tau$, so $\lambda_p(\tau) \le p < 1 - q/2$. So τ has exactly one red vertex, and at least two blue vertices else $\lambda_p(\tau) \le \lambda_p(\tau(1,1)) = 1 - pq < 1 - q/2$. Hence, by Theorem 6.5, $\rho_3 \subset \tau$, a contradiction.

We have now shown that if $p \ne 0, \frac{1}{2}, 1$ then the only optimal types are σ and $\tau(2,0)$, and so $\kappa_p(C_6^*) = \max\{1 - pq, 1 - p/(1 + 2p)\}$.

7.5   Split graphs

A *split graph* is a graph whose vertex set can be covered by a clique and an independent set — that is, it is precisely a member of the class $\mathcal{P}(\tau(1,1))$. This structure makes it possible to say quite a lot about their extremal functions. On the other hand, graphs in $\mathcal{P}(\tau(2,0))$ and $\mathcal{P}(\tau(0,2))$ are very hard to analyse. The reason for this disparity seems to lie in Theorem 6.5, which says that if τ is p-core and has vertices of both colours then $\tau(1,1) \subset \tau$, so every split graph is τ-colourable. This fact, plus the symmetry of the definition, opens up the analysis of split graphs.

A specific example of a split graph is $K_s + \overline{K_t}$. The edit distance of $K_1 + \overline{K_3}$ was computed by Alon and Stav [6], and the same was done for $K_s + \overline{K_t}$ in general by Balogh and Martin [10]. Let $H = \mathrm{cc}(K_s + \overline{K_t})$, which we abbreviate to $H = K_s + \overline{K_t}$. The value of $\kappa_p(K_s + \overline{K_t})$ was given in [39, Example 5.5]. More generally, let $H = \mathrm{cc}(F)$ where F is a split graph. Write ω for the clique number and α for the independence number of F; note that either $\omega + \alpha = |H|$ or $\omega + \alpha = |H| + 1$.

Let τ be a p-core type with $\kappa_p(H) = \lambda_p(\tau)$. As already noted, the vertices of τ are all the same colour. Suppose they are red. No vertex can have degree $|H| - \omega$ in green else H would be τ-colourable, since the edges of τ are blue or green. Hence, by Lemma 7.1, $\lambda_p(\tau) \le \max\{q, 1 - q/(|H| - \omega)\}$. Likewise if all vertices are blue we obtain $\lambda_p(\tau) \le \max\{p, 1 - p/(|H| - \alpha)\}$. Therefore $\kappa_p(H) \le \max\{1 - p/(|H| - \alpha), 1 - q/(|H| - \omega)\}$.

If $\omega + \alpha = |H| + 1$ then equality holds, because H is neither $\tau(|H| - \omega, 0)$-colourable nor $\tau(0, |H| - \alpha)$-colourable. But if $\omega + \alpha = |H|$ the inequality can be strict; for example, it is easily shown that if F is the path of length 3 then $\kappa_p(H) = \max\{p, q\}$ (a special case of [39, Example 5.8]).

Marchant [37, 38] has found bounds on $\text{kex}_p(n, K_s + \overline{K_t})$ by a direct argument not related to the methods of this article. He showed

$$\kappa \binom{n}{2} \leq \text{kex}_p(n, K_s + \overline{K_t}) \leq \text{ex}_p(n, K_s + \overline{K_t}) \leq \kappa \binom{n+1}{2}.$$

where $\kappa = \max\{1 - p/s, 1 - q/(t-1)\}$. Richer [44], and independently Diwan and Mubayi [23], obtained similar bounds when $s = 1$ and $p = 1/t$.

7.5.1   Green split graphs

Let H be the graph in the previous example but where the red edges have been turned green: that is, the red graph is the same as before but the blue graph is K_{s+t}. Note that $\text{kex}_p(n, H) = \text{ex}_p(n, H)$ because H has no red edges, and so any white (i.e. missing) edges in an extremal graph can safely be replaced by red edges.

Marchant [37, 38] proved the same inequalities for H as those just mentioned for $K_s + \overline{K_t}$, but with instead $\kappa = \max\{1 - p/s, 1 - q/(s + t - 1)\}$. It follows that $\kappa_p(H) = \lambda_p(H) = \kappa$.

7.5.2   Stars and stability

Return now to the graphs $H = K_s + \overline{K_t}$, and take the case $s = 1$. Then $\kappa_p(H) = \mu_p(H) = \max\{q, 1 - q/(t-1)\}$. The p-core type $\tau = \tau(t-1, 0)$ satisfies $\lambda_p(\tau) = 1 - q/(t-1)$ and H is not τ-colourable, so τ satisfies Theorem 3.7 for H.

Now we can take any $(t-2)$-regular graph G and turn it into a type τ_G by painting its vertices red, its edges green, and painting the missing edges blue. The 2-coloured graph H is not τ_G-colourable for any such type τ_G. By taking the weight vector with $x_u = 1/|\tau_G|$ for every vertex u, we obtain $\lambda_p(\tau) \geq q - (1 - pt)/|\tau_G|$. In the specific case $p = 1/t$ this means $\lambda_{1/t}(\tau_G) \geq q$. On the other hand, Lemma 7.1 shows $\lambda_{1/t}(\tau_G) \leq \max\{q, 1 - q/(t-1)\} = q$. So $\lambda_{1/t}(\tau_G) = q = \kappa_{1/t}(H)$.

It follows that, if τ_G is p-core when $p = 1/t$, then τ_G satisfies Theorem 3.7 for H. A simple argument [39, Theorem 5.6], adapted from Sidorenko [51], shows that τ_G is indeed p-core provided G is connected and $p \leq 1/t$. For $t \geq 4$ there is a boundless supply of connected $(t-2)$-regular graphs G, and so Theorem 5.2 is verified.

In fact, τ_G can be p-core for larger values of p than $1/t$. When $t = 4$ and G is a cycle, a restful calculation involving Sidorenko's criterion (Theorem 6.4) shows that τ_G is p-core if and only if $p < 1/(2 + 2\cos(2\pi/n))$.

7.6   Short cycles

The cycles form one of the simplest classes of graphs. Let $H = \text{cc}(C_n)$ where C_n is a cycle of length n, abbreviated as usual to $H = C_n$. We ignore the case $n = 3$ since H is then monochromatic (see §2.4).

Alon and Stav evaluated $\text{ed}(\text{Forb}(C_4))$ and, using arguments similar to those elsewhere in this section, it is easy to show that $\kappa_p(C_4) = 1 - pq$.

The 5-cycle already needs work. Marchant [37, 38] proved by induction that

$$\kappa \binom{n}{2} - \frac{n}{2} \leq \text{kex}_p(n, C_5) \leq \text{ex}_p(n, C_5) \leq \kappa \binom{n+1}{2}, \quad \kappa = \max\left\{1 - \frac{p}{2}, 1 - \frac{q}{2}\right\}.$$

So $\kappa_p(C_5) = \mu_p(C_5) = \max\{1 - p/2, 1 - q/2\}$, the types being $\tau(2, 0)$ and $\tau(0, 2)$.

Marchant also gives a direct argument for C_7, but it is much more complex than that for C_5. He proves

$$\kappa\binom{n}{2} - \frac{n}{2} \le \text{kex}_p(n, C_7) \le \kappa\binom{n}{2} + 2n \qquad \mu\binom{n}{2} - \frac{n}{2} \le \text{ex}_p(n, C_7) \le \mu\binom{n+1}{2},$$

where

$$\kappa = \max\left\{1 - \frac{p}{2}, 1 - \frac{pq}{1+p}, 1 - \frac{q}{3}\right\} \quad \text{and} \quad \mu = \max\left\{1 - \frac{p}{2}, \frac{5}{6}, 1 - \frac{q}{3}\right\}.$$

Clearly $\kappa_p(C_7) = \kappa$ and $\mu_p(C_7) = \mu$. The relevant types for $\kappa_p(C_7)$ are $\tau(0,2)$, $\tau(2,1)$ and $\tau(3,0)$. The type $\tau(2,1)$ is optimal in the range $1/3 < p < 1/2$. When considering $\mu_p(C_7)$, the type $\tau(2,1)$ is superseded in the same range by the weak type consisting of 6 white vertices joined by green. Thus $H = C_7$ gives another example of a graph where $\kappa_p(H) < \mu_p(H)$ for some p.

In classical extremal graph theory the behaviour of even cycles is quite different from that of odd cycles, but such a dichotomy is not in evidence here. Using the methods of this article it is easy to show $\kappa_p(C_6) = \max\{1 - q/2, 1 - pq\}$, the types being $\tau(2,0)$ and $\tau(1,1)$. Because C_6 is $\tau(3,0)$-colourable, the following lemma comes in handy: its simple proof rests on Mantel's theorem [36] (the triangle case of Turán's theorem [55]).

Lemma 7.2 ([39, Lemma 5.11]) *Let τ be a type whose vertices are all red and which contains no green triangle. If $p \le 1/2$ then $\lambda_p(\tau) \le (p - q)/|\tau| + q + p/2 \le 1 - p/2$.*

Life gets tougher, though, for larger cycles. Martin [41] has managed to evaluate $\kappa_p(C_n)$ for $n \le 9$, and also $\kappa_p(C_{10})$ for most p. Hence he can compute the edit distance of $\text{Forb}(C_n)$ for $n \le 10$. Here are the values for $8 \le n \le 10$.

$$\begin{aligned}
\kappa_p(C_8) &= \max\{1 - pq/(1+p), 1 - q/3\} \\
\kappa_p(C_9) &= \max\{1 - p/2, 1 - q/4\} \\
\kappa_p(C_{10}) &= \max\{1 - pq/(1+2p), 1 - q/4\} \quad \text{for } p \ge 1/7
\end{aligned}$$

7.7 More than one forbidden graph

Classical extremal graph theory is profoundly shaped by the Erdős-Stone theorem [31], one consequence of which is that the extremal function for a class of graphs is the minimum of the extremal functions for the individual graphs, at least in the limit as $n \to \infty$. A similar phenomenon does not hold for hypergraphs or for multigraphs, though explicit examples are not so easy to come by, due in large part to the difficulty of evaluating the extremal functions themselves.

Theorem 5.1 shows that the phenomenon does not hold either for 2-coloured multigraphs in general. Here is a simple explicit example [39, Example 5.12].

Let H_1 be the graph of order 4 with $H_r = K_{1,3}$ and $H_b = K_4$, discussed in §7.5.1. Then $\kappa_p(H_1) = \max\{q, 1 - q/3\}$. Let H_2 be a green K_3; by Mantel's theorem [36] $\kappa_p(H_2) = 1/2 + \max\{p/2, q/2\}$.

Now put $\mathcal{H} = \{H_1, H_2\}$. Suppose $p < 1/2$ and $\kappa_p(\mathcal{H}) = \lambda_p(\tau)$, where τ is a p-core type such that neither H_1 nor H_2 is τ-colourable. If τ has a blue vertex then there can be no other vertex, else H_1 would be τ-colourable by Theorem 6.5; thus $\lambda_p(\tau) \le q$. If τ has only red vertices then no vertex has three green edges, and there is no triangle. Lemma 7.2 shows $\lambda_p(\tau) \le 3/4$ if $|\tau| \le 4$ and the proof of Lemma 7.1 shows the same if $|\tau| \ge 4$. Hence $\kappa_p(\mathcal{H}) \le \max\{q, 3/4\}$ if $p \le 1/2$.

So for $1/4 < p < 1/2$ we have $3/4 = \kappa_p(\{H_1, H_2\}) < \min\{\kappa_p(H_1), \kappa_p(H_2)\} = \min\{1 - q/3, 1 - p/2\}$.

7.8 Complete bipartite graphs

As mentioned in §7.5, whereas $\tau(1,1)$-colourable graphs can be easy to analyse, this is far from true for $\tau(2,0)$-colourable graphs (or, equivalently, $\tau(0,2)$-colourable graphs). We outline here what is known about complete bipartite graphs. Let $H = \mathrm{cc}(K_{s,t})$ with $s \le t$, abbreviated to $H = K_{s,t}$.

We observe at once that $K_{s,t}$ is not $\tau(s-1, 1)$-colourable, and so $\kappa_p(K_{s,t}) \ge 1 - pq/(1 + p(s-2))$. For $s = t = 2$ we have equality for all p, as mentioned in §7.6. The next lemma, proved via Theorem 6.5, gives information about when equality fails.

Lemma 7.3 ([39, Lemma 5.14]) *Let τ be a p-core type such that $K_{s,t}$, $s \le t$, is not τ-colourable and $\kappa_p(K_{s,t}) = \lambda_p(\tau)$. Then either $\tau = \tau(s-1,1)$ and $\kappa_p(H) = 1 - pq/(1 + p(s-2))$, or all the vertices of τ are red. If $p > 1/2$ then $\tau = \tau(s-1,1)$ if $t = s$ and $\tau = \tau(t-1,0)$ otherwise.*

Hence the value of $\kappa_p(K_{s,t})$ is known for $p \ge 1/2$. But for $p < 1/2$ the evaluation of $\kappa_p(K_{s,t})$ is difficult and presents interesting features, even for $s = 2$. Different kinds of behaviour are illustrated by the following three examples.

7.8.1 $H = K_{2,t}$
The case $K_{2,2}$ has already been mentioned, and Martin and McKay [42] proved that $\tau(1,1)$ is optimal for $K_{2,3}$ for all $p < 1/2$. However, it was shown in [39, Example 5.16] that $\tau(1,1)$ is non-optimal for $K_{2,t}$ if $t \ge 4$ and $1/3 < p < 1/2$, though no clue was offered as to the correct value.

Martin and McKay [42] managed to find the value of $\kappa_p(K_{2,4})$ for all p. For $p < 1/5$ the optimal type is $\tau(1,1)$ and for $p > 1/3$ it is $\tau(3,0)$. But remarkably, the optimal type τ for $1/5 < p < 1/3$ comes from a $(15,6,1,3)$ strongly regular graph, the so-called "generalized quadrangle" $GQ(2,2)$. The type τ is obtained from $\mathrm{cc}(GQ(2,2))$ by painting the 15 vertices red. It can then be calculated that $\kappa_p(K_{2,4}) = \max\{1 - pq, 7(1+q)/15, 1 - q/3\}$.

7.8.2 $H = K_{3,3}$
This graph was already highlighted for attention by Richer [44] and by Diwan and Mubayi [23] (see §8). The problem of finding $\mathrm{ed}(\mathrm{Forb}(K_{3,3}))$ was raised as an interesting case by Alon and Stav (see [6]); the value was found by Balogh and Martin [10].

Let τ be a p-core type for which $K_{3,3}$ is not τ-colourable and $\kappa_p(K_{3,3}) = \lambda_p(\tau)$. We know by Lemma 7.3 that if $\tau \ne \tau(2,1)$ then $p < 1/2$ and all the vertices of τ are red. In the latter case τ cannot contain a green triangle, so Lemma 7.2 shows $\lambda_p(\tau) \le 1 - p/2 < 1 - pq/(1+p) = \lambda_p(\tau(2,1))$ if $p > 1/3$. Hence $\tau(2,1)$ is optimal for $p > 1/3$ and $\kappa_p(K_{3,3}) = 1 - pq/(1+p)$ in this range. The function $1 - pq/(1+p)$

has a local minimum at $\sqrt{2} - 1 > 1/3$ and so, by Theorems 2.8 and 4.4, we obtain Balogh and Martin's result [10] that $\mathrm{ed}(\mathrm{Forb}(K_{3,3})) = 3 - 2\sqrt{2}$.

In [39, §5.4] it is shown that $\kappa_p(K_{3,3}) = 1 - pq/(1 + p)$ continues to hold for $p \geq 1/9$. In the same way that we used $\tau(3,0) \not\subseteq \tau$ to show the equation for $p \geq 1/3$, the proof for $p \geq 1/9$ involves finding two classes of types that cannot be subtypes of τ. Unlike any of the examples discussed up to now, the argument is long and makes heavy use of extensions via Lemma 3.6 to find these types.

However, it turns out that $\kappa_p(K_{3,3}) = 1 - pq/(1 + p)$ is not true for all p. For the equation to fail, Lemma 7.3 says τ must be of the form τ_G for some graph G, as described in §7.5.2. To avoid $K_{3,3}$ being τ-colourable, G must contain neither K_3 nor $K_{3,3}$. A little calculation shows that, if G has average degree d, then there is some value of p for which $\lambda_p(\tau_G) > 1 - pq/(1 + p)$ provided $(d + 3)^2 > 8|G|$. This means G must be quite dense for a $K_{3,3}$-free graph — random graphs are not good enough — but Brown's graphs [18] work. As a result, $\kappa_p(K_{3,3}) > 1 - pq/(1 + p)$ for $p \leq 1/124$. The types here are large: the smallest one, which works in the range $1/219 \leq p \leq 1/124$, is the Brown graph for the prime 19 and it has 13718 vertices.

These facts suggest it is unlikely that $\kappa_p(K_{3,3})$ can be evaluated for all p without first finding the ordinary graph extremal function for $K_{3,3}$.

7.8.3 $\mathbf{H} = \mathbf{K_{7,7}}$ The type $\tau(6,1)$ has p-value $1 - pq/(1 + 5p)$. This is minimized when $p = (\sqrt{6} - 1)/5$ and the minimum value is $(18 + 2\sqrt{6})/25 < 11/12$. Let G be $K_{6,6}$ with a 1-factor removed. Then $K_{7,7}$ is not τ_G-colourable, and $\lambda_p(\tau_G) \geq 11/12$ for all p. It follows that $\min_p \kappa_p(K_{7,7})$ is not the same as $\min_p(\lambda_p(\tau(6,1)))$, so the edit distance for $\mathrm{Forb}(K_{7,7})$ is not realised by $\lambda_p(\tau(6,1))$. The edit distance in this case remains unknown: it is not $11/12$ because a better value is obtained by taking G to be K_{13} and removing 5 independent edges and a path of length 2.

Similar remarks apply to $K_{t,t}$ for all $t \geq 7$.

8   Richer's conjectures

Motivated by Theorem 4.6, Richer [44] made an interesting conjecture about $\min_p \kappa_p(H)$, perhaps reminiscent of the Erdős-Stone theorem [31].

Conjecture 8.1 (Richer [44]) *Let H be a 2-coloured complete graph with no green edges. Then*

$$\min_p \mu_p(H) \leq 1 - \frac{1}{|H|} .$$

A stronger conjecture would be $\min_p \mu_p(H) \leq 1 - 1/(|H| - 1)$, which is more in line with extremal graph theory. However, this inequality does not always hold. Suppose $H = K_{t,t}$, as in §7.8, or, more generally, $|H| = 2t$ and the components of H_r are complete equipartite bipartite graphs. Then H is not $\tau(t - 1, 1)$-colourable so $\min_p \mu_p(H) \geq \min_p \lambda_p(\tau(t - 1, 1)) = 1 - 1/(t + 2\sqrt{t - 1})$. This is greater than $1 - 1/(|H| - 1)$ for $t \in \{2, 3, 4\}$.

Nevertheless, the stronger conjecture might hold if $|H| \notin \{4, 6, 8\}$.

A closely related conjecture was made by Diwan and Mubayi [23], who conjectured that a 2-coloured multigraph G of order n that does not contain H satisfies $\min\{e(G_r), e(G_b)\} \leq (1 - 1/(|H| - 1))n^2/2$, apart from the special cases just cited. Certainly $\min\{e(G_r), e(G_b)\} \le w_p(G)$ so the asymptotic version of Diwan and Mubayi's conjecture is implied by the strong form of Conjecture 8.1.

Since $\kappa_p(H) \le \mu_p(H)$, Conjecture 8.1 implies $\min_p \kappa_p(H) \le 1 - 1/|H|$. This inequality is already known, though. Axenovich, Kézdy and Martin [8] proved that $\text{ed}(\text{Forb}(H_r)) \ge 1/2t$, where $t = \max\{a + b : H$ is not $\tau(a, b)$-colourable$\}$. By Theorem 4.5 this is equivalent to $\min_p \kappa_p(H) \le 1 - 1/2t$. In fact, $\kappa_{1/2}(H) \le 1 - 1/2t$. To see this, note that if $\lambda_{1/2}(\tau) > 1 - 1/2t$ then the proportion of edges of τ that are green must exceed $1 - 1/t$, in which case τ contains a green K_{t+1} by Turán's theorem. This means $\tau(a, b) \subset \tau$ for some a, b with $a + b = t + 1$, so H is τ-colourable. (This is essentially the proof in [8].)

Richer [44] proved Conjecture 8.1 for $|H| \le 5$ and proved that $\min_p \mu_p(H) \le 1 - 3/(5|H| - 5)$ in general. This was improved by Marchant [37, 38].

Theorem 8.2 (Marchant [37, 38]) *Let H be a 2-coloured complete graph with no green edges. Then*

$$\min_p \mu_p(H) \le 1 - \frac{2}{3|H| - 3}.$$

Richer gave a second conjecture, weaker than the first. To state it we must define certain weak types analogous to the types $\tau(a, b)$.

Definition 8.3 The weak type $\tau(a, b, c)$ has a red, b blue and c white vertices, all edges being green.

Hence $\tau(a, b) = \tau(a, b, 0)$. By an argument similar to that of Example 6.3, it is easily shown that $\tau(a, b, c)$ is p-core for all p and $\lambda_p(\tau(a, b, c)) = 1 - pq/(ap + bq + cpq)$.

Richer's second conjecture was the following theorem, which he proved but with $|H| + 8$ in place of H. By developing his ideas Marchant gave a full proof.

Theorem 8.4 (Marchant [37, 38]) *Let H be a 2-coloured complete graph with no green edges. Then there exists p, $0 \le p \le 1$, such that*

$$\lambda_p(\tau(a, b, c)) \le 1 - \frac{1}{|H|}$$

for every weak type $\tau(a, b, c)$ for which H is not $\tau(a, b, c)$-colourable.

Thus if Conjecture 8.1 is to fail then it must do so for an H for which the optimal type at the minimizing probability does not have just green edges.

9 Further remarks

We close with comments on a couple of further related issues.

9.1 Szemerédi's Regularity Lemma — is it needed?

Apart from the applications mentioned in §4, the proofs of the theorems in this article make no use of Szemerédi's Regularity Lemma, one of the most fundamental tools in graph theory [54].

On the other hand, the proofs of the applications use Szemerédi's lemma in an essential way. This is described, as far as the applications in §4.1.1 and §4.1.2 are concerned, in Bollobás's article [12, §5].

When a result is proved using Szemerédi's lemma, the question is sometimes asked whether there is a proof which avoids using the lemma. This can be because it is felt that a sledgehammer is being used to crack a nut, though it is equally reasonable to view the lemma as a basic and natural fact with an elementary proof. But another reason to avoid its use is that it introduces constants that are mind-numbingly large, as Gowers [34] showed.

An example of a theorem in which Szemerédi's Lemma was heavily used is that of Bollobás and Nikiforov [13], giving the following strengthening of the theorem of Prömel and Steger in §4.1.1.

Theorem 9.1 (Bollobás and Nikiforov [13]) *Let F be a graph, $\mathcal{P} = Forb(F)$ and $t = \max\{a + b : \mathcal{P}(\tau(a, b) \subset \mathcal{P}\}$. Then for every $\epsilon > 0$ there exists $\delta > 0$ such that, amongst any $2^{(1 - 1/t + \epsilon)\binom{n}{2}}$ graphs with vertex set $\{1, \ldots, n\}$, there is one containing at least $\delta n^{|F|}$ induced copies of F.*

The proof given in [13] uses Szemerédi's lemma multiple times: it is a combination of the Prömel-Steger theorem [43] and the removal lemma of Alon, Fisher, Krivelevich and Szegedy [3], both of which rely heavily on Szemerédi's lemma for their proofs.

The question, then, is asked in [13] whether Theorem 9.1 can be proved without Szemerédi's lemma. The answer is that it can.

First of all, the main theorem of §4.1.1, which includes the Prömel-Steger theorem, was proved by Alekseev [2] *inter alia*, and Alekseev's proof, unlike that of [15], does not use Szemerédi's lemma. It rests instead on the well-known shattering lemma proved by Sauer [47], by Shelah [50] and by Vapnik and Chervonenkis [56].

Thus Szemerédi's lemma is excised from one half of the proof in [13], and the job is completed by avoiding the removal lemma entirely and using a theorem of Saxton [48]. The statement of Saxton's theorem is reminiscent of Erdős and Simonovits's supersaturation theorem [29, Theorem 1]. In particular it is not necessary to know the number of graphs with no induced copy of F, and as such it extends in the sense of §4.1.2 to the case $p \neq 1/2$.

Theorem 9.2 (Saxton [48]) *Let F be a graph, let $\mathcal{P} = Forb(F)$ and let $0 < p < 1$. Then for every $\epsilon > 0$ there exists $\delta > 0$ such that, if $\mathcal{G}$ is a class of graphs with vertex set $\{1, \ldots, n\}$ and $\Pr[G(n, p) \in \mathcal{G}] \geq 2^{\epsilon\binom{n}{2}} \Pr[G(n, p) \in \mathcal{P}]$, then some graph in $\mathcal{G}$ contains at least $\delta n^{|F|}$ induced copies of F.*

So Szemerédi's lemma is not needed in Theorem 9.1, nor in §4.1.1. Is it needed in the more general §4.1.2? It appears that the results of [16] might be proved instead by means of extensions. The results of §4.3 might also be amenable to the same treatment. We do not offer details here, though.

That leaves the edit distance application in §4.2. When making their study of edit distance Alon and Stav [5] noticed that there was a very close connection between what they were doing and what was done in [15], but there were subtle differences which prevented them using the results from [15] directly. Curiously, this disconnection resurfaces here: whilst it looks as though extensions will work in §4.1.2, Szemerédi's lemma seems at present to be intrinsic to the study of edit distance, as it is to the study of property testing (as might be inferred from [3]. For more on this subject, including hypergraph extensions, see [4, 45, 9]).

9.2 The relationship between κ_p and μ_p

So far we have concentrated on the parameter κ_p rather than on μ_p, because of its greater importance in applications. On the other hand, μ_p might appear the more natural parameter. What is the relationship between these two?

Given that μ_p is the unconstrained extremal function and κ_p is the same but where the underlying graph is complete, the relationship between μ_p and κ_p is somewhat reminiscent of the relationship between Turán's theorem and Ramsey's theorem. Ramsey-Turán theory was an attempt to explore that relationship: we do not describe it in detail here but refer the reader instead to the survey by Simonovits and Sós [53].

Given a graph H, Ramsey-Turán asks for the value of the function

$$\mathrm{ex}(H, n, \alpha) = \max\{e(G) : |G| = n, H \not\subseteq G, \alpha(G) \leq \alpha\}$$

where $\alpha(G)$ is the independence number of G. Clearly $\mathrm{ex}(H, n, n)$ is the ordinary extremal function and the idea is that for small values of α the study of $\mathrm{ex}(H, n, \alpha)$ approaches Ramsey theory in some way. One of the most studied problems was to find

$$\lim_{\epsilon \to 0} \lim_{n \to \infty} \mathrm{ex}(H, n, \epsilon n) \binom{n}{2}^{-1}.$$

In like manner, given a 2-coloured multigraph H, we might define

$$\mathrm{ex}_p(H, n, \alpha) = \max\{w_p(G) : |G| = n, H \not\subseteq G, \alpha(G) \leq \alpha\}$$

where the independence number $\alpha(G)$ of the 2-coloured multigraph G is the independence number of the underlying graph G_u. We could then define

$$\mu_p^\epsilon(H) = \lim_{n \to \infty} \mathrm{ex}_p(\mathcal{H}, n, \epsilon n) \binom{n}{2}^{-1} \quad \text{and} \quad \mu_p^0(H) = \lim_{\epsilon \to 0} \mu_p^\epsilon(H).$$

If G is a complete 2-coloured multigraph G then $\alpha(G) = 1$, and so $\kappa_p(H) \leq \mu_p^\epsilon(H)$ for all $\epsilon > 0$. Thus $\kappa_p(H) \leq \mu_p^0(H)$. It seems natural to ask whether $\kappa_p(H) = \mu_p^0(H)$ holds.

Clearly there are graphs H for which $\kappa_p(H) = \mu_p^0(H)$, because $\kappa_p(H) \leq \mu_p^\epsilon(H) \leq \mu_p(H)$ holds for all H and we have seen plenty of examples where $\kappa_p(H) = \mu_p(H)$. But there are also graphs H for which $\kappa_p(H) < \mu_p^0(H)$. An example due to Saxton is as follows.

Let $H = \mathrm{cc}(tK_t)$: that is, H consists of t red cliques of order t with blue edges in between. Given $\epsilon > 0$ there is a triangle-free graph of order n and independence number at most ϵn which is sparse — that is, it has $o(n)$ edges (Erdős [24]). Let F be the graph consisting of k copies of this graph with all edges in between the copies (that is, F is complete k-partite with a copy of this graph inserted into each class); then F contains no complete subgraph of order greater than $2k$. Let G be the 2-coloured graph obtained by painting every edge of F green. Then $\alpha(G) \leq \epsilon|G|$ and $w_p(G) = (1 - 1/k + o(1))\binom{|G|}{2}$. Moreover if $k = \lfloor (t^2 - 1)/2 \rfloor$ then $H \not\subseteq G$. This shows that $\mu_p^0(H) \geq 1 - 2/(t^2 - 2)$. The construction is the same as that used by Erdős and Sós [30] when studying the Ramsey-Turán function for K_{2k+1}.

On the other hand, $\kappa_p(H) \leq 1 - \min\{p, q\}/(2t - 2)$. For otherwise there is some $\delta > 0$ and a sequence (G_n) of 2-coloured complete graphs not containing H with $|G_n| = n$ and $w_p(G) \geq (1 - \min\{p, q\}(1/(2t - 2) - \delta))\binom{n}{2}$. Let F_n be the graph consisting of the green edges of G_n. Then F_n has at least $(1 - 1/(2t - 2) + \delta)\binom{n}{2}$ edges. By the Erdős-Stone theorem [31], for large n the graph F_n contains a complete $(2t-1)$-partite subgraph J with $R(t)$ vertices in each class, where $R(t)$ is the ordinary two colour Ramsey number. Consider the subgraph of G_n spanned by the vertices of J. The edges of J are green and, since G_n is complete, each class of J spans a monochromatic copy of K_t, by the definition of $R(t)$. At least t classes contain monochromatic copies of the same colour. If these t copies of K_t are all red they span a copy of H in G_n, and the same is true too if they are all blue, because the edges of G_n between the K_ts are all green.

References

[1] V.E. Alekseev, Hereditary classes and coding of graphs (in Russian), *Probl. Cybern.* **39** (1982), 151–164.

[2] V.E. Alekseev, On the entropy values of hereditary classes of graphs, *Discrete Math. Appl.* **3** (1993), 191-199.

[3] N. Alon, E. Fischer, M. Krivelevich and M. Szegedy, Efficient testing of large graphs, *Combinatorica* **20** (2000), 451–476.

[4] N. Alon and A. Shapira, A characterization of the (natural) graph properties testable with one-sided error, *Proc. 46th IEEE Symp. Foundations of Computer Science* (2005), 429–438.

[5] N. Alon and U. Stav, What is the furthest graph from a hereditary property?, *Random Structures and Algorithms* **33** (2008), 87–104.

[6] N. Alon and U. Stav, The maximum edit distance from hereditary graph properties, *J. Combinatorial Theory (Ser. B)* **98** (2008), 672–697.

[7] N. Alon and U. Stav, Stability-type results for hereditary properties, *J. Graph Theory* **62** (2009), 62–83.

[8] M. Axenovich, A. Kézdy and R. Martin, On the editing distance of graphs, *J. Graph Theory* **58** (2008), 123–138.

[9] T. Austin and T. Tao, Testability and repair of hereditary hypergraph properties, *Random Structures and Algorithms* **36** (2010), 373–463.

[10] J. Balogh and R. Martin, Edit distance and its computation, *Electronic Journal of Combinatorics* **15** (2008), R20.

[11] J. Beck, *Combinatorial games. Tic-tac-toe theory*, Encyclopedia of Mathematics and its Applications **114**, Cambridge University Press, Cambridge (2008).

[12] B. Bollobás, Hereditary and monotone properties of combinatorial structures, in *Surveys in combinatorics 2007* (eds. A. Hilton and J. Talbot), *London Mathematical Society Lecture Note Series*, **346**, Cambridge University Press, Cambridge (2007), pp. 1–39.

[13] B. Bollobás and V. Nikiforov, The number of graphs with large forbidden subgraphs, *European J. Combinatorics* **31** (2010), 1964–1968.

[14] B. Bollobás and A. Thomason, Generalized chromatic numbers of random graphs, *Random Structures and Algorithms* **6** (1995), 353–356.

[15] B. Bollobás and A. Thomason, Hereditary and monotone properties of graphs, in *The Mathematics of Paul Erdős II* (eds. R.L. Graham and J. Nešetřil), *Algorithms and Combinatorics* **14**, Springer-Verlag, Berlin (1997), pp. 70–78.

[16] B. Bollobás and A. Thomason, The structure of hereditary properties and colourings of random graphs, *Combinatorica* **20** (2000), 173–202.

[17] J.A. Bondy and Zs. Tuza, A weighted generalization of Turán's theorem, *J. Graph Theory* **25** (1997), 267–275.

[18] W.G. Brown, On graphs that do not contain a Thomsen graph, *Canad. Math. Bull.* **9** (1966), 281–285.

[19] W.G. Brown, P. Erdős and M. Simonovits, Extremal problems for directed graphs, *J. Combinatorial Theory (Ser. B)* **15** (1973), 77–93.

[20] W.G. Brown, P. Erdős and M. Simonovits, Inverse extremal digraph problems, in *Finite and Infinite Sets, Eger 1981, Colloq. Math. Soc. János Bolyai* **37**, North Holland, Amsterdam (1985), pp. 119–156.

[21] W.G. Brown, P. Erdős and M. Simonovits, Algorithmic solution of extremal digraph problems, *Trans. Amer. Math. Soc.* **292** (1985), 421–449.

[22] W.G. Brown and M. Simonovits, Extremal multigraph and digraph problems, in *Paul Erdős and his mathematics, II (Budapest, 1999), Bolyai Soc. Math. Stud.* **11**, János Bolyai Math. Soc., Budapest (2002), pp. 157–203.

[23] A. Diwan and D. Mubayi, Turán's theorem with colors (preprint).

[24] P. Erdős, Graph theory and probability, II, *Canad. J. Math.* **13** (1961), 346–352.

[25] P. Erdős, Some recent results on extremal problems in graph theory (Results), in *Theory of Graphs (Internat. Sympos., Rome, 1966)*, Gordon and Breach, New York; Dunod, Paris (1967), pp. 117–123.

[26] P. Erdős, On some new inequalities concerning extremal properties of graphs, in *Theory of Graphs (Proc. Colloq. Tihany, 1966)*, Academic Press, New York (1968), 279–319.

[27] P. Erdős and R. Rado, A combinatorial theorem, *J. London Math. Soc.* **25** (1950), 249–255.

[28] P. Erdős and M. Simonovits, A limit theorem in graph theory, *Studia Sci. Math. Hungar.* **1** (1966), 51–57.

[29] P. Erdős and M. Simonovits, Supersaturated graphs and hypergraphs, *Combinatorica* **3** (1983), 181–192.

[30] P. Erdős and V.T. Sós, Some remarks on Ramsey's and Turán's theorem, in *Combinatorial theory and its applications, II (Proc. Colloq. Balatonfüred, 1969),* North Holland, Amsterdam (1970), 395–404.

[31] P. Erdős and A.H. Stone, On the structure of linear graphs, *Bull. Amer. Math. Soc.* **52** (1946), 1087–1091.

[32] Z. Füredi and A. Kündgen, Turán problems for integer-weighted graphs, *J. Graph Theory* **40** (2002), 195–225.

[33] Z. Füredi and M. Simonovits, Triple systems not containing a Fano configuration, *Combinatorics, Probability and Computing* **14** (2005), 467–484.

[34] W.T. Gowers, Lower bounds of tower type for Szemerédi's uniformity lemma, *Geom. Funct. Anal.* **7** (1997), 322–332.

[35] V.I. Levenshtein, Binary codes capable of correcting deletions, insertions, and reversals, *Soviet Physics Doklady* **10** (1966), 707–710.

[36] W. Mantel, Problem 28, *Wiskundige Opgaven* **10** (1907), 60–61.

[37] E. Marchant, (in preparation).

[38] E. Marchant, Ph.D. thesis, University of Cambridge, (2010).

[39] E. Marchant and A. Thomason, Extremal graphs and multigraphs with two weighted colours, in *Fete of Combinatorics and Computer Science* (eds. G.O.H. Katona, A. Schrijver and T. Szönyi), *Bolyai Soc. Math. Stud.* **20**, Springer, Berlin, in press.

[40] E. Marchant and A. Thomason, The structure of hereditary properties and 2-coloured multigraphs, *Combinatorica*, in press.

[41] R. Martin, The edit distance function and localization (preprint).

[42] R. Martin and T. McKay, On the edit distance from $K_{2,t}$-free graphs I: Cases $t = 3, 4$ (preprint).

[43] H.-J. Prömel and A. Steger, Excluding induced subgraphs III: a general asymptotic, *Random Structures and Algorithms* **3** (1992), 19–31.

[44] D.C. Richer, Ph.D. thesis, University of Cambridge, (2000).

[45] V. Rödl and M. Schacht, Generalizations of the removal lemma (to appear).

[46] V. Rödl and A. Sidorenko, On the jumping constant conjecture for multigraphs, *J. Combinatorial Theory (Ser. A)* **69** (1995), 347–357.

[47] N. Sauer, On the density of families of sets, *J. Combinatorial Theory (Ser. A)* **13** (1972), 145–147.

[48] D. Saxton, Supersaturation for hereditary properties (in preparation).

[49] E.R. Scheinerman, Generalized chromatic numbers of random graphs, *SIAM J. Discrete Math.* **5** (1992), 74–80.

[50] S. Shelah, A combinatorial problem: stability and order for models and theories in infinitary languages, *Pac. J. Math.* **41** (1972), 247–261.

[51] A.F. Sidorenko, Boundedness of optimal matrices in extremal multigraph and digraph problems, *Combinatorica* **13** (1993), 109–120.

[52] M. Simonovits, A method for solving extremal problems in graph theory, stability problems, in *Theory of Graphs (Proc. Colloq. Tihany, 1966)*, Academic Press, New York (1968), 279–319.

[53] M. Simonovits and V. T. Sós, Ramsey-Turán theory, *Discrete Mathematics* **229** (2001), 293-340.

[54] E. Szemerédi, Regular partitions of graphs, in *Problèmes combinatoires et théorie des graphes, Orsay, 1976* (eds. J.-C. Bermond, J.-C. Fournier, M. las Vergnas and D. Sotteau), *Colloq. Internat. CNRS* **260**, CNRS, Paris (1978), pp. 399–401.

[55] P. Turán, On an extremal problem in graph theory (in Hungarian), *Mat. Fiz. Lapok* **48** (1941), 436–452.

[56] V.N. Vapnik and A.Ya. Chervonenkis, On the uniform convergence of relative frequencies of events to their probabilities, *Th. Prob. and Applics.* **16** (1971), 71–79.

DPMMS, CMS, Wilberforce Road, Cambridge CB3 0WB
a.g.thomason@dpmms.cam.ac.uk

Random geometric graphs

Mark Walters

Abstract

Suppose that we place n points in a square, uniformly at random, and form a graph by joining two if they are within some distance r of each other. What does the resulting graph look like? And how does it vary as r changes?

Alternatively, we could form a graph by placing the n points uniformly at random as before, but this time joining each point to its k nearest neighbours.

Both of these models are random but have some structure coming from the underlying planar topology. This structure gives them a very different behaviour from that of the 'ordinary' random graphs.

We survey the known results about these models and some other closely related models, and summarise some of the techniques used in proving such results.

1 Introduction

Most combinatorialists would define the start of random graph theory by the seminal papers of Erdős and Rényi ([17, 18, 19, 20]) published in 1959–61, which introduced the standard random graph model. However, roughly simultaneously (1961), Gilbert [29] introduced a different random graph model which has only recently risen to prominence. In this model the vertices have some (random) geometric layout and the edges are determined by the position of the vertices. We shall call graphs formed in this way *random geometric graphs*.

We want to place points uniformly throughout the plane in such a way that they occur with a positive 'density'. This is the idea of a Poisson process. For the reader's convenience we very briefly recall the important features of such a process. (Alternatively, see e.g., Section 6.8 of [34] for background on Poisson processes, or [44] for an in-depth study of their properties.)

Definition A Poisson process $\mathcal{P}$ of density one in $\mathbf{R}^2$ is a random subset of $\mathbf{R}^2$ defined by the following two properties:

1. the number of points in any (measurable) set A is Poisson distributed with mean the (Lebesgue) measure of A,

2. if A, B are disjoint (measurable) subsets of $\mathbf{R}^2$ then the number of points in A is independent from the number of points in B.

Gilbert's original model was defined as follows: pick points in $\mathbf{R}^2$ according to a Poisson process $\mathcal{P}$ of density one and join two if their distance is less than some parameter r. We see that the expected number of neighbours of a point is exactly the expected number of points of $\mathcal{P}$ within distance r, which is the expected number of points in a disc of area $A = \pi r^2$. In other words the expected degree of a vertex is exactly A. Thus, although we have defined the graph in terms of the radius r, it is more natural to parametrise in terms of the area $A = \pi r^2$; we write $G(A)$ for the resulting graph. We shall, however, reserve the symbol r for the corresponding radius.

Having constructed this graph we can ask all the normal graph theoretic questions: does the graph have a giant (infinite) component, is it connected, what is the chromatic number? Some of these are applicable to the model as it stands, but many of them only really make sense for finite graphs. For example, it is easy to see that the graph is not connected for any r (somewhere there will be an isolated vertex), and that the chromatic number is infinite for any $r > 0$. Thus, we introduce a finite version of the model: we restrict the model to a square S_n of area n (so side length $\sqrt{n}$) which ensures the expected number of vertices is n. We call the resulting graph $G(n, A)$. We note that the graph $G(n, A)$ does not (usually) have exactly n vertices, but the expected number of vertices is n and, since the standard deviation of the number of vertices is only $\sqrt{n}$, it is (highly) unlikely to be very far from n.

We shall not be interested in results for fixed n, but rather in the asymptotic behaviour as n goes to infinity. More precisely, we have $A = A(n)$ and we will look at the properties of the sequence of graphs $G(n, A(n))$. However, we will suppress the notation where this is no ambiguity and speak of the graph $G(n, A)$.

We briefly remark that rather than placing points according to a Poisson process we could place exactly n points uniformly at random inside the square S_n. Since, conditional on a Poisson process having n points in a region A, they are uniformly distributed over that region (see e.g., [44]) the resulting graph (usually called the Binomial model) is very similar to the Gilbert model. It is, however, more difficult to analyse because knowing what happens in one small part of the box affects the behaviour throughout the box. This is analogous to the difference between $G(n, m)$ and $G(n, p)$ in ordinary random graphs. Owing to the increased complexity these dependencies introduce, and the fact that it does not naturally generalise to an infinite graph on the whole plane, we shall not consider the Binomial model in this paper.

One practical application of this model is to wireless networks (see the next section). In this application the Gilbert model is very inefficient. For example, to make it likely that the network is connected the area needs to be quite large (logarithmic; see Section 2.4) and since, in a wireless network, power is roughly proportional to area the power consumption is high (so battery life is reduced etc.) Indeed things are even worse: there is a lot of interference.

It is natural to consider whether we could alleviate these problems by reducing the power and range in regions of high density. We do this by adjusting the range to that of the kth nearest neighbour. More precisely, we take a Poisson process $\mathcal{P}$ of density one in the plane (respectively the square S_n) and place an undirected edge between each point and its (a.s. unique) k nearest neighbours. Call the resulting graph $K(k)$ (respectively $K(n, k)$). We consider this model in Section 3.

1.1 Motivation

Although these models are natural objects for 'pure mathematical' study, and in this paper we view them from that perspective, they also have 'real world' applications.

First, we imagine a network of radio transmitters, each of some fixed range[1] r, spread across the plane at random. The resultant communication network is exactly the graph defined by Gilbert. Indeed, this was a scenario Gilbert mentioned when originally defining his model. In the years following the original paper, there were several papers about the model which had this application in mind (see e.g., [45, 72, 37, 73, 42, 62, 55]). Recently there has been a rapid proliferation of wireless devices (mobile phones, wireless computing etc) and this has provided a significant impetus to the field. Indeed, there are far too many papers to give even a representative sample; we just mention Penrose's book [60] which includes many of the results of the first half of this paper (and very much more), the book of Meester and Roy [51] which concentrates on percolation rather than more general properties of random geometric graphs, and the survey paper of Balister, Bollobás and Sarkar [3] which has a rather different emphasis from this paper (for example, they give significant attention to other models of random geometric graphs such as Voronoi percolation). In addition, many of the above sources deal with the higher dimensional analogues of these models.

There is another rather more surprising application: to statistics. Indeed, this application has historically been the more important. A common statistical question is to ask whether a collection of (possibly multi-variate) data comes from a single distribution or multiple distributions. One technique is to treat the data as points in an appropriate dimension space and ask whether they are "randomly" distributed. For example if the data is bi-variate we can plot each point in the plane (or a subset of the plane), construct a graph as above, and ask whether the resulting graph is 'typical'. See, for example [26, 27, 32, 66, 68, 79, 40, 77]. (A related example, which is probably rather more familiar, is that of constructing a minimal spanning tree on the point set, and asking whether it is 'typical'.)

The second motivation, in particular, suggests that we should be interested not just in the properties of (graphs formed from) uniformly distributed point sets but other underlying densities (e.g., normally distributed points). These have been studied but, for simplicity, we shall not consider more general distributions: see the cited references and, in particular, Penrose's book [60] for details.

1.2 Notation and conventions

As mentioned in the introduction, when dealing with the graphs $G(n, A)$ and $K(n, k)$ we will be proving asymptotic results. Thus, we introduce one convenient abbreviation. For an event B depending on $G(n, A)$ we say that B occurs *with high probability* (abbreviated whp) if

$$\lim_{n \to \infty} P(B(G(n, A))) = 1,$$

and similarly for $K(n, k)$. We reserve 'almost sure' (a.s.) for events which genuinely have probability one such as no two points of the Poisson process having the the same distance from a third. To avoid clutter in our results, where appropriate, we will implicitly ignore such events (for example, we shall write 'is' to mean 'is with probability 1').

---

[1] even here the area is a natural measure: it corresponds to the power of each transmitter

On occasion we will want to talk about a geometric graph on some particular set of points $\mathcal{P}$: we use the notation $G(\mathcal{P}, A)$ to denote the Gilbert model with parameter A formed on the point set $\mathcal{P}$.

We will be making frequent reference to the Poisson distribution and we shall use $\mathrm{Po}(\lambda)$ to denote a Poisson random variable with mean λ.

We will use standard notation for asymptotic bounds: o, O, Ω and ω. Formally, for two functions f and g we say that $f = o(g)$ (respectively ($f = O(g)$, $f = \Omega(g)$ and $f = \omega(g)$) if $f/g \to 0$ (respectively is bounded above, is bounded away from zero, tends to infinity). If $f = O(g)$ and $f = \Omega(g)$ we write $f = \Theta(g)$.

2 The Gilbert Model

2.1 Degrees

We start by looking at the maximum and minimum degrees. In the infinite graph it is easy to see (by considering infinitely many disjoint discs of radius r) that there are vertices of any finite degree, but no vertices of infinite degree (since the probability a given vertex has infinite degree is zero and there are only countably many vertices).

Hence we turn to the finite graph $G(n, A)$. Before discussing more exact results we give a simple but important example: that of when the minimum degree δ stops being zero (obviously for small r the minimum degree is zero and for large r it is non-zero).

By the definition of the Poisson process, the probability a particular vertex is isolated is e^{-A} (for convenience we ignore vertices near the boundary of S_n, see Section 2.12 for details). The expected number of vertices is n so the expected number of isolated vertices is ne^{-A} which strongly suggests that $A = \log n$ is the threshold: that is if $A = (1 - \varepsilon)\log n$ then $\delta = 0$ whp, and if $A = (1 + \varepsilon)\log n$ then $\delta > 0$ whp.[2]

We see that when asking about the maximum and minimum degree we are essentially asking (at least for $A \ll n$) what is the maximum (minimum) of n independent Poisson random variables with mean $A = \pi r^2$; i.e., Δ is approximately the value of k such that $\mathbb{P}(\mathrm{Po}(\pi r^2) \geq k) = 1/n$ and δ is the value of k such that $\mathbb{P}(\mathrm{Po}(\pi r^2) \leq k) = 1/n$. Arguing along these lines one can prove (Chapter 6 of Penrose [60])

Theorem 2.1 *Let $G = G(n, A)$ as above and let Δ and δ be the maximum (respectively minimum) degree of G. Then the following all hold whp.*

1. If $A < n^{-\varepsilon}$ for some $\varepsilon > 0$ then Δ is bounded and $\delta = 0$.

2. If $A = o(\log n)$ but A is not less than $n^{-\varepsilon}$ for any $\varepsilon > 0$, then $\delta = 0$ and

$$\Delta = (1 + o(1))\frac{\log n}{\log((\log n)/A)}.$$

In particular Δ is $o(\log n)$ and $\omega(A)$.

---

[2] We have just looked at the expectation; for a formal proof we would need something more such as checking that the variance is small.

3. If $A = c\log n$ for some constant c then $\Delta = \Theta(\log n)$ and

(a) If $c < 1$ then $\delta = 0$,

(b) if $c > 1$ then $\delta = \Theta(\log n)$.

4. If $A = \omega(\log n)$ then δ and Δ are both $(1 + o(1))A$.

The first regime is not very interesting: the area is rapidly going to zero so the average degree is rapidly going to zero. The fourth regime is called the super-connectivity regime and this regime tends to be comparatively easy to analyse: every vertex has essentially the same degree and indeed in some heuristic sense every 'relevant' set of area A has about A points in it.

The middle two regimes are of the most interest: for reasons that will become apparent they are called the sub-connectivity and connectivity regimes respectively.[3]

We include one observation: if $A = A(n)$ satisfies $A = o(\log n)$ then

$$\Delta(G(n, \lambda A)) = (1 + o(1))\Delta(G(n, A))$$

for $\lambda > 0$; that is in the sub-connectivity regime multiplying the area by a constant does not change the maximum degree significantly. On the other hand it is immediate that if $A = \omega(\log n)$ then

$$\Delta(G(n, \lambda A)) = \lambda(1 + o(1))\Delta(G(n, A))$$

i.e., the maximum degree is proportional to the area in the super-connectivity regime.

Much tighter bounds on the degree are known. In fact, and this is a feature we will see of many 'local events', throughout the sub-connectivity regime the maximum degree is very tightly constrained: it must take one of two possible values (again see Chapter 6 of Penrose [60]).

Theorem 2.2 *Suppose that $G = G(n, A)$ with $A = A(n) = o(\log n)$ then there exists a sequence j_n such that $\Delta(G(n, A)) \in \{j_n, j_n + 1\}$ whp.*

In this case, the numbers j_n are known explicitly: suppose that k satisfies $\mathbb{P}(\text{Po}(A) > k) > 1/n$ but $\mathbb{P}(\text{Po}(A) > k + 1) < 1/n$. Then j_n is either k or $k - 1$. (Indeed, it is known exactly when each of these possibilities occurs.)

2.2 Clique number

Next we consider the clique number $\text{cl}(G)$. We start with the most trivial of observations: $\text{cl}(G) \leq \Delta(G) + 1$. Also, for any point set $\mathcal{P}$ we have $\text{cl}(G(\mathcal{P}, \pi r^2)) \geq \Delta(G(\mathcal{P}, \pi(r/2)^2)) + 1$ since, by the triangle inequality, any two points within a disc of radius $r/2$ are joined. We remark that, in the sub-connectivity regime, this is sufficient to give the asymptotic behaviour of the clique number.

We observe that in this argument the fact that the disc of radius $r/2$ was centred at a point of the process was irrelevant: if any disc of radius $r/2$ contains k points then these k points form a clique. Of course there are other ways for a large clique

---

[3]We remark that this terminology is not completely standard. For example, in his book [60] Penrose includes the first regime in the sub-connectivity regime.

to occur: we just need lots of points in *some* set of diameter r. Trivially this set need not fit into a disc of radius $r/2$. However, the following theorem, which is a slight strengthening of Theorem 6.15 of [60], shows that the largest clique is almost contained in a disc of radius $r/2$.

Theorem 2.3 *For any fixed $\varepsilon > 0$ the largest clique in $G(n, A)$ is, whp, contained in a disc of radius $(1 + \varepsilon)r/2$.*

Proof The Bieberbach isodiametric inequality (see e.g. page 32 of [47]) states that any set of diameter r has area at most $\pi r^2/4$. To prove Theorem 2.3 we need a stronger bound when we add the condition that the set is not contained in a disc of radius $(1 + \varepsilon)r/2$; i.e., that the circumradius is at least $(1 + \varepsilon)r/2$. The following lemma is exactly what we require (Hernández Cifre [41] based on Scott [69]).

Lemma 2.4 *Suppose that U is a (measurable) set with diameter d and circumradius R. Then the area of U is at most*

$$\frac{3}{2} \left[d^2 \left(\frac{\pi}{3} - \arccos \frac{\sqrt{3}R}{d} \right) - \sqrt{3}R \left(R - \sqrt{d^2 - 3R^2} \right) \right].$$

Applying this lemma in our situation we see that there is a $\delta > 0$ such that any set of diameter r and circumradius at least $(1 + \varepsilon)r/2$ has area at most $(1 - \delta)\pi r^2/4$.

The idea is that such sets have noticeably smaller area than a disc of radius $r/2$ and, hence, do not contain as many points. However, to make this argument work we need to bound the number of such sets in some way. To do this we use the following tessellation argument.

We tile the square S_n with tiles of side length ηr for some small constant η. Suppose that $G(n, A)$ contains a k-clique H which is not contained in a disc of radius $(1 + \varepsilon)r/2$. By definition H has diameter at most r. Let $\widetilde{H}$ be the collection of tiles that contain a vertex of H. The diameter of $\widetilde{H}$ is at most $r(1 + 2\sqrt{2}\eta)$ and the circumradius of $\widetilde{H}$ is at least $(1 + \varepsilon)r/2$. Hence, provided η was chosen small enough, Lemma 2.4 implies $|\widetilde{H}| \leq (1 - \delta/2)\pi r^2/4$.

How many possible choices for $\widetilde{H}$ are there? Well there are $O(n/r^2)$ choices for the first tile. Every other tile must lie within r of that tile, so there are only a fixed number (less than $4/\eta^2$) of possible choices for each tile. Since the total number of tiles in $\widetilde{H}$ is at most a constant, there are $O(n/r^2)$ choices for the collection $\widetilde{H}$.

We see that the size of the largest clique not contained in a disc of radius $(1 + \varepsilon)r/2$ is bounded above by the maximum of $O(n/r^2)$ Poisson random variables with mean $(1 - \delta/2)\pi r^2/4$, whereas $\mathrm{cl}(G)$ is bounded below by the maximum of n/r^2 independent Poisson random variables with mean $\pi r^2/4$ (consider disjoint discs of radius $r/2$). The result follows by simple computation. □

We remark that the bounds at the end of this proof are essentially the same as the bounds for the maximum degree in the graph $G(n, r/2)$ (i.e., the maximum of $O(n/r^2)$ random variables with mean $\pi r^2/4$). Thus we expect the maximum degree in the graph with radius $r/2$ to be essentially the same as the clique number in the graph with radius r. The following theorem shows this is true (we omit the proof: see Theorem 6.15 of [60])

Theorem 2.5 *Suppose that $r \to \infty$. Then, whp,*

$$\mathrm{cl}(G(n, A)) = (1 + o(1))\Delta(G(n, r/2)).$$

Exactly as with the maximum degree (see the previous section), in the sub-connectivity regime, the clique number is even more tightly controlled (Müller [53]).

Theorem 2.6 *Suppose that $A = o(\log n)$. Then there exist a sequence of numbers j_n such that $\mathrm{cl}(G(n, A)) \in \{j_n, j_{n+1}\}$ whp.*

In fact Müller proves rather more, he shows that two point concentration occurs for a large class of local properties.

2.3 The Giant Component

The next question we consider is 'when does this graph have an infinite component?' This is exactly a 'percolation' question and our initial ideas will be based on comparing $G(A)$ with percolation on a square lattice.

We use the idea of tessellation: tile the plane with squares of side length $r/\sqrt{5}$: this is chosen so that any two points in adjacent squares (i.e., squares sharing a side) are joined. There is a natural correspondence from the tiles to vertices in $\mathbf{Z}^2$ and we use this correspondence to construct a site percolation model on $\mathbf{Z}^2$ as follows. Declare a vertex of $\mathbf{Z}^2$ open if the corresponding square in the tessellation contains at least one point. Obviously, by the definition of a Poisson process, the sites are open independently each with probability $1 - \exp(-r^2/5)$.

Now suppose that there is an infinite component C in this site percolation on $\mathbf{Z}^2$. We can 'lift' this back to an infinite component in $G(A)$; indeed, the set of all points in tiles corresponding to sites in C form an infinite connected set in $G(n, A)$. (Note this set need not be a *component* as there may be points in other tiles that are joined to this infinite connected set.)

Hence, if

$$\mathbb{P}(\text{a site in } \mathbf{Z}^2 \text{ is open}) = 1 - \exp(-r^2/5) > p_c(\mathbf{Z}^2, \text{ site}),$$

where $p_c(\mathbf{Z}^2, \text{site})$ denotes the critical probability for site percolation on the square lattice, then the graph $G(A)$ has an infinite component. Using the best rigorous bounds for the critical probability $p_c(\mathbf{Z}^2, \text{ site}) \leq 0.679492$ (Wierman [76]), we see that this guaranteed to be true if $r > 2.39$ or equivalently if $A > 17.9$.

Conversely tile the plane with squares of side length r (so no points in tiles that are not adjacent or diagonally adjacent are joined). Once again we use the correspondence between tiles and $\mathbf{Z}^2$, and again we declare a site in $\mathbf{Z}^2$ to be open if it contains a point of the process. Hence, we have

$$\mathbb{P}(\text{a site in } \mathbf{Z}^2 \text{ is open}) = 1 - \exp(-r^2).$$

Suppose that $G(A)$ contains an infinite component. Then the set of all vertices in $\mathbf{Z}^2$ which contain a point of this infinite component are a connected subset of the graph on $\mathbf{Z}^2$ formed by including not only the normal edges but also the diagonal edges. Moreover since no tile contains infinitely many points of the Poisson process (a.s.) this connected set must be infinite.

Thus, we have shown that if there is no infinite component in the corresponding square lattice (with diagonals) then there is no infinite component in $G(A)$. Hence if

$$\mathbb{P}(\text{a site in } \mathbf{Z}^2 \text{ is open}) < p_c(\mathbf{Z}^2 \text{ with diagonal edges, site}), \qquad (2.1)$$

then $G(A)$ does not contain an infinite component (almost surely). Since the lattices $\mathbf{Z}^2$, and $\mathbf{Z}^2$ with diagonals are matching pairs (that is they are dual in a site percolation sense)

$$p_c(\mathbf{Z}^2 \text{ with diagonal edges, site}) = 1 - p_c(\mathbf{Z}^2, \text{site}).$$

(a combination of the results of Fisher [22], Fisher and Essam [23], and Kesten [43]; see Chapter 11 of Grimmett [33] for an exposition).

Combining these results we see that equation 2.1 holds if $\exp(-r^2) > p_c(\mathbf{Z}^2, \text{ site})$ which is guaranteed by $A < 1.214$. Thus, if $A < 1.214$ then $G(A)$ a.s. has no infinite component.

This technique of tiling and comparing the model with a known percolation model is very widely applicable but, as in this example, tends to give quite weak bounds. One way of viewing this method is that we approximate the graph G by a new graph $\widehat{G}$ where points of the process are joined whenever their corresponding tiles are joined in the corresponding lattice. In the first case we chose the tiling such that $\widehat{G}$ is a subgraph of G and in the second case such that G is a subgraph of $\widehat{G}$. In each case, we were able to apply known results to $\widehat{G}$ and deduce results about G.

There are two places where this argument is far from tight. First, our approximation of the graph $G(n, A)$ by a graph on tiles is not a very close approximation. For example in the first case we had to choose the tile side length so that every pair of points in adjacent tiles were joined so frequently points in tiles at distance two are joined; i.e., G has many more edges than $\widehat{G}$.

Secondly, we had to use the rigorous bound (0.679492) for $p_c(\mathbf{Z}^2, \text{site})$ which is quite a long way from the 'answer' given by computer simulations of 0.592746 (Ziff [80]). In other words we approximated by a model that itself does not have good (rigorous) bounds.

There are some ways we can improve this tessellation method (see below) but these improvements are not necessary for two of its key uses. First, it is useful for initial 'back of the envelope' calculations to work out the approximate order of magnitude of thresholds (e.g., are they constant or logarithmic etc). Secondly, and more importantly, it is useful to rule out 'unlikely cases'. We shall give some examples later but for the moment let us consider an analogous case in the normal random graph model $G(n, p)$. When proving that the threshold for connectivity is $p = (\log n)/n$ the proof has two parts: one ruling out small components (including isolated vertices) and one very easy part ruling out two large components where crude arguments are sufficient.

One way to improve the above bounds is by using different lattices: in particular we can tessellate the plane with hexagons rather than squares, so the corresponding lattice is triangular. This is one lattice where the critical probability for site percolation is known exactly: it is $1/2$ (Kesten [43]). Applying the same argument as above shows that for $A > 10.9$ there is an infinite component: indeed this was essentially proved by Gilbert in the original paper [29].[4]

We remark that we could use much smaller tiles in our tessellation: this obviously makes $\widehat{G}$ a much closer approximation to G, but at the cost of giving us a much more unusual graph on the underlying lattice so we no longer have good bounds on p_c. However, in some circumstances, this sort of tessellation is useful.

There has only been a small improvement to this bound of 10.9 (Hall [38]; Philips, Panwar and Tantawi [62] claim better but give no proof).

Theorem 2.7 *The critical area A_c for percolation in the model $G(n, A)$ satisfies $A_c < 10.6$.*

The improvement was obtained by taking slightly different regions: hexagons with the corners rounded rather than true hexagons. These rounded hexagons are still disjoint but now do not cover the whole plane. The size of the underlying true hexagons needs to be a little larger to get probability at least $1/2$ that each region contains a point, but the maximum distance between points in adjacent regions is reduced.

However significantly better lower bounds are known. These use a different technique: that of comparison with a branching process. Since this technique is very standard we only sketch the argument.

Suppose that we 'explore' the Poisson process as follows. We start at a point of the process v. Let $A_0 = \{v\}$. Then look in the disc of radius r about that point for other points. Let A_1 be the set of such points. Then look in discs of radius r about each of the points of A_1 and place all new points found in A_2. Repeat. This process can naturally be coupled with a dominating branching process in which each node has number of children Poisson distributed with mean $|A|$. Hence, if $|A| < 1$ the branching process dies almost surely. The branching process dominates the original exploration so we see that the probability v is in an infinite component is zero. Since this is true for every point in the Poisson process we see that $G(A)$ has no infinite component almost surely.

This bound can be improved by the following observation: after the first step each new point found must be in 'unexplored space'; in particular it cannot be in the disc about the current point's parent. Thus the unexplored area is at most $(\pi/3 + \sqrt{3}/2)r^2$. Hence, by a simple calculation, the branching process almost surely dies if $A < 1.64$ (see Gilbert's original paper [29][5]).

Observe that the above bound on the unexplored area is only attained if the child is actually at distance r from its parent: further improvements can be made by giving each vertex found a 'type' depending on its distance from its parent and comparing with a typed branching process. Using this method Hall [38] obtained:

Theorem 2.8 *The critical area A_c for percolation in the model $G(n, A)$ satisfies $A_c > 2.18$.*

2.3.1 Computer Simulations It is of course easy to simulate $G(A)$ with a computer, or more accurately, to simulate finite subsections of this model and 'extrapolate' from the behaviour on these finite regions to that of the infinite plane. Indeed,

---

[4]At the time p_c was not known to be $1/2$ and Gilbert proved his bound conditional on $p_c = 1/2$.

[5]a typographic error causes it to be incorrectly stated as 1.75 in the paper

Gilbert did this in his original 1961 paper obtaining an estimate of $A_c = 3.2$ (careful programming of a three million dollar IBM7090 mainframe allowing him to simulate 3000 points).

Since then many others have estimated A_c including [64, 70, 28, 67]. Most recently Quintanilla, Torquato and Ziff [63], using significant computer time and a very careful choice of algorithm, obtained the current best estimate of 4.51223 ± 0.00005.

In almost all cases the extrapolation from finite regions to the infinite plane is not justified: more precisely it is usually clear that the method works asymptotically but the speed of convergence is unclear.

One exception is the following 'in-between' result of Balister, Bollobás, Sarkar and Walters [9] giving a rigorous 99.99% confidence interval of $[4.508, 4.515]$ (this strange sounding statement is explained below). As this method has been used to prove some completely rigorous bounds for some related percolation models we briefly describe it.

The key step is a rigorous extrapolation from finite regions to the infinite model. We tessellate the plane with (large) square tiles and we make the obvious identification between tiles and vertices in $\mathbf{Z}^2$ as usual. This time, however, our comparison will be with a bond percolation process on $\mathbf{Z}^2$.

We take an event in the original model on $\mathbf{R}^2$ that depends on the Poisson process inside two neighbouring tiles *and nothing outside of these two tiles*. If the event occurs we declare the corresponding bond in the corresponding $\mathbf{Z}^2$ lattice open.

As a practical example let us give one such event: namely the event that the largest component in $G(A)|_{T_1}$ is joined to the largest component in $G(A)|_{T_2}$ *inside* $T_1 \cup T_2$. We shall denote this event by $T_1 \leftrightarrow T_2$.

Of course, these bonds are not independent. However, bonds which do not have a tile in common are independent. More precisely, we call a bond percolation model *1-independent* if for any two collections of bonds B_1 and B_2 that do not share an endpoint, the bonds in B_1 are independent of the bonds in B_2.

We use the following result of [9].

Theorem 2.9 *Suppose that* $\mathbb{P}$ *is a 1-independent bond measure on* $\mathbf{Z}^2$ *and that each edge is present with probability at least* 0.8639. *Then, a.s., there is an infinite component: i.e., the model percolates.*

Applying this to our example we see that if

$$\mathbb{P}(T_1 \leftrightarrow T_2) > 0.8639 \qquad (2.2)$$

then we have an infinite component in the 1-independent model on $\mathbf{Z}^2$ which, by the definition of this event, lifts back to an infinite connected set in the original model. The important feature is that this event looks at one fixed finite region of the plane (namely, two neighbouring tiles) and, just from the behaviour in this finite region, we can rigorously deduce the behaviour in the whole plane.

In most cases (see Section 3.3 for an exception) this finite region is still too large for us to prove anything rigorous, and we resort to Monte-Carlo methods for this finite event. However, by using this method there is exactly one possible source of error: we could be very unlucky in the random numbers.[6]

Let us consider this more precisely. We fix the tiling of the plane and define a random variable A_u in some way, and then prove rigorously that with probability at least 99.99% equation 2.2 holds when the area A on which the event $T_1 \leftrightarrow T_2$ implicitly depends is taken to be A_u. Then $[0, A_u]$ is a (random as always) 99.99% one-sided confidence interval for A_c.

How do we get such a random variable? We fix some value A', try a large number (e.g. 200) of random configurations with $A = A'$ and count how many times the event $T_1 \leftrightarrow T_2$ occurs. If it occurs more than 190 times we set our random variable $A_u = A'$ and otherwise we set $A_u = \infty$. This 'works' because if $\mathbb{P}(T_1 \leftrightarrow T_2) < 0.8639$ it is very unlikely to occur 190 times in 200 trials (easy to check by summing the binomial distribution). Of course, if $A_u = \infty$ this bound gives no information so we want to pick A' close to A_c but still make sure the event $T_1 \leftrightarrow T_2$ occurs at least 190 times. However, for the validity of the method it does not matter how we pick the value A'.

Having defined our confidence interval we obtained the 'upper bound' given earlier of 4.515 by evaluating A_u once. Using the same idea we obtained the 'lower bound' of 4.508.

2.4 Connectivity

Now let us turn to connectivity. As we saw in Section 2.1 the infinite graph $G(A)$ is never connected since it will always have minimum degree zero (i.e., it will contain isolated vertices). Indeed, we saw there that if $A < \log n$ then $\delta = 0$, and thus the graph $G(n, A)$ is not connected.

Turning to the upper bound, first let us see what a simple tessellation argument gives. Tile the square S_n with tiles of side length[7] $r/\sqrt{5}$. If every such square contains a point of the process then the resulting graph is obviously connected. Hence if,

$$\mathbb{P}(\text{there is an empty square}) \leq 5n/r^2 \exp(-r^2/5) \to 0$$

the graph $G(n, A)$ is connected whp. This corresponds to $r^2/5 \approx \log n$ or $A \approx 5\pi \log n$. Combining these two we see that the threshold for connectivity is $A = \Theta(\log n)$.

However, very much tighter results are known. Before we state them we need a little more notation. For a fixed point set $\mathcal{P}$ and a monotone graph property Π define $\mathcal{H}(\mathcal{P}, \Pi)$, the *hitting area* for Π, by

$$\mathcal{H}(\mathcal{P}, \Pi) = \inf\{A \colon G(\mathcal{P}, A) \text{ has property } \Pi\}$$

---

[6]In practice, there are two other possible errors: first, as with any other computer method, there could be a bug in our computer program, and secondly our random numbers may not be random (e.g., they may come from a pseudo-random number generator). The latter problem can be avoided by using a hardware random number generator.

[7]Strictly we need to change the size slightly to guarantee that the whole square S_n is divisible by the tile size. This effect is tiny and we ignore it.

Theorem 2.10 (Penrose [58]) *For almost all point sets $\mathcal{P}$ we have*

$$\mathcal{H}(\mathcal{P},\ connected) = \mathcal{H}(\mathcal{P},\ no\ isolated\ vertices)$$

This theorem is telling us exactly what the obstruction for connectivity is: it is the existence of isolated vertices. The threshold for isolated vertices is essentially trivial to calculate: indeed we did such a calculation for the lower bound mentioned above. We do, however, need to be careful about points close to the boundary but we defer discussion of this technicality until Section 2.12.

Since the events 'vertex v is isolated' are local events they are nearly independent. Hence the number of such isolated vertices is approximately Poisson (see e.g., [12] or [1]) (more precisely the number of isolated vertices converges in distribution to a Poisson random variable as n tends to infinity). Using this fact it is easy to give an explicit formula for the probability that there are no isolated vertices, which by Theorem 2.10 implies G is connected.

Theorem 2.11
$$\mathbb{P}(G(n,\log n + c)\ is\ connected) = e^{-e^{-c}}.$$

(A similar result was proved independently by Gupta and Kumar [35] a few years after Penrose.)

The proof splits into two pieces, first we show that the graph is very unlikely to contain two large components, and then we show that, in fact, all components must be single vertices.

Lemma 2.12 *Suppose that $c > 0$. Then there is a C such that, whp, the graph $G(n, c\log n)$ does not contain two components each of (Euclidean) diameter at least $C\sqrt{\log n}$.*

The ideas used in proving this lemma are used in many other proofs so we give a fuller explanation than elsewhere.

We use a tessellation argument. Tile the square S_n with tiles of side length $r/\sqrt{20}$. This is chosen so that points in neighbouring tiles have distance at most $r/2$. The argument consists of the following four steps:

1. each component must meet many tiles

2. all tiles adjacent to the component are empty

3. there must be many such tiles by the edge isoperimetric inequality for the grid

4. the probability that these neighbouring tiles are all empty is small.

Before we discuss each of these steps we note that the first two steps are trivial and the third easy, but the final step is rather more subtle so will discuss it in rather more detail.

Suppose that U and V are two components each of diameter at least $C\sqrt{\log n}$. Let $\tilde{U}$ be the set of tiles that meet a vertex or an edge of U which we identify with the corresponding subset of $\mathbf{Z}^2$. Obviously, since $\tilde{U}$ includes all tiles meeting an edge of U, $\tilde{U}$ is a connected[8] subset of the lattice $\mathbf{Z}^2$. Define $\tilde{V}$ similarly.

---

[8]we ignore the zero probability event that an edge goes through the corner of a tile

Step 1 Since each component has diameter at least $C\sqrt{\log n}$, it meets at least $C\sqrt{\log n}/(r/\sqrt{20}) = C\sqrt{20\pi/c}$ tiles. Let $c_1 = \sqrt{20\pi/c}$.

Step 2 First some notation: for any collection A of tiles define ∂A, the *boundary of A* to be the set of tiles in A^c adjacent to a tile in A.

Now, observe that the minimum distance between edges of two distinct components is at least $r/2$. Indeed suppose not. This minimum distance must be obtained between a vertex and an edge, and the edge has length at most r so the distance between the vertex and one of the ends is at most r, so the two components are joined which is a contradiction.

Hence, every tile in $\partial \widetilde{U}$ is empty since if any tile in $\partial \widetilde{U}$ contains a point of the process it would be at most $r/2$ from an edge of $\widetilde{U}$ and, thus, joined to U.

Step 3 One would expect that the boundary of a set is (at least) roughly the square root of the size of the set itself, and this is exactly what the edge isoperimetric inequality for the grid (Bollobás and Leader [14]) states.

Applying it to $\widetilde{U}$ shows that $\partial \widetilde{U}$ contains at least

$$\min\{2\sqrt{|\widetilde{U}|}, 2\sqrt{|\widetilde{U}^c|}\}$$

tiles. By Step 1 $|\widetilde{U}|$ is at least $c_1 C$ and since $\widetilde{U}^c \supset \widetilde{V}$ we see that $|\widetilde{U}^c|$ is also at least $c_1 C$. Therefore $\partial \widetilde{U}$ contains at least $2\sqrt{c_1 C}$ tiles.

Step 4 One might think that this is sufficient: but whilst it is very unlikely that a particular collection of $2\sqrt{c_1 C}$ tiles is empty there are a vast number of possible collections of tiles. To complete Step 4 we need to reduce this number in some way. We would like to say that the collection of boundary tiles has some special form. A natural candidate is that the boundary is connected (since it sort of looks like we can walk around it). However, a moment's reflection leads us to consider an annulus: this has a boundary consisting of two pieces: the inner and the outer boundary. (Also, but unimportantly, we need to be able to step round corners: that is the boundary pieces are connected in the lattice with diagonals included.)

We make use of the following fact, which in topological terms is saying that the square is unicoherent (see e.g., [16]).

Lemma 2.13 *If A and A^c are connected subsets of the lattice $\mathbf{Z}^2$ then ∂A is connected in the lattice with diagonals.*

There is no reason to believe that $\widetilde{U}^c$ is connected (e.g., $\widetilde{U}$ could be an annular component) so we use a standard complementing trick. Let $\widetilde{U}_1$ be the component of $\widetilde{V}^c$ that contains $\widetilde{U}$. Then $\widetilde{U}_1$ is connected by definition and $\widetilde{U}_1^c$ is also connected (since every other component of $\widetilde{V}^c$ is connected to $\widetilde{V}$ outside of $\widetilde{U}_1^c$).

Applying the edge isoperimetric inequality and Lemma 2.13 to $\widetilde{U}_1$ rather than $\widetilde{U}$ we see that $\partial \widetilde{U}_1$ contains at least $2\sqrt{c_1 C}$ tiles and is connected in the lattice with diagonals.

To complete the proof we just use the following standard bound on the number of connected subgraphs of the lattice (see e.g. Problem 45 of [13]):

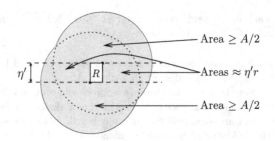

Figure 1: The rectangle R and the empty area (shaded).

Lemma 2.14 *Suppose that G is a graph with maximum degree Δ and that $v \in V(G)$. The number of connected subgraphs of G with n vertices that contain v is at most $(e\Delta)^n$.*

We have seen that $\partial \widetilde{U}_1$ is a (diagonally) connected set of at least $2\sqrt{c_1 C}$ empty tiles. The number of diagonally connected sets of size u containing a particular tile is at most $(8e)^u$ so the total number of diagonally connected sets of size at least u is at most $n(8e)^u$. Hence the probability that there is a diagonally connected set of u tiles all of which are empty is is at most

$$n(8e)^u \exp(-ur^2/20),$$

which tends to zero if $u > 20\pi/c$ (recall $\pi r^2 = A = c \log n$).

Hence, provided that we choose $C > 100\pi^2/c^2 c_1$ we see that, whp, there do not exist two components of diameter at least $C\sqrt{\log n}$.

Lemma 2.15 *Suppose that C is as in Lemma 2.12 with $c = 0.9$. Then whp, the graph $G(n, \log n - \frac{1}{2}\log\log n)$ does not contain a component consisting of more than one point and diameter at most $C\sqrt{\log n}$.*

(In fact the bound can be tightened a little: the lemma holds for $A = \log n - \log \log n$; see Penrose [60].)

Suppose that H is such a component. First we show that H cannot be too small; in particular we show that whp, the diameter of H is at least $\eta = (\log \log n)^2/\sqrt{\log n}$. Indeed, suppose not. Let x be a point of H. We see that $B(x, \eta)$, the ball of (Euclidean) radius η centred at x, would contain at least one other point and the region $B(x, r) \setminus B(x, \eta)$ would be empty. The probability that this event occurs at a particular vertex is at most

$$(1 - \exp(\pi\eta^2)) \exp(-(A - \pi\eta^2)) \leq 2\pi\eta^2 \exp(-A) = O\left(\frac{\sqrt{\log n}(\log \log n)^4}{n \log n}\right) = o(1/n).$$

Since there are n points, the probability that this event occurs at any vertex is $o(1)$, so whp there is no such point and, thus, no such H.

Now suppose that H has diameter $d > \eta$. Let J be the convex hull of H. Let t_1 and t_2 be horizontal tangents to J above and below J respectively and let t_3 and t_4

be vertical tangents to J to the left and right of J. These four tangents describe a circumscribed rectangle R containing J (see Figure 1). We see that R has diameter at least η. Let η' be the length of the longer side, so $\eta' \geq \eta/\sqrt{2}$. Let x_i for $1 \leq i \leq 4$ be the points where t_i intersects J (so each $x_i \in H$).

It is easy to see that the area of $\bigcup_{i=1}^{4} B(x_i, r) \setminus R$ is at least

$$A + (1 + o(1))2r\eta' \geq A + (1 + o(1))2r\eta/\sqrt{2} > A + \eta\sqrt{\log n} = A + (\log\log n)^2$$

and that this region is empty. The probability that four particular points have this region empty is at most

$$\exp(-(A + \eta r)) \leq \exp(-\log n + \tfrac{1}{2}\log\log n - (\log\log n)^2) = o\left(\frac{1}{n(\log n)^3}\right).$$

How many ways of choosing four such points are there? There are n choices for x_1 but, having chosen x_1, there are only $O(\log n)$ choices for the other three (they must be within $C\sqrt{\log n}$ of x_1). Hence, the total number of ways of choosing four such points is $O(n(\log n)^3)$. Therefore, the probability that there exist four points x_1, x_2, x_3, x_4 within $C\sqrt{\log n}$ of each other with $\bigcup_{i=1}^{4} B(x_i, r) \setminus R$ empty tends to zero and we see that, whp, no such rectangle exists and thus, that no such component exists.

Lemmas 2.12 and 2.15 together with the earlier observation that isolated vertices occur all the way up to $A = \log n$ suffice to prove Theorem 2.10.

2.5 Higher Connectivity

The same techniques as in the previous section can be used to prove bounds for κ-connectivity for $\kappa \geq 2$.

Theorem 2.16 (Penrose [59]) *For any fixed* κ

$$\mathcal{H}(\mathcal{P},\ minimum\ degree\ \kappa) = \mathcal{H}(\mathcal{P},\ \kappa\text{-}connected).$$

There is one slight change, the obstruction moves to the boundary of S_n; that is, for $\kappa > 1$ a vertex of degree at most κ is at least as likely to occur near the boundary of S_n as in the centre. We discuss this further in Section 2.12. For the moment we only state what this gives as a threshold.

Theorem 2.17 *Suppose that* $\kappa \geq 2$ *and* $A = \log n + (2\kappa - 3)\log\log n + c$. *Then*

$$\mathbb{P}(G(n, A)\ is\ \kappa\text{-}connected) \to f(c)$$

where $f(c)$ *is a function of* c *with* $\lim_{c \to -\infty} f(c) = 0$ *and* $\lim_{c \to \infty} f(c) = 1$.

(The function f is known but complicated.)

In other words the threshold for κ-connectivity is $A = \log n + (2\kappa - 3)\log\log n$. We remark that the reason this is not correct for $\kappa = 1$ is the difference in the boundary behaviour.

2.6 Hamiltonicity and Matchings

Hamiltonicity is another standard graph property to consider. Moreover, its intrinsic interest is increased by the greater difficulty in finding the threshold for Hamiltonicity in 'ordinary' $G(n, p)$ random graphs (see e.g., [12]). It does not, however, seem a natural property from a wireless network perspective.

An obvious necessary condition for Hamilitonicity is that the graph be 2-connected (which by Theorem 2.16 is whp the same as the graph having minimum degree two). It is natural to ask, as Penrose did, whether this is also a sufficient condition in the sense that

$$\mathcal{H}(\mathcal{P}, \text{ Hamiltonian}) = \mathcal{H}(\mathcal{P}, \text{ 2-connected}). \tag{2.3}$$

Progress on this was slow, Petit [61] proved that $A/\log n \to \infty$ implies the existence of a Hamilton cycle, and this was improved by Díaz, Mitsche and Pérez [15] to $A = (1 + \varepsilon)\log n$. Whilst this was a good step towards (2.3) the condition already implies that the graph $G(n, A)$ is $\varepsilon'\log n$ connected for some $\varepsilon' > 0$.

Then three groups proved it independently, resulting in the two papers Balogh, Bollobás, Krivelevich, Müller and Walters [11] and a little later Müller, Pérez and Wormald [54]. The latter authors also proved that as soon as the graph is 2κ-connected then there exist κ disjoint Hamilton cycles. Since the argument is rather involved we shall only sketch some of the key steps.

First we tessellate S_n with squares of side length $\sqrt{\log n}/c$ for some large constant c. We form the tessellation graph $\widehat{G}$ by joining two points if the centres of their tiles are at most $(c - 2)\sqrt{\frac{\log n}{\pi}}$ apart. Let r be the hitting radius for 2-connectivity. Since, the threshold for 2-connectivity of G is $A = \log n + \log\log n$ we may assume $\pi r^2 > \log n$ and thus that $\widehat{G}$ is a subgraph of G. Further, since c is large the tessellation graph $\widehat{G}$ is a fairly close approximation to G.

Almost all of these tiles contain many points: formally, $1 + o(1)$ of the tiles contain at least 100 points. Moreover, since almost all the tiles contain at least 100 points, there is a vast connected component in $\widehat{G}$: indeed it contains $1 + o(1)$ of all tiles. We call the tiles in this component the easy tiles and the remaining (few) tiles difficult.

Since there are so few difficult tiles they exists in several well separated small 'clumps'. Using the 2-connectedness of the graph it is possible to construct a Hamilton path for each clump (this is the complicated part of the proof, see [11] for details). Having dealt with all the difficult squares is is easy to join all these short paths and all the vertices in easy tiles together into a Hamilton cycle in G.

It may seem surprising that we can use the approximate tessellation graph to prove a precise hitting time result like Theorem 2.3. The reason this works is that when dealing with the intricate part of the results (dealing with a clump of difficult tiles) we do not use the tessellation graph but the 2-connectedness directly. We use the tessellation graph to identify small locally difficult parts of the graph and then to join these local parts together. Thus, by using the tessellation graph we reduce the problem from an intractable global problem to a more manageable local problem.

2.7 Chromatic number

Next we turn to the chromatic number. We saw earlier that the clique number $\mathrm{cl}(G)$ is well determined, and obviously the clique number is a lower bound for the chromatic number. The following theorem describes the situation accurately.

Theorem 2.18 (McDiarmid [48]) *Suppose that* $G = G(n, A)$ *as usual.*

1. *If* $A = o(\log n)$ *then*

$$\chi(G) = (1 + o(1))\mathrm{cl}(G) = (1 + o(1))\Delta(G).$$

2. *If* $A = \omega(\log n)$ *then*

$$\chi(G) = (1 + o(1))\frac{\sqrt{3}}{2\pi}A = (1 + o(1))\frac{\sqrt{3}}{2\pi}\Delta(G) = (1 + o(1))\frac{2\sqrt{3}}{\pi}\mathrm{cl}(G).$$

The case $A = o(\log n)$ of the theorem is simple: obviously

$$\mathrm{cl}(G(n, A)) \le \chi(G(n, A))) \le \Delta(G(n, A))$$

and in this range $\mathrm{cl}(G(n, A)) = (1 + o(1))\Delta(G(n, A))$ (see Sections 2.1 and 2.2).

For $A = \omega(\log n)$ we obtain a lower bound as follows. We bound the size of the maximal independent set. Suppose that W is an independent set in $G(n, A)$. Since the discs $B(w, r/2)$, for $w \in W$, must be disjoint we instantly have a bound of

$$|W| < \frac{n + 4r\sqrt{n} + r}{\pi r^2/4} = (1 + o(1))\frac{4n}{\pi r^2}$$

(the second two terms in the numerator of the first fraction are to deal with effects near the boundary). However, using Thue's Theorem (see e.g., [56] or [65]) on disc packing we can improve this to

$$|W| < \frac{\pi}{2\sqrt{3}}(1 + o(1))\frac{4n}{\pi r^2}$$

which gives a bound of

$$\chi \ge (1 + o(1))\frac{\sqrt{3}}{2\pi}\pi r^2 = (1 + o(1))\frac{\sqrt{3}}{2\pi}A.$$

The upper bound uses the following deterministic result: let $T(d)$ denote the unit triangular lattice but with edges between any two vertices at (Euclidean) distance d. Then $\chi(T(d)) = (1 + o(1))d^2$ (see e.g., [50]).

We tile the plane with hexagons with centres at distance εr so that the centres are the vertices of an εr scaled copy of the unit triangular lattice; thus, if we join any two of these vertices whenever they are at Euclidean distance at most r we get an εr scaled copy of $T(1/\varepsilon)$. Suppose that h is the maximum number of points in any of these hexagons. Since we are in the super-connectivity regime $h = (1 + o(1))$ times the area of one of these hexagons (i.e. $(\sqrt{3}/2)\varepsilon^2 r^2$). Hence,

$$\chi \le (1 + o(1))\frac{\sqrt{3}}{2}\varepsilon^2 r^2\chi(T(1/\varepsilon)) = (1 + o(1))\frac{\sqrt{3}}{2\pi}A$$

The other bounds follow immediately from Sections 2.1 and 2.2. We note that $\frac{2\sqrt{3}}{\pi} \approx 1.103$, so even in the super-connectivity regime the chromatic number is only a little bigger than the obvious lower bound of the clique number.

The behaviour for $A = \Theta(\log n)$ is also known (McDiarmid and Müller [49]): the chromatic number increases from $(1 + o(1))\mathrm{cl}(G)$ to $(1 + o(1))\frac{2\sqrt{3}}{\pi}\mathrm{cl}(G)$ as the constant in the 'Θ' increases from zero to infinity.

Although the chromatic number is not a 'local property', for $A = o(\log n)$ Theorem 2.18 says it is essentially the same as the clique number. This suggests that the chromatic number of a random geometric graph is very close to a local property and, indeed, once again we have two point concentration (Müller [53]).

Theorem 2.19 *Suppose that $A = o(\log n)$. Then there exist a sequence of numbers j_n such that $\chi(G(n, A)) \in \{j_n, j_{n+1}\}$ whp.*

2.8 Coverage

Since the graphs we are interested in come with an actual embedding we can ask some questions which are not strictly graph theoretical, but involve the embedding itself. One example is the following (Gilbert [30]; for an earlier heuristic argument see Moran and Fazekas de St Groth [52]).

Theorem 2.20 *Suppose that $\mathcal{P}$ is a Poisson distributed point set in S_n. Then the discs $B(x, r)$ for $x \in \mathcal{P}$ cover all of S_n whp if and only if $A - \log n - \log \log n \to +\infty$.*

One way of viewing this result is that it is saying that *wherever* we add a new point it will be connected into the network: that is, whp the network is completely robust to the addition of new vertices. (In a sense this is analogous to the graph property of 2-connectedness which requires that the graph remains connected if *any* vertex is deleted, not just a typical vertex).

We make a brief remark about the proof and why the stated threshold arises. Suppose that we draw discs of radius r about each of the points of the process. Gilbert's crucial observation is the following lemma. Define a *crossing point* to be the intersection point of the boundary of two of the discs in S_n or the intersection point of the boundary of a disc and the boundary of S_n.

Lemma 2.21 *The square S_n is covered by discs if and only if every crossing point is in the interior of some (other) disc (and we have at least one crossing point).*

This lemma reduces the problem of checking coverage at the uncountably many points in S_n to that of checking a manageable finite subset.

The number of such intersection points is of the order of nA (there are n discs each with roughly $4A$ intersections), and each intersection point has an e^{-A} chance of being uncovered. Hence, if $A = \log n + \log \log n$ the expected number of uncovered intersection points is

$$e^{-A}nA = e^{-(\log n + \log \log n)}n(\log n + \log \log n) \approx 1$$

i.e., we are around the threshold. Of course this is only a heuristic but it easy to make precise.

Similar results are known in higher dimensions, and for the same related question where we require every point in S_n to be covered multiple times (Hall [39]). For example, he showed that, if we want to cover every point in the *torus* T_n some constant s times, the threshold is $A = \log n + s \log \log n$. (The result in the torus T_n is simpler than in the square S_n since it avoids boundary effects; see Section 2.12.) In other words, the threshold for covering every point s times is only a tiny bit larger than the threshold for covering once. Balister, Bollobás, Sarkar and Walters [7] proved the surprising fact that at this threshold, not only is T_n s-covered, but the points can be partitioned into s classes each of which covers T_n (more precisely, whp the hitting time for these two events are the same).

2.9 Higher Dimensions

The Gilbert model can be defined in any dimension, and many of the above results can be proved in this more general setting. However, the higher-dimensional case is less studied for two reasons: first, one of the main applications (namely wireless networks) is naturally two dimensional. Secondly the extra difficulties in high dimensions tend not to be graph theoretic but more topological; for example arguments can no longer use planarity. Also, the boundary becomes both more significant and more complicated (there are many different dimension boundary faces): see Section 2.12.

We mention one result: that of the critical volume needed for percolation (Penrose [57]).

Theorem 2.22 *Let* $A_c(d)$ *denote the critical volume for percolation in d-dimensions. Then* $A_c(d) \to 1$ *as* $d \to \infty$.

This is what one would expect; the branching process argument implies that 1 is a lower bound in all dimensions, and as the dimension increases the balls become more and more disjoint so the percolation process becomes more and more like the branching process.

2.10 Other norms and neighbourhoods

Throughout this section we have defined neighbourhoods in terms of the Euclidean norm. Much of the above can be done in other norms but this tends to be of less interest: the natural applications use the Euclidean norm. In most cases the results are fairly similar (although, of course, details like the value of the critical area A_c differ).[9] There is one exception: the l_∞ and l_1 norms can behave differently. There are two (related reasons) for this; first, the balls are not strictly convex, and secondly the balls tessellate space: that is they fit together perfectly.

The model can also be extended to non-convex neighbourhoods. For example each transmitter may be able to transmit to the annular region of all points between distance r and $(1+\varepsilon)r$. This has an interesting property: that the area-threshold for an infinite component tends to 1 as ε goes to 0 (Balister, Bollobás and Walters [8]

---

[9]As far as we are aware this has not been proved, but the values given by computer simulations are significantly different.

or Franceschetti, Booth, Cook, Meester and Bruck [24]). Recall that, by comparison with a branching process, one is a lower bound for regions of *any* shape.

Of course, broadcasting to annular regions has little (if any) real world application, so this result raises the question of whether there is any real world shape for which the same threshold holds. In [10] Balister, Bollobás and Walters proved that randomly oriented directional transmitters have this property. (Formally, the neighbourhoods are randomly oriented sectors of a disc.)

In contrast, there is essentially no change in the connectivity results (e.g., Theorem 2.11) for different norms or shapes: indeed, if the expected number of neighbours of a node is less than $\log n$, then whp, at least one node will be isolated.

2.11   Sharpness of Monotone Properties

For any monotone graph property Π it is natural to ask how sharp the transition is. More precisely, we let $r(p)$ denote the value of r such that $G(n, A)$ has property Π with probability p, and we ask for bounds on the *width* $w(\varepsilon) = r(1 - \varepsilon) - r(\varepsilon)$.

In [31] Goel, Rai and Krishnamachari proved the following very general result about sharpness. (Note they used a different normalisation so their result looks rather different.)

Theorem 2.23 *Let* $G = G(n, A)$ *and* Π *be a monotone graph property. Then* $w(\varepsilon) = O((\log n)^{3/4})$. *Further, there is a monotone graph property with constant width (depending on* ε*).*

Before discussing the idea of the proof we briefly comment on this result. At first glance the upper and lower bounds on the width appear quite close: a log factor apart. However, as we have seen, most of the interesting graph properties have thresholds which are $O(\log n)$ for area, i.e., $O(\sqrt{\log n})$ for radius, and the above theorem does not say anything in this case. It is, however, a much sharper transition than is guaranteed to occur in ordinary Erdős-Rényi random graphs (Friedgut and Kalai [25]).

Their proof technique is interesting. For two n-points sets $\mathcal{P}$ and $\mathcal{Q}$ in S_n and a matching between them, we define the matching distance to be the longest edge in the matching. We define the bottleneck matching distance from $\mathcal{P}$ to $\mathcal{Q}$ to be the minimum matching distance over all possible matchings from $\mathcal{P}$ to $\mathcal{Q}$.

They rediscovered the following result (first proved by Leighton and Shor [46])

Theorem 2.24 *Let* $\mathcal{P}$ *and* $\mathcal{Q}$ *be two random uniformly distributed n-points sets in* S_n. *Then whp the bottleneck matching distance from* $\mathcal{P}$ *to* $\mathcal{Q}$ *is* $O((\log n)^{3/4})$.

It is easy to see that if two point sets $\mathcal{P}$ and $\mathcal{Q}$ have bottleneck matching distance d then the graphs $G(\mathcal{P}, \pi r^2) \subset G(\mathcal{Q}, \pi(r + 2d)^2)$. Hence, Theorem 2.24 roughly says that, for some constant C, the graph $G(n, \pi r^2)$ is whp a subgraph of $G(n, \pi(r + C(\log n)^{3/4})^2)$. Hence, if $G(n, \pi r^2)$ has some chance of having property Π then $G(n, (r + C(\log n)^{3/4})^2)$ has it whp.

They also proved the analogous result in higher dimensions (using a higher dimensional version of Theorem 2.24 first proved by Shor and Yukich [71]). In fact, in higher dimensions the result is rather tighter: the width is $O((\log n)^{1/d})$; i.e., sharpness covers all of the super-connectivity regime. (It still does not say anything useful in the connectivity or sub-connectivity regimes however).

2.12 Boundary Effects

We conclude this section with a brief discussion of the boundary effects that we ignored earlier. We see that the arguments in Section 2.1 are not correct near the boundary: vertices near the boundary of S_n do not have a neighbourhood of area A since some of the disc will lie outside S_n. Indeed, a point on the edge of S_n would only have neighbourhood area $A/2$, and a point in a corner of S_n only $A/4$.

To see how this can impact our arguments suppose that we look at the threshold for connectivity, so $A \approx \log n$. The probability that a 'central' vertex is isolated is e^{-A} and the are approximately n central vertices. In comparison the probability that a vertex near the boundary of S_n is isolated is around $e^{-A/2}$ (since half the area is outside S_n) and there are order $\sqrt{n}$ such vertices. Hence, to a first approximation, when the expected number of isolated central vertices is one, the expected number of isolated boundary vertices is also one.

If we do the calculation for boundary vertices a little more accurately we find that the area for a boundary vertex is actually about $A/2 + O(r)$ (since it is not 'on the edge' but some constant distance from it) and this means that central obstructions do indeed dominate. [We remark that the corners are not relevant, although the probability a vertex near a corner is isolated is much larger $(e^{-A/4})$ there are very few of them $O(\log n)$.]

However, for closely related problems, the boundary can be crucial. Indeed, if we look for minimum degree κ (or equivalently, by Theorem 2.16, κ-connectivity) we find that the threshold for central vertices is $\log n + (\kappa - 1) \log \log n$ whereas the threshold for boundary vertices is $\log n + (2\kappa - 3) \log \log n$. Hence the threshold for minimum degree κ is

$$\max\left(\log n + (\kappa - 1)\log \log n, \ \log n + (2\kappa - 3)\log \log n\right).$$

We see that the first term is bigger for $\kappa = 1$, they are equal for $\kappa = 2$ and that the second term is bigger for all $\kappa \geq 3$ (which is why Theorem 2.17 is not true for $\kappa = 1$). This tells us that for $\kappa = 1$ the obstruction (an isolated vertex) is central, for $\kappa = 2$ the obstruction (a vertex of degree 1) can be central and it can be near the boundary, and that for $\kappa \geq 3$ the obstruction (a vertex of degree $\kappa - 1$) is near the boundary.

The boundary also dominates the connectivity/isolated vertex threshold in three or more dimensions; of course, in higher dimensions there are many different types (dimensions) of boundary. A similar argument to that given above for two dimensions shows that the obstruction for connectivity (an isolated vertex) will occur near a two-dimensional face.

As we have seen the boundary effects cannot always be ignored, and they frequently require special treatment in proofs. There are two closely related models which can help in avoiding these boundary effects, and thus are amenable to simpler proofs.

The first one is obvious: work on a torus T_n rather than S_n. Obviously this removes all the boundary. It is often a good place to work when sketching initial proofs. However, it does have one undesirable consequence: arguments often mention things like the 'left most point' of a component, and it is unclear what, if anything, that would mean on a torus.

The second option is to work in the square S_n, but to allow points outside of S_n as well. For example, when considering coverage of S_n (see Section 2.8) we consider points throughout the plane (in fact we only care about points within distance r of S_n) and ask whether they cover S_n. By doing this we make the probability that a point of S_n is covered independent of which point is chosen, and again this can simplify the arguments.

Of course whether results about these simpler models are of interest depends on the circumstances!

3   The k-nearest neighbour model

Recall from the introduction that we defined the k-nearest neighbour model by taking a Poisson process $\mathcal{P}$ of density one and placing an undirected edge between each point and its k nearest neighbours. We called the resulting graph $K(k)$ or $K(n, k)$ as appropriate.

The results in the previous section showed that the reason a large radius was necessary to obtain a connected network was to avoid isolated vertices (or possibly other small components). Since the k-nearest neighbour model obviously has minimum degree at least k, there is hope that the average disc size (so total power) needed to ensure connectivity will be smaller.

There are other minor variants on this model (for example joining points if they are both k-nearest neighbours of each other) which we discuss in Section 3.4. However, the model defined here is the most widely studied.

3.1   Neighbourhood Radius

When we discussed the Gilbert model we started by looking at the minimum and maximum degrees; that is not a very interesting question for the k-nearest neighbour model. Indeed, every vertex has 'out-degree' k, and it is easy to see that every vertex has degree between k and $6k$ (we give more precise bounds in Section 3.5.3). However this is not really the analogous question: in the Gilbert model we fixed the radius and asked about the degree. In this model we are fixing the degree (strictly the out-degree) so we should ask for bounds on the 'radius', that is the longest edge and the shortest non-edge; we denote these by R and r respectively, and we let A and a be the corresponding areas (so $A = \pi R^2$ and $a = \pi r^2$). The following theorem follows easily from bounding the Poisson distribution (see e.g., [4]; alternatively it can be deduced by 'inverting' Theorem 2.1).

Theorem 3.1 *Suppose that G is the graph $K(n, k)$ and that a and A are as above.*

1. If $k = o(\log n)$ then $a = o(k)$ and $A = (1 + o(1)) \log n$.

2. If $k = \Theta(\log n)$ then a and A are both $\Theta(\log n)$.

3. If $k = \omega(\log n)$ then a and A are both $(1 + o(1))k$.

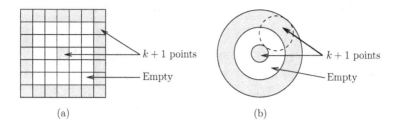

(a)   (b)

Figure 2: Two configurations which disconnect the graph $K(n,k)$. In (a) each shaded square contains (at least) $k+1$ points and each white square is empty (contains no points). The centre tile is disconnected in $K(n,k)$. The figure in (b) shows a rotationally symmetric variant (see Section 3.2.1).

Obviously $G(n,a)$ is a subgraph of $K(n,k)$ which in turn is is a subgraph of $G(n,A)$, and so this theorem, together with results for the Gilbert model, can be used to prove results for the k-nearest neighbour model, but with a loss of accuracy coming from the difference between a and A.

In particular we see that if $k = \omega(\log n)$ then $K(n,k)$ is essentially the same as $G(n,A)$ (formally for any $\varepsilon > 0$ we have $G(n,(1-\varepsilon)k) \subset K(n,k) \subset G(n,(1+\varepsilon)k)$).

Of course, much tighter results could be proved, including explicit bounds in place of the Θ and o; however, in most cases the above theorem is sufficient.

3.2 Connectivity

Obviously, by construction, this model has no isolated vertices; indeed the minimum degree is at least k. However, as in the Gilbert model it is easy to see that the infinite graph $K(k)$ is never connected (Xue and Kumar [78]). Indeed, using a tessellation, if we have the configuration in Figure 2 consisting of one tile with at least $k+1$ points surrounded by 24 tiles with no points surrounded by 24 more tiles each containing at least $k+1$ points) we can see that there are no edges from points in the central square to the rest of G. This has positive probability so will occur somewhere in the plane.

Indeed, if we look for such a configuration where each tile has area k we see that the probability each shaded tile contains at least $k+1$ points is about $1/2$, and the probability all the remaining tiles are empty is e^{-24k}. Thus the total event has probability about $2^{-25}e^{-24k} = \Theta(e^{-24k})$. The tessellation contains $n/(49k)$ disjoint 7×7 boxes where this event could occur, so heuristically (but easily rigorisably) if $k = (\frac{1}{24} - \varepsilon) \log n$ the event will occur somewhere; i.e., the threshold for $K(n,k)$ to be connected is at least $\frac{1}{24} \log n$ (see [78] again).

On the other hand if we again consider a tessellation and suppose that every tile contains at least one and at most M points, then the $49M$-nearest neighbour graph is connected: indeed, with this choice of k every point in one tile is joined to every point in a neighbouring tile. If we use tiles of area, say, $2 \log n$, then the it is easy to check that whp every tile contains a point. Moreover, provided that $M = C \log n$ for a sufficiently large constant C we see that no tile contains more than M points

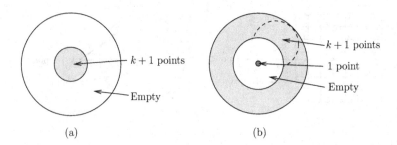

(a)   (b)

Figure 3: Two configurations. In (a) the central disc has radius one third of the
outer disc; thus the central points have no out edges to the rest of G (i.e., they are an
out-component); in (b) the central point has zero in-degree (i.e., is an in-component).

and, thus, whp the graph $K(n, 49C \log n)$ is connected.

We briefly remark as an alternative that we can see from Theorem 3.1 that for
c sufficiently large

$$K(n, c \log n) \supset G(n, (1 + \varepsilon) \log n)$$

for some $\varepsilon > 0$ and thus is connected whp by Penrose's result for the Gilbert model
(Theorem 2.11 from the previous section). See González-Barrios and Quiroz [32],
or [78].

Once again, we see that threshold for connectivity is $\Theta(\log n)$.

Having made this observation it is natural to ask what is the actual threshold.
More formally, we have seen that if c is a small constant then the graph $K(n, c \log n)$
is not connected whp, and if c is a large constant then $K(n, c \log n)$ is connected
whp. What is the critical value $c = c_*$ where the transition from not connected to
connected occurs? However, it does not even appear to be obvious that there is such
a critical value.

Before discussing some of the current proof techniques for this model we summarise the 'state of the art'. The best current bounds are given in the following theorem.

Theorem 3.2 *If $k < 0.3043 \log n$ then the graph $K(n, k)$ is not connected whp.
Conversely, if $k > 0.4125 \log n$ then $K(n, k)$ is connected whp.*

The lower bound was proved by Balister, Bollobás, Sarkar and Walters in [4] and
the upper bound by Walters [75] improving on the upper bound of $1/\log 7 \approx 0.5139$
proved in [4].

It is interesting that just above the threshold the graph is connected but, in the
underlying directed graph, there are vertices with zero in-degree and components
with zero out-degree. Indeed, the two constructions in Figure 3 are likely to occur
for $k < \frac{1}{\log 9} \log n \approx 0.455 \log n$ and $k < 0.5739 \log n$ respectively: see Section 3.2.1
for further details.

It is now known that there is a critical constant (Balister, Bollobás, Sarkar and
Walters [5]).

Theorem 3.3 *There is a constant c_* such that if $c < c_*$ then $K(n, \lfloor c \log n \rfloor)$ is not connected whp, and if $c > c_*$ then $K(n, \lfloor c \log n \rfloor)$ is connected whp.*

Having found that there is indeed a critical value for this transition it is natural to ask how 'rapid' the transition from not-connected to connected is. Indeed, this question has been widely studied for many transitions in the normal random graph $G(n, p)$. This was answered by Falgas-Ravry and Walters in [21].

Theorem 3.4 *There exist constants $C > 0$ and $\alpha > 0$ such that for any $\varepsilon = \Omega(n^{-\alpha})$ with*

$$\mathbb{P}(K(n, k) \text{ is connected}) \geq \varepsilon$$

we have

$$\mathbb{P}(K(n, k + \lfloor C \log(1/\varepsilon) \rfloor) \text{ is connected}) > 1 - \varepsilon.$$

In other words, the 'window' for connectivity is of width roughly a constant (c.f. Theorem 2.11 which shows the same behaviour in the Gilbert model).

As we saw in Section 2.12, arguments about the graph $G(n, A)$ have to deal with boundary effects, often giving rise to extra cases that need to be considered. The same is true in the k-nearest neighbour model although, intuitively, the boundary should be less of problem as the range is automatically increased (i.e., we still join to k points). The results in [4] suggested that points near the boundary are not an obstruction to connectivity but were not quite strong enough to prove this. More recently Walters [75] proved:

Theorem 3.5 *Suppose $k > 0.272 \log n$. Then whp all vertices within distance $\log n$ of the boundary of S_n are contained in the giant component.*[10]

The utility of this result arises not from the exact bound stated but in that it is smaller than the lower bound on c_* given in Theorem 3.2. That is if k is around $c_* \log n$ then the graph has no small boundary components. Thus, proofs around the connectivity threshold no longer need to consider the extra cases caused by the existence of small components near the boundary.

3.2.1 Lower bounds There is only one natural way to try and prove a lower bound: come up with a disconnecting configuration (like the example in Figure 2a discussed at the start of this section) that is likely to occur. In fact, the proof of Theorem 3.3 indicates that there is necessarily some such configuration.

It is reasonable to believe, but unproven, that the extremal configuration is rotationally symmetric. This leads us to consider the configuration defined as follows (see Figure 2b): let r be such that $\pi r^2 = k$. Consider three concentric discs D_1, D_3 and D_5 of radius r, $3r$ and $5r$ respectively. Suppose that there are at least $k + 1$ points in D_1, no points in $D_3 \setminus D_1$ and that any disc D of radius $2r$ centred on the boundary of D_3 contains at least $k + 1$ points (necessarily outside D_3). Then the graph $K(n, k)$ is not connected: indeed the points J in D_1 are disconnected from the rest of G, since the first two conditions ensure there are no out-edges from J to $G \setminus J$ and the third condition ensures that there are no in-edges from a point in $G \setminus J$ (necessarily outside D_3) to J.

---

[10]For $k = \Theta(\log n)$ there is a giant component consisting of $1 - o(1)$ of all vertices: see Section 3.3.

How likely is this event? Well the probability that there are at least $k+1$ points in D_1 is about $1/2$ and the probability D_3/D_1 is empty is $e^{-8\pi r^2} = e^{-8k}$. Finally, the probability that every disc D contains at least $k+1$ points is $1 - o(1)$ since the area of $D \setminus D_3$ is significantly greater than πr^2 so the probability that it does not contain at least $k+1$ points is tiny. (There are infinitely many discs to check, but by reducing the radius slightly we can reduce to a fixed finite number of discs; alternatively it is easy to prove using a very fine tessellation.)

Overall we see that the probability of this event is about $e^{-8k}/2$. We can place $\Theta(n/\log n)$ disjoint discs of radius $5r$ in S_n; in each of these disc the above disconnecting event occurs with probability $e^{-8k}/2$ and, since the discs are disjoint these events are independent. Hence, if $k = (1/8 - \varepsilon)\log n$ the probability G is connected is at most

$$(1 - e^{-8k}/2)^{n/\log n} = \left(1 - \frac{1}{2n^{1-8\varepsilon}}\right)^{n/\log n} \to 0;$$

that is, we have shown that $c_* \geq 1/8$.

There are some ways we can improve this bound: we can shrink the whole configuration. That makes the first and third condition less likely, but significantly increases the probability of the second condition (the empty area is much smaller). If we optimise this scaling we obtain a bound of $0.2739 \log n$ (see discussion in the proof of Theorem 5 of [4]).

However, this construction can be further improved. Rather than asking for uniform density in $D_5 \setminus D_3$ we can have a varying density which is largest near the boundary of D_3. (This helps as points near the boundary of D_3 are in more of the discs D occurring in the third condition.) Optimising this varying density for a rotationally symmetric configuration is relatively easy (Theorem 5 of [4]) and this gave the lower bound in Theorem 3.2.

3.2.2 Proof techniques In this section we discuss two proof techniques which seem to have wide applicability to the k-nearest model. For both of these it would be nice to have one underlying result which can be applied in each case; unfortunately, however, the details have to be adapted to work in each individual situation.

First we observe that using a very similar argument to Lemma 2.12 we can prove that the graph does not have two large components (see [4]). Formally

Lemma 3.6 *Suppose that $k = \Omega(\log n)$. Then there is a C such that, whp, the graph $K(n, k)$ does not contain two components each of order at least $C \log n$.*

In other words, around the threshold for connectivity all the interesting obstructions are small: within a constant multiple of the smallest they could possibly be.

The first technique we will discuss is that of the circumscribed hexagon (introduced in [4]). Suppose that G' is a small component in $K(n, k)$ which is not near the boundary of S_n. We can circumscribe a hexagon H around G': formally we consider the six tangents to the convex hull of G' which are inclined at angles 0, $\frac{\pi}{3}$, and $\frac{2\pi}{3}$ to the horizontal. These tangents form the hexagon H containing G', as shown in Figure 4. (Note this hexagon could be degenerate: some sides could have length zero.) By definition each tangent t_i intersects G' in a point $P_i \in V(G')$ (some of the

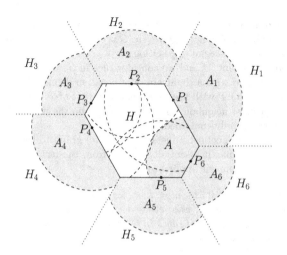

Figure 4: The circumscribed hexagon H

P_i may coincide). Let D_i be the k-nearest neighbour disc for P_i. Consider P_1: since G' is a component all neighbours of P_1 lie in G' so, in particular, lie in the hexagon H. Hence, all k neighbours of P_1 lie in $D_1 \cap H$ and $D_1 \setminus H$ is empty. We shall apply this observation to each of the six points $P_1, \dots, P_6$.

The exterior angle bisectors of H divide the exterior of H into six (disjoint) regions $H_1, \dots, H_6$ as shown in Figure 4. Let $A_i = D_i \cap H_i$ and let A be the smallest of the sets $D_i \cap H$. We see that each A_i is empty and that A contains k points; that is there are k points in $A \cup \bigcup_{i=1}^{6} A_i$ and they all lie in A. Moreover, for each i,

$$|A_i| = |D_i \cap H_i| \geq |D_i \cap H| \geq |A|.$$

so

$$|A| \leq \frac{1}{7} \left| A \cup \bigcup_{i=1}^{6} A_i \right|. \tag{3.1}$$

For any collection of sets $A, A_1, \dots, A_6$ satisfying (3.1) the probability that there are exactly k points in the union and they all lie in A is at most 7^{-k}.

We have established that, if there is a small component, then there are seven regions $A, A_1, \dots, A_6$ in this special layout. How many ways of choosing these regions are there? The regions are defined by the 6 points giving the hexagon, together with the 6 points defining the k-nearest neighbour radii. There are n ways of choosing the first point, but having chosen that point there are only $O(\log n)$ ways of choosing the remaining points (since they all lie within distance $O(\sqrt{\log n})$ of the first point chosen.). Hence the number of ways of choosing these 12 points is $O(n(\log n)^{11})$. Thus, provided that $n(\log n)^{11} 7^{-k} \to 0$, whp no such hexagon exists: i.e., there is no small component. This gives the upper bound $0.51 \log n$.

Many other results also use this technique e.g., [5, 21, 75]. It is useful as it provides a reasonable handle on the structure of these small components.

The second technique is a type of tessellation argument. We demonstrate the technique for one specific example: the event G is connected. Hence we assume that we are in the connectivity regime $k = \Theta(\log n)$.

We fix a large constant M and we consider tiles in S_n of side length $M\sqrt{\log n}$. Fix a specific tile T and with a slight abuse of notation define $\frac{1}{2}T$ to be the square of half the side length of T with the same centre as that of T. Define A_T to be the event that the k-nearest neighbour graph G_T formed on the points of $\mathcal{P}$ that lie in T has a component entirely contained in $\frac{1}{2}T$.

We wish to relate the connectedness of the entire graph to the events A_T. We start with a trivial lemma: it essentially follows from the fact that G has no long edges (and a similar argument applied to G_T).

Lemma 3.7 *Suppose that $0.3\log n < k < \log n$ and that $G = K(n,k)$. Then, provided M is sufficiently large, whp, every tile T in S_n of side length $M\sqrt{\log n}$ has the following property. The set of edges in G which meet a vertex in $\frac{1}{2}T$ is the same as the corresponding set in G_T.*

In other words there are no edges from outside T into $\frac{1}{2}T$ and the extra edges formed when we ignore the points outside T do not reach into $\frac{1}{2}T$. In particular, G has a component wholly contained in $\frac{1}{2}T$ if and only if G_T does; i.e., if and only if the event A_T occurs.

We can place $n/(M^2\log n)$ disjoint tiles in S_n, and the Poisson process in each tile is independent, so

$$\mathbb{P}(G \text{ is not connected}) \geq \mathbb{P}(A_T \text{ occurs for any tile } T) = 1 - (1 - \mathbb{P}(A_T))^{n/M\log n}.$$

On the other hand, suppose we cover S_n with a collection of (overlapping) tiles $\mathcal{T}$ such that every set of diameter at most $(M/8)\sqrt{\log n}$ is contained in the square $\frac{1}{2}T$ for some $T \in \mathcal{T}$ (which we can do using $16n/(M\log n)$ tiles). Now, if G is not connected then, assuming M is sufficiently large, by Lemma 3.6 we see that one of the components has diameter at most $M/8\sqrt{\log n}$ and thus that the event A_T occurs for some $T \in \mathcal{T}$. Hence

$$\mathbb{P}(G \text{ is not connected}) \leq \frac{16n}{M\log n}\mathbb{P}(A_T).$$

Roughly speaking, these two inequalities show that the transition from not connected to connected corresponds to the transition from $\mathbb{P}(A_T) \gg \frac{\log n}{n}$ to $\mathbb{P}(A_T) \ll \frac{\log n}{n}$. In other words, we can focus on the relatively simple event A_T, and deduce the global behaviour in S_n. For example, Theorems 3.3 and 3.4 were proved using two (quite different) careful analyses of the event A_T.

3.3 The Giant Component

As in the Gilbert model we ask when does the (infinite) k-nearest neighbour model contain an infinite component. Using a tessellation argument similar to that in Section 2.3 we can prove that if k is some large constant (of the order of a few hundred) then the model percolates (Häggström and Meester [36] or rather later Teng and Yao [74]). Thus we see that there is some critical threshold k_c such that percolation occurs if and only if $k \geq k_c$.

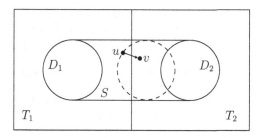

Figure 5: The 'bond event' in the proof of Theorem 3.8

However, it is not immediately obvious that even just connecting to the (one) nearest neighbour does not give an infinite component. To see that this is the case suppose that there is an infinite path in the graph. Along this path the edge lengths either increase or decrease monotonically. If we head in the decreasing direction then either the path terminates (the final vertex's nearest neighbour is the previous vertex on the path) or it does not. The former case must have probability zero since otherwise a positive fraction of vertices would terminate an infinite path.

The latter case is a little more difficult to rule out. Start from one edge of the path, say length r, and pass along the path in the direction of decreasing edge length. At each vertex there is a chance at least $\exp(-\pi r^2)$ that the disc of radius r about the vertex contains no new points of the process (i.e., the path terminates). Since there are infinitely many vertices in the path where it could fail it will fail somewhere (a.s.).

We have shown that $k_c \geq 2$ and that k_c is at most a few hundred. Since the critical constant k_c is inherently an integer there is some hope that the exact value could be found, although the wide gap in the bounds is not promising.

Recently Balister and Bollobás [2] proved the first 'good' bound for k_c.

Theorem 3.8 *The critical constant k_c for 2-dimensional k-nearest neighbour percolation is at most 11.*

We give some of the key steps of the proof. The proof compares the percolation model with a 1-independent bond percolation on $\mathbf{Z}^2$. First we consider a rectangle B as shown in Figure 5 with side lengths s and $2s$ (s to be optimised later) made up of two adjacent squares T_1 and T_2. We consider two discs D_1 and D_2 of radius r (again to be optimised later) centred in the left square and right square of the rectangle respectively. We shall declare the corresponding bond in $\mathbf{Z}^2$ open if the following holds: that each vertex in D_1 can be joined to a vertex in D_2 *regardless* of the process outside B, and the same from D_2 to D_1, and that D_1 and D_2 are non-empty.

With this definition of open it is easy to see that an infinite component in this one-independent model on $\mathbf{Z}^2$ 'lifts' back to an infinite component in G. By Theorem 2.9 we 'just' need to show the probability of the event above is at least 0.8639; of course the main step in the argument is the proof of this bound, but we have reduced the problem to the analysis of a finite region.

We very briefly sketch the proof of the bound. Let L be the line between the centres of D_1 and D_2, and let S be the 'sausage' shaped convex hull of $D_1 \cup D_2$. The aim is to show that every point u in the $S \setminus D_2$ has a neighbour v 'further along' towards D_2 in the sense that the disc centred on the line L with v on its left boundary is nearer D_2 than the corresponding disc through u is (see Figure 5 again).

The authors show that with $r \approx 3.1$ and $s \approx 11.6$ the probability that there is any point in $S \setminus D_2$ that does not have such a neighbour is less than 0.065291 and, thus, that the probability the edge is open (i.e., paths exist in both directions) is at least 0.869 > 0.8639 as required. Showing this involves some unpleasant calculations, including bounding some complicated integrals.

Computer simulations (e.g., Section 7 of [4]), including the confidence interval method with confidence better than 10^{-100} (see [2]), indicate that, in fact, $k_c = 3$. Indeed, these simulations suggest that for $k = 3$ the giant component contains 98.5% of all vertices! That is 3 is 'much bigger' than the critical constant k_c.

We remark that in sufficiently high dimensions Häggström and Meester [36] proved that the critical constant is equal to two.

3.4 Related Models

3.4.1 Directed graphs The k-nearest neighbour model naturally gives a directed graph (which we previously treated as a simple graph) so we can ask directed graph questions, for example is the graph directed connected? The best result is the following (Balister, Bollobás, Sarkar and Walters [4]).

Theorem 3.9 *Let $\vec{K}$ be the directed k-nearest neighbour graph $\vec{K}(n,k)$. If $k > 0.7209 \log n$ then $\vec{K}$ is whp not directed connected, whereas if $k < 0.9967 \log n$ then $\vec{K}$ is connected whp. In particular the threshold for directed connectivity is strictly less than $\log n$.*

Note that, in the directed graph model, there can be vertices with zero indegree: and indeed this is where the lower bound comes from (using a density varying argument similar to that in Section 3.2.1). The authors conjecture that this is in fact extremal.

3.4.2 A different undirected graph In most of this section we formed an undirected graph by placing an undirected edge whenever an edge occurs with either orientation. An alternative is to place an edge only when both orientations occur.

This is actually a fairly natural model from a real-world perspective: most wireless transmission protocols require acknowledgements from the recipient of the information (so if there is no acknowledgement the information can be resent). Whilst, the acknowledgement need not follow the reverse of the original route it may be convenient for it to do so.

Despite this, not many results have been proved about this model. The paper of Balister and Bollobás [2] is an exception: they proved that for $k = 15$ there is percolation in this model.

As far as we are aware, there are no results about the connectivity of this graph (apart from the obvious $\Theta(\log n)$ bound that can be read out of the arguments at the start of Section 3.2). However, since the obstructions for small out-components and small in-components look very different it is reasonable to believe that any obstruction to connectivity in this model is actually an 'in-obstruction' or an 'out-obstruction'. Hence we make the following conjecture.

Conjecture 3.10 *Suppose that k is such that $\vec{K}(n,k)$ is connected whp. Then the above model is connected whp.*

3.5 Other results

3.5.1 Coverage As for the Gilbert model one can ask when do the k-nearest neighbour discs cover the whole of S_n. In [4] Balister, Bollobás, Sarkar and Walters proved that the threshold is essentially the same as that for having a vertex of in-degree zero: i.e., the bounds in Theorem 3.9 apply in this case too.

3.5.2 Higher connectivity As we saw there is no 'nice' obstruction for connectivity in the k-nearest neighbour model and the upper and lower bounds for the critical constant c_* are still a significant distance apart. Thus, it is not reasonable to hope for a good bound for the thresholds for higher order connectivity. However, there are good bounds on the gap between connectivity and higher order connectivity (Falgas-Ravry and Walters [21] improving on Balister, Bollobás, Sarkar and Walters [6]):

Theorem 3.11 *There is a constant c such that for any $k = k(n)$ for which $K(n,k)$ is connected whp, the graph $K(n, k + \lfloor cs \log \log n \rfloor)$ is s-connected whp.*

Essentially this result is proved by showing that the probability that $K(n,k)$ is 2-connected is not too much smaller than that of being connected, and then applying the sharpness result Theorem 3.4.

3.5.3 Degree, Clique Size and Chromatic Number These turn out to be easy. The minimum degree of G is exactly k. Indeed, take the vertex v with the k-nearest neighbour disc of largest radius. By definition no point outside this disc sends any edges in to v so v has degree exactly k. Moreover, exactly the same argument shows that every subgraph H of G has minimum degree at most k; i.e., G is k-degenerate. It follows immediately that $\chi(G) \leq k+1$ (curiously this does not seem to have been observed before).

For $k < \frac{1}{\log 9} \log n$ then the construction in Figure 3a shows that G has a clique of size $k + 1$ whp and, thus, that $\mathrm{cl}(G) = \chi(G) = k + 1$.

On the other hand for $k = \omega(\log n)$ then $K(n,k)$ and $G(n,A)$ (with $A = k$) are almost the same graph so by the corresponding results for the Gilbert model (Theorems 2.5 and 2.18) the clique size will be $(1 + o(1))k/4$, and the chromatic number will be $(1 + o(1))\frac{\sqrt{3}}{2\pi}k$.

The maximum degree is at most $6k$: indeed, any 60 degree sector about a vertex can contain at most k neighbours (consider the furthest neighbour in the sector). With the extra observation that the 120 degree sector about the furthest neighbour contains at most $k-1$ other neighbours this bound reduces to $5k$.[11] Moreover, this is easily obtainable provided that k is at most $c\log n$ for some small constant c (place k points in small regions near the corners of a pentagon about the vertex).

Finally, for $k = \omega(\log n)$ the graph $K(n, k)$ is approximately $G(n, A)$ with $A = k$, and the maximum degree will be $(1 + o(1))k$.

3.5.4 Hamilton Cycles This is rather less well understood than in the Gilbert model (see Section 2.6) where we saw that the obvious necessary condition for the existence of a Hamilton cycle (2-connectivity) is in fact sufficient. Balogh, Bollobás, Krivelevich, Müller and Walters [11] proved the following weaker result.

Theorem 3.12 *Suppose that* $G = K(n,k)$ *is* $5 \cdot 10^7$ *connected whp. Then* G *is Hamiltonian.*

In Section 2.6 we used a tessellation argument with a fine tessellation. This meant that the tessellation graph was a good approximation to the genuine graph and that was sufficient to prove the result. In the k-nearest neighbour graph the natural tessellation graph, even with small tiles, is not a close approximation (since the k-nearest neighbour discs vary in size significantly) and that is the reason why the authors needed the extra connectivity in the result above. However, they conjectured that 2-connectivity is sufficient: that is, as soon as the graph becomes 2-connected it is Hamiltonian.

3.5.5 Higher dimensions An easy modification of the argument at the beginning of Section 3 shows that in any dimension the threshold for connectivity is $k = \Theta(\log n)$, and the proof of Theorem 3.3 could be modified to show the existence of a critical constant $c_*(d)$ in any dimension d. Thus, we can ask what is the asymptotic behaviour of $c_*(d)$ as $d \to \infty$.

The hexagon argument (see Section 3.2.2) is inherently two dimensional, but we can use a variant: applying the argument to a (hyper)-cube instead of the hexagon. This yields that $c_*(d) = O(1/\log(d))$.

In particular we see that $c_*(d)$ tends to zero as d increases. However, the speed of this convergence is not known. The lower bounds discussed in Section 3.2.1 can be modified to work in high dimensions giving a bound $c_*(d) = \Omega(1/d)$, but there is a significant gap between this and the upper bound mentioned above.

3.5.6 Other norms Any norm could be used in defining the k-nearest neighbour model, but we are not aware of any significant results in other norms.

References

[1] N. Alon and J. H. Spencer. *The probabilistic method*. Wiley-Interscience Series in Discrete Mathematics and Optimization. John Wiley & Sons Inc., Hoboken, NJ, third edition, 2008.

---

[11] We are ignoring the zero probability event that two points have the same distance from a third point.

[2] P. Balister and B. Bollobás. Percolation in the k-nearest neighbor graph. Preprint.

[3] P. Balister, B. Bollobás, and A. Sarkar. Percolation, connectivity, coverage and colouring of random geometric graphs. In *Handbook of large-scale random networks*, volume 18 of *Bolyai Soc. Math. Stud.*, pages 117–142. Springer, Berlin, 2009.

[4] P. Balister, B. Bollobás, A. Sarkar, and M. Walters. Connectivity of random k-nearest-neighbour graphs. *Adv. in Appl. Probab.*, 37(1):1–24, 2005.

[5] P. Balister, B. Bollobás, A. Sarkar, and M. Walters. A critical constant for the k-nearest-neighbour model. *Adv. in Appl. Probab.*, 41(1):1–12, 2009.

[6] P. Balister, B. Bollobás, A. Sarkar, and M. Walters. Highly connected random geometric graphs. *Discrete Appl. Math.*, 157(2):309–320, 2009.

[7] P. Balister, B. Bollobás, A. Sarkar, and M. Walters. Sentry selection in wireless networks. *Adv. in Appl. Probab.*, 42(1):1–25, 2010.

[8] P. Balister, B. Bollobás, and M. Walters. Continuum percolation with steps in an annulus. *Ann. Appl. Probab.*, 14(4):1869–1879, 2004.

[9] P. Balister, B. Bollobás, and M. Walters. Continuum percolation with steps in the square or the disc. *Random Structures Algorithms*, 26(4):392–403, 2005.

[10] P. Balister, B. Bollobás, and M. Walters. Random transceiver networks. *Adv. in Appl. Probab.*, 41(2):323–343, 2009.

[11] J. Balogh, B. Bollobás, M. Krivilevich, T. Müller, and M. Walters. Hamilton cycles in random geometric graphs. *Ann. Appl. Probab.*, to appear.

[12] B. Bollobás. *Random graphs*, volume 73 of *Cambridge Studies in Advanced Mathematics*. Cambridge University Press, Cambridge, second edition, 2001.

[13] B. Bollobás. *The art of mathematics*. Cambridge University Press, New York, 2006. Coffee time in Memphis.

[14] B. Bollobás and I. Leader. Edge-isoperimetric inequalities in the grid. *Combinatorica*, 11(4):299–314, 1991.

[15] J. Díaz, D. Mitsche, and X. Pérez. Sharp threshold for Hamiltonicity of random geometric graphs. *SIAM J. Discrete Math.*, 21(1):57–65 (electronic), 2007.

[16] J. Dugundji. *Topology*. Allyn and Bacon Inc., Boston, Mass., 1966.

[17] P. Erdős and A. Rényi. On random graphs. I. *Publ. Math. Debrecen*, 6:290–297, 1959.

[18] P. Erdős and A. Rényi. On the evolution of random graphs. *Magyar Tud. Akad. Mat. Kutató Int. Közl.*, 5:17–61, 1960.

[19] P. Erdős and A. Rényi. On the evolution of random graphs. *Bull. Inst. Internat. Statist.*, 38:343–347, 1961.

[20] P. Erdős and A. Rényi. On the strength of connectedness of a random graph. *Acta Math. Acad. Sci. Hungar.*, 12:261–267, 1961.

[21] V. Falgas-Ravry and M. Walters. Sharpness in the k-nearest neighbours random geometric graph model. *ArXiv e-print 1101.3083 math.PR*, Jan. 2011.

[22] M. E. Fisher. Critical probabilities for cluster size and percolation problems. *J. Mathematical Phys.*, 2:620–627, 1961.

[23] M. E. Fisher and J. W. Essam. Some cluster size and percolation problems. *J. Mathematical Phys.*, 2:609–619, 1961.

[24] M. Franceschetti, L. Booth, M. Cook, R. Meester, and J. Bruck. Continuum percolation with unreliable and spread-out connections. *J. Stat. Phys.*, 118(3-4):721–734, 2005.

[25] E. Friedgut and G. Kalai. Every monotone graph property has a sharp threshold. *Proc. Amer. Math. Soc.*, 124(10):2993–3002, 1996.

[26] J. H. Friedman and L. C. Rafsky. Multivariate generalizations of the Wald-Wolfowitz and Smirnov two-sample tests. *Ann. Statist.*, 7(4):697–717, 1979.

[27] J. H. Friedman and L. C. Rafsky. Graph-theoretic measures of multivariate association and prediction. *Ann. Statist.*, 11(2):377–391, 1983.

[28] E. T. Gawlinski and E. H. Stanley. Continuum percolation in two dimensions: Monte carlo tests of scaling and universality for non-interacting discs. *Journal of Physics A: Mathematical and General*, 14(8):L291, 1981.

[29] E. N. Gilbert. Random plane networks. *J. Soc. Indust. Appl. Math.*, 9:533–543, 1961.

[30] E. N. Gilbert. The probability of covering a sphere with N circular caps. *Biometrika*, 52:323–330, 1965.

[31] A. Goel, S. Rai, and B. Krishnamachari. Monotone properties of random geometric graphs have sharp thresholds. *Ann. Appl. Probab.*, 15(4):2535–2552, 2005.

[32] J. M. González-Barrios and A. J. Quiroz. A clustering procedure based on the comparison between the k nearest neighbors graph and the minimal spanning tree. *Statist. Probab. Lett.*, 62(1):23–34, 2003.

[33] G. Grimmett. *Percolation*, volume 321 of *Grundlehren der Mathematischen Wissenschaften [Fundamental Principles of Mathematical Sciences]*. Springer-Verlag, Berlin, second edition, 1999.

[34] G. R. Grimmett and D. R. Stirzaker. *Probability and random processes*. Oxford University Press, New York, third edition, 2001.

[35] P. Gupta and P. R. Kumar. Critical power for asymptotic connectivity in wireless networks. In *Stochastic analysis, control, optimization and applications*, Systems Control Found. Appl., pages 547–566. Birkhäuser Boston, Boston, MA, 1999.

[36] O. Häggström and R. Meester. Nearest neighbor and hard sphere models in continuum percolation. *Random Structures Algorithms*, 9(3):295–315, 1996.

[37] B. Hajek. Adaptive transmission strategies and routing in mobile radio networks. *Proceedings of the Conference on Information Sciences and Systems*, pages 373–378, 1983.

[38] P. Hall. On continuum percolation. *Ann. Probab.*, 13(4):1250–1266, 1985.

[39] P. Hall. On the coverage of k-dimensional space by k-dimensional spheres. *Ann. Probab.*, 13(3):991–1002, 1985.

[40] N. Henze. A multivariate two-sample test based on the number of nearest neighbor type coincidences. *Ann. Statist.*, 16(2):772–783, 1988.

[41] M. A. Hernández Cifre. Optimizing the perimeter and the area of convex sets with fixed diameter and circumradius. *Arch. Math. (Basel)*, 79(2):147–157, 2002.

[42] T.-C. Hou and V. Li. Transmission range control in multihop packet radio networks. *IEEE Transactions on Communications*, 34(1):38–44, 1986.

[43] H. Kesten. *Percolation theory for mathematicians*, volume 2 of *Progress in Probability and Statistics*. Birkhäuser Boston, Mass., 1982.

[44] J. F. C. Kingman. *Poisson processes*, volume 3 of *Oxford Studies in Probability*. The Clarendon Press Oxford University Press, New York, 1993. Oxford Science Publications.

[45] L. Kleinrock and J. Silvester. Optimum transmission radii for packet radio networks or why six is a magic number. In *NTC '78; National Telecommunications Conference, Volume 1*, volume 1, pages 4–+, 1978.

[46] T. Leighton and P. Shor. Tight bounds for minimax grid matching with applications to the average case analysis of algorithms. *Combinatorica*, 9(2):161–187, 1989.

[47] J. E. Littlewood. *Littlewood's miscellany*. Cambridge University Press, Cambridge, 1986. Edited and with a foreword by Béla Bollobás.

[48] C. McDiarmid. Random channel assignment in the plane. *Random Structures Algorithms*, 22(2):187–212, 2003.

[49] C. McDiarmid and T. Müller. On the chromatic number of random geometric graphs. *Combinatorica*, to appear.

[50] C. McDiarmid and B. Reed. Colouring proximity graphs in the plane. *Discrete Math.*, 199(1-3):123–137, 1999.

[51] R. Meester and R. Roy. *Continuum percolation*, volume 119 of *Cambridge Tracts in Mathematics*. Cambridge University Press, Cambridge, 1996.

[52] P. A. P. Moran and S. Fazekas de St Groth. Random circles on a sphere. *Biometrika*, 49:389–396, 1962.

[53] T. Müller. Two-point concentration in random geometric graphs. *Combinatorica*, 28(5):529–545, 2008.

[54] T. Müller, X. Pérez, and N. Wormald. Disjoint hamilton cycles in the random geometric graph. *J. Graph Theory*, to appear.

[55] J. Ni and S. Chandler. Connectivity properties of a random radio network. *IEE Proceedings - Communications*, 141(4):289–296, 1994.

[56] J. Pach and P. K. Agarwal. *Combinatorial geometry*. Wiley-Interscience Series in Discrete Mathematics and Optimization. John Wiley & Sons Inc., New York, 1995. A Wiley-Interscience Publication.

[57] M. D. Penrose. Continuum percolation and Euclidean minimal spanning trees in high dimensions. *Ann. Appl. Probab.*, 6(2):528–544, 1996.

[58] M. D. Penrose. The longest edge of the random minimal spanning tree. *Ann. Appl. Probab.*, 7(2):340–361, 1997.

[59] M. D. Penrose. On k-connectivity for a geometric random graph. *Random Structures Algorithms*, 15(2):145–164, 1999.

[60] M. D. Penrose. *Random geometric graphs*, volume 5 of *Oxford Studies in Probability*. Oxford University Press, Oxford, 2003.

[61] J. Petit. *Layout Problems*. PhD thesis, Universitat Politècnica de Catalunya, Barcelona, Spain, 2001.

[62] T. Philips, S. Panwar, and A. Tantawi. Connectivity properties of a packet radio network model. *IEEE Transactions on Information Theory*, 35(5):1044–1047, 1989.

[63] J. Quintanilla, S. Torquato, and R. M. Ziff. Efficient measurement of the percolation threshold for fully penetrable discs. *J. Phys. A*, 33(42):L399–L407, 2000.

[64] F. D. K. Roberts. A Monte Carlo solution of a two-dimensional unstructured cluster problem. *Biometrika*, 54:625–628, 1967.

[65] C. A. Rogers. *Packing and covering*. Cambridge Tracts in Mathematics and Mathematical Physics, No. 54. Cambridge University Press, New York, 1964.

[66] F. J. Rohlf. Generalization of the gap test for the detection of multivariate outliers. *Biometrics*, 31(1):93–101, 1975.

[67] M. Rosso. Concentration gradient approach to continuum percolation in two dimensions. *Journal of Physics A: Mathematical and General*, 22(4):L131, 1989.

[68] M. F. Schilling. Multivariate two-sample tests based on nearest neighbors. *J. Amer. Statist. Assoc.*, 81(395):799–806, 1986.

[69] P. R. Scott. Sets of constant width and inequalities. *Quart. J. Math. Oxford Ser. (2)*, 32(127):345–348, 1981.

[70] C. H. Seager and G. E. Pike. Percolation and conductivity: A computer study. ii. *Phys. Rev. B*, 10(4):1435–1446, Aug 1974.

[71] P. W. Shor and J. E. Yukich. Minimax grid matching and empirical measures. *Ann. Probab.*, 19(3):1338–1348, 1991.

[72] J. A. Silvester. *On the spatial capacity of packet radio networks.* PhD thesis, California Univ., Los Angeles., 1980.

[73] H. Takagi and L. Kleinrock. Optimal transmission ranges for randomly distributed packet radio terminals. *IEEE Transactions on Communications*, 32(3):246–257, 1984.

[74] S.-H. Teng and F. F. Yao. k-nearest-neighbor clustering and percolation theory. *Algorithmica*, 49(3):192–211, 2007.

[75] M. Walters. Small components in k-nearest neighbour graphs. *ArXiv e-print 1101.2619 math.PR*, Jan. 2011.

[76] J. C. Wierman. Substitution method critical probability bounds for the square lattice site percolation model. *Combin. Probab. Comput.*, 4(2):181–188, 1995.

[77] M. A. Wong and T. Lane. A kth nearest neighbour clustering procedure. *J. Roy. Statist. Soc. Ser. B*, 45(3):362–368, 1983.

[78] F. Xue and P. R. Kumar. The number of neighbors needed for connectivity of wireless networks. *Wireless Networks*, 10:169–181, 2004.

[79] C. T. Zahn. Graph-theoretical methods for detecting and describing gestalt clusters. *IEEE Trans. Comput.*, 20(1):68–86, 1971.

[80] R. M. Ziff. Spanning probability in 2D percolation. *Physical Review Letters*, 69:2670–2673, Nov. 1992.

Queen Mary, University of London, London E1 4NS, UK
M.Walters@qmul.ac.uk

Transversals in latin squares: a survey

Ian M. Wanless

Abstract

A latin square of order n is an $n \times n$ array of n symbols in which each symbol occurs exactly once in each row and column. A transversal of such a square is a set of n entries containing no pair of entries that share the same row, column or symbol. Transversals are closely related to the notions of complete mappings and orthomorphisms in (quasi)groups, and are fundamental to the concept of mutually orthogonal latin squares.

Here we survey the literature on transversals and related notions. We cover (1) existence and enumeration results, (2) generalisations of transversals including partial transversals and plexes, (3) the special case when the latin square is a group table, (4) a connection with covering radii of sets of permutations, (5) transversals in arrays that generalise the notion of a latin square in various ways.

1 Introduction

By a *diagonal* of a square matrix we will mean a set of entries that contains exactly one representative from each row and column. A *transversal* is a diagonal in which no symbol is repeated. A *latin square* of order n is an $n \times n$ array of n symbols in which each symbol occurs exactly once in each row and in each column. The majority of this survey[1] looks at transversals (and their generalisations) in latin squares. In a transversal of a latin square every symbol must occur exactly once, although in §10 we will consider transversals of more general matrices where this property no longer holds.

Historically, interest in transversals arose from the study of orthogonal latin squares. A pair of latin squares $A = [a_{ij}]$ and $B = [b_{ij}]$ of order n are said to be *orthogonal mates* if the n^2 ordered pairs (a_{ij}, b_{ij}) are distinct. It is simple to see that if we look at all n occurrences of a given symbol in B, then the corresponding positions in A must form a transversal. Indeed,

Theorem 1.1 *A latin square has an orthogonal mate iff it has a decomposition into disjoint transversals.*

For example, below there are two orthogonal latin squares of order 8. Subscripted letters are used to mark the transversals of the left hand square which correspond

---

[1]The present survey extends and updates an earlier survey [123] on the same theme.

to the positions of each symbol in its orthogonal mate (the right hand square).

$$
\begin{array}{llllllll}
1_a & 2_b & 3_c & 4_d & 5_e & 6_f & 7_g & 8_h \\
7_b & 8_a & 5_d & 6_c & 2_f & 4_e & 1_h & 3_g \\
2_c & 1_d & 6_a & 3_b & 4_g & 5_h & 8_e & 7_f \\
8_d & 7_c & 4_b & 5_a & 6_h & 2_g & 3_f & 1_e \\
4_f & 3_e & 1_g & 2_h & 7_a & 8_b & 5_c & 6_d \\
6_e & 5_f & 7_h & 8_g & 1_b & 3_a & 2_d & 4_c \\
3_h & 6_g & 2_e & 1_f & 8_c & 7_d & 4_a & 5_b \\
5_g & 4_h & 8_f & 7_e & 3_d & 1_c & 6_b & 2_a
\end{array}
\qquad
\begin{array}{llllllll}
a & b & c & d & e & f & g & h \\
b & a & d & c & f & e & h & g \\
c & d & a & b & g & h & e & f \\
d & c & b & a & h & g & f & e \\
f & e & g & h & a & b & c & d \\
e & f & h & g & b & a & d & c \\
h & g & e & f & c & d & a & b \\
g & h & f & e & d & c & b & a
\end{array}
\qquad (1.1)
$$

It was conjectured by no less a mathematician than Euler [54] that orthogonal latin squares of order n exist iff $n \not\equiv 2 \bmod 4$. This conjecture was famously disproved by Bose, Shrikhande and Parker who in [16] showed instead that:

Theorem 1.2 *There is a pair of orthogonal latin squares of order n iff $n \notin \{2, 6\}$.*

More generally, there is interest in sets of *mutually orthogonal latin squares* (MOLS), that is, sets of latin squares in which each pair is orthogonal in the above sense. The literature on MOLS is vast (start with [31, 39, 40, 87]) and provides ample justification for studying transversals. In the interests of keeping this survey to a reasonable size, we will not discuss MOLS except as far as they bear directly and specifically on questions to do with transversals. While Theorem 1.1 remains the original motivation for studying transversals, subsequent investigations have shown that transversals are interesting objects in their own right. Despite this, a number of basic questions about their properties remain unresolved. In 1995, Alon *et al.* [6] bemoaned the fact that "There have been more conjectures than theorems on latin transversals in the literature." While there are still some frustratingly simply conjectures that remain unresolved, the progress in the last five years has finally rendered the lament from [6] untrue. Much of that progress has resulted from the discovery of a new tool called the "Delta Lemma".

2   The Delta Lemma

The deceptively simple idea behind the Delta Lemma occurred to two sets of researchers simultaneously and independently in 2005, leading eventually to the publications [49, 57]. Variants of the Lemma have also been used in [21, 37, 48, 50, 51, 101, 124].

To use the Delta Lemma it is useful to think of a latin square as being a set of *entries*, each of which is a (row, column, symbol) triple. It is convenient to index the rows, columns and symbols of a latin square of order n with $\mathbb{Z}_n$, in which case the square can be viewed as a subset of $\mathbb{Z}_n \times \mathbb{Z}_n \times \mathbb{Z}_n$. The latin property insists that distinct entries agree in at most one coordinate.

In its simplest form the Delta Lemma is this:

Lemma 2.1 *Let L be a latin square of order n indexed by $\mathbb{Z}_n$. Define a function $\Delta : L \to \mathbb{Z}_n$ by $\Delta(r, c, s) = r + c - s$. If T is a transversal of L then, modulo n,*

$$
\sum_{(r,c,s) \in T} \Delta(r, c, s) = \begin{cases} 0 & \text{if } n \text{ is odd,} \\ \frac{1}{2}n & \text{if } n \text{ is even.} \end{cases} \qquad (2.1)
$$

The proof is a triviality, since by definition r, c and s take every value in $\mathbb{Z}_n$ once in T. Yet the simplicity of the result belies its power. The function Δ can be thought of as measuring the difference of a latin square from the cyclic group. It is uniformly zero on the addition table of $\mathbb{Z}_n$, which leads to an immediate corollary:

Theorem 2.2 *The addition table of $\mathbb{Z}_n$ has no transversals when n is even.*

This fact was proved by Euler [54], making it one of the first theorems ever proved about transversals[2]. Variants of the Delta Lemma can be used to show that many other groups lack transversals. We will revisit the question of which groups have a transversal in §6.

As Δ measures the difference of a latin square from $\mathbb{Z}_n$ it is at its most powerful when applied to latin squares where most entries agree with $\mathbb{Z}_n$. In many such cases, the few entries that have $\Delta \neq 0$ can readily be seen to have restrictions on the transversals that include them. This approach was used in [124] to show:

Theorem 2.3 *For every order $n > 3$ there exists a latin square which contains an entry that is not included in any transversal.*

Given Theorem 1.1, an immediate corollary is:

Theorem 2.4 *For every order $n > 3$ there is a latin square that has no orthogonal mate.*

The even case of this result was already known in Euler's day (Theorem 2.2), and the case of $n \equiv 1 \bmod 4$ was shown by Mann [92] in 1944 (see Theorem 4.6). However, despite prominence as an open problem [14, §3.3], [15, X.8.13], [46], [79, p.181] and [118], the $n \equiv 3 \bmod 4$ case resisted until the discovery of the Delta Lemma. With that history spanning back to the 18th century, it is remarkable that within 5 years the Delta Lemma has provided no fewer than four different proofs of Theorem 2.4, underscoring that it is the right tool for the job.

The first two proofs [57, 124] were simultaneous. Evans obtained Theorem 2.4 using a version of the Delta Lemma but without showing Theorem 2.3. In Evans' version of the Delta Lemma rows, columns and symbols have indices chosen from $\mathbb{Z}_m$ for some m smaller than the order of the square. Obviously, this necessitates some duplication of indices, but nevertheless, for any assignment of indices, there is a single value that must be the sum of the Δ function along any transversal.

Two further proofs of Theorem 2.4, both via Theorem 2.3, are given in [48, 51]. Although some of the earlier proofs are really quite neat, Egan's proof in [48] deserves recognition as the proof from "The Book". Here it is:

Proof [of Theorem 2.3] In light of Theorem 2.2, we need only consider odd $n > 3$.

---

[2]Euler used the name "formule directrix" for a transversal. Subsequently, in some statistical literature (e.g. [62]) a transversal was called a directrix.

Define a latin square $L = [L_{ij}]$ of order n, indexed by $\mathbb{Z}_n$, by

$$L_{ij} = \begin{cases} 1 & \text{if } (i,j) \in \{(0,0),(1,n-1)\}, \\ 0 & \text{if } (i,j) \in \{(1,0),(2,n-1)\}, \\ j+2 & \text{if } i = 0 \text{ and } j \in \{1,3,5,\ldots,n-2\}, \\ j & \text{if } i = 2 \text{ and } j \in \{1,3,5,\ldots,n-2\}, \\ i+j & \text{otherwise.} \end{cases}$$

To check that the entry $(1,0,0)$ is not in any transversal T of L, observe that $\Delta(1,0,0) = 1$. Any other entries which might lie in T, namely the ones that do not share any coordinate with $(1,0,0)$, have Δ value in $\{-2,0,2\}$. Since the entries with $\Delta = -2$ all share a row, at most one of them can be in any transversal. Likewise for the entries with $\Delta = +2$. As $n > 3$, it is impossible to satisfy (2.1). $\qquad\square$

As a coda to this proof, we observe that a similar argument shows that the entry $(1, n-1, 1)$ is not in any transversal.

Evans [57] demonstrates that his variant of the Delta Lemma can be used to explain a number of classical results about transversals. In addition to the above, the Delta Lemma can be (and in many cases was) used to prove the Theorems numbered 3.1, 3.4, 3.5, 4.2, 4.3, 4.5, 4.6, 8.3, 8.4, 8.7, 8.8, 8.9, 8.10, 10.7, 10.8 and 10.11 in this survey. For such a simple device it is immensely powerful!

3   Entries not in transversals

As shown by Theorem 2.3, some latin squares have an entry that is not in any transversal. In extreme cases, such as Theorem 2.2, the latin square has no transversals at all. We now look at some further results of this nature.

A latin square of order mq is said to be of q-step type if it can be represented by a matrix of $q \times q$ blocks A_{ij} as follows

$$\begin{matrix} A_{11} & A_{12} & \cdots & A_{1m} \\ A_{21} & A_{22} & \cdots & A_{2m} \\ \vdots & \vdots & \ddots & \vdots \\ A_{m1} & A_{m2} & \cdots & A_{mm} \end{matrix}$$

where each block A_{ij} is a latin subsquare of order q and two blocks A_{ij} and $A_{i'j'}$ contain the same symbols iff $i + j \equiv i' + j' \bmod m$. The following classical theorem is due to Maillet [90] (and was rediscovered by Parker [99]).

Theorem 3.1 *Suppose that q is odd and m is even. No q-step type latin square of order mq possesses a transversal.*

As we will see in §6, this rules out many group tables having transversals. In particular, as we saw in Theorem 2.2, no cyclic group of even order has a transversal. By contrast, there is no known example of a latin square of odd order without transversals.

Conjecture 3.2 *Each latin square of odd order has at least one transversal.*

τ	Order n							
	2	3	4	5	6	7	8	9
0		1	1	1	2	54	267 932	19 270 833 530
1						11	13 165	18 066
2						26	1 427	1 853
3						12	253	54
4	1					12	508	21
5						6	89	7
6					1	8	65	7
7						3	33	1
8						4	48	1
9							25	
10						1	27	1
11						1	9	
12				1	2	6	9	
13						1	2	
14							2	
16			1		1	1	27	
18							1	
20							1	
28							1	
36					6	1		
64							33	
Total	1	1	2	2	12	147	283 657	19 270 853 541

Table 1: Species of order $n \leqslant 9$ according to their number of transversal-free entries.

This conjecture is known [94] to be true for $n \leqslant 9$. It is attributed to Ryser [104] and has been open for forty years. In fact, Ryser's original conjecture was somewhat stronger: for every latin square of order n, the number of transversals is congruent to n mod 2. In [11], Balasubramanian proved the even case.

Theorem 3.3 *In any latin square of even order the number of transversals is even.*

Despite this, it has been noted in [3, 25, 113] (and other places) that there are many counterexamples of odd order to Ryser's original conjecture. Hence the conjecture has now been weakened to Conjecture 3.2 as stated.

Latin squares of moderate order are typically blessed with many transversals, although it is clear that some rare cases have restrictions. One measure of the restrictions on transversals is $\tau(L)$, the number of transversal-free entries in a latin square L. The value of τ for latin squares of order up to 9 is shown in Table 1, from [51]. The entries in the table are counts of the number of species[3]. This table, together with tests on random latin squares of larger order suggests that almost all latin squares of large order have a transversal through every entry (i.e. $\tau = 0$). Nevertheless, we have the following result, which was implicitly shown in [124] and

---

[3]A *species* or *main class*, is an equivalence class of latin squares each of which has essentially the same structure. See [39, 87] for the definition.

explicitly stated in [51].

Theorem 3.4 *For all $n \geqslant 4$, there exists a latin square L of order n with $\tau(L) \geqslant 7$.*

This result is likely to be far short of best possible. Clearly, τ can be as large as n^2 for even n, by Theorem 2.2. For odd n we also know the following, from [51].

Theorem 3.5 *For all odd $m \geqslant 3$ there exists a latin square of order $3m$ that contains an $(m-1) \times m$ latin subrectangle consisting of entries that are not in any transversal.*

In this example $\tau/n^2 \geqslant m(m-1)/(3m)^2 \sim 1/9$, so at least a constant fraction of the entries are transversal free as $n = 3m \to \infty$. This raises the following interesting question [51]:

Question 3.6 *Is $\liminf\limits_{n\to\infty} \max\limits_{L} \frac{1}{n^2}\tau(L) > 0$, where L ranges over squares of order n?*

4 Disjoint transversals

Motivated by Theorem 1.1, we next consider sets of disjoint transversals. Such a set will be described as *maximal* if it is not a subset of a strictly larger set of disjoint transversals. For a given latin square L we consider two measures of the number of disjoint transversals in L. Let $\lambda = \lambda(L)$ be the largest cardinality of any set of disjoint transversals in L, and let $\alpha = \alpha(L)$ be the smallest cardinality of any maximal set of disjoint transversals in L. Clearly $0 \leqslant \alpha \leqslant \lambda \leqslant n$. We will also be interested in $\beta(n)$ and $\mu(n)$, which we define to be the minimum of $\alpha(L)$ and $\lambda(L)$, respectively, among all latin squares L of order n.

Example 4.1 A latin square of order n has $\lambda = n$ iff it has an orthogonal mate, by Theorem 1.1. For $n = 6$ there is no pair of orthogonal squares (see Theorem 1.2), but we can get close. Finney [62] gives the following example which contains 4 disjoint transversals indicated by the subscripts a, b, c and d.

$$
\begin{array}{cccccc}
1_a & 2 & 3_b & 4_c & 5 & 6_d \\
2_c & 1_d & 6 & 5_b & 4_a & 3 \\
3 & 4_b & 1 & 2_d & 6_c & 5_a \\
4 & 6_a & 5_c & 1 & 3_d & 2_b \\
5_d & 3_c & 2_a & 6 & 1_b & 4 \\
6_b & 5 & 4_d & 3_a & 2 & 1_c \\
\end{array}
$$

This square has $\lambda = 4$, and $\alpha = 3$ (the different shadings show a maximal set of 3 disjoint transversals).

Table 2 shows the species of order $n \leqslant 9$, counted according to their maximum number λ of disjoint transversals. Table 3 shows the species of order $n = 9$ categorised according to their values of λ and α. The data in both tables was computed in [51].

Evidence for small orders (such as that in Table 2) led van Rees [118] to conjecture that, as $n \to \infty$, a vanishingly small proportion of latin squares have orthogonal

λ	n = 2	3	4	5	6	7	8	9
0	1	0	1	0	6	0	33	0
1	-	0	0	1	0	1	0	0
2	0	-	0	0	2	5	7	0
3	-	1	-	0	0	24	46	3
4	-	-	1	-	4	68	712	23
5	-	-	-	1	-	43	71 330	142 915
6	-	-	-	-	0	-	209 505	61 613
7	-	-	-	-	-	6	-	18 922 150 935
8	-	-	-	-	-	-	2 024	-
9	-	-	-	-	-	-	-	348 498 052
Total	1	1	2	2	12	147	283 657	19 270 853 541

Table 2: Latin squares of order $n \leqslant 9$ with λ disjoint transversals.

	λ						
α	3	4	5	6	7	9	Total
1	0	7	36 000	0	0	0	36 007
2	2	4	6 765	528	873	5	8 177
3	1	12	100 150	61 085	18 786 989 798	340 588 766	19 127 739 812
4		0	0	0	135 160 264	7 909 243	143 069 507
5			0	0	0	32	32
6				0	0	5	5
7					0	1	1
Total	3	23	142 915	61 613	18 922 150 935	348 498 052	19 270 853 541

Table 3: Species of order 9 categorised according to λ and α.

mates. However, the trend seems to be quite the reverse (see [93, 124]), although no rigorous way of establishing this has yet been found.

For even orders, [50] showed that λ can achieve many different values[4]:

Theorem 4.2 *For each even* $n \geqslant 6$ *and each* $j \equiv 0 \bmod 4$ *such that* $0 \leqslant j \leqslant n$, *there exists a latin square* L *of order* n *with* $\lambda(L) = j$.

In [51] it was shown that $\lambda = 1$ is also achievable for even $n \geqslant 10$, as a corollary of:

Theorem 4.3 *For all even* $n \geqslant 10$, *there exists a latin square of order* n *that has transversals, but in which every transversal coincides on a single entry*[5].

It is not possible for a latin square of order n to have $\lambda = n - 1$. Theorems 4.2 and 4.3, together with small order examples, led the authors of [51] to conjecture that for large even orders, all other values of λ are achievable:

Conjecture 4.4 *For all even* $n \geqslant 10$ *and each* $m \in \{0, 1, \ldots, n - 3, n - 2, n\}$ *there exists a latin square of order* n *such that* $\lambda(L) = m$.

---

[4]We will see in Theorem 6.1 that the situation is markedly different for group tables.

[5]If n is a multiple of 16 then every transversal in the construction includes two specific entries.

For odd orders n, there is not even a conjecture as to which values of λ can be achieved, except that Conjecture 3.2 predicts that λ must be positive. This leads naturally to a discussion of $\mu(n)$, the minimum value of λ among the latin squares of order n. Clearly, $\mu(n) = 0$ for all even n by Theorem 2.2, so we are concerned with the case when n is odd. If Conjecture 3.2 is true, then $\mu(n) \geqslant 1$ for all odd n. Our best general upper bound currently is:

Theorem 4.5 *If $n > 3$ then $\mu(n) \leqslant \frac{1}{2}(n+1)$.*

This result was first explicitly stated in [51], although it follows immediately from the only known proof[6] of Theorem 2.4 that does not go via Theorem 2.3. The $n \equiv 3 \bmod 4$ case of Theorem 4.5 was implicitly shown by Evans [57], 62 years after the $n \equiv 1 \bmod 4$ had been shown by Mann [92], who proved:

Theorem 4.6 *Let L be a latin square of order $4k + 1$ containing a latin subsquare S of order $2k$. Let U be the set of entries in L that do not share a row, column or symbol with any element of S. Then every transversal of L contains an odd number of elements of U.*

In Theorem 4.6, simple counting shows that U has $2k + 1$ elements and hence $\lambda(L) \leqslant 2k + 1 = (n+1)/2$.

There appears to be room to improve on the upper bound for $\mu(n)$ stated in Theorem 4.5. The known values of $\mu(n)$ for odd n are $\mu(1) = \mu(5) = \mu(7) = 1$ and $\mu(3) = \mu(9) = 3$. It would be of interest to determine if $\mu(n) < \frac{1}{2}n$ for $n > 3$. In particular [51]:

Question 4.7 *Is $\mu(n)$ bounded as $n \to \infty$?*

Next we consider $\beta(n)$, the minimum value of α among the latin squares of order n. Although we know little about the size of $\mu(n)$ for odd n, we can narrow $\beta(n)$ down to a very small set of possible values. A result in [50] proved that

$$\beta(n) \leqslant \begin{cases} 1 & \text{if } n \equiv 1 \bmod 4, \\ 3 & \text{if } n \equiv 3 \bmod 4. \end{cases}$$

Even this strong restriction leaves some potential for improvement. The known values of $\beta(n)$ for odd n are $\beta(3) = 3$ and $\beta(1) = \beta(5) = \beta(7) = \beta(9) = 1$. If Conjecture 3.2 is correct then $\beta(n) \geqslant 1$ for all odd n, but at this stage it is still plausible that equality holds for odd $n > 3$.

Finally, we remark that if transversals are not disjoint then they intersect. It is obvious that two transversals of an $n \times n$ latin square can never share exactly $n - 1$ or $n - 2$ entries. However, [28] shows that they can intersect in any other way:

Theorem 4.8 *For all odd $n > 5$ and every integer $t \in \{0, 1, 2, \ldots, n - 3, n\}$ there exist two transversals of the addition table of $\mathbb{Z}_n$ that intersect in exactly t entries.*

---

[6]Theorem 2.4 is a direct corollary of Theorem 4.5.

5 Partial transversals

We have seen in Theorems 2.2 and 3.1 that not all latin squares have transversals, which prompts the question of how close we can get to finding a transversal in such cases. We define a *partial transversal*[7] of *length* k to be a set of k entries, each selected from different rows and columns of a latin square such that no two entries contain the same symbol. A partial transversal is *completable* if it is a subset of some transversal, whereas it is *non-extendible* if it is not contained in any partial transversal of greater length.

Since not all squares of order n have a partial transversal of length n (i.e. a transversal), the best we can hope for is to find one of length $n - 1$. Such partial transversals are called *near transversals*. The following conjecture has been attributed to Brualdi (see [39, p.103]) and Stein [111] and, in [52], to Ryser. For generalisations of it, in terms of hypergraphs, see [2].

Conjecture 5.1 *Every latin square has a near transversal.*

A claimed proof of this conjecture by Deriyenko [42] contains a fatal error, as mentioned in [40, p.40] and discussed in detail in [25]. More recently, a paper [77] appeared in the maths arXiv claiming to prove Conjecture 5.1. However, the paper was subsequently withdrawn when it was discovered that the proof was invalid. By copying the method of Theorem 3.3, Akbari and Alireza [3] managed to show that the number of non-extendible near transversals in any latin square is divisible by 4. Unfortunately in many cases that number can be zero, as we will see in Corollary 6.6.

The best reliable lower bound to date states that there must be a partial transversal of length at least $n - O(\log^2 n)$. This was shown by Shor [108], and the implicit constant in the 'big O' was very marginally improved by Fu *et al.* [64]. Subsequently Hatami and Shor [73] discovered an error in [108] (duplicated in [64]) and corrected the constant to a higher one. Nonetheless, the important thing remains that the bound is $n - O(\log^2 n)$. This improved on a number of earlier bounds including $\frac{2}{3}n + O(1)$ (Koksma [83]), $\frac{3}{4}n + O(1)$ (Drake [45]) and $n - \sqrt{n}$ (Brouwer *et al.* [17] and Woolbright [126]).

It has also been shown in [25] that every latin square possesses a diagonal in which no symbol appears more than twice. An earlier claimed proof of this result [22, Thm 8.2.3] is incomplete.

Conjecture 5.1 has been open for decades and has now gained a degree of notoriety. A much simpler problem is to consider the shortest possible length of a non-extendible partial transversal. It is easy to see the impossibility of a non-extendible partial transversal having length strictly less than $\frac{1}{2}n$, since there would not be enough 'used' symbols to fill the submatrix formed by the 'unused' rows and columns. However, for all $n > 4$, non-extendible partial transversals of length $\lceil \frac{1}{2}n \rceil$ can easily be constructed using a square of order n which contains a subsquare S of order $\lfloor \frac{1}{2}n \rfloor$ and a partial transversal containing the symbols of S but not using any of the same rows or columns as S.

The antithesis of non-extendibility is for a partial transversal to be *completable* in the sense that it is a subset of some transversal. Theorems 3.4, 3.5 and 4.3 all furnish

---

[7]In some papers (e.g. [64, 73, 108]) a partial transversal of length k is defined slightly differently to be a diagonal on which k different symbols appear.

examples where even some partial transversals of length 1 fail to be completable. An interesting open question concerns the completability of short partial transversals in cyclic groups. Grüttmüller [68] defined $C(k)$ to be the smallest odd integer such that it is possible to complete every partial transversal of length k in $\mathbb{Z}_n$ for any odd $n \geqslant C(k)$. He showed that $C(1) = 1$ and $C(2) = 3$. It is not proved that $C(k)$ even exists for $k \geqslant 3$, but if it does then Grüttmüller [69] showed that $C(k) \geqslant 3k - 1$. He also provided computational evidence to suggest that this bound is essentially best possible. Further evidence that $C(3) = 9$ was given by Cavenagh et al. [27], who proved:

Theorem 5.2 *For any prime $p > 7$, every partial transversal of length 3 in the addition table of $\mathbb{Z}_p$ is completable.*

To complement this result, [110] gives a method for completing short partial transversals in $\mathbb{Z}_n$ when n has many different prime factors.

6   Finite Groups

By using the symbols of a latin square to index its rows and columns, each latin square can be interpreted as the Cayley table of a quasigroup [39]. In this section we consider the important special case when that quasigroup is associative; in other words, it is a group. The extra structure in this case allows for much stronger results. For example, let L_G be the Cayley table of a finite group G. Suppose that we know of a transversal of L_G that comprises a choice from each row i of an element g_i. Let g be any fixed element of G. Then if we select from each row i the element $g_i g$ this will give a new transversal. Moreover, as g ranges over G the transversals so produced will be mutually disjoint. Hence:

Theorem 6.1 *If the Cayley table of a finite group has a single transversal then it has a decomposition into disjoint transversals.*

In other words, using the notation of §4, the only two possibilities if $|G| = n$ are that $\lambda(L_G) = 0$ or $\lambda(L_G) = n$.

6.1   Complete Mappings and Orthomorphisms

Much of the study of transversals in groups has been phrased in terms of the equivalent concepts of complete mappings and orthomorphisms[8]. Mann [91] introduced complete mappings for groups, but the definition works just as well for quasigroups. It is this: a permutation θ of the elements of a quasigroup $(Q, \oplus)$ is a *complete mapping* if $\eta : Q \mapsto Q$ defined by $\eta(x) = x \oplus \theta(x)$ is also a permutation. The permutation η is known as an *orthomorphism* of $(Q, \oplus)$, following terminology introduced in [78]. All of the results of this paper could be rephrased in terms of complete mappings and/or orthomorphisms because of our next observation.

Theorem 6.2 *Let $(Q, \oplus)$ be a quasigroup and L_Q its Cayley table. Then $\theta : Q \mapsto Q$ is a complete mapping iff we can locate a transversal of L_Q by selecting, in each row*

---

[8]These two concepts are so closely related that some references (e.g. [75, 107]) confuse them.

x, the entry in column $\theta(x)$. Similarly, $\eta : Q \mapsto Q$ is an orthomorphism iff we can locate a transversal of L_Q by selecting, in each row x, the entry containing symbol $\eta(x)$.

There are also notions of near complete mappings and near orthomorphisms that correspond naturally to near transversals [12, 40, 55].

Orthomorphisms and complete mappings have been used to build a range of different combinatorial designs and algebraic structures including MOLS [12, 55, 91], generalized Bhaskar Rao designs [1], diagonally cyclic latin squares [27, 121], left neofields [12, 40, 55], Bol loops [96] and atomic latin squares [122]. This wide applicability and the intimate connection with transversals, as demonstrated by Theorem 6.2, justifies a closer look at orthomorphisms (or equivalently, at complete mappings). Various special types of orthomorphisms have been considered, with the focus often on orthomorphisms with a particularly nice algebraic structure. Such orthomorphisms have the potential to be exploited in a variety of applications, so we now examine them in some detail.

An orthomorphism of a group is *canonical* [12, 13, 40, 110] (also called *normalized* [31] or *standard* [75]) if it fixes the identity element[9]. In the following we will suppose that θ is an orthomorphism in a group G with identity ε. For the sake of simplicity we will assume that G is abelian and θ is canonical, although in some of the following categories these restrictions may be relaxed if so desired.

1. *Linear orthomorphisms*: θ is linear if $\theta(x) = \lambda x$ for some fixed $\lambda \in G$. Clark and Lewis [30] show that the number of such linear orthomorphisms of $\mathbb{Z}_n$ is

$$\prod_{p|n} p^{a-1}(p-2),$$

   where the product is over prime divisors of n and $a = a(p, n)$ is the greatest integer such that p^a divides n.

2. *Quadratic orthomorphisms*: Suppose G is the additive group of a finite field F and let $\square$ denote the set of non-zero squares in F. If there are constants $\lambda_1, \lambda_2 \in F$ such that θ can be defined as $x \mapsto \lambda_1 x$ for $x \in \square$ and $x \mapsto \lambda_2 x$ for $x \notin \square$, then we say θ is a quadratic orthomorphism. Linear orthomorphisms are a special case of quadratic orthomorphisms for which $\lambda_1 = \lambda_2$. Note that $\square$ is an index 2 subgroup of the multiplicative group of F, and the non-squares form a coset of $\square$.

3. *Cyclotomic orthomorphisms*: These generalise the quadratic orthomorphisms. Take any non-trivial subgroup H of the multiplicative group of F and choose a multiplier λ_i for each coset of H. Multiply every element in the coset (sometimes called a cyclotomy class) by the chosen multiplier. If the resulting map is an orthomorphism then we say it is a cyclotomic orthomorphism. See [55] for more information on cyclotomic orthomorphisms, including the special cases of linear and quadratic orthomorphisms.

---

[9]Some references (e.g. [55]) define all "orthomorphisms" to be canonical, but that is undesirable since it leaves no easy way to talk about the orthomorphisms which are not canonical.

4. *Regular orthomorphisms*: Let θ' be the restriction of θ to $G \setminus \{\varepsilon\}$. The orthomorphism θ is k-regular if the permutation θ' is regular in the sense that
   it permutes all elements of $G \setminus \{\varepsilon\}$ in cycles of length k. This notion was
   introduced in [63] and later studied in [107]. The special case when θ' has a
   single cycle of length $|G| - 1$ corresponds to the idea of an R-sequencing of G
   (see [40, Chap.3]).

5. *Involutory orthomorphisms*: If $\theta = \theta^{-1}$ then $\big\{\{x, \theta(x)\} : x \in G \setminus \{\varepsilon\}\big\}$ is a
   starter[10] in G. Conversely every starter in G defines an orthomorphism of G
   that is its own inverse.

6. *Strong orthomorphisms*: A permutation that is both an orthomorphism and a
   complete mapping is called a strong orthomorphism [8] (alternatively, a strong
   complete mapping [59] or a strong permutation [74]). They exist in an abelian
   group G if and only if the Sylow 2-subgroups and Sylow 3-subgroups of G
   are either trivial or non-cyclic [59]. Strong orthomorphisms are connected to
   strong starters, see [31, 74] for definitions and details.

7. *Polynomial orthomorphisms*: If G is the additive group of a ring R and there
   exists any polynomial $p(x)$ over R such that $\theta(x) = p(x)$ for all $x \in G$ then
   we say that θ is a polynomial orthomorphism. For example, linear orthomorphisms are polynomial, as is any orthomorphism of a finite field. In fact,
   [97, 119] any orthomorphism of a field of order $q \geqslant 4$ is realised by a polynomial
   of degree at most $q - 3$. In contrast, [110] showed that for any odd composite
   n there is a non-polynomial orthomorphism of $\mathbb{Z}_n$. The polynomials of small
   degree that produce orthomorphisms are classified in [97]. Note that quadratic
   orthomorphisms are polynomial, but are not produced by quadratic polynomials (indeed no orthomorphism is produced by a quadratic polynomial).

8. *Compound orthomorphisms*: Let d be a divisor of n. An orthomorphism θ of
   $\mathbb{Z}_n$ is defined to be d-compound if $\theta(i) \equiv \theta(j)$ whenever $i \equiv j \bmod d$. This
   notion was introduced in [110] where compound orthomorphisms were used,
   among other things, for completing partial orthomorphisms.

9. *Compatible orthomorphisms*: An orthomorphism θ of $\mathbb{Z}_n$ is compatible if it
   is d-compound for all divisors d of n. Every polynomial orthomorphism is
   necessarily compatible. The converse holds only for certain values of n, as
   characterised in [110]. The same paper contains a formula for the number
   of compatible orthomorphisms of $\mathbb{Z}_n$ expressed in terms of the number of
   orthomorphisms of $\mathbb{Z}_p$ for prime divisors p of n.

Having seen in Theorem 6.2 that transversals, orthomorphisms and complete
mappings are essentially the same thing, we will adopt the practice of expressing
our remaining results in terms of transversals even when the original authors used
one of the other notions.

---

[10]A *starter* is a pairing of the non-zero elements of an additive group such that every non-zero
element can be written as the difference of the two elements in some pair. Starters are useful for
creating numerous different kinds of designs [31].

6.2 Which groups have transversals?

We saw in §3 that the question of which latin squares have transversals is far from settled. However, if we restrict our attention to group tables, the situation is a lot clearer.

Consider these five propositions for the Cayley table L_G of a finite group G:

(i) L_G has a transversal.

(ii) L_G can be decomposed into disjoint transversals.

(iii) There exists a latin square orthogonal to L_G.

(iv) There is some ordering of the elements of G, say $a_1, a_2, \ldots, a_n$, such that $a_1 a_2 \cdots a_n = \varepsilon$, where ε denotes the identity element of G.

(v) The Sylow 2-subgroups of G are trivial or non-cyclic.

The fact that (i), (ii) and (iii) are equivalent comes directly from Theorem 1.1 and Theorem 6.1. Paige [98] showed that (i) implies (iv). Hall and Paige [72] then showed[11] that (iv) implies (v). They also showed that (v) implies (i) if G is a soluble, symmetric or alternating group. They conjectured that (v) is equivalent to (i) for all groups.

It was subsequently noted in [41] that both (iv) and (v) hold for all non-soluble groups, which proved that (iv) and (v) are equivalent. A much more direct and elementary proof of this fact was given in [117]. Thus the Hall-Paige Conjecture could be rephrased as the statement that all five conditions (i)–(v) are equivalent.

For decades there was incremental progress, as the Hall-Paige Conjecture was shown to hold for various groups, including the linear groups $GL(2, q)$, $SL(2, q)$, $PGL(2, q)$ and $PSL(2, q)$ (see [56] and the references therein). Then a very significant breakthrough was obtained by Wilcox [125] who reduced the problem to showing it for the sporadic simple groups (of which the Mathieu groups have already been handled in [36]). Evans [58] then showed that the only possible counterexample was Janko's group J_4. Finally, in unpublished work Bray claims to have showed that J_4 has a transversal, thereby proving the important theorem:

Theorem 6.3 *Conditions* (i), (ii), (iii), (iv), (v) *are equivalent for all finite groups.*

An interesting first step towards finding non-associative analogues of Theorem 6.3 was taken by Pula [100]. Regarding the non-associative analogue of condition (iv) above, it was shown in [19] that for all $n \geqslant 5$ there exists a loop[12] of order n in which every element can be obtained as a product of all n elements in some order and with some bracketing.

While Theorem 6.3 settles the question of which groups have a transversal[13], it remains an interesting open question as to whether Conjecture 5.1 holds for groups. A related concept is the idea of a sequenceable group. A group of finite order n is

---

[11]As shown in [120], the fact that (i) implies (v) is actually a special case of Theorem 3.1.

[12]A *loop* is a quasigroup with an identity element [39].

[13]From now on we will sometimes refer to groups having transversals (or near transversals etc.) when strictly speaking it is the Cayley table of the group that has these structures.

called *sequenceable* if its elements can be labelled in an order $a_1, a_2, \ldots, a_n$ such that the products $a_1, a_1a_2, a_1a_2a_3, \ldots, a_1a_2 \cdots a_n$ are distinct. This idea was introduced by Gordon [67] who showed that abelian groups are sequenceable iff they have a non-trivial cyclic Sylow 2-subgroup (in other words if condition (v) above fails). Since then, many non-abelian groups have been shown to be sequenceable as well (see [40, Chap 3] or [31, p.350] for details). The importance of this idea for our purposes is that the entries $(a_1a_2 \cdots a_i, a_{i+1}, a_1a_2 \cdots a_{i+1})$ for $i = 1, 2, \ldots, n-1$ form a near transversal of a sequenceable group. Hence we have this folklore result:

Theorem 6.4 *If a finite group is sequenceable then it has a near transversal.*

The converse of Theorem 6.4 is false. For example, the dihedral groups of order 6 and 8 have near transversals but are not sequenceable. All larger dihedral groups are sequenceable [88]. Indeed, it has been conjectured by Keedwell [80] that all non-abelian groups of order at least 10 are sequenceable.

For abelian groups, an important result was proved by Hall [71]. Recast into the form most useful to us, it is this:

Theorem 6.5 *Let L_G be the Cayley table of an abelian group G of finite order n, with identity 0. Suppose $b_1, b_2, \ldots, b_n$ is a list of (not necessarily distinct) elements of G. A necessary and sufficient condition for L_G to possess a diagonal on which the symbols are $b_1, \ldots, b_n$ (in some order), is that $\sum b_i = 0$.*

Since the sum of the elements in an abelian group is the identity if condition (v) above holds, and is the unique involution in the group otherwise, we have:

Corollary 6.6 *If a finite abelian group has a non-trivial cyclic Sylow 2-subgroup then it possesses non-extendible near transversals, but no transversals. Otherwise it has transversals but no non-extendible near transversals.*

This corollary has been rediscovered several times, most recently by Stein and Szabó [113]. They also show that for p prime, $\mathbb{Z}_p$ has no diagonal with exactly two distinct symbols on it. Again, this is a direct corollary of Theorem 6.5. We will see yet a third result that follows easily from Theorem 6.5 in Theorem 10.8.

6.3   How many transversals does a group have?

We turn next to the question of how many transversals a given group may have. In this subsection and the next, we will be concerned with estimating the number of transversals, as well as demonstrating that it must satisfy certain congruences. Analogous questions for more general latin squares will be considered in §7.

Using theoretical methods it seems very difficult to find accurate estimates for the number of transversals in a latin square (unless, of course, that number is zero). This difficulty is so acute that there are not even good estimates for z_n, the number of transversals of the cyclic group of order n. Clark and Lewis [30] conjecture that $z_n \geqslant n(n-2)(n-4) \cdots 3 \cdot 1 = n! \, o(\sqrt{e/n}^n)$ for odd n, while Vardi [116] makes a stronger prediction:

Conjecture 6.7 *There exist real constants* $0 < c_1 < c_2 < 1$ *such that*

$$c_1^n n! \leqslant z_n \leqslant c_2^n n!$$

for all odd $n \geqslant 3$.

Vardi makes this conjecture[14] while considering a variation on the toroidal n-queens problem. The toroidal n-queens problem is that of determining in how many different ways n non-attacking queens can be placed on a toroidal $n \times n$ chessboard. Vardi considered the same problem using semiqueens in place of queens, where a semiqueen is a piece which moves like a toroidal queen except that it cannot travel on right-to-left diagonals. The solution to Vardi's problem provides an upper bound on the toroidal n-queens problem. The problem can be translated into one concerning latin squares by noting that every configuration of n non-attacking semiqueens on a toroidal $n \times n$ chessboard corresponds to a transversal in a cyclic latin square L of order n, where $L_{ij} \equiv i - j \bmod n$. Note that the toroidal n-queens problem is equivalent to counting diagonals which simultaneously yield transversals in L and L', where $L'_{ij} = i + j \bmod n$.

Cooper and Kovalenko [34] were the first to prove the upper bound in Conjecture 6.7 by showing $z_n = o(0.9154^n n!)$, and this was subsequently improved to $z_n = o(0.7072^n n!)$ in [84]. In Theorem 7.2 we will see a stronger bound, that applies to all latin squares, not just to cyclic groups.

Finding a lower bound of the form given in Conjecture 6.7 is still an open problem. However, [32, 103, 107] do give some lower bounds, each of which applies only for some n. The following better bound was found in [28], although it is still a long way short of proving Vardi's Conjecture:

Theorem 6.8 *If n is odd and sufficiently large then* $z_n > (3.246)^n$.

Estimates for the rate of growth of z_n are given by Cooper *et al.* [33], who arrived at a value around $0.39^n n!$ and Kuznetsov [85, 86] who favours the slightly smaller $0.37^n n!$. Acting on a hunch, the present author proposes:

Conjecture 6.9

$$\lim_{n \to \infty} \frac{1}{n} \log(z_n/n!) = -1.$$

Of course, at this stage it is not even known that this limit exists.

6.4   Congruences and divisors

We next consider congruences satisfied by the number of transversals in a group table. An immediate corollary of the proof of Theorem 6.1 is that for any group the number of transversals through a given entry of the Cayley table is independent of the entry chosen. Hence (see Theorem 3.5 of [40]) we get:

Theorem 6.10 *The number of transversals in the Cayley table of a group G is divisible by $|G|$, the order of G.*

---

[14]Vardi's actual statement is not very concrete. Conjecture 6.7 is the present author's interpretation of Vardi's intention.

McKay *et al.* [94] also showed the following simple results, in the spirit of Theorem 3.3:

Theorem 6.11 *The number of transversals in any symmetric latin square of order n is congruent to n modulo 2.*

Corollary 6.12 *Let G be a group of order n. If G is abelian or n is even then the number of transversals in G is congruent to n modulo 2.*

Corollary 6.12 cannot be generalised to non-abelian groups of odd order, given that the non-abelian group of order 21 has 826 814 671 200 transversals.

Theorem 6.13 *If G is a group of order $n \not\equiv 1$ mod 3 then the number of transversals in G is divisible by 3.*

We will see below that the cyclic groups of small orders $n \equiv 1$ mod 3 have a number of transversals which is not a multiple of three.

Let z_n be the number of transversals in the cyclic group of order n and let $z'_n = z_n/n$ denote the number of transversals through any given entry of the cyclic square of order n. Since $z_n = z'_n = 0$ for all even n by Theorem 2.2 we shall assume for the following discussion that n is odd. The initial values of z'_n are known from [105] and [106]. They are

$$z'_1 = z'_3 = 1, \quad z'_5 = 3, \quad z'_7 = 19, \quad z'_9 = 225, \quad z'_{11} = 3\,441, \quad z'_{13} = 79\,259,$$
$$z'_{15} = 2\,424\,195, \quad z'_{17} = 94\,471\,089, \quad z'_{19} = 4\,613\,520\,889, \quad z'_{21} = 275\,148\,653\,115,$$
$$z'_{23} = 19\,686\,730\,313\,955, \quad z'_{25} = 1\,664\,382\,756\,757\,625.$$

Interestingly, if we take these numbers modulo 8 we find that this sequence begins 1,1,3,3,1,1,3,3,1,1,3,3,1. We know from Theorem 6.11 that z'_n is always odd for odd n, but it is an open question whether there is any deeper pattern modulo 4 or 8. The initial terms of z'_n mod 3 are 1,1,0,1,0,0,2,0,0,1,0,0,2. We know from Theorem 6.13 that z'_n is divisible by 3 when $n \equiv 2$ mod 3. In fact we can say more:

Theorem 6.14 *Let n be an odd number. If $n \geqslant 5$ and $n \not\equiv 1$ mod 3 then z'_n is divisible by 3. If n is a prime of the form $2 \times 3^k + 1$ then $z'_n \equiv 1$ mod 3.*

Theorem 6.14 is from [110]. In the same paper, the sequence z'_n mod n was completely determined:

Theorem 6.15 *If n is prime then $z'_n \equiv -2$ mod n, whereas if n is composite then $z'_n \equiv 0$ mod n.*

A nice fact about z_n is that it is the number of *diagonally cyclic latin squares* of order n. Equivalently, z_n is the number of quasigroups on the set $\{1, 2, \ldots, n\}$ which have the transitive automorphism $(123 \cdots n)$. Moreover, z'_n is the number of such quasigroups which are idempotent. See [20, 121] for more details and a survey of the many applications of such objects.

n	Number of transversals in groups of order n
3	3
4	0, 8
5	15
7	133
8	0, 384, 384, 384, 384
9	2 025, 2 241
11	37 851
12	0, 198 144, 76 032, 46 080, 0
13	1 030 367
15	36 362 925
16	0, 235 765 760, 237 010 944, 238 190 592, 244 744 192, 125 599 744, 121 143 296, 123 371 520, 123 895 808, 122 191 872, 121 733 120, 62 881 792, 62 619 648, 62 357 504
17	1 606 008 513
19	87 656 896 891
20	0, 697 292 390 400, 140 866 560 000, 0, 0
21	5 778 121 715 415, 826 814 671 200
23	452 794 797 220 965

Table 4: Transversals in groups of order $n \leqslant 23$.

6.5   Groups of small order

We now discuss the number of transversals in general groups of small order. For groups of order $n \equiv 2 \bmod 4$ there can be no transversals, by Theorem 6.3. For each other order $n \leqslant 23$ the number of transversals in each group is given in Table 4. The groups are ordered according to the catalogue of Thomas and Wood [115]. The numbers of transversals in abelian groups of order at most 16 and cyclic groups of order at most 21 were obtained by Shieh *et al.* [107]. The remaining values in Table 4 were computed by Shieh [105]. McKay *et al.* [94] then independently confirmed all counts except those for cyclic groups of order $\geqslant 21$, correcting one misprint in Shieh [105].

Bedford and Whitaker [13] offer an explanation for why all the non-cyclic groups of order 8 have 384 transversals. The groups of order 4, 9 and 16 with the most transversals are the elementary abelian groups of those orders. Similarly, for orders 12, 20 and 21 the group with the most transversals is the direct sum of cyclic groups of prime order. It is an open question whether such a statement generalises.

Question 6.16 *Is it true that a direct sum of cyclic groups of prime order always has at least as many transversals as any other group of the same order?*

By Corollary 6.12 we know that in each case covered by Table 4 (except the non-abelian group of order 21), the number of transversals must have the same parity as the order of the square. It is remarkable though, that the groups of even order have a number of transversals which is divisible by a high power of 2. Indeed, any 2-group of order $n \leqslant 16$ has a number of transversals which is divisible by 2^{n-1}. It would be interesting to know if this is true for general n. Theorem 6.10 does provide a partial answer, but there seems to be more to the story.

7   Number of transversals

In this section we consider the question of how many transversals a general latin square can have. We define $t(n)$ and $T(n)$ to be respectively the minimum and maximum number of transversals among the latin squares of order n.

In §6 we have already considered z_n, the number of transversals in the addition table of $\mathbb{Z}_n$, which is arguably the simplest case. Since $t(n) \leqslant z_n$, Theorem 2.2 tells us that $t(n) = z_n = 0$ for even n. It is unknown whether there is any odd n for which $t(n) = 0$, although Conjecture 3.2 asserts there is not. In any case, for lower bounds on $t(n)$ we currently can do no better than to observe that $t(n) \geqslant 0$, and to note that $t(1) = 1$, $t(3) = t(5) = t(7) = 3$ and $t(9) = 68$. A related question, for which no work seems to have been published, is to find an upper bound on $t(n)$ when n is odd.

Turning to the maximum number of transversals, we have $z_n \leqslant T(n)$ and hence Theorem 6.8 gives a lower bound on $T(n)$ for odd n. In fact, the bound applies for even n as well [28]:

Theorem 7.1 *Provided n is sufficiently large, $T(n) > (3.246)^n$.*

It is clear that $T(n) \leqslant n!$ since there are only $n!$ different diagonals. An exponential improvement on this trivial bound was obtained by McKay *et al.* [94], who showed:

Theorem 7.2 *For $n \geqslant 5$,*

$$15^{n/5} \leqslant T(n) \leqslant c^n \sqrt{n}\, n!$$

where $c = \sqrt{\frac{3-\sqrt{3}}{6}}\, e^{\sqrt{3}/6} \approx 0.6135$.

As a corollary of Theorem 7.2 we can infer that the upper bound in Conjecture 6.7 is true (asymptotically) with $c_2 = 0.614$. This also yields an upper bound for the number of solutions to the toroidal n-queens problem.

The lower bound in Theorem 7.2 is very simple and is weaker than Theorem 7.1. The upper bound took considerably more work, although it too is probably far from the truth.

The same paper [94] reports the results of an exhaustive computation of the transversals in latin squares of orders up to and including 9. Table 5 lists the minimum and maximum number of transversals over all latin squares of order n for $n \leqslant 9$, and the mean and standard deviation to 2 decimal places.

Table 5 confirms Conjecture 3.2 for $n \leqslant 9$. The following semisymmetric[15] latin squares are representatives of the unique species with $t(n)$ transversals for

---

[15]A latin square is semisymmetric if it is invariant under cyclically permuting the roles of rows, columns and symbols. See [39] for more details.

n	$t(n)$	Mean	Std Dev	$T(n)$
2	0	0	0	0
3	3	3	0	3
4	0	2	3.46	8
5	3	4.29	3.71	15
6	0	6.86	5.19	32
7	3	20.41	6.00	133
8	0	61.05	8.66	384
9	68	214.11	15.79	2241

Table 5: Transversals in latin squares of order $n \leqslant 9$.

$n \in \{5, 7, 9\}$. In each case the entries in the largest subsquares are shaded.

1	2	3	4	5
2	1	4	5	3
3	5	1	2	4
4	3	5	1	2
5	4	2	3	1

3	2	1	5	4	7	6
2	1	3	6	7	4	5
1	3	2	7	6	5	4
5	6	7	4	1	2	3
4	7	6	1	5	3	2
7	4	5	2	3	6	1
6	5	4	3	2	1	7

2	1	3	6	7	8	9	5	4
1	3	2	5	4	9	6	7	8
3	2	1	4	9	5	7	8	6
9	5	4	3	2	1	8	6	7
8	4	6	2	5	7	1	9	3
4	7	9	8	3	6	5	1	2
5	8	7	9	6	2	3	4	1
6	9	8	7	1	4	2	3	5
7	6	5	1	8	3	4	2	9

n	Lower Bound	Upper Bound
10	5 504	75 000
11	37 851	528 647
12	198 144	3 965 268
13	1 030 367	32 837 805
14	3 477 504	300 019 037
15	36 362 925	2 762 962 210
16	244 744 192	28 218 998 328
17	1 606 008 513	300 502 249 052
18	6 434 611 200	3 410 036 886 841
19	87 656 896 891	41 327 486 367 018
20	697 292 390 400	512 073 756 609 248
21	5 778 121 715 415	6 803 898 881 738 477

Table 6: Bounds on $T(n)$ for $10 \leqslant n \leqslant 21$.

In Table 6 we reproduce from [94] bounds on $T(n)$ for $10 \leqslant n \leqslant 21$. The upper bound is somewhat sharper than that given by Theorem 7.2, though proved by the same methods. The lower bound in each case is constructive and likely to be of the same order as the true value. When $n \not\equiv 2 \bmod 4$ the lower bound comes from the group with the highest number of transversals (see Table 4). When $n \equiv 2 \bmod 4$ the lower bound comes from a so-called turn-square, many of which were analysed in [94]. A *turn-square* is obtained by starting with the Cayley table of a group (typically

I. M. Wanless

a group of the form $\mathbb{Z}_2 \oplus \mathbb{Z}_m$ for some m) and "turning" some of the intercalates (that is, replacing a subsquare of order 2 by the other possible subsquare on the same symbols). For example,

$$
\begin{array}{ccccc|ccccc}
5 & 6 & 2 & 3 & 4 & 0 & 1 & 7 & 8 & 9 \\
6 & 2 & 3 & 4 & 0 & 1 & 7 & 8 & 9 & 5 \\
2 & 3 & 4 & 0 & 1 & 7 & 8 & 9 & 5 & 6 \\
3 & 4 & 0 & 1 & 2 & 8 & 9 & 5 & 6 & 7 \\
4 & 0 & 1 & 2 & 3 & 9 & 5 & 6 & 7 & 8 \\
\hline
0 & 1 & 7 & 8 & 9 & 5 & 6 & 2 & 3 & 4 \\
1 & 7 & 8 & 9 & 5 & 6 & 2 & 3 & 4 & 0 \\
7 & 8 & 9 & 5 & 6 & 2 & 3 & 4 & 0 & 1 \\
8 & 9 & 5 & 6 & 7 & 3 & 4 & 0 & 1 & 2 \\
9 & 5 & 6 & 7 & 8 & 4 & 0 & 1 & 2 & 3 \\
\end{array}
\tag{7.1}
$$

has 5504 transversals. The 'turned' entries have been shaded. The study of turn-squares was pioneered by Parker (see [18] and the references therein) in his unsuccessful quest for a triple of MOLS of order 10. He noticed that turn-squares often have many more transversals than is typical for squares of their order, and used this as a heuristic in the search for MOLS.

It is has long been suspected that $T(10)$ is achieved by (7.1). This suspicion was strengthened by McKay *et al.* [93] who examined several billion squares of order 10, including every square with a non-trivial symmetry, and found none had more than 5504 transversals. Parker was indeed right that the square (7.1) is rich in orthogonal mates[16]. However, using the number of transversals as a heuristic in searching for MOLS is not fail-safe. For example, the turn-square of order 14 with the most transversals (namely, 3 477 504) does not have any orthogonal mates [94]. Meanwhile there are squares of order n with orthogonal mates but which possess only the bare minimum of n transversals (the left hand square in (1.1) is one such).

Nevertheless, the number of transversals does provide a useful species invariant for squares of small orders where this number can be computed in reasonable time (see, for example, [82] and [120]). It is straightforward to write a backtracking algorithm to count transversals in latin squares of small order, though this method currently becomes impractical if the order is much over 20. See [75, 76, 107] for some algorithms and complexity theory results[17] on the problem of counting transversals.

8 Generalised transversals

There are several ways to generalise the notion of a transversal. We have already seen one of them, namely the partial transversals in §5. In this section we collect results on another generalisation, namely plexes.

---

[16]It has 12 265 168 orthogonal mates [89], which is an order of magnitude greater than Parker estimated.

[17]An unfortunate feature of the analysis in [75] of the complexity of counting transversals in cyclic groups is that it hinges entirely on a technicality as to what constitutes the input for the algorithm. The authors consider the input to be a single integer that specifies the order of group. However, their conclusions would be very different if the input was considered to be a Cayley table for the group in question, which in the context of counting transversals is a more natural approach.

A *k-plex* in a latin square of order n is a set of kn entries which includes k representatives from each row and each column and of each symbol. A transversal is a 1-plex.

Example 8.1 The shaded entries form a 3-plex in the following square:

$$
\begin{array}{cccccc}
1 & 2 & 3 & 4 & 5 & 6 \\
2 & 1 & 4 & 3 & 6 & 5 \\
3 & 5 & 1 & 6 & 2 & 4 \\
4 & 6 & 2 & 5 & 3 & 1 \\
5 & 4 & 6 & 2 & 1 & 3 \\
6 & 3 & 5 & 1 & 4 & 2
\end{array}
$$

The name k-plex was coined in [120] only fairly recently. It is a natural extension of the names duplex, triplex, and quadruplex which have been in use for many years (principally in the statistical literature, such as [62]) for 2, 3 and 4-plexes.

The entries not included in a k-plex of a latin square L of order n form an $(n - k)$-plex of L. Together the k-plex and its complementary $(n - k)$-plex are an example of what is called an *orthogonal partition* of L. For discussion of orthogonal partitions in a general setting see Gilliland [66] and Bailey [10]. For our purposes, if L is decomposed into disjoint parts $K_1, K_2, \ldots, K_d$ where K_i is a k_i-plex then we call this a $(k_1, k_2, \ldots, k_d)$-partition of L. A case of particular interest is when all parts have the same size. We call a $(k, k, \ldots, k)$-partition more briefly a *k-partition*. For example, the marked 3-plex and its complement form a 3-partition of the square in Example 8.1. By Theorem 1.1, finding a 1-partition of a square is equivalent to finding an orthogonal mate.

Some results about transversals generalise directly to other plexes, while others seem to have no analogue. Theorem 3.3 and Theorem 6.1 seem to be in the latter class, as observed in [94] and [120] respectively. However, Theorem 6.10 does generalise to the following [50]:

Theorem 8.2 *Let m be the greatest common divisor of positive integers n and k. Suppose L is the Cayley table of a group of order n. The number of k-plexes in L is a multiple of n/m.*

Also, Theorems 3.1 and 6.3 showed that not every square has a transversal, and similar arguments work for any k-plex where k is odd [120]:

Theorem 8.3 *Suppose that q and k are odd integers and m is even. No q-step type latin square of order mq possesses a k-plex.*

Theorem 8.4 *Let G be a group of finite order n with a non-trivial cyclic Sylow 2-subgroup. The Cayley table of G contains no k-plex for any odd k but has a 2-partition and hence contains a k-plex for every even k in the range $0 \leqslant k \leqslant n$.*

The situation for even k is quite different to the odd case. Rodney [31, p.143] conjectured that every latin square has a duplex. He subsequently strengthened this conjecture, according to Dougherty [44], to the following:

Conjecture 8.5 *Every latin square has the maximum possible number of disjoint duplexes. In particular, every latin square of even order has a 2-partition and every latin square of odd order has a* $(2,2,2,\ldots,2,1)$*-partition.*

Conjecture 8.5 was stated independently in [120]. It implies Conjecture 3.2 and also that every latin square has k-plexes for every even value of k up to the order of the square. Thanks to [50], Conjecture 8.5 is now known to be true for all latin squares of orders $\leqslant 9$. It is also true for all groups[18], as can be seen by combining Theorem 6.3 and Theorem 8.4.

If a group has a trivial or non-cyclic Sylow 2-subgroup then it has a k-plex for all possible k. Otherwise it has k-plexes for all possible even k but for no odd k. It is worth noting that other scenarios occur for latin squares which are not based on groups. For example, the square in Example 8.1 has no transversal but clearly does have a 3-plex. It is conjectured in [120] that there exist arbitrarily large latin squares of this type.

Conjecture 8.6 *For all even $n > 4$ there exists a latin square of order n which has no transversal but does contain a 3-plex.*

Another possibility was shown by a family of squares constructed in [49].

Theorem 8.7 *For all even n there exists a latin square of order n that has k-plexes for every odd value of k between $\lfloor n/4 \rfloor$ and $\lceil 3n/4 \rceil$ (inclusive), but not for any odd value of k outside this range.*

Interestingly, there is no known example of odd integers $a < b < c$ and a latin square which has an a-plex and a c-plex but no b-plex.

The union of an a-plex and a disjoint b-plex of a latin square L is an $(a+b)$-plex of L. However, it is not always possible to split an $(a+b)$-plex into an a-plex and a disjoint b-plex. Consider a duplex which consists of $\frac{1}{2}n$ disjoint intercalates (latin subsquares of order 2). Such a duplex does not contain a partial transversal of length more than $\frac{1}{2}n$, so it is a long way from containing a 1-plex.

We say that a k-plex is *indivisible* if it contains no c-plex for $0 < c < k$. The duplex just described is indivisible[19]. Indeed, for every k there is an indivisible k-plex in some sufficiently large latin square. This was first shown in [120], but "sufficiently large" in that case meant at least quadratic in k. This was improved to linear as a corollary of our next result, from [21, 50]. An *indivisible partition* is a partition of a latin square into indivisible plexes.

Theorem 8.8 *For every $k \geqslant 2$ and $m \geqslant 2$ there exists a latin square of order mk with an indivisible k-partition.*

In particular, for all even $n \geqslant 4$ there is a latin square of order n composed of two indivisible $\frac{1}{2}n$-plexes. Egan [48] recently showed an analogous result for odd orders.

---

[18]For an alternative proof that Cayley tables of groups have at least one duplex, see [117].

[19]In contrast, it is known that any 3-plex that forms a latin trade is divisible; in fact it must divide into 3 disjoint transversals. See [26] for details.

Theorem 8.9 *If* $n = 2k + 1 \geqslant 5$ *then there is a latin square of order* n *with an indivisible* $(k, k, 1)$-*partition and an indivisible* $(k, k + 1)$-*partition.*

The previous two theorems mean that some squares can be split in "half" in a way that makes no further division possible. This is slightly surprising given that latin squares typically have a vast multitude of partitions into various plexes. For a detailed study of the indivisible partitions of latin squares of order up to 9, see [50].

It is an open question for what values of k and n there is a latin square of order n containing an indivisible k-plex. However, Bryant *et al.* [21] found the answer when k is small relative to n.

Theorem 8.10 *Let* n *and* k *be positive integers satisfying* $5k \leqslant n$. *Then there exists a latin square of order* n *containing an indivisible* k-*plex.*

It is also interesting to ask how large k can be relative to n. Define $\kappa(n)$ to be the maximum k such that some latin square of order n contains an indivisible k-plex. From Theorems 8.8 and 8.9 we know $\kappa(n) \geqslant \lceil n/2 \rceil$. Even though the numerical evidence (e.g. from [50]) suggests that latin squares typically contain many plexes, we are currently unable to improve on the trivial upper bound $\kappa(n) \leqslant n$. A proof of even the weak form of Conjecture 8.5 would at least show $\kappa(n) < n$ for $n > 2$. The values of $\kappa(n)$ for small n, as calculated in [50], are shown in Table 7.

n	1	2	3	4	5	6	7	8	9
$\kappa(n)$	1	2	1	2	3	4	5	5	6 or 7

Table 7: $\kappa(n)$ for $n \leqslant 9$.

So far we have examined situations where we start with a latin square and ask what sort of plexes it might have. To complete the section we consider the reverse question. We want to start with a potential plex and ask what latin squares it might be contained in. We define[20] a k-*protoplex* of order n to be an $n \times n$ array in which each cell is either blank or filled with a symbol from $\{1, 2, \ldots, n\}$, and which has the properties that (i) no symbol occurs twice within any row or column, (ii) each symbol occurs k times in the array, (iii) each row and column contains exactly k filled cells. We can then sensibly ask whether this k-protoplex is a k-plex. If it is then we say the partial latin square is *completable* because the blank entries can be filled in to produce a latin square. Donovan [43] asks for which k and n there exists a k-protoplex of order n that is not completable. The following partial answer was shown in [120].

Theorem 8.11 *If* $1 < k < n$ *and* $k > \frac{1}{4}n$ *then there exists a* k-*protoplex of order* n *that is not completable.*

Grüttmüller [69] showed a related result by constructing, for each k, a non-completable k-plex of order $4k - 2$ with the additional property that the plex is the union of k disjoint transversals. Daykin and Häggkvist [38] and Burton [23] independently conjectured that if $k \leqslant \frac{1}{4}n$ then every k-protoplex is completable. It

---

[20]Our definition of protoplex is very close to what is known as a "k-homogeneous partial latin square", except that such objects are usually allowed to have empty rows and columns [26].

seems certain that for k sufficiently small relative to n, every k-protoplex is completable. This has already been proved when $n \equiv 0 \bmod 16$ in [38]. Alspach and Heinrich [7] ask more specifically whether there exists an $N(k)$ with the property that if k transversals of a partial latin square of order $n \geqslant N(k)$ are prescribed then the square can always be completed. Grüttmüller's result just mentioned shows that $N(k) \geqslant 4k - 1$, if it exists. A related result due to Burton [23] is this:

Theorem 8.12 *For* $k \leqslant \frac{1}{4}n$ *every* k-*protoplex of order* n *is contained in some* $(k + 1)$-*protoplex of order* n.

An interesting generalisation of plexes was recently introduced by Pula [101]. A k-plex can be viewed as a function from the entries of a latin square to the set $\{0, 1\}$, such that the function values add to k along any row, any column or for any symbol. Pula generalises this idea to a k-*weight*, which he defines exactly the same way, except that the function is allowed to take any integer value. He shows that the Delta Lemma still works with this more general definition and uses it to obtain analogues of several classical results, including Theorem 3.1. Perhaps more tantalisingly, he shows that Ryser's Conjecture (Conjecture 3.2) and the weak form of Rodney's Conjecture (Conjecture 8.5) are simple to prove for k-weights; every latin square has a 2-weight and all latin squares of odd order have a 1-weight. However, the analogue of Conjecture 5.1 for 1-weights is still an open question.

9 Covering radii for sets of permutations

Several years ago, a novel approach to Conjecture 3.2 and Conjecture 5.1 was opened up by Andre Kézdy and Hunter Snevily in unpublished notes. These notes were then utilised in the writing of [25], from which the material in this section is drawn. To explain the Kézdy-Snevily approach, we need to introduce some terminology.

Consider the *symmetric group* S_n as a metric space equipped with *Hamming distance*. That is, the distance between two permutations $g, h \in S_n$ is the number of points at which they disagree (n minus the number of fixed points of gh^{-1}). Let P be a subset of S_n. The *covering radius* $\mathrm{cr}(P)$ of P is the smallest r such that the balls of radius r with centres at the elements of P cover the whole of S_n. In other words every permutation is within distance r of some member of P, and r is chosen to be minimal with this property. The next result is proved in [24] and [25].

Theorem 9.1 *Let* $P \subseteq S_n$ *be a set of permutations. If* $|P| \leqslant n/2$, *then* $\mathrm{cr}(P) = n$. *However, there exists* P *with* $|P| = \lfloor n/2 \rfloor + 1$ *and* $\mathrm{cr}(P) < n$.

This result raises an obvious question. Given n and s, what is the smallest set S of permutations with $\mathrm{cr}(S) \leqslant n - s$? We let $f(n, s)$ denote the cardinality of the smallest such set S. This problem can also be interpreted in graph-theoretic language. Define the graph $G_{n,s}$ on the vertex set S_n, with two permutations being adjacent if they agree in at least s places. Now the size of the smallest dominating set in $G_{n,s}$ is $f(n, s)$.

Theorem 9.1 shows that $f(n, 1) = \lfloor n/2 \rfloor + 1$. Since any two distinct permutations have distance at least 2, we see that $f(n, n - 1) = n!$ for $n \geqslant 2$. Moreover, $f(n, s)$ is a monotonic increasing function of s (by definition).

The next case to consider is $f(n,2)$. Kézdy and Snevily made the following conjecture:

Conjecture 9.2 *If n is even, then $f(n,2) = n$; if n is odd, then $f(n,2) > n$.*

The Kézdy–Snevily Conjecture has several connections with transversals [25].

Theorem 9.3 *Let S be the set of n permutations corresponding to the rows of a latin square L of order n. Then S has covering radius $n - 1$ if L has a transversal and has covering radius $n - 2$ otherwise.*

Corollary 9.4 *If there exists a latin square of order n with no transversal, then $f(n,2) \leqslant n$. In particular, this holds for n even.*

Hence Conjecture 9.2 implies Conjecture 3.2, as Kézdy and Snevily observed. They also showed:

Theorem 9.5 *Conjecture 9.2 implies Conjecture 5.1.*

In other words, to solve the longstanding Ryser and Brualdi conjectures it may suffice to answer this: How small can we make a subset $S \subset S_n$ which has the property that every permutation in S_n agrees with some member of S in at least two places?

In Corollary 9.4 we used latin squares to find an upper bound for $f(n,2)$ when n is even. For odd n we can also find upper bounds based on latin squares. The idea is to choose a latin square with few transversals, or whose transversals have a particular structure, and add a small set of permutations meeting each transversal twice. For $n \in \{5,7,9\}$, we now give a latin square for which a single extra permutation (shaded) suffices, showing that $f(n,2) \leqslant n + 1$ in these cases.

1	2	3	4	5
2	1	4	5	3
3	5	1	2	4
4	3	5	1	2
5	4	2	3	1
1	3	4	2	5

1	2	3	4	5	6	7
2	3	1	5	4	7	6
3	1	2	6	7	4	5
4	5	6	7	1	2	3
5	4	7	1	6	3	2
6	7	4	2	3	5	1
7	6	5	3	2	1	4
3	2	1	7	6	5	4

1	3	2	4	6	5	7	9	8
2	1	3	5	4	6	8	7	9
3	2	1	7	9	8	4	6	5
4	6	5	9	8	7	1	3	2
5	4	6	8	7	9	3	2	1
6	5	4	2	1	3	9	8	7
7	9	8	1	3	2	5	4	6
8	7	9	3	2	1	6	5	4
9	8	7	6	5	4	2	1	3
5	4	6	1	3	2	9	8	7

In general, we have the following [25]:

Theorem 9.6 $f(n,2) \leqslant \frac{4}{3}n + O(1)$ *for all n.*

The reader is encouraged to seek out [25] and the survey by Quistorff [102] for more information on covering radii for sets of permutations.

10   Generalisations of latin squares

There are a number of different ways in which the definition of a latin square can be relaxed. In this section we consider such generalisations, and what can be said about their transversals. In this context, it is worth clarifying exactly what we mean by a transversal. For a rectangular matrix we will always assume that there are no more rows than there are columns. In such a case, a *diagonal* will mean a selection of cells that includes one representative from each row and at most one representative from each column. As before, a *transversal*[21] will mean a diagonal in which no symbol is repeated. Of course, in situations where the number of symbols in the matrix exceeds the number of rows, this no longer means that every symbol must occur within the transversal. A *partial transversal of length ℓ* will be a selection of ℓ cells no two of which share their row, column or symbol. In an $m \times n$ matrix, a *near transversal* is a partial transversal of length $m - 1$.

A matrix is said to be *row-latin* if it has no symbol that occurs more than once in any row. Similarly, it is *column-latin* if no symbol is ever repeated within a column. Stein [111] defines an $n \times n$ matrix to be an *equi-n-square* if each of n symbols occurs n times within the matrix. In this terminology, a latin square is precisely a row-latin and column-latin equi-n-square. Stein [111] was able to show a number of interesting results including these:

Theorem 10.1 *In an equi-n-square there is a row or column that contains at least $\sqrt{n}$ distinct symbols.*

Theorem 10.2 *An equi-n-square has a partial transversal of length at least*

$$n\left(1 - \frac{1}{2!} + \frac{1}{3!} - \cdots \pm \frac{1}{n!}\right) \approx (1 - 1/e)n \approx 0.63n.$$

Theorem 10.3 *Suppose a positive integer q divides $n > 2$. If each of n^2/q symbols occurs q times in an $n \times n$ matrix then there is a partial transversal of length exceeding $n - \frac{1}{2}q$.*

Stein [111] also makes the following conjectures, some of which are special cases of others in the list:

1. An equi-n-square has a near transversal.

2. Any $n \times n$ matrix in which no symbol appears more than $n - 1$ times has a transversal.

3. Any $(n - 1) \times n$ matrix in which no symbol appears more than n times has a transversal.

4. Any $(n - 1) \times n$ row-latin matrix has a transversal.

5. For $m < n$, any $m \times n$ matrix in which no symbol appears more than n times has a transversal.

---

[21]Usage in this article follows the majority of the latin squares literature. For more general matrices the word 'transversal' has been used (for example, in [5, 6, 111, 114]) to mean what we are calling a diagonal. Other papers call diagonals 'sections' [113, 61] or '1-factors' [3]. When 'transversal' is used to mean diagonal, 'latin transversal' is used to mean what we are calling a transversal.

6. Any $(n-1) \times n$ matrix in which each symbol appears exactly n times has a transversal.

7. For $m < n$, any $m \times n$ matrix in which no symbol appears more than $m + 1$ times has a transversal.

Example 10.4 Drisko [47] gives counterexamples to Stein's Conjecture 5, whenever $m < n \leqslant 2m - 2$. The construction is simply to take $m - 1$ columns that have the symbols in order $[1, 2, 3, \ldots, m]^T$ and the remaining columns to be $[2, 3, \ldots, m, 1]^T$. It is easy to argue directly that the resulting matrix has no transversal[22]. Taking $m = n - 1$ we see immediately that Stein's Conjectures 3, 6 and 7 also fail.

A weakened form of Stein's Conjecture 5 can be salvaged. Stein's response to Example 10.4 was to propose a new conjecture [112] that all column-latin matrices have a near transversal. For "thin" matrices we can do even better. In Drisko's original paper [47] he proved the following result, which was subsequently generalised to matroids by Chappell [29], and later proved in a slightly different way by Stein [112]:

Theorem 10.5 *Let $n \geqslant 2m - 1$. Any $m \times n$ column-latin matrix has a transversal.*

Häggkvist and Johansson [70] showed that every large enough and thin enough latin rectangle[23] not only has a transversal, but has an orthogonal mate:

Theorem 10.6 *Suppose $0 < \varepsilon < 1$. If n is sufficiently large and $m < \varepsilon n$ then every $m \times n$ latin rectangle can be decomposed into transversals.*

m	$n = 2$	3	4	5	6	7
2	1	5	7	9	11	13
3		2	3	7	8	10
4			3	4	5	8
5				3	5	
6					4	

Table 8: $L(m, n)$ for small m, n.

Stein and Szabó [113] introduce the function $L(m, n)$ which they define as the largest integer i such that there is a transversal in every $m \times n$ matrix that has no symbol occurring more than i times. Table 8, reproduced from [113], shows the value of $L(m, n)$ for small m and n. The examples that show $L(5, 5) < 4$ and $L(6, 6) < 5$ yield counterexamples to Stein's Conjecture 2. So of the seven conjectures listed above, only Conjectures 1 and 4 remain. In addition to the values in Table 8, Stein and Szabó show that $L(2, n) = 2n - 1$ for $n \geqslant 3$ and $L(3, n) = \lfloor (3n - 1)/2 \rfloor$ for $n \geqslant 5$ and prove the lower bound $L(m, n) \geqslant n - m + 1$. Akbari *et al.* [4] showed that $L(m, n) = \lfloor (mn - 1)/(m - 1) \rfloor$ if $m \geqslant 2$ and $n \geqslant 2m^3$. Parker is attributed in [112, 113] with a construction proving the following result, which also follows from Example 10.4:

---

[22]This can also easily be proved by a variant of the Delta Lemma.

[23]A *latin rectangle* is row-latin and column-latin and also must use the same symbols in each row.

Theorem 10.7 *If $n \leqslant 2m - 2$ then $L(m, n) \leqslant n - 1$.*

Various results have been shown using the probabilistic method. Erdős and Spencer [53] showed that $L(n, n) \geqslant (n - 1)/16$. Fanaï [60] took the direct generalisation to rectangular matrices, showing $L(m, n) \geqslant \frac{1}{8}n(n - 1)/(m + n - 2) + 1$. Alon *et al.* [6] showed that there is a small $\varepsilon > 0$ such that if no symbol occurs more than $\varepsilon 2^m$ times in a $2^m \times 2^m$ matrix, then the matrix can be decomposed into transversals. They claim the same method can be adapted to show the existence of a number of pairwise disjoint transversals in matrices of a similar type, but whose order need not be a power of 2.

Akbari and Alireza [3] define $l(n)$ to be the smallest integer such that there is a transversal in every $n \times n$ matrix that is row-latin and column-latin and contains at least $l(n)$ different symbols. They show that $l(1) = 1$, $l(2) = l(3) = 3$, $l(4) = 6$, and $l(5) \geqslant 7$. Theorem 2.2 shows that $l(n) > n$ for all even n. Establishing a meaningful upper bound on $l(n)$ remains an interesting open challenge. The authors of [3] conjecture that $l(n) - n$ is not bounded as $n \to \infty$. They also prove that $l(2^a - 2) > 2^a$ for $a \geqslant 3$, which follows from:

Theorem 10.8 *Let a, b be any two elements of an elementary abelian 2-group G, of order $2^m \geqslant 4$. Let M be the matrix of order $2^m - 2$ formed from the Cayley table of G by deleting the rows and columns indexed by a and b. Then M has no transversal.*

Although Akbari and Alireza did not do so, it is simple to derive this result from Theorem 6.5, given that the sum of all but two elements of an elementary abelian group can never be the identity. A related new result by Arsovski [9] is this:

Theorem 10.9 *Any square submatrix of the Cayley table of an abelian group of odd order has a transversal.*

This result was originally conjecture by Snevily [109]. He also conjectured that in Cayley tables of cyclic groups of even order, the only submatrices without a transversal are subgroups of even order or "translates" of such subgroups. This conjecture (which remains open) does not generalise directly to all abelian groups of even order, as Theorem 10.8 shows.

Theorem 10.9 was first proved[24] for prime orders by Alon [5] and for all cyclic groups by Dasgupta *et al.* [35]. Gao and Wang [65] then showed it is true in arbitrary abelian groups for submatrices whose order is less than $\sqrt{p}$, where p is the smallest prime dividing the order of the group. In related work, Fanaï [61] showed that there is a transversal in any square submatrix of the addition table of $\mathbb{Z}_n$ (for arbitrary n)[25], provided the rows selected to form the submatrix are sufficiently close together relative to n. He also showed existence of a partial transversal in certain submatrices satisfying a slightly weaker condition.

Finally, we briefly consider arrays of dimension higher than two. A *latin hypercube* of order n is an $n \times n \times \cdots \times n$ array filled with n symbols in such a way that no symbol is repeated in any *line* of n cells parallel to one of the axes. A 3-dimensional

---

[24]In fact Alon showed a stronger result that allows rows to be repeated when selecting the submatrix. His proof uses the fascinating combinatorial nullstellensatz. For related work, see [81].

[25]Although, given [35], this result is only of interest when n is even.

latin hypercube is a *latin cube* and a 2-dimensional latin hypercube is a latin square. Furthermore, any 2-dimensional "slice" of a latin hypercube is a latin square. By a transversal of a latin hypercube of order n we mean a selection of n cells no two of which agree in any coordinate or share the same symbol. The literature contains some variation on the definitions of both latin hypercubes and transversals thereof [95]. Perhaps the most interesting work in this area is by Sun [114], who conjectured that all latin cubes have transversals[26]. Generalising this conjecture (and Conjecture 3.2) on the basis of the examples catalogued in [95], we propose:

Conjecture 10.10 *Every latin hypercube of odd dimension or of odd order has a transversal.*

The restriction to odd dimension or odd order is required. Let $Z_{n,d}$ denote the d-dimensional hypercube whose entry in cell $(x_1, x_2, \ldots, x_d)$ is $x_1 + x_2 + \cdots + x_d \bmod n$. Then, by direct generalisation of Theorem 2.2, we have:

Theorem 10.11 *If n and d are even, there are no transversals in $Z_{n,d}$.*

Sun [114] showed that if d is odd and n is arbitrary then $Z_{n,d}$ has a transversal. In fact, any $k \times k \times \cdots \times k$ subarray of $Z_{n,d}$ has a transversal, where $1 \leqslant k \leqslant n$.

11 Concluding Remarks

We have only been able to give a brief overview of the fascinating subject of transversals in this survey. Space constraints have forced the omission of much worthy material, including proofs of most of the theorems quoted. However, even this brief skim across the surface has shown that many basic questions remain unanswered and much work remains to be done.

The subject is peppered with tantalising conjectures. Even the theorems in many cases seem to be far from best possible, leaving openings for future improvements. It is hoped that this survey may motivate and assist such improvements.

References

[1] R. J. R. Abel, D. Combe, G. Price and W. D. Palmer, Existence of generalized Bhaskar Rao designs with block size 3, *Discrete Math.* **309** (2009), 4069–4078.

[2] R. Aharoni and E. Berger, Rainbow matchings in r-partite r-graphs, *Electron. J. Combin.* **16** (2009), R119.

[3] S. Akbari and A. Alireza, Transversals and multicolored matchings, *J. Combin. Des.* **12** (2004), 325–332.

[4] S. Akbari, O. Etesami, H. Mahini, M. Mahmoody and A. Sharifi, Transversals in long rectangular arrays, *Discrete Math.* **306** (2006), 3011–3013.

[5] N. Alon, Additive Latin transversals, *Israel J. Math.* **117** (2000), 125-130.

[6] N. Alon, J. Spencer and P. Tetali, Covering with latin transversals, *Discrete Appl. Math.* **57** (1995), 1–10.

---

[26]Sun defined a transversal to be n cells no two of which lie in the same line. However, this seems too broad a definition. It includes some examples where, say, all cells in the transversal share the same first coordinate. However, Sun's results all relate to transversals as we have defined them.

[7] B. Alspach and K. Heinrich, Matching designs, *Australas. J. Combin.* **2** (1990), 39–55.

[8] B. A. Anderson, Sequencings and houses, *Congr. Numer.* **43** (1984), 23–43.

[9] B. Arsovski, A proof of Snevily's conjecture, *Israel J. Math.*, to appear.

[10] R. A. Bailey, Orthogonal partitions in designed experiments (Corrected reprint), *Des. Codes Cryptogr.* **8** (1996), 45–77.

[11] K. Balasubramanian, On transversals in latin squares, *Linear Algebra Appl.* **131** (1990), 125–129.

[12] D. Bedford, Orthomorphisms and near orthomorphisms of groups and orthogonal Latin squares: a survey, *Bull. Inst. Combin. Appl.* **15** (1995), 13–33.

[13] D. Bedford and R. M. Whitaker, Enumeration of transversals in the Cayley tables of the non-cyclic groups of order 8, *Discrete Math.* **197/198** (1999), 77–81.

[14] T. Beth, D. Jungnickel and H. Lenz, *Design theory*, Cambridge University Press, Cambridge, 1986.

[15] T. Beth, D. Jungnickel and H. Lenz, *Design theory* (2nd ed., Vol II), Cambridge University Press, Cambridge, 1999.

[16] R. C. Bose, S. S. Shrikhande and E. T. Parker. Further results on the construction of mutually orthogonal latin squares and the falsity of Euler's conjecture, *Canad. J. Math.* **12** (1960), 189–203.

[17] A. E. Brouwer, A. J. de Vries and R. M. A. Wieringa, A lower bound for the length of partial transversals in a latin square, *Nieuw Arch. Wisk. (3)* **26** (1978), 330–332.

[18] J. W. Brown and E. T. Parker, More on order 10 turn-squares, *Ars Combin.* **35** (1993), 125–127.

[19] J. Browning, P. Vojtěchovský and I. M. Wanless, Overlapping latin subsquares and full products, *Comment. Math. Univ. Carolin.* **51** (2010), 175–184.

[20] D. Bryant, M. Buchanan and I. M. Wanless, The spectrum for quasigroups with cyclic automorphisms and additional symmetries, *Discrete Math.* **309** (2009), 821–833.

[21] D. Bryant, J. Egan, B. M. Maenhaut and I. M. Wanless, Indivisible plexes in latin squares, *Des. Codes Cryptogr.* **52** (2009), 93–105.

[22] R. A. Brualdi and H. J. Ryser, *Combinatorial matrix theory*, Encyclopedia of Mathematics and its Applications **39**, Cambridge University Press, 1991.

[23] B. Burton, *Completion of partial latin squares*, honours thesis, University of Queensland, 1996.

[24] P. J. Cameron and C. Y. Ku, Intersecting families of permutations, *European J. Combin.* **24** (2003), 881–890.

[25] P. J. Cameron and I. M. Wanless, Covering radius for sets of permutations, *Discrete Math.* **293** (2005), 91–109.

[26] N. J. Cavenagh, A uniqueness result for 3-homogeneous Latin trades, *Comment. Math. Univ. Carolin.* **47** (2006), 337–358.

[27] N. J. Cavenagh, C. Hämäläinen and A. M. Nelson, On completing three cyclically generated transversals to a Latin square, *Finite Fields Appl.* **15** (2009), 294–303.

[28] N. J. Cavenagh and I. M. Wanless, On the number of transversals in Cayley tables of cyclic groups, *Disc. Appl. Math.* **158** (2010), 136–146.

[29] G. G. Chappell, A matroid generalization of a result on row-Latin rectangles, *J. Combin. Theory Ser. A* **88** (1999), 235–245.

[30] D. Clark and J. T. Lewis, Transversals of cyclic latin squares, *Congr. Numer.* **128** (1997), 113–120.

[31] C. J. Colbourn and J. H. Dinitz (eds), *Handbook of combinatorial designs* (2nd ed.), Chapman & Hall/CRC, Boca Raton, 2007.

[32] C. Cooper, A lower bound for the number of good permutations, Data Recording, Storage and Processing (Nat. Acad. Sci. Ukraine) **2** (2000), 15–25.

[33] C. Cooper, R. Gilchrist, I. Kovalenko and D. Novakovic, Deriving the number of good permutations, with applications to cryptography, *Cybernet. Systems Anal.* **5** (2000), 10–16.

[34] C. Cooper and I. M. Kovalenko, The upper bound for the number of complete mappings, *Theory Probab. Math. Statist.* **53** (1996), 77–83.

[35] S. Dasgupta, G. Károlyi, O. Serra and B. Szegedy, Transversals of additive Latin squares, *Israel J. Math.* **126** (2001), 17–28.

[36] F. Dalla Volta and N. Gavioli, Complete mappings in some linear and projective groups, *Arch. Math. (Basel)* **61** (1993), 111–118.

[37] P. Danziger, I. M. Wanless and B. S. Webb, Monogamous latin squares, *J. Combin. Theory Ser. A* **118** (2011), 796–807.

[38] D. E. Daykin and R. Häggkvist, Completion of sparse partial latin squares, *Graph theory and combinatorics*, 127–132, Academic Press, London, 1984.

[39] J. Dénes and A. D. Keedwell, *Latin squares and their applications*, Akadémiai Kiadó, Budapest, 1974.

[40] J. Dénes and A. D. Keedwell, *Latin squares: New developments in the theory and applications*, Ann. Discrete Math. **46**, North-Holland, Amsterdam, 1991.

[41] J. Dénes and A. D. Keedwell, A new conjecture concerning admissibility of groups, *European J. Combin.* **10** (1989), 171–174.

[42] I. I. Deriyenko, On a conjecture of Brualdi (in Russian), *Mat. Issled.* No. 102, Issled. Oper. i Kvazigrupp, (1988), 53–65, 119.

[43] D. Donovan, The completion of partial Latin squares, *Australas. J. Combin.* **22** (2000), 247–264.

[44] S. Dougherty, Planes, nets and codes, *Math. J. Okayama Univ.* **38** (1996), 123–143.

[45] D. A. Drake, Maximal sets of latin squares and partial transversals, *J. Statist. Plann. Inference* **1** (1977), 143–149.

[46] D. A. Drake, G. H. J. van Rees and W. D. Wallis, Maximal sets of mutually orthogonal latin squares, *Discrete Math.* **194** (1999), 87–94.

[47] A. A. Drisko, Transversals in row-Latin rectangles, *J. Combin. Theory Ser. A* **84** (1998), 181–195.

[48] J. Egan, Bachelor latin squares with large indivisible plexes, *J. Combin. Designs*, doi:10.1002/jcd.20281 to appear.

[49] J. Egan and I. M. Wanless, Latin squares with no small odd plexes, *J. Combin. Designs* **16** (2008), 477–492.

[50] J. Egan and I. M. Wanless, Indivisible partitions of latin squares, *J. Statist. Plann. Inference* **141** (2011), 402–417.

[51] J. Egan and I. M. Wanless, Latin squares with restricted transversals, submitted.

[52] P. Erdős, D. R. Hickerson, D. A. Norton and S. K. Stein, Unsolved problems: Has every latin square of order n a partial latin transversal of size $n - 1$? *Amer. Math. Monthly* **95** (1988), 428–430.

[53] P. Erdős and J. Spencer, Lopsided Lovász local lemma and latin transversals, *Discrete Appl. Math.* **30** (1991), 151–154.

[54] L. Euler, Recherches sur une nouvelle espéce de quarrés magiques, *Verh. Zeeuwsch. Gennot. Weten. Vliss.* **9** (1782), 85–239. Eneström E530, Opera Omnia OI7, 291–392.

[55] A. B. Evans, *Orthomorphism Graphs of Groups*, Springer, 1992.

[56] A. B. Evans, The existence of complete mappings of SL$(2, q)$, $q \equiv 3$ modulo 4, *Finite Fields Appl.* **11** (2005), 151–155.

[57] A. B. Evans, Latin squares without orthogonal mates, *Des. Codes Cryptog.* **40** (2006), 121–130.

[58] A. B. Evans, The admissibility of sporadic simple groups, *J. Algebra* **321** (2009), 105–116.

[59] A. B. Evans, The existence of strong complete mappings, submitted.

[60] H.-R. Fanaï, A note on transversals, *Ars Combin.* **91** (2009), 83–85.

[61] H.-R. Fanaï, Existence of partial transversals, *Linear Algebra Appl.* **432** (2010), 2608–2614.

[62] D. J. Finney, Some orthogonal properties of the 4 × 4 and 6 × 6 latin squares. *Ann. Eugenics* **12** (1945), 213–219.

[63] R. J. Friedlander, B. Gordon and P. Tannenbaum, Partitions of groups and complete mappings, *Pacific J. Math.* **92** (1981), 283–293.

[64] H.-L. Fu, S.-C. Lin and C.-M. Fu, The length of a partial transversal in a latin square, *J. Combin. Math. Combin. Comput.* **43** (2002), 57–64.

[65] W. D. Gao and D. J. Wang, Additive Latin transversals and group rings, *Israel J. Math.* **140** (2004), 375–380.

[66] D. C. Gilliland, A note on orthogonal partitions and some well-known structures in design of experiments, *Ann. Statist.* **5** (1977), 565–570.

[67] B. Gordon, Sequences in groups with distinct partial products, *Pacific J. Math.* **11** (1961), 1309–1313.

[68] M. Grüttmüller, Completing partial Latin squares with two cyclically generated prescribed diagonals, *J. Combin. Theory Ser. A* **103** (2003), 349–362.

[69] M. Grüttmüller, Completing partial Latin squares with prescribed diagonals, *Discrete Appl. Math.* **138** (2004), 89–97.

[70] R. Häggkvist and A. Johansson, Orthogonal Latin rectangles, *Combin. Probab. Comput.* **17** (2008), 519–536.

[71] M. Hall, A combinatorial problem on abelian groups, *Proc. Amer. Math. Soc.* **3** (1952), 584–587.

[72] M. Hall and L. J. Paige, Complete mappings of finite groups, *Pacific J. Math.* **5** (1955), 541–549.

[73] P. Hatami and P. W. Shor, Erratum to: A lower bound for the length of a partial transversal in a latin square, *J. Combin. Theory Ser. A* **115** (2008), 1103–1113.

[74] J. D. Horton, Orthogonal starters in finite abelian groups, *Discrete Math.* **79** (1990), 265–278.

[75] J. Hsiang, D. F. Hsu and Y. P. Shieh, On the hardness of counting problems of complete mappings, *Disc. Math.* **277** (2004), 87–100.

[76] J. Hsiang, Y. Shieh and Y. Chen, The cyclic complete mappings counting problems, *PaPS: Problems and problem sets for ATP workshop in conjunction with CADE-18 and FLoC 2002*, Copenhagen, (2002).

[77] L. Hu and X. Li, Color degree condition for large heterochromatic matchings in edge-colored bipartite graphs, http://arxiv.org/abs/math.CO/0606749.

[78] D. M. Johnson, A. L. Dulmage and N. S. Mendelsohn, Orthomorphisms of groups and orthogonal latin squares I, *Canad. J. Math.* **13** (1961), 356–372.

[79] D. Jungnickel, "Latin squares, their geometries and their groups. A survey", in *Coding theory and design theory. Part II. Design theory* (Ed. Dijen Ray-Chaudhuri), Springer-Verlag, New York, 1990 (pp. 166–225).

[80] A. D. Keedwell, Sequenceable groups, generalized complete mappings, neofields and block designs, Combinatorial mathematics X (Adelaide, 1982), 49–71, Lecture Notes in Math. **1036**, Springer, Berlin, 1983.

[81] A. E. Kézdy and H. S. Snevily, Distinct sums modulo n and tree embeddings, *Combin. Probab. Comput.* **11** (2002), 35–42.

[82] R. Killgrove, C. Roberts, R. Sternfeld, R. Tamez, R. Derby and D. Kiel, Latin squares and other configurations, *Congr. Numer.* **117** (1996), 161–174.

[83] K. K. Koksma, A lower bound for the order of a partial transversal in a latin square, *J. Combinatorial Theory* **7** (1969), 94–95.

[84] I. N. Kovalenko, Upper bound for the number of complete maps, *Cybernet. Systems Anal.* **32** (1996), 65–68.

[85] N. Yu. Kuznetsov, Applying fast simulation to find the number of good permutations, *Cybernet. Systems Anal.* 43 (2007), 830–837.

[86] N. Yu. Kuznetsov, Estimating the number of good permutations by a modified fast simulation method, *Cybernet. Systems Anal.* 44 (2008), 547–554.

[87] C. F. Laywine and G. L. Mullen, *Discrete mathematics using latin squares*, Wiley, New York, 1998.

[88] P. Li, Sequencing the dihedral groups D_{4k}, *Discrete Math.* **175** (1997), 271–276.

[89] B. M. Maenhaut and I. M. Wanless, Atomic latin squares of order eleven, *J. Combin. Des.* **12** (2004), 12–34.

[90] E. Maillet, Sur les carrés latins d'Euler, *Assoc. Franc. Caen.* **23** (1894), 244–252.

[91] H. B. Mann, The construction of orthogonal latin squares, *Ann. Math. Statistics* **13** (1942), 418–423.

[92] H. B. Mann, On orthogonal latin squares, *Bull. Amer. Math. Soc.* **50** (1944), 249–257.

[93] B. D. McKay, A. Meynert and W. Myrvold, Small latin squares, quasigroups and loops, *J. Combin. Des.* **15** (2007), 98–119.

[94] B. D. McKay, J. C. McLeod and I. M. Wanless, The number of transversals in a latin square, *Des. Codes Cryptogr.* **40** (2006), 269–284.

[95] B. D. McKay and I. M. Wanless, A census of small latin hypercubes, *SIAM J. Discrete Math.* **22** (2008), 719–736.

[96] H. Niederreiter and K. H. Robinson, Bol loops of order pq, *Math. Proc. Cambridge Philos. Soc.* **89** (1981), 241–256.

[97] H. Niederreiter and K. H. Robinson, Complete mappings of finite fields, *J. Austral. Math. Soc. Ser. A* **33** (1982), 197–212.

[98] L. J. Paige, Complete mappings of finite groups, *Pacific J. Math.* **1** (1951), 111–116.

[99] E. T. Parker, Pathological latin squares, *Proceedings of symposia in pure mathematics*, Amer. Math. Soc. (1971), 177–181.

[100] K. Pula, Products of all elements in a loop and a framework for non-associative analogues of the Hall-Paige conjecture, *Electron. J. Combin.* **16** (2009), R57.

[101] K. Pula, A generalization of plexes of latin squares, *Disc. Math.*, **311** (2011), 577–581.

[102] J. Quistorff, A survey on packing and covering problems in the Hamming permutation space, *Electron. J. Combin.* **13** (2006), A1.

[103] I. Rivin, I. Vardi and P. Zimmerman, The n-Queens Problem, *Amer. Math. Monthly* **101** (1994), 629–639.

[104] H. J. Ryser, Neuere Probleme der Kombinatorik, *Vortrage über Kombinatorik Oberwolfach*, 24–29 Juli (1967), 69–91.

[105] Y. P. Shieh, *Partition strategies for #P-complete problem with applications to enumerative combinatorics*, PhD thesis, National Taiwan University, 2001.

[106] Y. P. Shieh, private correspondence, (2006).

[107] Y. P. Shieh, J. Hsiang and D. F. Hsu, On the enumeration of abelian K-complete mappings, *Congr. Numer.* **144** (2000), 67–88.

[108] P. W. Shor, A lower bound for the length of a partial transversal in a latin square, *J. Combin. Theory Ser. A* **33** (1982), 1–8.

[109] H. S. Snevily, The Cayley addition table of $\mathbb{Z}_n$, *Amer. Math. Monthly* **106** (1999), 584–585.

[110] D. S. Stones and I. M. Wanless, Compound orthomorphisms of the cyclic group, *Finite Fields Appl.* **16** (2010), 277–289.

[111] S. K. Stein, Transversals of Latin squares and their generalizations, *Pacific J. Math.* **59** (1975), 567–575.

[112] S. K. Stein, Transversals in rectangular arrays, *Amer. Math. Monthly* **117** (2010), 424–433.

[113] S. K. Stein and S. Szabó, The number of distinct symbols in sections of rectangular arrays, *Discrete Math.* **306** (2006), 254–261.

[114] Z.-W. Sun, An additive theorem and restricted sumsets, *Math. Res. Lett.* **15** (2008), 1263-1276.

[115] A. D. Thomas and G. V. Wood, *Group tables*, Shiva Mathematics Series 2, Shiva Publishing, Nantwich, 1980.

[116] I. Vardi, *Computational Recreations in Mathematics*, Addison-Wesley, Redwood City, CA, 1991.

[117] M. Vaughan-Lee and I. M. Wanless, Latin squares and the Hall-Paige conjecture, *Bull. London Math. Soc.* **35** (2003), 1–5.

[118] G. H. J. van Rees, Subsquares and transversals in latin squares, *Ars Combin.* **29**B (1990), 193–204.

[119] D. Q. Wan, On a problem of Niederreiter and Robinson about finite fields, *J. Austral. Math. Soc. Ser. A* **41** (1986), 336–338.

[120] I. M. Wanless, A generalisation of transversals for latin squares, *Electron. J. Combin.* **9** (2002), R12.

[121] I. M. Wanless, Diagonally cyclic latin squares, *European J. Combin.* **25** (2004), 393–413.

[122] I. M. Wanless, Atomic Latin squares based on cyclotomic orthomorphisms, *Electron. J. Combin.* **12** (2005), R22.

[123] I. M. Wanless, Transversals in latin squares, *Quasigroups Related Systems* **15** (2007), 169–190.

[124] I. M. Wanless and B. S. Webb, The existence of latin squares without orthogonal mates, *Des. Codes Cryptog.* **40** (2006), 131–135.

[125] S. Wilcox, Reduction of the Hall-Paige conjecture to sporadic simple groups, *J. Algebra* **321** (2009), 1407–1428.

[126] D. E. Woolbright, An $n \times n$ latin square has a transversal with at least $n - \sqrt{n}$ distinct symbols, *J. Combin. Theory Ser. A* **24** (1978), 235–237.

School of Mathematical Sciences
Monash University
Vic 3800, Australia
ian.wanless@monash.edu